ACS SYMPOSIUM SERIES **242**

Polymers in Electronics

Theodore Davidson, EDITOR
Xerox Corporation

Based on a symposium sponsored by
the Division of Organic Coatings
and Plastics Chemistry
at the 185th Meeting
of the American Chemical Society,
Seattle, Washington,
March 20–25, 1983

American Chemical Society, Washington, D.C. 1984

Library of Congress Cataloging in Publication Data
Polymers in electronics.
 (ACS symposium series, ISSN 0097-6156; 242)

 Bibliography: p.
 Includes index.

 1. Polymers and polymerization—Congresses.
 2. Electronics—Materials—Congresses.

 I. Davidson, Theodore, 1939-   . II. American
Chemical Society. Division of Organic Coatings and
Plastics Chemistry. III. American Chemical Society.
Meeting (185th: 1983: Seattle, Wash.) IV. Series.
TK7871.15.P6P63   1984   621.381′028   83-25782
ISBN 0-8412-0823-9

Copyright © 1984

American Chemical Society

All Rights Reserved. The appearance of the code at the bottom of the first page of each chapter in this volume indicates the copyright owner's consent that reprographic copies of the chapter may be made for personal or internal use or for the personal or internal use of specific clients. This consent is given on the condition, however, that the copier pay the stated per copy fee through the Copyright Clearance Center, Inc., 21 Congress Street, Salem, MA 01970, for copying beyond that permitted by Sections 107 or 108 of the U.S. Copyright Law. This consent does not extend to copying or transmission by any means—graphic or electronic—for any other purpose, such as for general distribution, for advertising or promotional purposes, for creating a new collective work, for resale, or for information storage and retrieval systems. The copying fee for each chapter is indicated in the code at the bottom of the first page of the chapter.

The citation of trade names and/or names of manufacturers in this publication is not to be construed as an endorsement or as approval by ACS of the commercial products or services referenced herein; nor should the mere reference herein to any drawing, specification, chemical process, or other data be regarded as a license or as a conveyance of any right or permission, to the holder, reader, or any other person or corporation, to manufacture, reproduce, use, or sell any patented invention or copyrighted work that may in any way be related thereto. Registered names, trademarks, etc., used in this publication, even without specific indication thereof, are not to be considered unprotected by law.

PRINTED IN THE UNITED STATES OF AMERICA

Second printing 1985

Author Index 587
Subject Index 588

30. Thermal Expansion Coefficients of Leadless Chip Carrier Compatible
    Printed Wiring Boards 379
    Z. N. Sanjana, R. S. Raghava, and J. R. Marchetti

POLYMERS FOR SPECIAL FUNCTIONS

31. Piezoelectric Poly(vinylidene fluoride) in Small-Bore, Thick-Walled
    Tubular Form: Continuous Production and Properties 399
    P. Pantelis

32. Polymeric Reactions in Magnetic Recording Media 409
    Lesley J. Goldstein

CONDUCTIVE POLYMERS

33. Fabrication of Conductive Polyimide-Gate Transistors 423
    David R. Day

34. Redox Properties of Conjugated Polymers: A Successful Correlation
    of Theory and Experiment 433
    R. R. Chance, D. S. Boudreaux, J. L. Bredas, and R. Silbey

35. A Novel Phase of Organic Conductors: Conducting Polymer Solutions 447
    J. E. Frommer, R. L. Elsenbaumer, and R. R. Chance

36. Poly-p-phenylene Selenide and Poly-p-phenylene Telluride: Characterization
    and Assessment as Active Elements 461
    L. A. Acampora, D. L. Dugger, T. Emma, J. Mohammed, M. F. Rubner,
    L. Samuelson, D. J. Sandman, and S. K. Tripathy

37. Electrochemical Synthesis and Characterization
    of Poly(2,2'-bithiophene) 473
    M. A. Druy, R. J. Seymour, and S. K. Tripathy

38. The Influence of Microstructure on the Properties of Polyacetylene/
    Polybutadiene Blends 487
    S. K. Tripathy and M. F. Rubner

39. Ethylene–Propylene–Diene Terpolymer/Polyacetylene and Styrene–Diene
    Triblock Copolymer/Polyacetylene Blends: Characterization and Stability
    Studies 497
    Kang I. Lee and Harriet Jopson

40. Conductive Hybrids Based on Polyacetylene: Copolymers and Blends 507
    M. E. Galvin, G. F. Dandreaux, and G. E. Wnek

41. Electrically Conductive Polymer Composites
    of 7,7,8,8-Tetracyanoquinodimethane (TCNQ) Salt Dispersion:
    Influence of Charge-Transfer Interaction and Film Morphology 515
    Oh-Kil Kim

42. Synthesis and Properties of Conducting Films by Plasma Polymerization
    of Tetramethyltin 533
    R. K. Sadhir and W. J. James

43. Plasma Polymerized Organometallic Thin Films: Preparation
    and Properties 555
    R. K. Sadhir, H. E. Saunders, and W. J. James

44. Polyacetylene, $(CH)_x$: An Electrode-Active Material in Aqueous
    and Nonaqueous Electrolytes 575
    R. B. Kaner, A. G. MacDiarmid, and R. J. Mammone

13. The Radiation Degradation of Poly(2-methyl-1-pentene sulfone): Radiolysis Products 153
    T. N. Bowmer and M. J. Bowden
14. Novolac Based Positive Electron Beam Resist Containing a Polymeric Dissolution Inhibitor: Preparation and Exposure Characteristics 167
    Hiroshi Shiraishi, Asao Isobe, Fumio Murai, and Saburo Nonogaki
15. Molecular Design for Cross-linking Negative Resists: Relationship Between Sensitivity and Component Ratio in Copolymer Resists 177
    Katsumi Tanigaki, Yoshitake Ohnishi, and Shozo Fujiwara
16. Molecular Design for Cross-linking Negative Resists: Optimum Design for Poly(chloromethylstyrene-co-2-vinylnaphthalene) 191
    Yoshitake Ohnishi, Katsumi Tanigaki, and Akihiro Furuta
17. Poly(butadiene-co-glycidyl methacrylate) as Negative Electron Resist 201
    B. Bednář, J. Devátý, J. Králíček, and J. Zachoval
18. Plasma Developable Electron Resists 213
    Juey H. Lai

POLYIMIDES

19. Solution Characterization of Polyamic Acids and Polyimides 227
    P. Metzger Cotts and W. Volksen
20. Ultrapure Polyimides: Synthesis and Applications 239
    J. Duran and N. S. Viswanathan
21. Photosensitive Polyimide Siloxane 259
    Gary C. Davis

ENCAPSULANTS

22. Catalysts for Epoxy Molding Compounds in Microelectronic Encapsulation 273
    Winston C. Mih
23. Thermogravimetric Analysis of Silicone Elastomers as Integrated Circuit Device Encapsulants 285
    Ching-Ping Wong
24. Removable Polyurethane Encapsulants 305
    K. B. Wischmann
25. Improvements to Microcircuit Reliability by the Use of Inhibited Encapsulants 313
    F. W. Ainger, J. Brettle, I. Dix, and M. T. Goosey

POLYMERS FOR PRINTED WIRING

26. Photopolymer Dielectrics: The Characterization of Curing Behavior for Modified Acrylate Systems 325
    R. D. Small, J. A. Ors, and B. S. H. Royce
27. Morphology of Rubber-Modified Photopolymers 345
    J. A. Ors and J. B. Enns
28. UV Solder Masks as Insulators for Printed Circuit Boards 367
    Neil S. Fox
29. UV-Curable Conformal Coatings 373
    C. R. Morgan, D. R. Kyle, and R. W. Bush

CONTENTS

Preface ix

PHOTORESISTS

1. **Applications of Photoinitiated Cationic Polymerization to the Development of New Photoresists** 3
   J. V. Crivello

2. **Applications of Photoinitiators to the Design of Resists for Semiconductor Manufacturing** 11
   Hiroshi Ito and C. Grant Willson

3. **Semiempirical Calculations of Electronic Spectra: Use in the Design of Mid-UV Sensitizers** 25
   R. D. Miller, D. R. McKean, T. L. Tompkins, N. Clecak, C. Grant Willson, J. Michl, and J. Downing

4. **Mid-UV Photosensitization of Diazoquinone Positive Photoresists** 41
   W. T. Babie, M-F. Chow, and W. M. Moreau

5. **Deep UV Photolithography with Composite Photoresists Made of Poly(olefin sulfones)** 55
   H. Hiraoka and L. W. Welsh, Jr.

6. **Method for a Comparative Study of Positive Photoresist Lithographic Performance** 65
   C. C. Walker and J. N. Helbert

7. **Dependence of Dissolution Rate on Processing and Molecular Parameters of Resists** 79
   A. C. Ouano

8. **Effect of Composition on Resist Dry-Etching Susceptibility: Novel Vinyl Polymers** 91
   J. N. Helbert, M. A. Schmidt, C. Malkiewicz, E. Wallace, Jr., and C. U. Pittman, Jr.

ELECTRON BEAM RESISTS

9. **Resists for Electron Beam Lithography** 103
   Toshiaki Tamamura, Saburo Imamura, and Shungo Sugawara

10. **Chain-Scission Yields of Methacrylate Copolymers Under Electron Beam Radiation** 119
    C. C. Anderson, P. D. Krasicky, F. Rodriguez, Y. Namaste, and S. K. Obendorf

11. **Poly(tetrafluorochloropropyl methacrylate) as Positive Electron Resist** ... 129
    B. Bednář, J. Devátý, J. Králíček, and J. Zachoval

12. **Radiation-Induced Degradation of Poly(2-methyl-1-pentene sulfone): Kinetics and Mechanism** 135
    M. J. Bowden, D. L. Allara, W. I. Vroom, J. Frackoviak, L. C. Kelley, and D. R. Falcone

FOREWORD

The ACS SYMPOSIUM SERIES was founded in 1974 to provide a medium for publishing symposia quickly in book form. The format of the Series parallels that of the continuing ADVANCES IN CHEMISTRY SERIES except that in order to save time the papers are not typeset but are reproduced as they are submitted by the authors in camera-ready form. Papers are reviewed under the supervision of the Editors with the assistance of the Series Advisory Board and are selected to maintain the integrity of the symposia; however, verbatim reproductions of previously published papers are not accepted. Both reviews and reports of research are acceptable since symposia may embrace both types of presentation.

ACS Symposium Series

M. Joan Comstock, *Series Editor*

Advisory Board

Robert Baker
U.S. Geological Survey

Martin L. Gorbaty
Exxon Research and Engineering Co.

Herbert D. Kaesz
University of California—Los Angeles

Rudolph J. Marcus
Office of Naval Research

Marvin Margoshes
Technicon Instruments Corporation

Donald E. Moreland
USDA, Agricultural Research Service

W. H. Norton
J. T. Baker Chemical Company

Robert Ory
USDA, Southern Regional
  Research Center

Geoffrey D. Parfitt
Carnegie Mellon University

Theodore Provder
Glidden Coatings and Resins

James C. Randall
Phillips Petroleum Company

Charles N. Satterfield
Massachusetts Institute of Technology

Dennis Schuetzle
Ford Motor Company
  Research Laboratory

Davis L. Temple, Jr.
Mead Johnson

Charles S. Tuesday
General Motors Research Laboratory

C. Grant Willson
IBM Research Department

PREFACE

POLYMERS ARE USED IN MANY AREAS OF ELECTRONICS. The purpose of the symposium upon which this book is based was to bring forward the latest work in photon and electron beam resists, polymers for device applications, polyimides, polymers for printed wiring, and conductive polymers. Rapid development is apparent in areas where polymer chemistry intersects semiconductor device fabrication, dielectric materials, information storage media, and organic conductors. Many times, fundamental research can give rise to entirely new materials and processes. For example, research on photoinitiation is applicable to resists, barrier coatings, encapsulants, and printed wiring board technologies. These rather diverse technologies are based firmly in polymer chemistry; this fact underscores the vitality of the science of macromolecules and the versatility of polymers in electronic applications.

Support from the Xerox Corporation enabled the Editor to organize this symposium. Special thanks go to Stephen F. Pond of the Xerox Webster Research Center. Thanks also go to the session chairpersons, the referees who reviewed each chapter of this volume, the staff of the ACS Books Department, and the contributors for their efforts.

THEODORE DAVIDSON
Xerox Corporation
Webster Research Center
Rochester, NY 14644

PHOTORESISTS

Applications of Photoinitiated Cationic Polymerization to the Development of New Photoresists

J. V. CRIVELLO

Corporate Research & Development Center, General Electric Company, Schenectady, NY 12301

> The recent development of several new classes of highly efficient photoinitiators for cationic polymerization makes possible their application in the design of novel photoresists. The concepts on which these imaging processes are based are set forth in this article.

Although both positive and negative working photoresists based on photoinduced condensation and free-radical chemistry are well-known, cationic polymerization chemistry has received little attention for the fabrication of photoresists. The recent development of several new classes of practical photoinitiators for cationic polymerization has now made it possible to utilize this chemistry in a number of ways to produce highly sensitive photoresists (1-6). The facile synthesis of onium salts I-III together with their ready structural modification to manipulate

$$Ar_2I^+ \ X^- \qquad Ar_3S^+ \ X^- \qquad Ar\text{-}\overset{O}{\overset{\|}{C}}\text{-}CH_2\text{-}\overset{+}{S}R_2 \ X^-$$

$$\text{I} \qquad\qquad \text{II} \qquad\qquad \text{III}$$

their spectral absorption characteristics make them very attractive for photoresist applications. Using the above photoinitiators together with monomers which undergo cationic polymerization, acid catalyzed polymerization and polymers which undergo acid catalyzed depolymerization, one can prepare a variety of unique and novel photoresists. In the present paper, several examples of these concepts are demonstrated.

Results and Discussion

<u>Mechanism of Photolysis.</u> When diaryliodonium (I) and triarylsulfonium (II) salts are irradiated at wavelengths from 200-300 nm,

0097-6156/84/0242-0003$06.00/0
© 1984 American Chemical Society

they undergo irreversible photolysis with rupture of a carbon-iodine or carbon-sulfur bond as shown in Equations 1 and 2.

$$Ar_2I^+ X^- \xrightarrow{h\nu} Ar\cdot + ArI^{+\cdot} X^- \qquad (1)$$

$$Ar_3S^+ X^- \xrightarrow{h\nu} Ar\cdot + Ar_2S^{+\cdot} X^- \qquad (2)$$

Interaction of the respective cation-radicals with solvent or monomer, R-H, results in the release of a proton and in the formation of protonic acids as depicted in Equations 3-7.

$$ArI^{+\cdot} + R-H \longrightarrow ArI^+-H\ X^- + R\cdot \qquad (3)$$

$$ArI^+-H\ X^- \longrightarrow ArI + HX \qquad (4)$$

$$Ar_2S^{+\cdot} + R-H \longrightarrow Ar_2S^+-H\ X^- + R\cdot \qquad (5)$$

$$Ar_2S^+-H\ X^- \longrightarrow Ar_2S + HX \qquad (6)$$

$$2\ Ar_2S^{+\cdot} + H_2O \longrightarrow Ar_2SO + Ar_2S + 2H^+ \qquad (7)$$

Dialkylphenacylsulfonium salts (III), in contrast, undergo reversible dissociation on photolysis with generation of ylids and protonic acids (6).

$$Ar-\overset{O}{\underset{\|}{C}}-CH_2-\overset{+}{S}R_2\ X^- \underset{\Delta}{\overset{h\nu}{\rightleftarrows}} Ar-\overset{O}{\underset{\|}{C}}-CH_2-S\overset{R}{\underset{R}{\diagdown}} + HX \qquad (8)$$

$$\downarrow$$

$$Ar-\overset{O}{\underset{\|}{C}}-CH=S\overset{R}{\underset{R}{\diagdown}}$$

Although onium salts I-III are efficiently photolyzed on irradiation, they display little tendency toward thermal decomposition even when exposed to elevated temperatures (150°C) for long periods of time (1-2 hours). This excellent thermal stability allows resists containing these photoinitiators to be subjected to long bake cycles without danger of premature degradation of their sensitivity.

<u>Photosensitization</u>. The chief absorption bands for onium salts I-III lie below 230 nm in the ultraviolet region. Photolysis at longer wavelengths is, therefore, ineffective. However, the photolysis of these compounds can be readily photosensitized to respond to long wavelength uv and to visible light. The mechanism of photosensitization involves an electron transfer process

between the photosensitizer and the onium salt as shown in Equations 9-12 for diaryliodonium salts (7,8).

$$P \xrightarrow{h\nu} P^* \qquad (9)$$

$$P^* + Ar_2I^+ X^- \longrightarrow [P\cdots Ar_2I^+ X^-]^* \qquad (10)$$

$$[P\cdots Ar_2I^+ X^-]^* \longrightarrow P^{+\cdot} X^- + Ar_2I\cdot \qquad (11)$$

$$Ar_2I\cdot \longrightarrow Ar\cdot + ArI \qquad (12)$$

Typical photosensitizers for diaryliodonium salts are condensed ring aromatic hydrocarbons, diaryl ketones, and acridinium dyes. Condensed ring aromatic hydrocarbons are particularly effective photosensitizers for triarylsulfonium salts. The use of photosensitizers in onium salt photolysis permits the photoimaging processes induced by these compounds to be optimally fitted to the specific irradiation source used for their exposure.

Photosensitizers can also be used to great advantage in multilevel photoresists. A typical system incorporates a long wavelength uv or visible light photosensitized resist on the top level and a deep uv sensitive resist on the bottom. Both resists may be based on cationic photoinitiators or one may be of this type while the other may be a conventional resist material.

Photoimaging Processes Based on Onium Salts

Cationic Polymerization. Protonic acids such as HBF_4, HPF_6, $HSbF_6$, etc., derived from the photolysis of onium salts I-III are well known initiators of cationic polymerization (9). In equations 13-15 is shown the proposed mechanism of cationic polymerization using a triarylsulfonium salt and a typical monomer, M.

Photolysis

$$Ar_3S^+ X^- \xrightarrow[R-H]{h\nu} Ar_2S + ArH + HX_{(S)} \qquad (13)$$

Initiation

$$M + HX_{(S)} \rightleftharpoons HM^+ X^- \qquad (14)$$

Propagation

$$HM^+ X^- + nM \longrightarrow H(M)_n M^+ X^- \qquad (15)$$

Virtually all known types of cationically polymerizable monomers are polymerized using onium salt photoinitiators. Included among these monomers are epoxides, cyclic ethers, mono and polyfunctional vinyl compounds, spiroesters, spirocarbonates and cyclic silicones. Using such monomers, a wide variety of negative working resists can be developed. The most successful of these to date have been photoresists based on epoxy resins (10). Solid multifunctional novolac-epoxy resins are particularly attractive precursors for use in the fabrication of photoresists because the presence of multiple epoxide functionality in these resins enhances their apparent photoresponse. Similarly, vinyl ether difunctional monomers are exceptionally reactive in photoinitiated

$$\text{(novolac-epoxy resin)} \xrightarrow[Ar_2I^+ X^-]{h\nu} \text{(cured polymer)} \qquad (16)$$

cationic polymerization, and photoresists based on these materials have also been investigated.

$$n \begin{array}{c} CH=CH_2 \\ | \\ O \\ | \\ R \\ | \\ O \\ | \\ CH=CH_2 \end{array} \xrightarrow[Ar_3S^+ X^-]{h\nu} \begin{array}{c} +CH-CH_2\!\!\!+_n \\ | \\ O \\ | \\ R \\ | \\ O \\ | \\ +CH-CH_2\!\!\!+_n \end{array} \qquad (17)$$

Acid Catalyzed Condensation Polymerizations. The strong protonic acids produced by the photolysis of onium salts I-III can also be employed to catalyze the condensation of phenolic, melamine, and urea formaldehyde resins. Very durable photoresists based on these inexpensive and readily available resins can be made. Such resists generally require a postbake prior to development to complete the condensation and to enhance image formation.

Acid Catalyzed Rearrangements. An example of a photoresist based on an acid catalyzed rearrangement is the diaryliodonium salt photoinduced cyclization of cis-1,4-polyisoprene shown in Equation 18. This facile cyclization which has been reported previously (11) by non-photochemical processes results in a polycyclic polymer whose physical properties and solubility characteristics are considerably different than the initial polymer. Exploitation of these differences in the exposed and unexposed regions of the polymer film permit their use as either positive or negative tone resists.

$$\text{(structure)} \xrightarrow[Ar_2I^+ X^-]{h\nu} \text{(structure)} \quad (18)$$

Another type of acid catalyzed rearrangement induced by diaryliodonium and triarylsulfonium salts was recently reported by Ito and Willson of IBM (12) and is depicted in Equation 19.

$$\begin{array}{c}-(CH_2-CH)_n- \\ | \\ \text{Ar} \\ | \\ O-\underset{\parallel}{C}-O-C(CH_3)_3 \\ O \end{array} \xrightarrow[Ar_3S^+ AsF_6^-]{\substack{h\nu \\ Ar_2I^+ AsF_6^-}} \begin{array}{c}-(CH_2-CH)_n- \\ | \\ \text{Ar} \\ | \\ OH \end{array} + CH=C\begin{array}{c}CH_3 \\ CH_3\end{array} + CO_2 \quad (19)$$

Imagewise exposure of the salts dissolved in the polymer film generates a local concentration of strong protonic acids. Upon baking at 100°C for a few seconds, the labile ester group is catalytically decomposed to isobutylene, carbon dioxide and poly-p-hydroxystyrene. Depending on the nature of the developing solvent, either negative or positive tone images are formed. Ito and Willson report that the resolution capability of this resist is about 0.75 μm.

Cationic Depolymerization. Many polymers containing heteroatoms in their main chain are known to undergo facile depolymerization when treated with a cationic initiator (13). Such polymers undergo reinitiation at the chain ends or along the backbone followed by a "backbiting" process which eliminates monomers and cyclic oligomers. Negative resists using onium salt photoinitiators can be devised based on this concept. An example is shown in Equations 20-23:

$$\sim O-CH_2-CH_2-O-CH_2-CH_2-O\sim \xrightarrow[Ar_3S^+ X^-]{h\nu} \sim O-CH_2-CH_2-\overset{H}{\underset{}{O^+}}-CH_2-CH_2-O\sim \quad (20)$$

$$\begin{array}{c}\sim O-CH_2-CH_2-\overset{H}{\underset{}{O^+}}-CH_2-CH_2 \\ \sim O-CH_2-CH_2-O-CH_2-CH_2\end{array} \longrightarrow \sim O-CH_2-CH_2-OH + \sim O-CH_2-CH_2-\overset{+}{O}\bigcirc O \quad (21)$$

$$\sim\sim O-CH_2-CH_2-O^+\!\!\bigcirc\!\!O \longrightarrow \sim\sim O-CH_2-CH_2^+ \;+\; O\!\bigcirc\!O \quad (22)$$

$$\sim\sim O-CH_2-CH_2^+ \longrightarrow \sim\sim O^+\!\!\bigcirc\!\!O \quad (23)$$

The strong protonic acids generated by the photolysis of triarylsulfonium salts provide a means of selectively depolymerizing those portions of a photoresist film exposed to light. Dimeric species are the major products of these depolymerization reactions, although in some cases higher cyclic oligomers are also produced. In most cases, the cyclic oligomers have sufficiently low boiling points that they volatilize either during exposure or during a postbake cycle. This feature makes these photoresists unique in that no solvent is required for their development, i.e., they are "dry developable".

Ito and Wilson (12) have extensively investigated another resist system of this type based on polyphthalaldehyde (Equation 24).

$$\left[\text{polyphthalaldehyde}\right]_n \xrightarrow[\text{Ar}_2\text{I}^+ \text{AsF}_6^-]{h\nu} n \;\text{(}o\text{-phthalaldehyde)} \quad (24)$$

This resist has excellent resolution capabilities and may be imaged with x-rays, by electron beam as well as with uv irradiation. Because phthalaldehyde is reasonably volatile, the resist is dry developable.

Acid Catalyzed Hydrolysis. When photosensitive onium salts are incorporated into a film of a hydrolytically sensitive polycarbonate and exposed to ultraviolet irradiation, the acid which is produced catalyzes chain scission at random sites along the backbone (Equation 25).

$$\sim\sim O\text{-Ar-Ar-}O\text{-}\overset{O}{\overset{\|}{C}}\text{-}O\text{-Ar-Ar-}O\sim\sim \xrightarrow[\text{H}_2O]{\overset{h\nu}{\text{Ar}_3\text{S}^+\text{X}^-}}$$

$$\sim\sim O\text{-Ar-Ar-}OH \;+\; HO\text{-Ar-Ar-}O\sim\sim \;+\; CO_2 \quad (25)$$

During development, one takes advantage of the higher dissolution rate of the degraded polymer vis-à-vis the intact polymer to generate a positive image of the mask. This concept is general in its application and has been extended to produce positive resists using a number of polyesters, polycarbonates, polysulfonates, polyacetals, and polyazomethines.

Simultaneous Radical and Acid Catalyzed Condensation Polymerization. As shown in Equations 1-7, the photolysis of diaryliodonium and triarylsulfonium salts produces in addition to strong protonic acids, a variety of radical fragments. These photoinitiators are, therefore, capable of initiating free radical polymerizations. A number of hybrid imaging systems which take advantage of both radical and acidic species formed from the photolysis of these salts have been designed. For example, Equation 26 illustrates one such system based on simultaneous radical and acid catalyzed condensation polymerizations which has been explored in our laboratory.

$$\begin{array}{c}CH_2OH\\|\\NH\\|\\CO\\|\\CH=CH_2\end{array} + (CH_2-CH(OH))_n \xrightarrow[Ar_3S^+X^-]{h\nu} (CH_2-CH)_n\ O\ CH_2\ NH\ CO\ (CH-CH_2)_m + H_2O \quad (26)$$

Aqueous polyvinyl alcohol solutions containing 5-20% N-methylolacrylamide and a triarylsulfonium salt are film cast onto a substrate and dried. On irradiation, the free radical polymerization of the N-methylolacrylamide proceeds rapidly.

Simultaneously, the two polymers undergo acid catalyzed crosslinking through the alcohol and methylol groups with the elimination of water. These photoresists are especially attractive since they are inherently low in cost, exhibit good photosensitivity and can be developed in neutral aqueous solutions. These systems, like the others described in this paper, can be dye sensitized to broaden their response to both long wavelength uv and to visible light.

Conclusions

Photosensitive onium salts provide the means for the development of novel positive and negative photoresists based on the concepts advanced in this paper. Work is currently proceeding both at this laboratory and in others to explore the unique opportunities for photoresists based on these materials as latent photochemical sources of strong protonic acids.

Literature Cited

1. Crivello, J. V.; Lam, J. H. W. Macromolecules 1977, 10 (6) 1307.
2. Crivello, J. V.; Lam, J. H. W. J. Polym. Sci., Symposium No. 56 1976, 383.
3. Smith, G. H. Belg. Patent 828 841, Nov. 7, 1975.
4. Crivello, J. V.; Lam, J. H. W. J. Polym. Sci., Polym. Chem. Ed. 1979, 17, 977.
5. Nemcek, J. Belg. Patent 833 472, March 16, 1976; Chem. Abstr. 1976, 85, 34024f.
6. Crivello, J. V.; Lam, J. H. W. J. Polym. Sci., Polym. Chem. Ed. 1979, 17, 2877.
7. Pappas, S. P.; Jilek, J. H. J. Photogr. Sci. & Eng. 1979, 23 (3) 141.
8. Crivello, J. V.; Lee, J. L. Macromolecules 1981, 14, 1141.
9. Kennedy, J. P.; Marechal, E. "Carbocationic Polymerization"; John Wiley: New York, 1982; p. 56.
10. Schlessinger, S. I. Polym. Eng. and Sci. 1974, 14 (7), 513.
11. Golub, M. A. "High Polymers"; Kennedy, J. P.; Tornquist, E. G. M., Eds.; Wiley (Interscience): New York, 1969; Vol. 23, pt. II, p. 939.
12. Ito, H.; Willson, C. G. Technical Papers, Society of Plastics Engineers Meeting 1982 p. 331.
13. Goethals, E. J. "The Formation of Cyclic Oligomers in the Cationic Polymerization of Heterocycles"; ADVANCES IN POLYMER SCIENCE Vol. 23, Springer Verlag: New York, 1977; p. 103.

RECEIVED September 2, 1983

Applications of Photoinitiators to the Design of Resists for Semiconductor Manufacturing

HIROSHI ITO and C. GRANT WILLSON

IBM Research Laboratory, San Jose, CA 95193

> Acids generated via radiolysis of cationic photoinitiators, such as aryldiazonium, diaryliodonium, and triarylsulfonium metal halides can induce a variety of chain reactions in polymers, including (1) cross-linking, (2) catalytic deprotection, and (3) catalytic depolymerization. All of these reactions can be exploited in the design of resist materials. When formulated with the onium salt photoinitiators, (1) epoxy resins undergo efficient cross-linking reactions to provide negative tone resists; (2) poly(p-t-butoxy carbonyloxystyrene) is converted to poly(p-hydroxystyrene) in the exposed areas to provide either a positive or negative resist tone depending on the developer employed; and (3) polyaldehydes are completely reverted to monomer upon exposure and vaporize during the exposure to provide self-developing positive resists. The extremely high sensitivity of these systems is achieved through *chemical amplification*.

The continuing trend toward higher circuit density in microelectronic devices has motivated research efforts in a variety of high resolution lithography techniques, including electron beam, x-ray, and deep UV (200-300 nm) technologies. Each of these new technologies demands a highly sensitive resist system in order to be of practical use. We have sought resist systems designed to meet the high sensitivity requirements of these microlithography techniques, and have found that certain cationic photoinitiators provide a basis for the design of several such systems. These systems achieve their high sensitivity as a result of designing the polymer matrix to provide *chemical amplification* of the initial photoreaction.

In each system, the primary photo-event is dissociation of the cationic photoinitiator to produce an acid. This reaction proceeds with a quantum efficiency that is characteristic of the particular initiator. The photogenerated acid then interacts with a carefully chosen polymer matrix to initiate a chain reaction, or acts as a catalyst, such that a single molecule of photogenerated acid serves to initiate a cascade of bond making or breaking reactions. The effective quantum efficiency of the overall process is the product of the photolysis reaction efficiency times the length of the chain reaction (or the catalytic chain length). This multiplicative response constitutes gain or *chemical amplification* of the initial photo-event and provides for very high effective quantum efficiency (1-5).

The acid generating photoinitiators that provide the basis for our studies are onium salts that have been described in the work of Schlesinger (6,7) and Watt (8) and by Crivello (9) who pioneered their use as initiators for photocuring of coatings. The initiators include aryldiazonium salts that generate Lewis acids upon photolysis and diaryliodonium and triarylsulfonium salts that generate strong Bronsted acids (Scheme 1).

$$ArN_2^+ MX_n^- \xrightarrow{h\nu} ArX + N_2 + MX_{n-1}$$

$$Ar_2I^+ MX_n^- \xrightarrow{h\nu} ArI + HMX_n + \text{others}$$

$$Ar_3S^+ MX_n^- \xrightarrow{h\nu} Ar_2S + HMX_n + \text{others}$$

$MX_n = BF_4, PF_6, AsF_6, SbF_6$, etc. Scheme 1. Photochemistry of onium salts.

The acids that are generated by photolysis of these salts have been exploited in the cross-linking of polymers and oligomers containing oxirane or thiirane rings (6-12). Very sensitive, negative tone resists have been developed on the basis of these reactions (13). We have found that these photogenerated acids are also capable of catalyzing the scission of carbon-heteroatom linkages in certain polymers. These cleavage reactions have been utilized to design both positive and negative resists.

Results and Discussion

Negative Mode (Cross-linking). Dubois et al. (13) have described the preparation of a sensitive negative resist based on onium salt sensitization of copolymers of 2,3-epithiopropyl methacrylate and methyl methacrylate and Crivello and co-workers have used a similar reaction to generate low resolution images in epoxy resins (14). The sensitivity of these systems derives from the chain reaction character of the ring opening polymerization (cross-linking) that is responsible for their function. This is an example of chemical amplification. We were therefore tempted to explore high resolution image generation in epoxy resins.

Celanese Epi-Rez SU-8 was chosen for these studies. This resin is a polyfunctional epoxy novolac resin (Scheme 2) made from bisphenol-A. It is optically

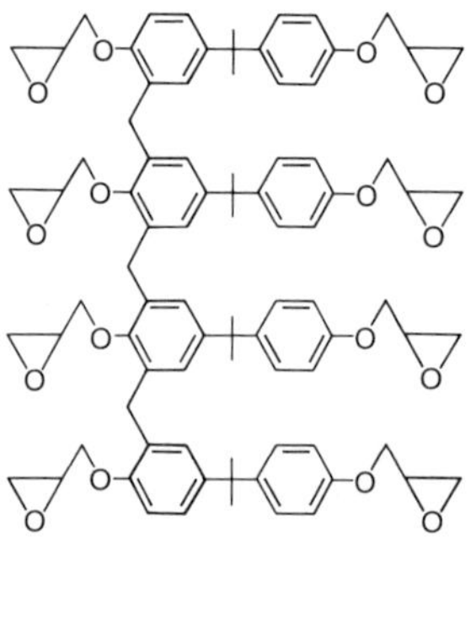

Scheme 2. Epi-Rez SU-8 (Celanese).

dense and strongly absorbing below 300 nm hence, we chose diazonium salts that absorb above 300 nm as initiators. p-n-Hexyloxybenzenediazonium metal halides were prepared for this purpose. They have an absorption maximum near the high intensity 313 nm mercury emission with $\varepsilon = 12{,}000\text{-}17{,}000$ and are soluble in organic solvents that will also dissolve SU-8. The SU-8 resin (4 g) was dissolved in 2-butanone (6 ml) to which was added the diazonium hexafluorophosphate (0.2 g). Resist films were prepared by spin coating the resulting solution on silicon wafers. The films were baked at 85° for 10 minutes to give 6 μ thick coatings. Because the diazonium salt begins to undergo thermolysis at 97°C (TGA), the baking conditions had to be carefully controlled. The onium salt-epoxy resist films were exposed through a patterned mask to 313 nm radiation, then developed in a mixture of $CHCl_3$ and hexanes (2:1 by volume) for 90 seconds, rinsed with hexanes for 10 seconds, then with CH_2Cl_2 for 15 seconds. Scanning electron micrographs of the resulting negative resist patterns are provided as Figure 1.

The patterns generated in the SU-8 resist show excellent image quality and slightly undercut profiles. However, the swelling phenomena that always occur during the development of negative resists which function on the basis of cross-linking reactions became apparent in smaller features, and limit the resolution of this system. The lower resolution limit depends on film thickness, developer and rinse composition, and on a variety of other process variables. No attempt was made to determine the ultimate resolution limits of the system but, it is clearly possible to generate images with minimum features in the 1 to 2 μ range with aspect ratios near 1.

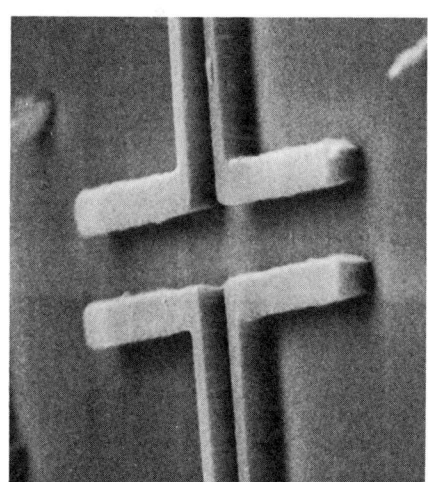
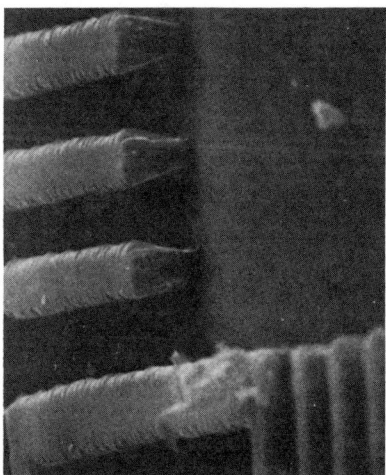

Figure 1. Negative patterns in SU-8.

Positive or Negative Mode (Catalytic Deprotection). We have recently focussed our attention on the design of positive tone resist systems that incorporate chemical amplification. One such new class of positive resist materials is based on the photochemical generation of a substance that catalyzes the cleavage of polymer pendant groups. Here, the effective quantum efficiency of the system is the product of the quantum yield of the photochemical event and the number of side chain transformations achieved by one molecule of the catalyst. This multiplicative response constitutes gain of the sort we seek. One polymer upon which the new system can be based is poly(p-t-butoxycarbonyloxystyrene) (PBOCST) (15) (Scheme 3). The initiators for this system are again various onium salts that undergo efficient photolysis to produce strong acids.

PBOCST is thermally stable to ca. 200°C. Above 200°C, the polymer loses about 45% of its weight as carbon dioxide and isobutene (15). Diphenyliodonium and triphenylsulfonium hexafluoroarsenates are thermally stable to ca. 250° and 300°, respectively. (16,17) Consequently, the resist formulated from PBOCST and these salts is stable to the baking conditions required for formation of high quality spin coated films, and the formulations have a long shelf life when stored at room temperature under yellow light.

The t-butoxycarbonyl group (t-BOC) and t-butyl esters have been widely employed as protecting groups in peptide synthesis (18). These structures are labile toward strong acids. In fact, while studying the cationic polymerization of p-t-butoxycarbonyloxy-α-methylstyrene, we noticed that a substantial amount of the t-BOC protecting group was cleaved during polymerizations initiated with BF_3OEt_2 or PF_5 (19). This acidolysis reaction can be conveniently monitored by either UV or IR spectroscopy.

Figure 2a shows the IR spectrum of a 1 μ thick PBOCST film containing 20% w/w of diphenyliodonium hexafluoroarsenate. The film was prepared by spin coating and was then baked at 100°C for 30 minutes. The thermal stability of the resist system is evidenced by the observation that no measurable phenolic hydroxyl band is generated during the baking process. After exposure to UV light, only a slight spectral change is observed, indicating that exposure alone is not an efficient means of deprotection. However, upon baking the exposed film at 100° for one minute, the reaction was brought to completion as evidenced by disappearance of the characteristic t-BOC carbonyl band at 1755 cm^{-1} and the appearance of a strong phenolic OH stretching absorption (Figure 2c). The deprotection reaction can also be followed by UV spectroscopy since the long wavelength absorption band of poly(p-hydroxystyrene) (PHOST) is both more intense and red shifted from its precursor, PBOCST (Figures 3 and 4). The acid produced by photolysis of the onium salt catalyzes thermolysis (acidolysis) of the t-BOC group via an A_{AL}-1 type mechanism (20). The reaction is slow at room temperature but very fast at 100°C, a temperature well below that required to thermolyze the protecting group in the absence of an acid catalyst. It should be noted that the photogenerated acid is not consumed in the t-BOC cleavage reaction. It serves only as a catalyst. Therefore, one mole of acid can cleave many t-BOC side chain groups, thus providing the desired chemical amplification of the initial photoreaction.

After exposure and post baking, the exposed areas of the t-BOC resist films are converted into PHOST. This is a polar, even acidic polymer that is soluble in a range of polar solvents such as alcohols or aqueous base, in which the precursor, PBOCST, is totally insoluble. Conversely, PBOCST is soluble in many nonpolar, lipophilic solvents in which the phenolic PHOST is insoluble. This large change in polarity that results

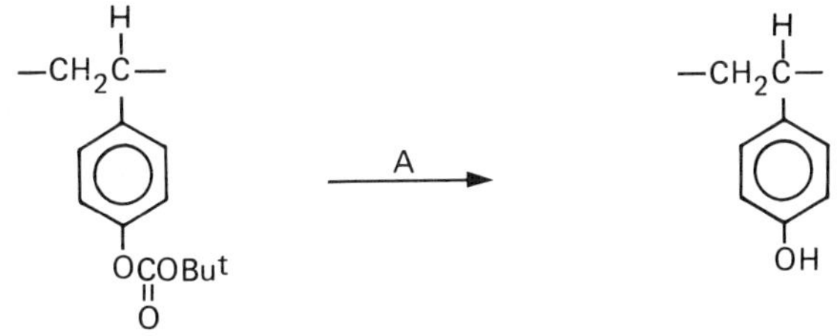

Scheme 3. Acidolysis of PBOCST.

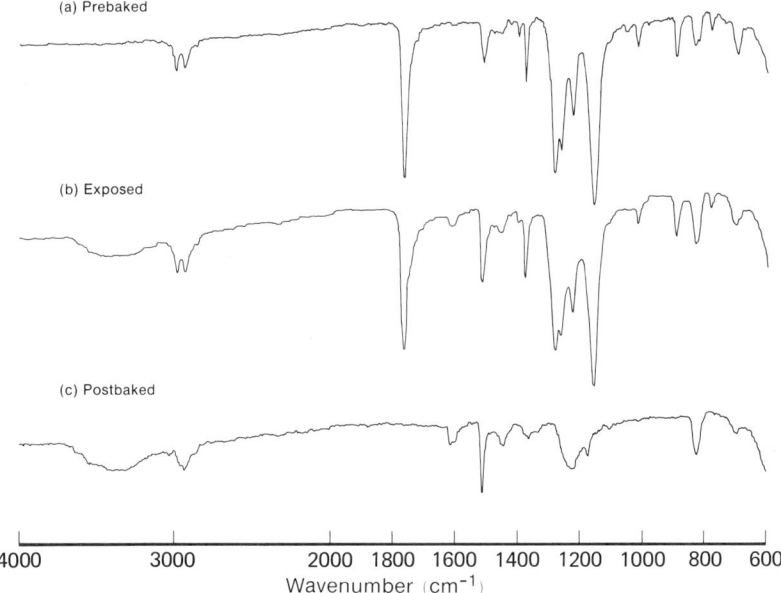

Figure 2. IR spectra of PBOCST/Ph_2IAsF_6: (a) before exposure, (b) after exposure, and (c) after postbake.

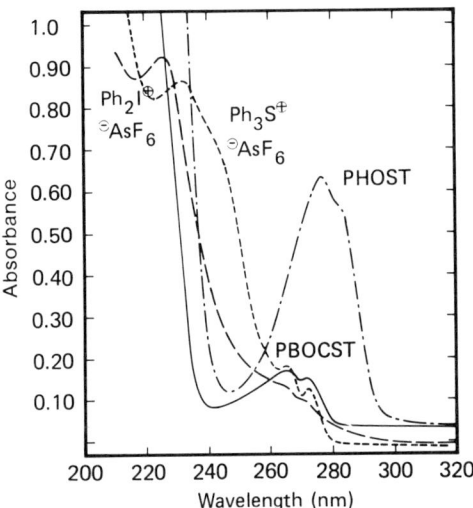

Figure 3. UV spectra of Ph_2IAsF_6, Ph_3SAsF_6, PBOCST, and PHOST.

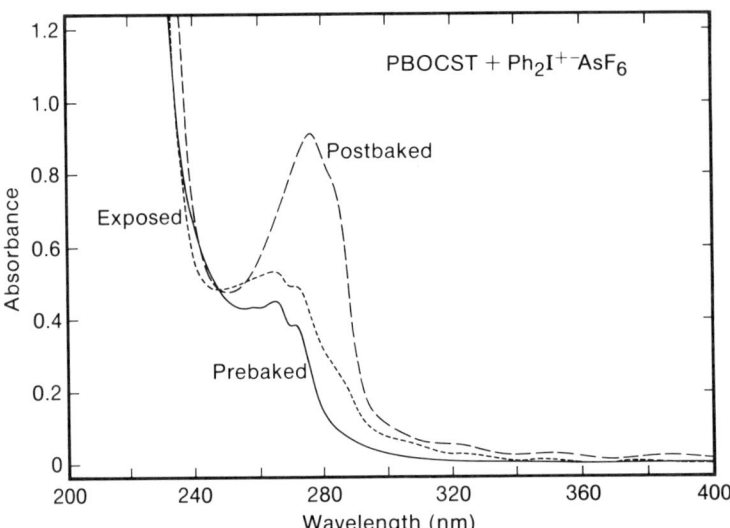

Figure 4. UV spectra of $PBOCST/Ph_2IAsF_6$ before and after exposure and after postbake.

from exposure and baking allows the exposed films to be developed either as negative or positive resists. Use of a nonpolar developer selectively dissolves the unexposed (PBOCST) areas of resist films to give negative tone images. Development with a polar solvent selectively dissolves the exposed (PHOST) areas of the films to give positive tone images. The solubility differentiation in the system is solely dependent on a change in the polarity of the repeating units, and thus its mechanism of action is quite different than that of either the familiar positive tone resists based on dissolution inhibitors, such as the diazonaphthoquinone/novolac systems, or resists that funcion on the basis of polymer main chain scission reactions, such as PMMA or the cross-linking negative systems. Since the phenolic, PHOST polymer is totally insoluble in solvents that dissolve PBOCST, negative tone images in this new resist do not suffer from resolution limitations due to swelling. No evidence of swelling is observed in either the positive or negative tone developers.

Resist systems that function on the basis of a change in the polarity of a side chain are few, and none apparently suffer from swelling phenomena. IBM researchers have reported a negative resist system that presumably functions on the basis of insolubilization by salt formation. This system is based on alkyl halide sensitized films of polystyrene bearing tetrathiofulvalene pendant groups (21). We have reported that poly(t-butyl methacrylate) is converted to poly(methacrylic acid) by reaction with acids generated via photolysis of onium salts (1,4) and recently, Hatada et al. reported a related system in which poly(α,α-dimethylbenzyl methacrylate) is transformed into poly(methacrylic acid) upon e-beam radiation which allows aqueous base development (22). In none of these cases is there reported any evidence of swelling during development.

The absorbance spectra of PBOCST and PHOST are provided as Figure 3. PBOCST is very transparent at 240-300 nm. The iodonium and sulfonium salts both absorb significantly in this spectral region but do not absorb significantly above 300 nm. Consequently, the PBOCST-onium salt resist is ideally suited to deep UV lithography (23). The resist can, however, be sensitized to longer wavelength through incorporation of dyes. The effect of polynuclear aromatics has been studied by Crivello (24,25) and by Pappas (26). These compounds have proven to be effective in the PBOCST system. Sensitization of resist formulations based on iodonium salts with pyrene renders them sensitive in the mid-UV (313 nm), and addition of perylene provides sensitivity in the near UV (365, 404, 436 nm). Figure 5 documents the conversion of PBOCST to PHOST by perylene sensitization of diphenyliodonium hexafluoroarsenate to 365 nm radiation. Preliminary investigations have shown that the PBOCST-onium salt systems are also sensitive to electron beam exposure. We have also demonstrated that polymers containing t-butyl ester side chains can be imaged via incorporation of onium salts. These include poly(t-butyl methacrylate), poly(t-butyl p-vinylbenzoate) and poly(t-butyl p-isopropenylphenyloxyacetate).

We have succeeded in producing excellent, high resolution images in the PBOCST-onium salt, deep UV resists. Films of PBOCST (M_n 30K) containing 20 wt% of diphenyliodonium hexafluoroarsenate were spin coated on silicon wafers from cyclohexanone then baked at 100° for 30 minutes to provide striation free 1.25 μ thick films. The resist was exposed through a quartz mask and a narrow band width, 254 nm filter using an Oriel Associates Inc. Model 780, Hg-Xe lamp. The films were baked at 100° for 10 seconds following exposure. Positive images were obtained by development in a mixture of Shipley MF 312 (aqueous tetramethylammonium hydroxide solution) and water (1:1 by volume) (Figure 6a) and negative images were produced by development in a mixture of CH_2Cl_2 and hexane (2:1 by volume) (Figure 6b). Other positive developers, such as Shipley AZ 2401 and isopropanol, work equally well.

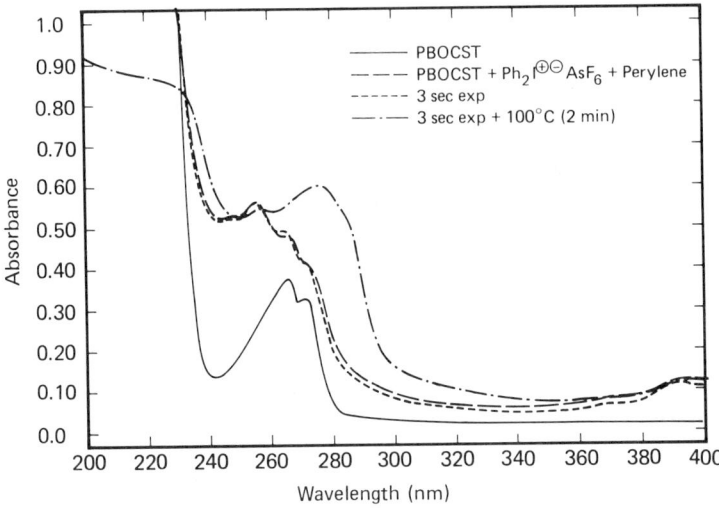

Figure 5. UV spectra of PBOCST/Ph$_2$IAsF$_6$/Perylene before and after exposure to 365 nm radiation and after postbake.

a   b

Figure 6. Positive (a) and negative (b) images in PBOCST/Ph$_2$IAsF$_6$.

Positive Mode (Catalytic Depolymerization). In the conceptual design of these systems, amplification is achieved by choosing polymers, which, when subjected to a single radiation induced bond scission, undergo a spontaneous depolymerization that results in complete reversion to monomer. The polymers upon which the new system is based are polyaldehydes. Aldehydes undergo anionic or cationic polymerization in a reversible, equilibrium reaction. Many of these systems have ceiling temperatures well below room temperature such that the polymerization must be carried out at cryogenic temperatures, and if the polymer is isolated, it rapidly depolymerizes to monomer. If, however, these polymers are end-capped by acylation or alkylation prior to isolation, they are often quite stable. Vogl et al. have extensively and intensively studied the equilibrium polymerization of aliphatic aldehydes (27). Most of these materials are intractable, insoluble substances because of their high crystallinity. Certain aromatic dialdehydes, such as phthalaldehyde and o-formylphenylacetaldehyde undergo cyclopolymerization with a ceiling temperature of approximately −40°C (28,29). After end-capping, these aromatic polyacetals are stable to 180°C, as evidenced by TGA analysis and, unlike their aliphatic counterparts, are soluble in common organic solvents. We have found polyphthalaldehyde to be particularly useful for our systems (Scheme 4).

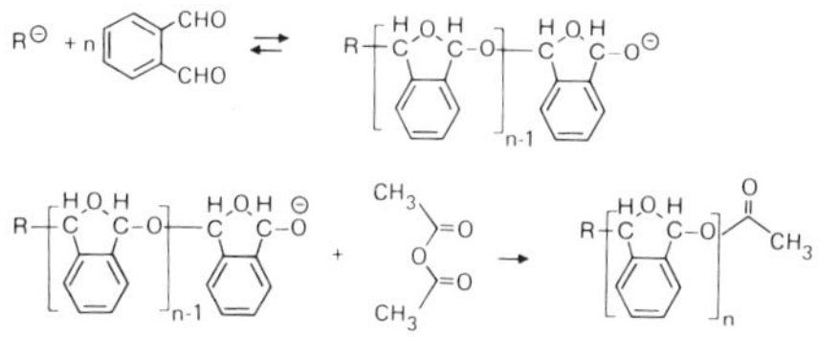

Scheme 4. Anionic polymerization of phthalaldehyde.

There have been earlier reports of the utility of polyaldehydes as imaging media. Solvent development of e-beam exposed polyphthalaldehyde has been reported (30). Aliphatic polyacetals with photosensitive end groups have been investigated as a photoimaging system in which the aldehyde monomer released from a polyacetal upon radiation followed by heating is used to cross-link certain polyamides (31). Sensitization of polyphthalaldehyde by addition of substances which generate acids upon radiation such as poly(vinyl chloride) or phenol derivatives has also been applied to an imaging technology on the basis of changes in optical density (32,33) and used as resist materials that produce images by heating and/or solvent development after exposure (34).

We have reported that unsensitized polyphthalaldehyde undergoes partial depolymerization when exposed to deep UV or e-beam radiation and that the sensitization of polyphthalaldehyde by copolymerization with o-nitrobenzaldehyde provides about 50% dry development upon heating after deep UV irradiation (3,5). Subsequently Hatada et al. reported that copolymers of aliphatic aldehydes are depolymerized to monomeric aldehydes upon exposure to e-beam or x-ray to provide highly sensitive self-developing positive resists (35). We have studied the sensitization of polyphthalaldehyde through incorporation of the onium salt photoinitiators in hope of achieving catalytic acidolysis of the acid-labile main chain acetal bonds (Scheme 5), a reaction of the sort observed by one of the authors (H.I.) during polymerization of anhydrosugars by reaction with the cationic initiator PF_5 at cryogenic temperatures (36). As expected, addition of triphenylsulfonium or diphenyliodonium hexafluoroarsenate (10 wt% to the polymer) has allowed imaging of 1 μ thick films of polyphthalaldehyde at 2-5 mJ/cm^2 of 254 nm radiation and at 1.0 μC/cm^2 of

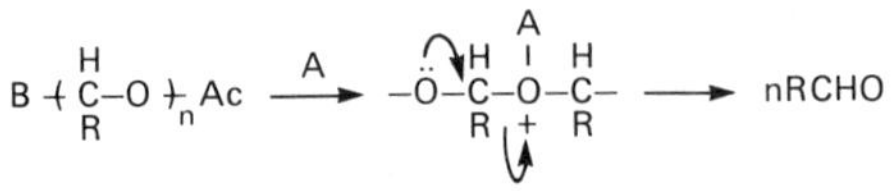

Scheme 5. Acidolysis and depolymerization of polyphthalaldehyde.

20 keV e-beam radiation. Clean relief patterns with resolution below 1 μ line width are generated by exposure alone even without heating. No development step is required.

In a typical experiment, polyphthalaldehyde was dissolved in bis-(2-methoxyethyl) ether or cyclohexanone, to which was added the onium salt at 10 wt% to the polymer. Films spin coated on Si wafers were baked at 100°C for 10 minutes and then image-wise exposed. Optical micrographs of the resist images generated upon UV, e-beam, and x-ray radiations are exhibited in Figures 7a, b, and c, respectively.

The UV sensitivity of the polyphthalaldehyde-onium salt system is dependent on the concentration and structure of the onium salts. At 10 wt% loading the sensitivity to narrow band width 254 nm radiation is 1-7 mJ/cm^2 and is insensitive to the structure of the salts. However, at 2 wt% loading of triphenylsulfonium or diphenyliodonium hexafluoroarsenates or p-hexyloxybenzenediazonium tetrafluoroborate much higher doses are required to achieve self-development and the sensitivity decreases with the salt structures in the sequence listed.

Formation of monomeric phthalaldehyde due to acidolysis and depolymerization of polyphthalaldehyde initiated by the photolysis of onium salts is clearly demonstrated by IR spectroscopy (Figure 8). Figure 8c shows vaporization of the aldehyde monomer upon heating. The monomeric phthalaldehyde is crystalline at room temperature but sublimes under vacuum without heating. We have successfully delineated self-developed images in a 2.8 μ thick film at a dose of 1.0 μC/cm^2 of 20 keV e-beam radiation, which were subsequently transferred to silicon oxide by treatment with either buffered HF or CF_4/O_2 plasma.

Dry developing resist materials that require no post-exposure development involving solvents have been sought for some time (37-41). The interest in such materials throughout the semiconductor industry stems from the potential they provide for higher throughput and higher device yields. The higher yields result from both the reduction in process steps and the fact that such materials provide the potential for carrying out exposure, development, and pattern transfer under high vacuum, thereby reducing the number of defects due to atmospheric contaminants and the development process. The polyphthalaldehyde/onium salt system is one of only two resists of which we are aware that undergo clean self-development without the need for post-exposure processing of any kind. The other resist is based on the copolymers of aliphatic aldehydes and was recently reported by Hatada et al. (35).

We are continuing to explore the limits of the PBOCST and polyphthalaldehyde resist systems, as well as other materials that incorporate chemical amplification.

Experimental

Materials. The epoxy resin used for the negative patterning was Epi-Rez SU-8 purchased from Celanese. Poly(p-t-butoxycarbonyloxystyrene) (PBOCST) was synthesized by radical polymerization of p-t-butoxycarbonyloxystyrene (BOCST), as described by Frechet et al. (15). Phthalaldehyde purchased from Aldrich Chemical Company was purified by recrystallization from CH_2Cl_2-hexanes and polymerized by

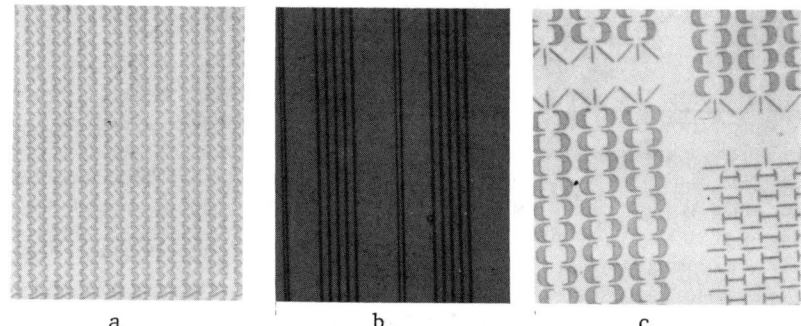

Figure 7. Self-developed images in polyphthalaldehyde: (a) 254 nm deep UV, Ph_3SAsF_6, (b) 20 keV e-beam, Ph_3AsF_6, and (c) Al-Kα x-ray, Ph_2IAsF_6.

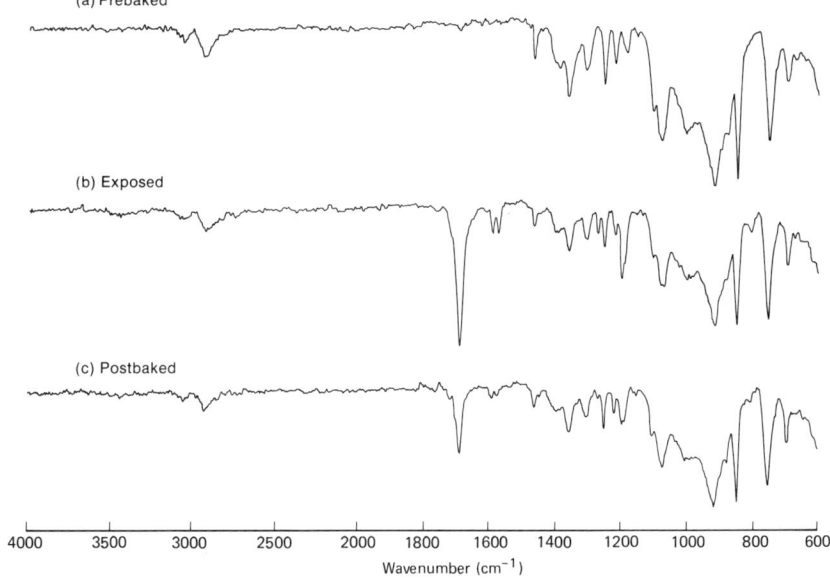

Figure 8. IR spectra of polyphthalaldehyde/Ph_2IAsF_6 before and after exposure to 254 nm radiation and after postbake.

anonic initiation with n-butyllithium, phenylmagnesium bromide, or potassium t-butoxide/18-crown-6 in THF at $-78°C$ under high vacuum with use of a break-seal technique. The phthalaldehyde polymerization was terminated by adding a cold mixture of acetic anhydride and pyridine at $-78°C$ to end-cap the polymer. The end-capped polymers are of high molecular weight (M_n=35,000-54,000 and M_w=43,000-100,000), stable to ca. 180°C, according to TGA analysis, soluble in common organic solvents, and can be spin cast from solution to provide clear, isotropic, noncrystalline films. The synthesis of the onium salt photoinitiators has been well documented in the literature (6-12).

Measurements. IR spectra were recorded on a Perkin Elmer 283 spectrometer and UV spectra on a Hewlett Packard Model 8450A UV/VIS spectrophotometer. Molecular weight measurements were made using a Wescan 230 or 231 recording membrane osmometer with toluene or THF as solvent while GPC measurements were made on a Waters Model 150 chromatograph equipped with six or eight μstyragel columns at 30° or 40°C in THF. TGA analyses were performed on a DuPont 951 or 1090 at a heating rate of 10°C/min.

Lithography. Electron beam exposures were carried out with an IBM vector scan e-beam exposure tool at 20 keV. X-ray exposures were carried out under vacuum by Al-Kα radiation, and UV exposures with a Cannon PLA 500, Oriel illuminator, Hybrid Technology Group Model 345-10, or Optical Associates Inc. Model 780 in contact mode.

Acknowledgment

Synthesis of polyphthalaldehyde, PBOCST, and early contributions to the chemical amplification concept by Professor Jean Frechet of the University of Ottawa are gratefully acknowledged. The authors thank S. MacDonald, R. Kwong and R. Cox for preparation of various diazonium salts, and J. Carothers, D. Mathias, C. Hrusa, and C. Cole for assistance in TGA, GPC, and ^{13}C NMR measurements. The iodonium and sulfonium salts used in the preliminary stages of this work were generously provided by V. Lee.

Literature Cited

1. Ito, H.; Willson, C. G.; Frechet, J. M. J. 1982 Symposium on VLSI Technology, Oiso, Japan, September 1982.
2. Frechet, J. M. J.; Ito, H.; Willson, C. G. Microcircuit 82, Grenoble, France, October 1982.
3. Willson, C. G.; Ito, H.; Frechet, J. M. J. Microcircuit 82, Grenoble, France, October 1982.
4. Ito, H.; Willson, C. G.; Frechet, J. M. J. SPE Regional Technical Conference on Photopolymers, Ellenville, New York, November 1982.
5. Ito, H.; Willson, C. G. SPE Regional Technical Conference on Photopolymers, Ellenville, New York, November 1982.
6. Schlesinger, S. I. Photog. Sci. Eng. 1974, 18, 387.
7. Schlesinger, S. I. Polym. Sci. Eng. 1974, 14, 513.
8. Watt, W. R. in "Epoxy Resin Chemistry"; Bauer, R. S., Ed.; ACS SYMPOSIUM SERIES No. 114, American Chemical Society: Washington, D.C., 1979, pp. 17-46.
9. Crivello, J. V. in "UV Curing: Science and Technology"; Pappas, S. P., Ed.; Technology Marketing Corporation: Stamford, Conneticut, 1978, pp. 23-77.
10. Ketley, A. D.; Tsao, J.-H. J. Rad. Curing 1979, 22.
11. Bal, T. S.; Cox, A.; Kemp, T. J.; Pinot de Moira, P. Polymer 1980, 21, 423.
12. Perkins, W. C. J. Rad. Curing 1981, 16.

13. Dubois, J. C.; Eranian, A.; Datmanti, E. Proceedings of the Symposium on Electron and Ion Beam Science and Technology, 8th International Conference, 1978, p. 940.
14. Crivello, J. V. ACS Organic Coatings and Applied Polymer Science Proceedings 1983, 48, 226.
15. Frechet, J. M.; Eichler, E.; Ito, H.; Willson, C. G. Polymer, in press.
16. Crivello J. V.; Lam, J. H. W. J. Org. Chem. 1978, 43, 3055.
17. Crivello, J. V.; Lam, J. H. W. J. Polym. Sci., Polym. Chem. Ed. 1980, 18, 2677.
18. Barton, J. W. in "Protective Groups in Organic Chemistry," McOmie, J. F. W., Ed; Plenum Publishing Co., Ltd.: London, 1973, Chapter 2.
19. Ito, H.; Willson, C. G.; Frechet J. M. J.; Farrall, J. M.; Eichler, E. Macromolecules 1983, 16, 510.
20. Ingold, C. K. "Structure and Mechanism in Organic Chemistry," Cornell University Press: Ithaca and London, 1953, Chapter 15.
21. Hofer, D. C.; Kaufman, F. B.; Kramer, S. R.; Aviram, A. Appl. Phys. Lett. 1980, 37, 314.
22. Hatada, K.; Tsubokura, Y.; Danjo, S.; Yuki, H.; Aritome, H.; Namba, S. Polymer Preprints Japan 1981, 30, 423.
23. Lin, B. J. J. Vac. Sci. Technol. 1975, 12, 1317.
24. Crivello, U. V.; Lam, J. H. W. J. Polym. Sci., Polym. Chem. Ed. 1978, 16, 2441.
25. Crivello J. V.; Lam, J. H. W. J. Polym. Sci., Polym. Chem. Ed. 1979, 17, 1059.
26. Pappas, S. P.; Jilek, J. H. Photog. Sci. Eng. 1979, 23, 140.
27. Vogl, O. J. Polym. Sci. 1960, 46, 241.
28. Aso, C.; Tagami, S.; Kunitake, T. J. Polym. Sci., Part A-1 1969, 7, 497.
29. Aso C.; Tagami, S. Macromolecules 1969, 2, 414.
30. Fech, J., Jr.; Poliniak E. S. U.S. Patent 3 940 507, 1976.
31. Chambers W. J.; Foss, R. P. U.S. Patent 4 108 839, 1979.
32. Limburg, W. W.; Stolka, M. U.S. Patent 3 915 704, 1975.
33. Limburg, W. W.; Marsh, D. G. U.S. Patent 3 917 483, 1975.
34. Nelson, R. W. U.S. Patent 3 984 253, 1976.
35. Hatada, K.; Kitayama, T.; Danjo, S.; Yuki, H.; Aritome, H.; Namba, S.; Nate, K.; Yokono, H. Polym. Bull. 1982, 8, 469.
36. Ito, H.; Schuerch, C. Macromolecules 1981, 14, 246.
37. Bowden M. J.; Thompson, L. F. Polym. Sci. Eng. 1974, 14, 525.
38. Taylor, G. N.; Wolf, T. M. J. Electrochem. Soc. 1980, 127, 2665.
39. Hiraoka, H. J. Electrochem. Soc. 1981, 128, 1065.
40. Tsuda, M.; Oikawa, S.; Kanai, W.; Yokota, A.; Hijikata, I.; Uehara, A.; Nakane, H. J. Vac. Sci. Technol. 1981, 19, 259.
41. Hattori, S.; Morita, S.; Yamada, S.; Tamano, J.; Ieda, M. SPE Regional Technical Conference on Photopolymers, Ellenville, New York, November 1982.

RECEIVED September 2, 1983

3

Semiempirical Calculations of Electronic Spectra
Use in the Design of Mid-UV Sensitizers

R. D. MILLER, D. R. MCKEAN, T. L. TOMPKINS, N. CLECAK, and C. GRANT WILLSON

IBM Research Laboratory, San Jose, CA 95193

J. MICHL and J. DOWNING

University of Utah, Salt Lake City, UT 84112

As device dimensions continue to shrink, pattern resolution becomes of critical importance. In this regard, the use of radiation sources with maximized output in the mid-UV region (300-350 nm) would permit increased resolution if an appropriate resist could be utilized. In this regard, commercial resist formulations function inefficiently in this region for a variety of reasons. Accordingly, we have used semiemipirical quantum mechanical techniques to calculate the electronic absorption spectra of two of the most commonly employed chromophores and to computationally assess the effect of a variety of substituents. The results of these calculations have been used to drive a synthetic program designed to produce an efficient sensitizer for use in the mid-UV.

The drive toward higher density circuitry in microelectronic devices has promoted interest in a variety of lithographic techniques designed to provide higher resolution than currently available using conventional tools which accomplish resist exposure utilizing radiation in the 350-450 nm region. The theoretical resolution attainable with a 1:1 optical projection system is proportional to the quantity λ/NA where λ is the wavelength of the incident light and NA is the numerical aperture of the lens system. In principle, one could obtain better resolution with conventional wavelengths by going to larger numerical aperatures but *only* at the expense of the depth of focus and the image field diameter (1,2). For this reason, improved resolution is best obtained by decreasing the wavelength of the incident irradiation. In this regard, Bruning (3) has already demonstrated significant improvement in resolution using 313 nm light in a 1:1 Perkin-Elmer projection printer. Tools equipped with mid-UV filters and mercury sources transmit primarily the mercury emission lines at 313 and 334 nm with the former as the major component.

With respect to realizing the improved resolution possible by the use of shorter wavelengths, tool design has progressed more rapidly than resist

0097-6156/84/0242-0025$08.00/0
© 1984 American Chemical Society

development. Most commercial resists show low sensitivity when utilized in the mid-UV region which adversely effects wafer throughput. This lack of sensitivity can be traced primarily to two features of existing positive resists: (1) low optical absorption of the sensitizer in the mid-UV region and/or (2) incomplete bleachability of the sensitizer chromophore. The former results in inefficient light absorption while the latter causes a degradation of the resist profile caused by overexposure in the top layers of the resist. The result of residual optical density caused, either by resin absorption or by the presence of an unbleachable sensitizer chromophore on the resist profile (4), is shown graphically by computer simulation (Figure 1). In this case, it is obvious that an increase in the residual optical density results in a serious degradation of the profile even at higher doses which simply lead to overexposure of the upper layers.

Although many types of compounds have been tested as sensitizers in phenolic host resins (Novolacs, Resols, *etc.*) (5), all commercial positive resists employ aromatic diazoquinones of some type which photochemically generate base soluble products via Wolff rearrangement initiated by the loss of nitrogen (6). A staggering variety of diazoketones have been synthesized and evaluated for lithographic purposes, but derivatives of 1 and 2 are most commonly employed (5).

Since most of these materials are described only in the patent literature unaccompanied by pertinent UV-visible spectral data, it was not possible to survey the effect of substituents on the absorption spectra of the basic chromophores in order to predict *a priori* what substituents (types and position) might generate sensitizers suitable for use in the mid-UV region. In light of this, we decided to calculate the theoretical electronic absorption spectra of 1 and 2 and to computationally assess the effect of a variety of substituents. We describe here the results of these studies and the use of the theoretical predictions to drive a synthetic program designed to produce viable mid-UV sensitizers.

Results

While semiempirical models which can be applied to molecules the size of <u>1</u> and <u>2</u> are necessarily only approximate, we were searching for trends rather than absolute values. In concept, the design of semiempirical quantum mechanical models of molecular electronic structure requires the definition of the electronic wavefunction space by a basis set of atomic orbitals representing the valence shells of the atoms which constitute the molecule. A specification of quantum mechanical operators in this function space is provided by means of parameterized matrices. Specification of the number of electrons in the system completes the information necessary for a calculation of electronic energies and wavefunctions if the molecular geometry is known. The selection of the appropriate functional forms for the parameterization of matrices is based on physical intuition and analogy to exact quantum mechanics. The numerical values of the parameters are obtained by fitting to selected experimental results, typically atomic properties.

Two models are particularly well suited for the calculation of the electronic spectra of conjugated π-electronic systems. Both are based on the principle of differential overlap and involve no multicenter repulsion integrals. The Pariser-Parr-Pople (PPP) model (7), whose basis set contains only atomic orbitals of π symmetry, is particularly good for $\pi\pi^*$ transitions in hydrocarbons and other molecules with a limited number of heteroatoms. Because of the unreliability of the PPP method in the presence of heteroatoms, we chose a second model which is available in two versions (CNDO/S (8) and INDO/S (9) for the calculations on <u>1</u> and <u>2</u>. This model utilizes a basis set which contains *all* valence shell orbitals and leads to less uncertainty in parameterization even when a large number of heteroatoms are present. This model works well for $\pi\pi^*$ transitions but often fails when applied to $n\pi^*$, $\sigma\pi^*$ and $\pi\sigma^*$ transitions. This should not be a serious limitation in the present application since the only transitions of sufficient intensity to be of interest are of the $\pi\pi^*$ type.

The calculations have been performed using the Zerner-Ridley parameterization of the INDO/S method (9). The geometries were determined by using regular polygons with sides of 1.40Å for the rings and standard CNDO geometries (10) for the substituents. Configuration interaction was included by using the 100 singly excited configurations produced by excitation of the ten highest occupied molecular orbitals to the ten lowest occupied molecular orbitals.

The results of the calculations on the diazoketones <u>1</u> and <u>2</u> are shown graphically in Figure 2. Although the calculated absorption maxima are not exactly in agreement with those observed, the assignment of the calculated transitions to the observed absorption bands is straightforward. In the case of diazoketone <u>1</u>, the first absorption peak near 390 nm corresponds to a transition calculated around 372 nm and the weak shoulder near 300 nm, which had not previously been recognized as a separate electronic transition, is strongly suggested to correspond to a weaker transition calculated at 300 nm.

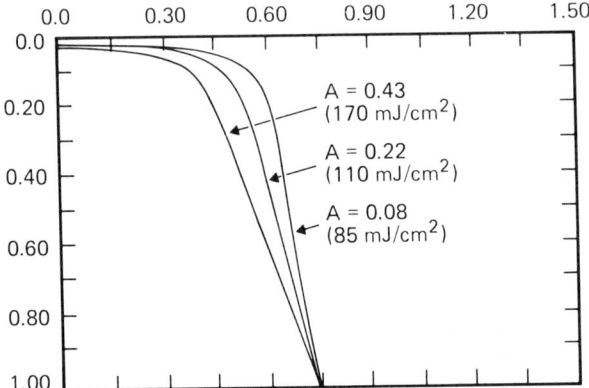

Figure 1. Sample resist profile: λ=313 nm; T_1=1.000; NA=0.167; sigma=0.071; DF=2.82; LW=1.50; SW=1.50; version 1.2; contour times=15, 30, 60, and 75; A=optical density (residual - resin + photoproducts at 313 nm); and dose=85, 110, and 170 mJ.

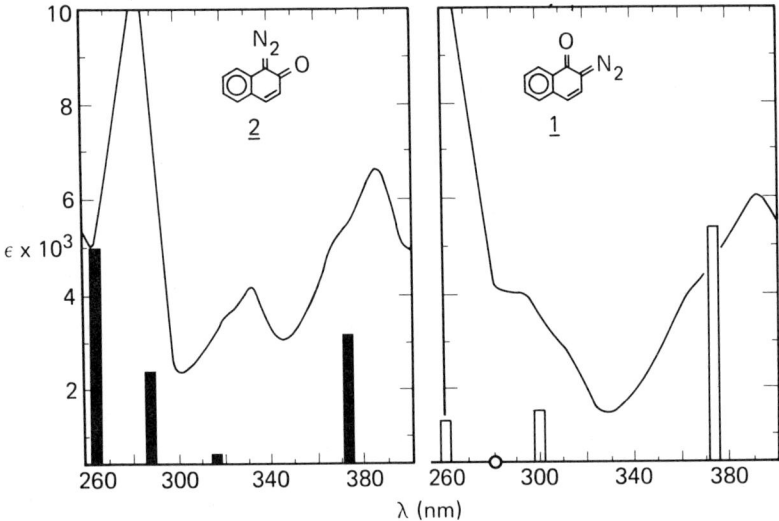

Figure 2. Experimental and calculated spectra of 1 and 2; vertical bars represent the position and relative intensities of calculated transitions.

More intense transitions appear at shorter wavelengths but are not of interest in the present study. For the diazoketone 2, the first peak at 385 nm clearly corresponds to the first calculated transition at 372 nm. The much weaker second peak at 330 nm is assigned to the weak calculated transition near 320 nm, and the intense peak at shorter wavelength is assigned to two intense overlapping transitions.

In order to obtain a qualitative understanding of the modes of action of the substituents, it is useful to decompose their overall action in an approximate way into independent inductive and resonance effects (11). As shown in Figure 3, these have different effects on orbital energies. The inductive effect of a substituent (also referred to as its sigma effect) modifies the effective electronegativity of the ring atom to which it is attached. The simplest example of such an effect is the replacement of a ring carbon atom by a ring nitrogen atom. To a first approximation, the effect of the electronegativity change $\Delta \alpha$ on the energy of a molecular orbital with a coefficient c_μ at the position of perturbation μ, is given by $\Delta \alpha \times c_\mu^2$. Electron-withdrawing inductive substituents (+I) stabilize each MO ($\Delta \alpha$ is negative), whereas electron-releasing inductive substituents (−I) destabilize the molecular orbitals ($\Delta \alpha$ is positive). The inductive effect of a substituent is thus seen to affect all orbitals in a way which discriminates only according to the coefficient of the orbital at the position of substitution, and not the orbital energy. This is physically reasonable, considering that the square of the orbital coefficient reflects the probability density for finding the electron located in this molecular orbital at the position of substitution.

Resonance (π-effects) of substituents are quite different in this respect. They are due to the extension of the π-electron system by the presence of the substituent. The overlap of the π-system of the ring with that of the substituent across the ring-substituent bond permits electron flow between the ring and the substituent. If the substituent contains only occupied π orbitals, for instance, a halogen substituent, only electron flow into the ring is possible. Such substituents act as π-electron donors (−E). If the π-symmetry atomic orbitals on the substituent are all empty, only electron flow from the ring is possible, and such substituents act as π-electron acceptors (+E), a classical example being the BH_2 substituent. Many substituents contain both occupied and empty molecular orbitals of π symmetry, suitable for interaction with the ring (*e.g.*, vinyl, nitro). Among these substituent orbitals, the highest occupied [HOMO(S)] and the lowest vacant [LUMO(S)] are usually more important. These substituents can act as net donors or acceptors depending on the relative strength of the two interactions. This is determined not only by the nature of the substituent, but also by the orbital energies of the ring itself. The electron flow into the ring occurs primarily into the lowest unoccupied ring molecular orbitals [LUMO(R)], and if these are very low in energy, the ring can provoke even reluctant substituents to act as donors. The electron flow into the substituent proceeds primarily from the highest occupied molecular orbital on the ring [HOMO(R)], and if this is very high in energy, the ring can force even a reluctant substituent to act as an acceptor. Conversely, the effect of the π

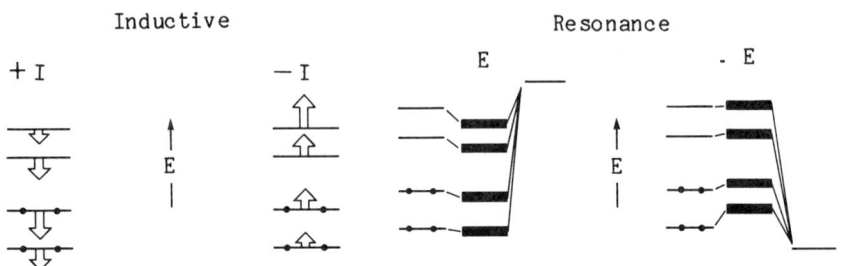

Figure 3. Perturbation substituent effects.

substituent on the ring molecular orbitals depends on their energies: the larger the separation between the energy of the substituent π orbital and the ring π orbital, the less interaction there will be if all else is constant. In this respect, the resonance effect is quite different from the inductive effect, which does not "discriminate" between the various ring orbitals. However, as in the case of inductive effect, the magnitude of the coefficient in the ring molecular orbital at the position of substitution is important in determining the extent of the resonance interaction. Thus, even a high energy ring orbital may be prevented from donating electron density into a substituent if the substituent is attached at the position of a node in the ring orbital.

Although we have so far concentrated our attention on the description of substituent effects in terms of net electron flow between the ring and the substituent in the ground state, this is generally not the most important interaction for electronic spectroscopy. Because of their relative energetic proximity, the occupied orbitals of the ring and of the substituent can interact very strongly with each other, as can the vacant orbitals of the ring and the substituent. These interactions, e.g., HOMO(R)-HOMO(S), do not lead to any net electron flow in and out of the ring in the first approximation, but they have the potential for changing orbital energies by a much larger amount than the previously discussed HOMO(R)-LUMO(S) and HOMO(S)-LUMO(R) interactions (Figure 3). As a matter of fact, the effect of the latter on orbital energies can frequently be neglected by comparison. The same factors which determine whether a substituent is a strong π donor or a strong π acceptor, also determine the magnitude of the shifts of the ring orbitals caused by the HOMO-HOMO and LUMO-LUMO interactions. Since these interactions permit a flow of charge between the ring and the substituent in the excited state when a hole has been generated in the HOMO(R) and an electron has been placed into the LUMO(R) of the ring system, it is hardly surprising that substituents which act as charge transfer donors in the ground state are also those which act as donors in the HOMO-LUMO excited state, and that their action is usually much stronger in the excited state. There is no simple quantitative relationship between the relative strength of the substituents in the two situations: as expressed in the HOMO-LUMO interactions important for ground state charge transfer on the one hand, and in the HOMO-HOMO and

LUMO-LUMO interactions important for charge transfer in the excited state and for electronic excitation energies on the other hand. The former are best expressed qualitatively in terms of the σ^+ constants, while the I_π constant of a substituent has been introduced to describe the latter (12). Few numerical values are available for this parameter. Yet, this is the property which will be of interest in our application where we emphasize the energetics of excited states and we shall return below to the problem of a quantitative description of the substituent effect.

Using the simple principles outlined above, we have estimated the effects to be expected for +I and –I, +E and –E substituents at all possible positions on the ring of diazoketones $\underline{1}$ and $\underline{2}$ from the knowledge of the squares of the calculated coefficients of the molecular orbitals using first-order perturbation theory. We have then chosen the cyano group as a typical +E substituent, the hydroxy group as representative of –E substituents, and aza nitrogen as representative of +I substituents, and have performed calculations for substituted derivatives of the diazoketones $\underline{1}$ and $\underline{2}$. The results strongly suggest that aza substitution in positions 4 and 5 of either diazoketone, and cyano substitution in positions 4, 5 or 6 in $\underline{1}$ and 5 or 6 in $\underline{2}$, as well as hydroxy substitution in position 4 of $\underline{1}$ and position 5 of $\underline{2}$ could be useful for our purposes.

A number of practical considerations caused us to favor derivatives of $\underline{1}$ over $\underline{2}$ in the search for a new mid-UV sensitizer: 1. they are extremely common and the synthetic methodology has been well developed, 2. the 2-diazonaphthalene-1-one $\underline{1}$ chromophore seems to show better thermal stability than the isomer $\underline{2}$ and finally 3. the photosensitivity of derivatives of $\underline{1}$ is reported to be somewhat higher than comparable derivatives of $\underline{2}$ (5). The last two features will be important ingredients in defining the ultimate sensitivity of a (sensitizer-host) resist system.

With regard to $\underline{1}$, the calculations clearly predict that inductive (+I) withdrawal in positions 4 and 5 affects mainly the first transition and the shift is to shorter wavelengths. On the other hand, examination of Figure 4 also shows that mesomeric (+E) electron withdrawl in position 5 significantly shifts the second transition of $\underline{1}$ to longer wavelengths and intensifies it. A similar but lesser shift is expected either for the substitution of electron withdrawing substituents in positions 4 and 6 or from an electron donating group at carbon 4.

In order to test the validity of the calculations, the aza substituted derivative 3-diazaquinalin-4-one $\underline{3}$ was prepared by standard techniques (13). As predicted, there is a blue shift in the longest wavelength transition to 362 nm which is in accord with the suggestion that an aza ring substituent is an inductive attractor. Although this shift is very significant, it is not large enough to produce the desired increase in absorption in the mid-UV region. It does, however, suggest a number of derivatives ($\underline{4-7}$) which should be spectrally suitable for use in the mid-UV based on the calculated substituent effects. Unfortunately, however, these materials possess other drawbacks such as synthetic inaccessibility, thermal instability, insolubility and lack of adequate photobleachability which limit their practical utility.

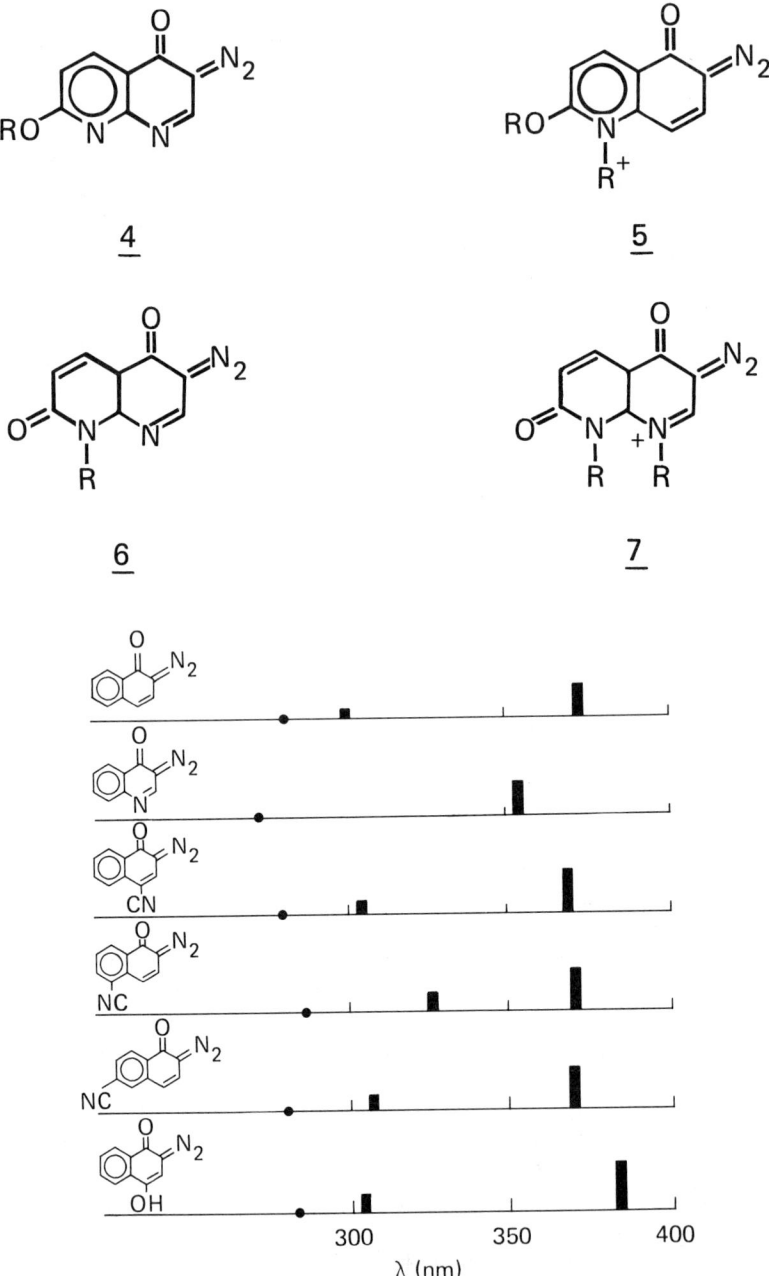

Figure 4. Calcualted transitions for selected derivatives of 2-diazonaphthalene-1-one 1. Vertical bars represent position and relative intensities of calculated transitions.

At this point, it seemed reasonable to investigate the other approach to spectral modification (*i.e.*, the shifting and intensification of the second and weaker transition of 1 into the mid-UV region). In this regard, it is helpful to recall that this may be accomplished either by the incorporation of a resonance attracting substituent at positions 4, 5 or 6 or by a donor at position 4 of 1. Since the red shift is predicted to be maximized for attracting substituents in the 5 position, derivatives of this type were selected for initial study. Furthermore, it should be noted that cyano substituents, while advantageous for calculations, are not particularly convenient synthetically, particularly in molecules containing additional reactive functionality such as in diazoketones. Sulfonyl derivatives of many types, are however, easily prepared, thermally stable and, photochemically inert. The sulfonyl substituent is potentially electronically ambiguous since it could conceivably behave as either an inductive withdrawer or as a hyperconjugative resonance attractor. The spectrum of a common aryl sulfonate sensitizer, 8 is shown in Figure 5. We were concerned that the presence of the benzophenone moiety in 8 might result in spectral effects which were not typical of simple 5-substituted aryl sulfonate esters. In this regard, we note, however, that the intensity of the mid UV transition suggests that it is clearly a $\pi\pi^*$ type and that the mid UV transition of the simple phenyl sulfonate derivative (Table II, R=OPh) is similarly strongly red shifted. Comparison of the spectrum of 8 with the study depicted in Fig. 4 clarifies the electronic role of sulfonyl substituents in modifying the absorption spectrum of the chromophore 1. The longest wavelength transition of 8 is not blue shifted from the parent 1 which implies that the substituent is not serving as a typical inductive withdrawer. The second transition in 1 is however, strongly red shifted and intensified in 8

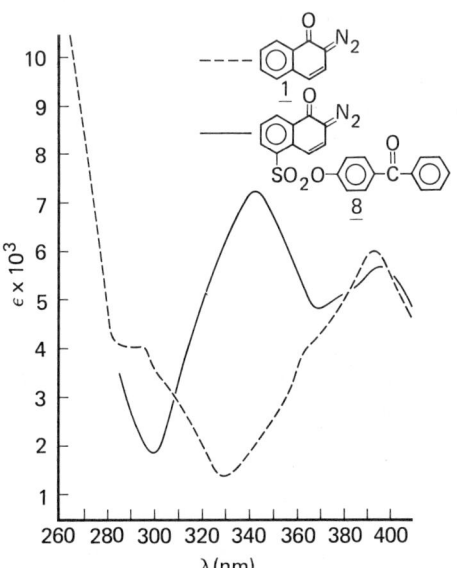

Figure 5. Absorption spectra of 1 and 8, solvent 5% dioxane/methanol. (Reproduced with permission from Ref. 16.)

suggesting that the aryl sulfonate group is behaving as a resonance attracting substituent. Since the electronic structure of the sulfonate group differs from that of a classical resonance attracting group (*i.e.*, it has no low-lying π^* antibonding orbitals), it may be considered as resonance electron attracting by hyperconjugation. In the case of substituents such as -SO_2X, the strength of the resonance interaction depends strongly on the nature of the σ^* bond orbital of the S-X bond, particularly on the energy of this orbital and also on the magnitude of its coefficient on the sulfur atom.

It has been demonstrated that, in accord with theoretical predictions, the presence of resonance attracting substituents in position 5 (either standard or hyperconjugative) of the basic chromophore 1 produces the desired red shift of the second electronic transition. The sulfonate group fulfills this function, in fact in the case of aryl derivatives such as 8, a bit too well for the strong red shift of the second transition in aryl sulfonates results in very little residual absorption in the mid-UV region particularly at 313 nm.

Having demonstrated the concept in principle, what remained was the selection of a substituent which produces a shift of the desired magnitude. For this one needs ideally a table of appropriate substituent constants for use in the prediction of the magnitude of induced spectral changes. To be generally useful, these constants should be relatively insensitive to the inductive nature of the substituent and reflect only the electromeric contribution of the group.

Here, our previous work on the magnetic circular dichroism of aromatic systems has proven useful (14). In mcd spectroscopy, the differential absorption of left-handed and right-handed circular polarized light propagating along the direction of magnetic field through the sample is measured as a function of photon energy. In molecules of low symmetry such as of interest to us here, each electronic absorption band has an analog in the mcd spectrum which can be positive or negative. The magnitude of the mcd band is expressed as the integrated area of the band in appropriate units and is referred to as the B term. It has been shown previously that the B term of the lowest energy $\pi\pi^*$ transition of aromatic molecules (L_b or L_a in Platt's nomenclature) is simply related to the orbital energies of the aromatic system. To a first approximation, the sign and magnitude of the B term depend on the relative size of the splitting of the two highest occupied molecular orbitals of the aromatic π system, and the extent of splitting in the lowest two empty molecular orbitals. The former is referred to as ΔHOMO and the latter as ΔLUMO. When the aromatic system is simple 4N+2 annulene, such as benzene, one of its two degenerate HOMO's and one of its two degenerate LUMO's will have a node at the position of substitution and its energy will be largely unaffected by the substituent. The other HOMO(R) and the other LUMO(R) of the ring have an antinode at this position and will be susceptible to shifting by the I and E effects of the substituent. Since the squares of the coefficient in HOMO(R) and in LUMO(R) at the position of substitution are equal, the I effect will change the energies of these two orbitals equally in first order perturbation theory and therefore will not affect the difference ΔHOMO-ΔLUMO. However, from the above discussion of resonance effects of substituents, it is clear that for −E substituents, ΔHOMO will be larger than ΔLUMO, and that the opposite will be true for +E substituents. The net strength of the substituent is related to the magnitude of the difference ΔHOMO-ΔLUMO and thus to the size of the mcd B term. The B terms have been measured for some common substituents over the years, and a few of the values for substituents of particular interest from a practical point of view in the present problem are listed in Table I (15).

Examination of this table shows that inductive effects make a minimal contribution. For resonance electron donating and electron attracting substituents, the nature and the magnitude of the expected effect is defined by the sign and the size of the B value. Therefore, -CN and -SO_2Cl substituents should produce a similar shift and -SO_2NH_2 should have the smallest effect of the electron attracting substituents tabulated. From the table and from intuition regarding the nature of hyperconjugative electron attraction, certain useful predictions can be made. One would predict that the mid UV transition of derivatives of I should be substituent sensitive and that alkyl sulfonamides and perhaps sulfonates should be most useful for our purposes. Table II shows a tabulation of the experimentally measured mid UV absorption maxima for a number of derivatives of I and demonstrates that this is indeed the case. The maxima are somewhat solvent sensitive and the values reported are those in methanol-5% dioxane. It is interesting that these substituents also have relatively little effect on the maximum of the longest wavelength transition, a feature also correctly predicted by the calculations. The practical result of the shift in the absorption maxima of the second band of the alkyl sulfonates and sulfonamides to ~330 nm is an increased extinction coefficient at *both* 313 and 334 nm which are the principle emission lines of mid-UV lithographic tools and suggests their potential utility as mid-UV sensitizers (Fig. 6). This utility is further enhanced by the observation that the bleachability of these materials in this region significantly exceeds that of the corresponding aryl derivatives.

In summary, we find that the application of semiempirical quantum mechanical calculations adequately predict the electronic spectra of the lithographically important chromophores 1 and 2 and describe the relative effects of substituents (both position and type) on the wavelengths of the important transitions. In this regard, it has been demonstrated that inductive attracting substituents in positions (4 and 5) result in blue shifts for the longest wavelength transition of 1 while leaving the other absorption bands relatively unperturbed. In contrast, we find that resonance electron attracting groups in positions 4, 5 and 6 or electron donors in position 4 have little influence on the first absorption band of 1 but red shift the second band into the mid-UV region. The magnitude of this substituent effect is accurately predicted by a consideration of substituent constants derived by mcd model studies. Consideration of all the necessary criteria for a mid-UV sensitizer such as spectral matching (*i.e.,* significant absorption at both 313 and 334 nm), thermal stability, solubility and compatibility with Novolac type resin hosts, *etc.* leads to the conclusion that derivatives of 1 substituted in position 5

Table I. Electronic substitutent constants determined from magnetic circular dichroism spectra of simple substituted benzene derivatives $^1A_{1g} \rightarrow {}^1B_{2u}$ transition.

	R	B
Inductive	$-NH_3^+$	0
	$-SO_3^-$	0.04
Donating	$-F$	Weak
	$-O^-$	-0.70
	$-NH_2$	-0.42
	$-OH$	-0.22
Attracting	$-NO_2$	$+0.60$
	$-CN$	$+0.38$
	$-SO_2Cl$	$+0.35$
	$-CHO$	$+0.31$
	$-CO_2^-$	$+0.22$
	$-SO_2NH_2$	$+0.18$

Table II. Substituent effects on the mid UV absorption of sulfonyl derivatives of $\underline{1}$.

R	λ_{max}(MeOH)nm	$\epsilon(\lambda_{max})$
benzimidazolyl	352	10,756
$-N_3$	349	9,000
$-OPh$	343	7,761
$-N(Me)Ph$	341	7,500
$-NMeOMe$	337	6,690
$-OC_8H_{17}$	334	6,947
$-N(Me)_2$	332	6,337

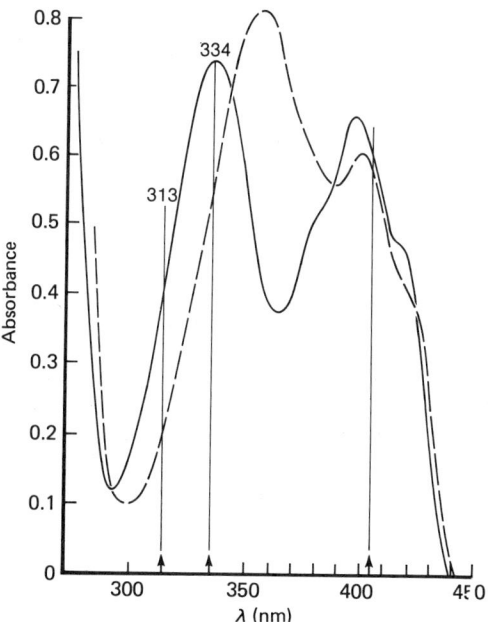

Figure 6. Spectra of a typical 5-alkyl (———) and a 5-aryl (---) sulfonate derivative of I in diglyme. (c=5 X 10^{-5}M.) (Reproduced with permission from Ref. 16.)

by alkylsulfonate or sulfonamide groups should lead to mid-UV sensitizers of improved performance.

For a variety of reasons, we chose to design our sensitizers around the alkyl sulfonates. We found that both 5-alkyl and aryl sulfonates of $\underline{1}$ can conveniently be prepared in excellent yields from the corresponding sulfonyl chloride through the use of catalytic amounts of sulfonyl transfer reagents such as 4-dimethylaminopyridine (DMAP) in the presence of triethylamine. This procedure is more convenient then the more traditional Schotten-Baumann conditions or those employing pyridine and allowed the preparation of sulfonate derivatives from both 1° and 2° aliphatic alcohols in high yield.

Unfortunately, the 5-alkylsulfonates while greatly improved relative to their aryl analogs (Fig. 6) are not ideally spectrally matched to the output of most mid UV tools. Resists formulated to have the optimal optical density at 313 nm are much too strongly absorbing at 334 nm. Ideally, one would prefer the optical densities of a mid UV resist to be comparable at both of these wavelengths. In this regard, we noticed that 4-alkyl and aryl sulfonate derivatives of $\underline{1}$ have a mid UV absorption band around 310 nm but very little absorption in the 334 nm region. It seems reasonable, therefore, that a sensitizer which was comprised of a combination of 4- and 5-sulfonate derivatives of 1 should produce the desired flat optical profile over the entire mid UV range. This result could be accomplished by preparing a mixed 4,5-disulfonate from an appropriately substituted aliphatic diol by sequential sulfonation using first 5-chlorosulfonyl-2-diazo-1-naphthalenone followed by treatments of the half ester with 4-chlorosulfonyl-2-diazo-1-naphthalenone. It was noted that the structure of the aliphatic diol had very little effect on the actual absorption spectrum of the sensitizer. For this reason, diols were selected which produced sensitizers with appropriate physical properties, *e.g.*, solubility in the Novalac-type resin-hosts, thermal stability, *etc*. Sensitizers of this type were then formulated into resists designed to optimize performance in the mid UV region (16). The absorption spectrum of such a resist as a thin film on a quartz substrate is shown in Fig. 7. Using resists of this type we were able to resolve 2.5 μm period grating arrays using 1:1 projection printing tools as demonstrated by the electron micrographs shown in Fig. 8. These results represent a satisfying conclusion to our studies and demonstrate the efficiency of a joint theoretical and experimental effort.

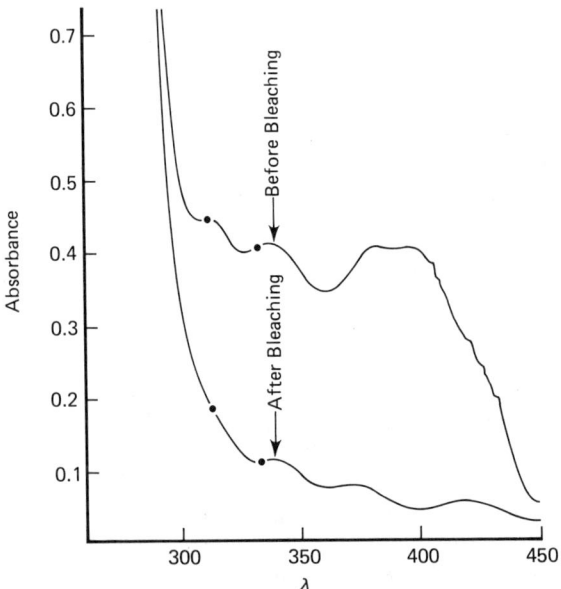

Figure 7. Unexposed and bleached spectra of experimental resist formulated from the 4,5-disulfonate of an aliphatic diol. (Reproduced with permission from Ref. 16.)

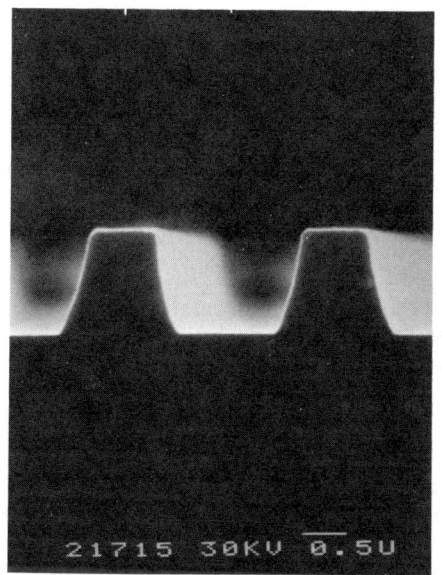 

Figure 8. SEMS of 2.5 pitch (1.25 X 1.25) gratings printed in an experimental mid-UV resist. (Reproduced with permission from Ref. 16.)

Acknowledgments

The authors gratefully acknowledge helpful discussions with D. Hofer particularly regarding the computer simulation of resist profiles. J. Michl and J. Downing also acknowledge partial financial support from the NSF (Grant No. CHE 81-21122).

References

1. Doane, D. A. *Solid State Technology* 101 (1981);
2. Bowden, M. D. *J. Electrochem. Soc.* **128**, 195 (1981).
3. Bruning, J. H. *J. Vac. Sci. Technol.* **16**, 1925 (1979).
4. Oldham, W. G.; Nandgaonkar, S. N.; Neureuther, A. R; Toole, M. O. *IEEE Trans. Electron Devices* **ED-26**, 717 (1979).
5. DeForest, W. S. "Photoresist"; McGraw-Hill Book Company, Inc.: New York, New York, 1975.
6. Pacanksy, J.; Lyerla, J. R. *IBM J. Res. Develop.* **23**, 42 (1979), and references cited therein.
7. Salem, L. "Molecular Orbital Theory of Conjugated Systems"; W. A. Benjamin, Inc.: New York, 1966.
8. Ellis, E. L.; Jaffe, J. H. "Semiempirical Methods of Electronic Structure Calculations"; Part B, G. A. Segal, Ed., Plenum Press: New York, 1977.
9. Zerner, M.; Ridley, J. *Theoret. Chim. Act.* **32**, 111 (1973); We thank the authors for providing us with a copy of the program.
10. Pople, J. A.; Beveridge, D. L. "Approximate Molecular Orbital Theory"; McGraw-Hill Inc.: New York, New York, 1970.
11. Dewar, M. J. S.; Dougherty, R. C. "The PMO Theory of Organic Chemistry"; Plenum Publishing Corporation: New York, New York 1975.
12. Murrell, J. N.; Kettle, S. F. A.; Tedder, J. M. "Valence Theory"; J. Wiley Inc.: New York, New York, 1970, p. 325, second edition.
13. Süs, O.; Glos, M. U.S. Patent 2 859 112 (1958).
14. Michl, J.; *J. Am. Chem. Soc.* **100**, 6801 (1978); *ibid.* **100**, 6812 (1978); *ibid.* **100**, 6819 (1978).
15. McCarville, M. E.; Thesis, Iowa State University, 1967.
16. Willson, G.; Miller, R. D.; McKean, D. R.; Clecak, N.; Tompkins, T.; Hofer, D.; Michl, J.; Downing, J.; "Design of a Positve Resist for Projection Lithography in the Mid UV," SPE Regional Technical Conference, Ellenville, New York, November 1982.

RECEIVED September 2, 1983

Mid-UV Photosensitization of Diazoquinone Positive Photoresists

W. T. BABIE, M-F. CHOW, and W. M. MOREAU

East Fishkill GTD, IBM Corporation, Rt. 52, Hopewell Junction, NY 12533

> Higher resolution projection printing requires utilization of shorter exposure wavelengths than 365–435nm. The spectral sensitivity, however, of diazonaphthoquinones (DQ) is lower at 313nm than at 365 and 405nm. At 313nm, the actinic absorption of diazoquinones is lower than the actinic peaks at 365 and 405nm. The enhancement of the spectral absorption and sensitivity can be achieved by the addition of pyrene photosensitizers. Pyrene (P) acts as a singlet energy transfer sensitizer via intermolecular resonance interaction. The photomechanism and application of sensitized PDQ to the mid-UV region will be discussed.

In order to enhance the resolution of 1:1 projection photolithography, shorter exposure wavelengths (240 to 330nm) than the currently employed wavelengths of 360 to 400nm are required.(1) In the 300-330nm region, however, commercial diazoquinone-novolak resists suffer from a lower sensitizer absorption and interference by a tailing novolak resin. Because of the undesirable absorption characteristics of the novolak photoresists and the lower output of mercury lamps in this region, we investigated the enhancement of the actinic sensitivity of diazoquinone-novolak resists for the mid-UV region.

The enhancement of spectral absorptivity at 313nm was accomplished by the addition of singlet photosensitizers to ester derivatives of diazonaphthoquinone (DQ). These photosensitizers absorb incident photons and transfer the energy to the DQ derivative which is then converted to the alkaline soluble product without the interference of the sensitizer. Theoretically, the mechanism of energy transfer in the solid phase involves dipole-dipole interaction because of the insignificant extent of diffusion of the sensitizer and acceptor. In order to be effective, the sensitizer must meet the following requirements: (1) Have a high ε (ca. $10^4 M^{-1} cm^{-1}$ at 313nm), (2) sustain a long singlet state lifetime τ (3) be insensitive to oxygen; and (4) possess a higher energy of excited singlet state than that of the DQ (ca. 58 kcal mol^{-1}). It is well known that pyrene and methyl pyrene are suitable sensitizers for this purpose.(2) The e, τ, and energy of pyrene are 1 x $10^4 M^{-1} cm^{-1}$, 300ns, and 75 kcal mol^{-1} respectively.

The Perkin Elmer projection printer is the tool most currently used in photolithography manufacturing. Because of the interference of photons at the resist-substrate interface and the absorption of the resist, it is always more difficult to develop the bottom of the resist image to size than to obtain a surface image. A double layer technique has been developed to resolve this problem. The top image layer is sensitized DQ and the bottom layer is composed of a DQ derivative.

Experimental

Material. The resist material photoactive compound (PAC), a diazonaphthoquinone derivative in a novolak resin matrix, was used as a control and also reformulated to obtain PAC/pyrene. Pyrene was obtained from Aldrich and was added to the PAC in this study in the quantity of 2.5% of the total solids content of the resist. The resists were mixed on a roller overnight and then filtered through a 0.8μm silver membrane filter.

DQ Disappearance. To study the effect of pyrene on DQ disappearance, n-propanol solutions in a quartz tube with I.D. of 4mm were irradiated with a 450W medium pressure Hg lamp through a 313nm interference filter. Typically, the optical densities at 313nm for DQ solutions with and without pyrene are 0.8 and 0.4 respectively. The disappearance of DQ was monitored at 405nm in a solution of sulfonyl chloride diazonaphthoquinone.

To observe the effect of pyrene on the disappearance of the DQ derivative in the solid film, samples were spin coated on quartz plates, prebaked at 85°c for 30 minutes, and then irradiated with a 250W Hg lamp with a 313nm interference filter under air. The disappearance of the DQ derivative was again monitored at 405nm by UV spectroscopy.

Lifetime of Pyrene. The lifetime of singlet pyrene was measured in n-propanol with and without the addition of DQ. The pyrene was excited by a nanosecond deuterium lamp. The emission of singlet pyrene was monitored at 420nm with a single photon counter. The lifetime was calculated by employing the method of least-square fitting to the data.

Lithographic Studies. Silicon wafers with ca. 0.5μm SiO_2 were coated with the resist materials to a thickness of 1.1μm and then prebaked at 85°C for 30 minutes. A Perkin-Elmer UV-3 exposure tool equipped with a UBK-7 filter was employed in this lithographic study. The output of the tool shows major emission peaks at 295, 313, and 334nm with 313nm being the dominant wavelength. The samples were exposed for 9 seconds using a 1mm slit width, aperture 1, and a scan speed of 720. The wafers were developed with AZ2401(3) developer diluted with deionized water to a composition of 1:3.5 $AZ2401/H_2O$ at 24.0°C. SEM micrographs were obtained of the developed images and line measurement data was taken.

A two-layer system was also studied employing a layer of the unsensitized PAC as the bottom layer and the sensitized PAC on top. The bottom layer was spin coated at 4000 rpm for 30 seconds and prebaked at 85°C for 30 minutes. This material was then blanket exposed with ca. 60mJ at 313nm. This layer was then covered with the sensitized PAC at a spin speed of 4300 rpm and then again prebaked at 85°C for 30 minutes. The wafers were then imaged on a UV-3 exposure tool through a UBK-7 filter employing a 1mm slit width, aperture 1 and a scan speed of 720. These wafers were subsequently developed using the same developer and conditions described previously. For the topology study, identical exposure and development conditions were employed with the exception of a scan speed of 600 in place of 720.

Results and Discussions

Photochemical Studies. The scheme below outlines the mechanism of energy transfer from pyrene (P) to the diazonaphthoquinone derivative (DQ) and its subsequent conversion to the alkaline soluble product.

$$P \xrightarrow{h\nu} P^*$$
$$P^* + DQ \longrightarrow P + DQ^*$$

Where R is an aryl substituent

Electronic energy transfer can take place by two mechanisms: exchange interaction and dipole-dipole interaction. Exchange interaction requires a collision between a donor (D) and acceptor (A) and diminishes as an exponential function of the distance between D and A. Dipole-dipole interaction can operate over a greater distance and decreases as a reciprocal function of the distance. Because of the different functions of the distance between D and A, electronic energy transfer can be expected to be conducted mainly via dipole-dipole interaction in a diffusion restricted medium, solid phase. A qualitative expression for the rate constant (k) of dipole-dipole interaction is shown in Equation 1:

$$k \propto \Phi_D \varepsilon_A \qquad (1)$$

where Φ_D and ε_A are the quantum yield of donor emission and the extinction coefficient of the acceptor in the region where the donor emits. Because of the high ε of pyrene in the mid-UV region, a small quantity of pyrene can absorb sufficient energy in this region. The energy transfer from pyrene to the diazonaphthoquinone is feasible in the solid phase because of the significant overlap between the emission spectra of pyrene and the diazonaphthoquinone derivative absorption. The UV spectra of the diazonaphthoquinone derivative (DQ) and the pyrene sensitized DQ derivative (PDQ) are shown in Figure 1.

Generally, the diazonaphthoquinone derivative consists of DQ and other aromatic portions which are connected by sulfonyl ester linkages. In order to avoid the interference of the aromatic portion on the energy transfer from pyrene to DQ in liquid solutions, a sulfonyl chloride diazonaphthoquinone was used in the study in n-propanol. In the absence of sulfonyl chloride DQ, the lifetime of pyrene in n-propanol was measured to be 280 ns.(4) In the presence of 6×10^{-5}M and 3×10^{-4}M sulfonyl chloride DQ, the lifetimes of pyrene were 155ns and 107ns respectively. The total quenching rate constant of sulfonyl chloride DQ was calculated to be ca. $3 \times 10^{10} M^{-1} s^{-1}$. When 3×10^{-4} sulfonyl chloride DQ in propanol was irradiated at 313nm, the disappearance of sulfonyl chloride DQ was 42% and 57% in the absence and presence of 2×10^{-4}M pyrene respectively. From the disappearance of sulfonyl chloride DQ, the rate constant of energy transfer was estimated to be ca. $2 \times 10^{10} M^{-1} s^{-1}$. Overall, the efficiency of energy transfer is approximately 70% of the total quenching process with a quenching rate constant of $3 \times 10^{10} M^{-1} s^{-1}$. The data is summarized in Table I.

PAC-pyrene was used for the study of the effect of pyrene on the conversion of DQ. When the film was irradiated at wavelengths where the pyrene does not absorb (>360nm), the disappearance of DQ was not affected by pyrene. This observation demon-

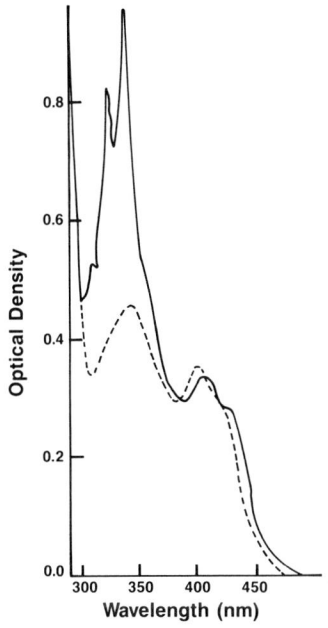

Figure 1

Absorbance spectra of DQ and PDQ. DQ is represented by dashed lines, PDQ by solid line.

strates that pyrene does not alter the photoreaction of DQ. When the film was irradiated at 313nm, pyrene increases the conversion of DQ by 21% and 31% in films with thicknesses of 1.3μm and 0.9μm respectively. The data is summarized in Table II. In summary, the addition of pyrene increases the conversion of DQ and improvement of photolithographic performance can be expected.

Table I. Conversions of sulfonyl chloride DQ as a function of pyrene in n-propanol. The solution was irradiated with 313nm. Optical pathlength was ca. 3 mm.

Sulfonyl Chloride DQ $t=0$	[Pyrene]	Conversion of Sulfonyl Chloride DQ
	0	42%
3×10^{-4}M	2×10^{-4}M	57%
	0	56%
1×10^{-4}M	7×10^{-5}M	64%

Table II. Conversions of DQ as a Function of Pyrene in PAC-pyrene Irradiated at 313nm

	PAC	PAC PYRENE	PAC	PAC PYRENE
Conversion of PAC	28%	34%	32%	42%
Relative Sensitivity	1	1.23	1	1.33
D 313nm $t=0$	0.45	0.80	0.30	0.50
Thickness	1.3μm		0.9μm	

Lithographic Studies. Wafers coated with PAC and PAC-pyrene were developed in 1:3.5 AZ2401 developer at 24.0°C after exposure on the Perkin-Elmer UV-3 tool. These results are summarized in Table III. PAC-pyrene can provide a sidewall angle of 65° and an exposed/unexposed dissolution rate ratio (R/Ro) of 10 with 1.5μm line/space geometries. In comparison, PAC provides sidewall angles of 46° and R/Ro = 2 at the same geometries. SEM micrographs are shown in Figure 2.

Table III. Summary of Lithographic Results

Resist	Expose Time	Scan Speed	Aperture	Develop Time	Δ Thickness at 1.5mm L/S (R/Ro)	Sidewall Angle	Dose of Bottom Coating
PAC	10"	720	1	130"	50% (2)	46°	–
PAC-Pyrene	10"	720	1	580"	10% (10)	65°	–
PAC-Pyrene	10"	720	1	240"	7% (15)	70°	50mJ
PAC							

Because of the photon diffraction standing waves in projection printing and the inherent absorption of the photoresist, the available energy is not homogeneously distributed in the resist. A qualitative picture of a 1μm thick resist with an optical density (OD) of 0.5 is illustrated in Figure 3a-d. From this figure, it is shown that there is just 32% of the incident dose available for absorption at the bottom of the resist layer. From this, it can be seen that a surface image is easier to obtain than to achieve a completely developed one. In order to develop the desired image, the sidewall angle and the contrast will be decreased significantly as shown in Figure 3c.

For example, let us assume that before the image is developed to the solid line in Figure 3c, suddenly the dissolution rate increases anisotropically normal to the surface plane. Instead of the shape obtained by the dotted line in Figure 3c, the image can be developed to the shape shown by the dotted line in Figure 3d. Overall, it is predicted that by increasing the dissolution rate in the vertical direction, the contrast and sidewall angle of the image can be increased significantly.

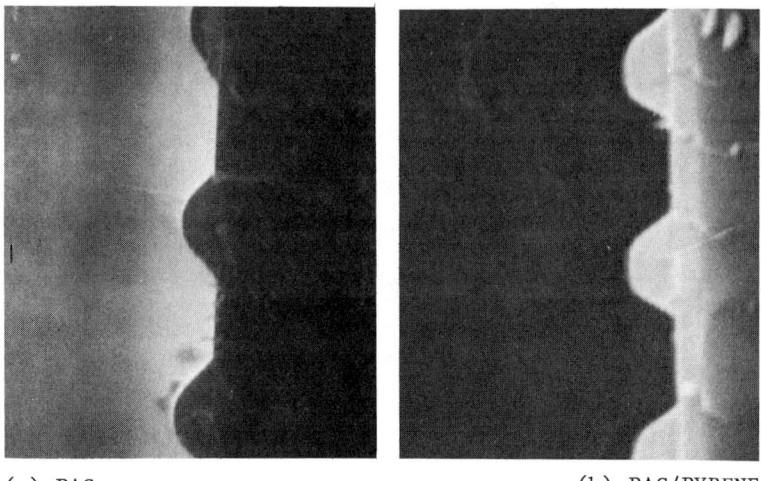

(a) PAC (b) PAC/PYRENE

Figure 2

Lithographic Study Controls of 1.5μm Line/1.5μm Space

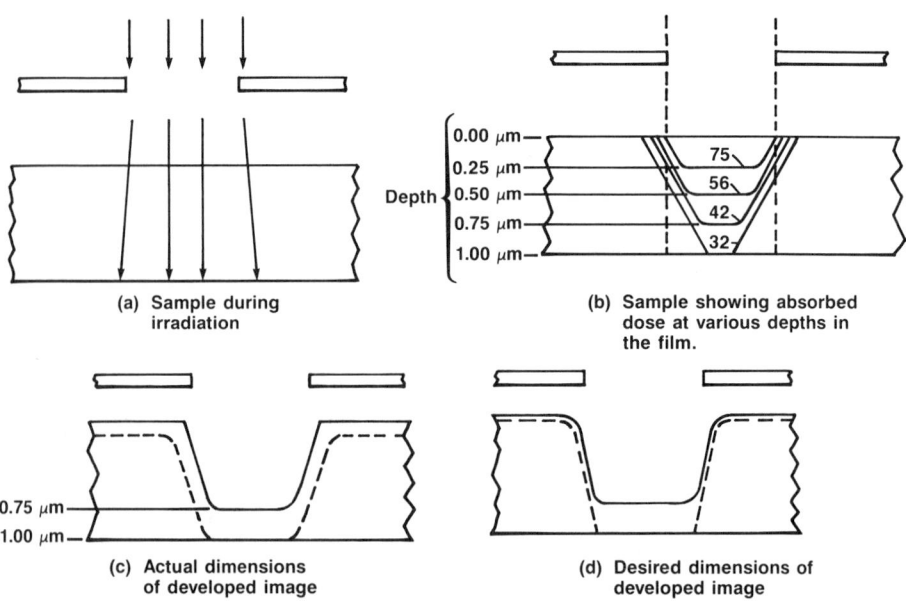

(a) Sample during irradiation

(b) Sample showing absorbed dose at various depths in the film.

(c) Actual dimensions of developed image

(d) Desired dimensions of developed image

Figure 3

Single 1.0μm Coating of PAC Displaying Available Dose as a Function of Depth into the Film

To accomplish this sudden change in dissolution rate during development of the image, a two layer system was employed, the top resist layer having a slower dissolution rate than the bottom layer. From Table III, it is shown that the PAC has a faster dissolution rate than its sensitized counterpart and that both resists are soluble in the same developer. To test our hypothesis, we employed the sensitized PAC as the top layer and the unsensitized PAC as the bottom one. Naturally, when the sensitized PAC is spin coated on top of the existing PAC, there will not be a sharp separation between the two layers. In fact, there is a concentration gradient of pyrene with the highest and lowest concentration being in the top and bottom layers respectively. Also, the dissolution rate of the double coated system will increase gradually as the concentration of the pyrene decreases as the developer penetrates deeper into the film. In order to have the dissolution rate change dramatically, the bottom coating was pre-exposed before the sensitized material was coated over the top of the existing material.

Wafers coated with the sensitized PAC over the unsensitized PAC were exposed on a UV-3 tool employing a UBK-7 filter. These results are summarized in Table III and Figures 4a-c. These results are completely consistent with our predictions; compared with single coating, the development time is reduced by approximately 50%. Also, sidewall angle is increased from 65° to 70° and the dissolution rate ratio R/R_o is increased from 10 to 15. Furthermore, the blanket pre-exposure of the bottom unsensitized layer can be omitted by irradiation of this material in solution before coating or by prebaking at a higher temperature to increase the dissolution rate by thermal degradation of the photoactive compound. The latter process may actually improve adhesion of the unsensitized DQ to the substrate.(5)

When a double layer system is employed, the uniformity of the films may be in question. To test the uniformity of this system, the linewidth tolerance of the image was measured. This tolerance data is summarized in Table IV. The data represents 15 sets of images from three wafers developed simultaneously. This includes five sets from each wafer with each set consisting of four geometries, 2.0μm, 1.75μm, 1.50μm and 1.25μm. All of the tolerances excluding those of the 1.25μm image are within the error of the dimensional measurements. This tolerance test demonstrates that the double coating can provide satisfactory film uniformity in terms of lithographic performance. Also, the average uniformity of the film thickness was better for the double layer system than for the uniformity of the single sensitized or unsensitized DQ layer as measured on a film thickness analyzer (FTA). Results are shown in Table V.

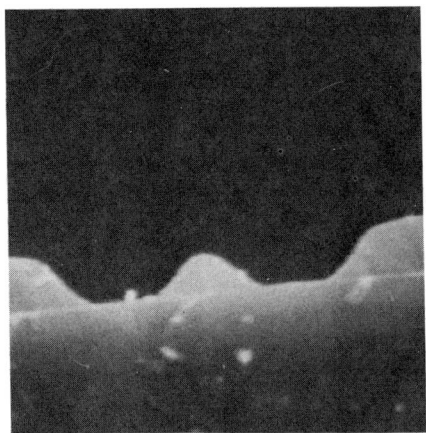

(a) PAC single layer control

(b) PAC/Pyrene single layer control

(c) PAC/Pyrene over PAC

Figure 4

Double Coating Employing PAC/Pyrene Over a Layer of PAC. 1.5μm Line/Space Arrays

Table IV. Tolerance of Geometries

Mask Geometry		Measured wafer geometry (The data presents 15 sets of geometries from 3 wafers developed simultaneously. Measurement was done by SEM.)			
		Single Coating PAC/Pyrene		Double Coating PAC/Pyrene-PAC	
Line	Space	Line	Space	Line	Space
1.43	0.90±0.03	1.5±0.2	1.2±0.2	1.4±0.1	1.1±0.1
1.63	1.14±0.04	1.62±0.06	1.45±0.09	1.70±0.07	1.36±0.07
1.94	1.38±0.04	1.94±0.07	1.70±0.07	1.94±0.06	1.62±0.06
2.19	1.63±0.05	2.2±0.1	1.9±0.1	2.21±0.05	1.88±0.03

Table V. Uniformity of Film Thickness

		Percentage of Variation			
		Wafer 1	Wafer 2	Wafer 3	Average
	PAC	5.1%	12.3%	12.4%	9.9%
PAC Pyrene	Single Coating	12%	3.2%	12.2%	9.1%
	Double Coating	7.8%	9.9%	4.6%	7.4%

Another concern is the problem of lithographic resolution over topology. It is required that the line and space width dimensions be constant when coating a resist material over topologies. The double coating system employed was spin coated on a 0.5μm depth topology wafer to yield a final thickness of 1.2μm. Scanning electron micrographs of the 2.0μm line/1.5μm space and 1.5μm line/1.0μm space geometries are shown in Figures 5a-d. The linewidth over the topology remained constant in the 2.0μm and 1.5μm geometries. These results demonstrate that there

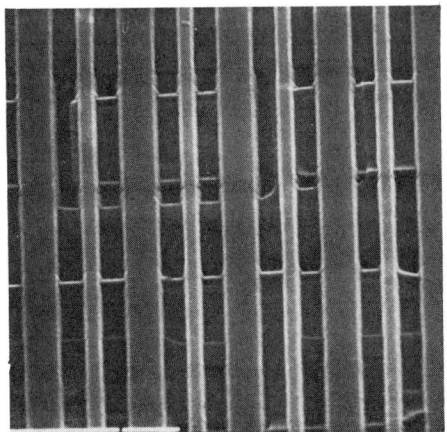 

(a & b) 2.0μm line/1.5μm space

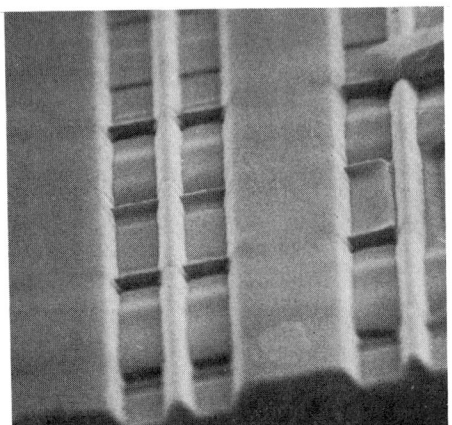 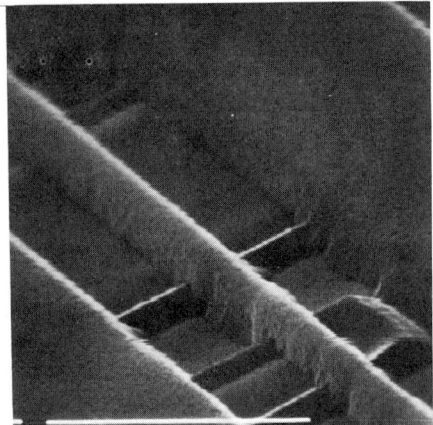

(c & d) 1.5μm line/1.0μm space

Figure 5

Double Layer PAC/Pyrene Over PAC Coatings, Spin Coated
Over 0.5μm Topology

is no problem of lithographic resolution over a 0.5µm depth of topology.

Conclusions

PAC-pyrene has shown a significant lithographic improvement over PAC. Further formulation and process study is required to optimize the lithographic performance of this system.

This system also demonstrates that the concept of energy transfer can enhance lithographic sensitivity significantly. Energy transfer is not limited by the particular donor (pyrene) and acceptor. The active energy acceptor can be as general as α-diazoketone, and pyrene can enhance the sensitivity of any photoresist in which an α-diazoketone is used as a photoactive compound, such as AZ resists.(3) Also, pyrene is but one example of a singlet sensitizer. Compounds which exhibit photophysical properties similar to pyrene can also be potential candidates for energy transfer in photoresists.

Acknowledgments

The authors would like to thank F. P. Hood and J. M. Yang of IBM East Fishkill for their critical review and comments. We would also like to thank Karen J. Kline, Angela M. Marcano and Debra J. Perez for their patience in helping prepare this manuscript.

Literature Cited

1. King, M. C., IEEE Trans. Elect. Devices 1979, 26, 711.
2. Murov, S., "Handbook of Photochemistry"; Marcel Dekker, New York, 1973.
3. Trademark of American Hoechst, Summit, NJ.
4. The lifetime of singlet pyrene was measured by single photon counting.
5. Horst, R.; Kaplan, L.; Merritt, D., U.S. Patent 3 873 813, 1977.

RECEIVED September 15, 1983

Deep UV Photolithography with Composite Photoresists Made of Poly(olefin sulfones)

H. HIRAOKA and L. W. WELSH, JR.

IBM Research Laboratory, San Jose, CA 95193

Poly(butene-1 sulfone), poly(methylcyclopentene sulfone), and a composite resist of poly(2-methyl pentene-1 sulfone) in a novolac resin are well known positive tone electron beam resists. These poly(olefin sulfones) do not have deep UV-sensitivity because they do not absorb UV-light down to 215 nm. When mixed with sensitizers, or used as a composite resist with novolac resins, they showed deep UV-sensitivity in a range of 100 to 250 mJ/cm^2; they provided positive tone polymer patterns in straight wet development or in dry development, and negative tone polymer patterns, in which case exposed resist films with a novolac resin were first postbaked at 115°C for 15 min, and then developed in an alkaline developer. Their photosensitivity may be derived from charge transfer complex formations because of observed color change of a novolac resin solution upon addition of poly(olefin sulfone) solution with newly appearing UV-absorption bands, or by simple energy transfer from the light absorbing resin or from photosensitizers which resulted in an unzipping degradation of poly(olefin sulfones) as evidence by gaseous products formed. Poly(olefin sulfones) alone did not give any gaseous photoproducts upon exposure of 254 nm light, but in presence of sensitizers like nitropyridine N-oxide or in a matrix resin poly(olefin sulfones) degrade upon 254 nm irradiation, yielding SO_2 and monomeric olefins.

Poly(olefin sulfones), like simple olefin sulfones or cyclic alkene sulfones, do not have any UV absorption down to 215 nm. This absence of UV absorption of poly(olefin sulfones) make them uninteresting for deep UV photolithographic applications. Several years ago, along with electron impact induced reactions, we reported vapor phase Hg(^{3}P) photosensitized reactions of methyl sulfone, tetramethylene sulfone and butadiene sulfone with 254 nm light irradiation, which resulted in efficient elimination of SO_2 to yield ethane, cyclobutane, and butadiene, respectively (1). In absence of excited Hg(^{3}P) vapor, these reactions did not occur because olefin sulfones do not have any UV absorption. Thermally these compounds are fairly stable; methyl sulfone and tetramethylene sulfone boil at 238° and 285°C, respectively. Clearly, the energy transfer from the excited mercury atoms Hg(^{3}P) to dissociative triplet states of olefin sulfones took place in vapor phase, resulting in efficient decomposition of olefin sulfones. We also reported an electron beam induced degradation study of poly(cycloalkene sulfones) together with poly(olefin sulfones) in a dose range of 10^{-5} coul/cm^2 of electrons with 25 keV energy (2). The elimination of a methylene group was significant particularly with poly(norbornylene sulfone) and in a less extent with poly(cyclopentene sulfone), giving rise to carbonyl sulfide formation; with poly(olefin sulfones) like poly(butene sulfones) unzipping reactions took place, yielding equal amounts of olefin and SO_2 (3).

Photosensitized degradation of poly(olefin sulfones) similar to the Hg(^{3}P) photosensitized reactions of olefin sulfones make them subject to photodegradation in easily accessible wavelength regions. Almost all poly(olefin sulfones) have been reported only as positive tone electron beam resists (4). As the only exception, poly(5-hexene-2-one sulfone) has been reported as a *positive* tone photoresist with or without a photosensitizer, benzophenone (5). Because this polymer has a carbonyl chromophore, its photosensitivity is clearly derived from the polymer structure itself.

We wish to report first dry developable photoresists with poly(olefin sulfones) with photosensitizers like pyridine N-oxide, and then we present our study on composite resists made of poly(olefin sulfones) with novolac resins or with poly(p-hydroxystyrenes).

Dry Developable Composite Photoresists with Poly(Olefin Sulfones)

Pyridine N-oxide and its derivatives have been studied as photocrosslinking agents for polymers like polystyrene, poly(methyl methacrylate) and others (6). They also act as crosslinking agents for novolac resins (7). Pyridine N-oxide has its maximum UV-absorption at 265 nm in ethanol, while p-nitropyridine N-oxide has its UV absorption peaks at 332 and 234 nm with almost equal intensity in ethanol.

We have found that poly(1-butene sulfone) with p-pyridine N-oxide as a sensitizer yielded positive tone polymer patterns in thermal development after UV light exposure through a conventional photo-mask with a cut-off wavelength of 300 nm. Poly(1-butene sulfone) was mixed with p-nitropyridine N-oxide (20 wt% in solid) and dissolved in nitromethane; the films were spin-coated onto silicon wafers, and prebaked at 100°C for 15 min. After UV exposure, with a medium pressure mercury lamp with an approximate dose of 100 mJ/cm^2, polymer images were developed by heating at 100°C for 7 min. The developed patterns are shown in Fig. 1. Pyridine N-oxide and benzophenone were also used as sensitizers; p-nitropyridine N-oxide, however, gave clear deeper images. Poly(2-methyl-1-pentene sulfone) can replace poly(1-butene sulfone) for positive tone photo-images; the resulted images, however, were inferior to poly(1-butene sulfone). Different from poly(1-butene sulfone), poly(2-butene sulfone) with an aromatic bisazide sensitizer yielded negative tone photo-images; here the sensitizer acted as a crosslinking agent.

The mass spectroscopic analysis of the gases formed in thermal development of the UV exposed poly(1-butene sulfone)/pyridine N-oxide revealed only 1-butene and SO_2 as the products, which indicated depolymerization of the polymer initiated by energy transfer form the sensitizer. These photosensitized poly(olefin sulfones) are not suitable for dry etching processes, and they are not reactive ion etching resistant. Resists made of poly(olefin sulfones) and novolac resins which will be described next are CF_4 plasma etch resistant with reasonable photosensitivities.

Composite Photoresists Made of Poly(Olefin Sulfones) and Novolac Resins

Matrix resists consisting of poly(2-methyl-1-pentene sulfone) and novolac resins have been reported as positive tone electron beam resists (8). UV absorption spectrum of the matrix resist of poly(2-methyl-1-pentene sulfone) in Varcum, which is cresol-formaldehyde novolac resin, is shown in Fig. 2 along with the spectra of the polysulfone and Varcum; clearly new absorption peaks appear in the near UV region; the absorption intensity per unit film thickness remains unchanged for the main absorption at 280 nm. This is also revealed by color changes when two separate solutions of poly(olefin sulfone) and novolac resin dissolved in AZ-thinner, which is 2-ethoxyethyl acetate, are mixed together. The similar color change and new UV absorption peaks appear in composite resists of poly(olefine sulfones) with poly(p-hydroxystyrenes), as shown in Fig. 3. These results indicate some complex formations between poly(olefin sulfones) and the resins, which may affect the energy transfer of absorbed UV light from the host resin to polysulfones.

The photo-energy transfer from the resin to polysulfones can be substantiated by a mass spectroscopic analysis of gaseous products from the composite resists exposed to 254 nm; the polysulfone alone does not give any gaseous products because it does not absorb UV light at this wavelength. Figure 4 shows the mass spectra of the gaseous products formed upon 254 nm exposure of the composite resist of poly(2-methyl-1-pentene sulfone) in Varcum resin. The energy transfer from the resin to the polysulfone caused the main chain scission of the polysulfone, giving rise to 2-methyl-1-pentene and SO_2 formation. Figure 5 shows the wide band ESCA data of the composite resist before and after UV light exposure in air. Reduced intensities of sulfur signals and increased intensity of the oxygen signal after UV light

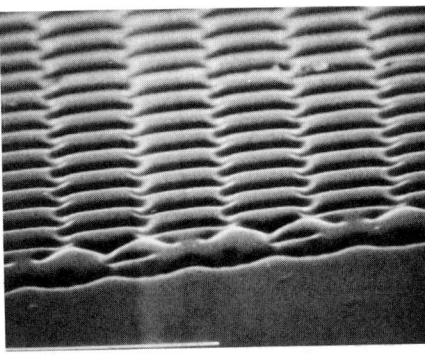 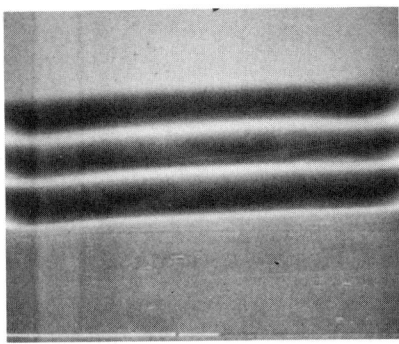

Figure 1. Dry developed images of poly(1-butene sulfone)/p-nitropyridine N-oxide.

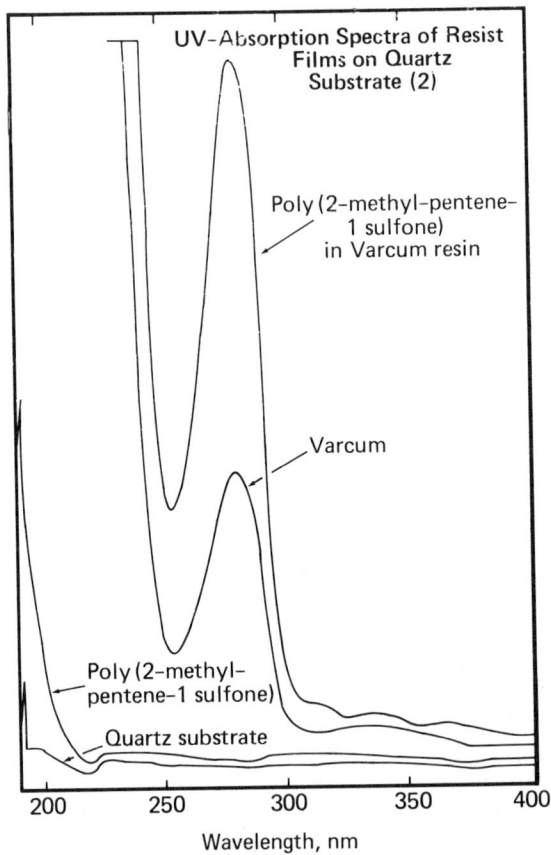

Figure 2. UV absorption spectra of composite resist.

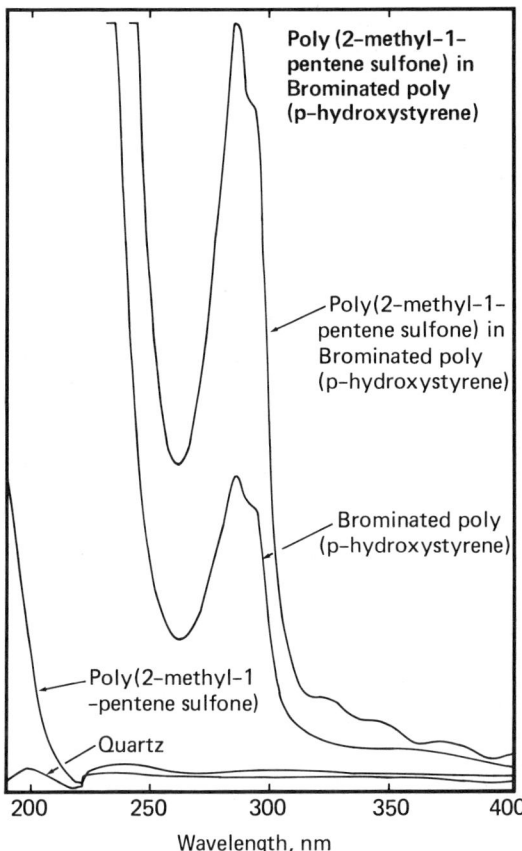

Figure 3. UV absorption spectra of PMPS in brominated poly(p-hydroxystyrene).

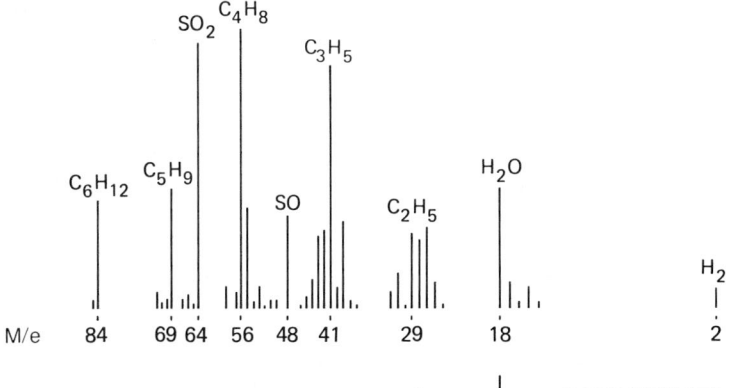

Figure 4. Gaseous products formed in 254 nm irradiation of PMPS in Varcum resin.

exposure in air are clearly demonstrated in the ESCA data. The change of the C_{1s} core level signal, shown in Fig. 6, indicates formation of carbonyl carbons at 289 eV and of carbons attached with hydroxy groups at 287 eV after a prolonged exposure to UV light.

The charge transfer complex absorption bands described above are not strong enough to be practically useful. For lithographic image generation, a quartz mask has to be used so that even the bottom part of the resist films absorbs 254 nm UV light, and then this energy is to be transferred to cause the degradation of the polyolefin sulfone. Because of this energy transfer only the most readily degradable polysulfone can be used as a composite photoresist; poly(2-methyl-1-pentene sulfone) yielded far better images than poly(1-butene sulfone), although their electron beam sensitivity is the same (4). The polymer patterns can be made either in positive mode or in negative mode depending upon the processing technique as described next. The fact that negative tone patterns were obtained in a KOH-based developer after heating at 130°C for 5 min indicates presence of trapped polymeric free radicals after the UV exposure which initiated crosslinking upon the postbake. Negative tone patterns are superior in quality than those in positive tone.

Figure 7 shows photo-images obtained with the composite resist made of poly(2-methyl-1-pentene sulfone) and Varcum; positive tone patterns are shown on the left, and negative tone patterns on the right. The resists films were spin-coated onto silicon wafers from an AZ-thinner solution of poly(2-methyl-1-pentene sulfone) (16.5 wt% in solid) and Varcum, and prebaked at 100°C for 20 min. The UV exposures were carried out with a medium pressure mercury lamp with a dose of 200 mJ/cm^2 of 254 nm light; it can be reduced to 100 mJ/cm^2 for full development.

For positive tone polymer patterns, the resist films were developed in a KOH-based developer without post-bake after the UV exposure. For negative tone polymer patterns, the resist films were developed after post-bake at 130°C for 5 min, following the UV exposure. We speculate that during this post-bake process, polymer free radicals generated in the main chain scission undergo cross-linking with the novolac resin. There are no film thickness loss observed after complete development with an adequate UV dose. Without the polysulfones, the novolac resin alone yielded very poor quality negative tone images with very thin remaining film thickness. The wall profiles of the negative tone patterns are shown in Fig. 8; vertical profiles or slightly undercutting profiles are obtained. The relation of the remaining film thickness after complete development *versus* 254 nm dose is shown in Fig. 9; (a) for the positive tone images, and (b) for the negative tone images with a composite resist of poly(2-methyl-1-pentene sulfone) and Varcum resin. The similar results are obtained with a composite resist made of brominated poly(p-hydroxystyrene) and poly(2-methyl-1-pentene sulfone) with an increased photosensitivity for negative tone images.

In all these studies, UV light in a region from 250 to 300 nm has been used to degrade poly(olefin sulfones) *via* energy transfer from the host resin in the case of composite resists with Varcum, and from sensitizers in the case of

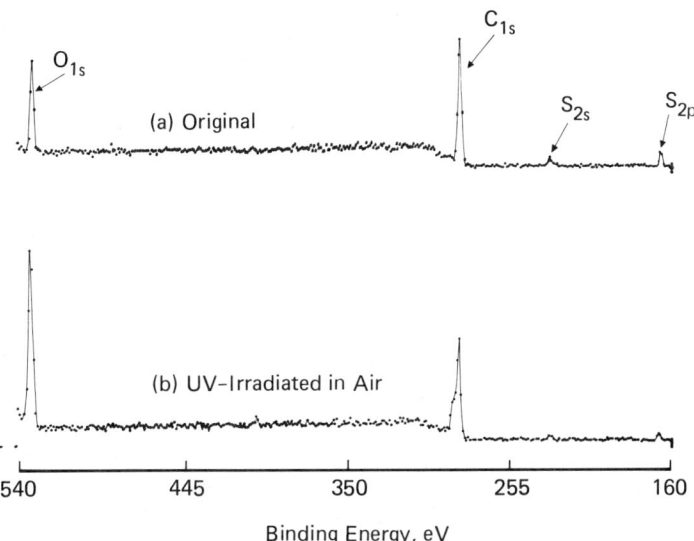

Figure 5. Wide band ESCA data of matrix resist films, PMPS in Varcum.

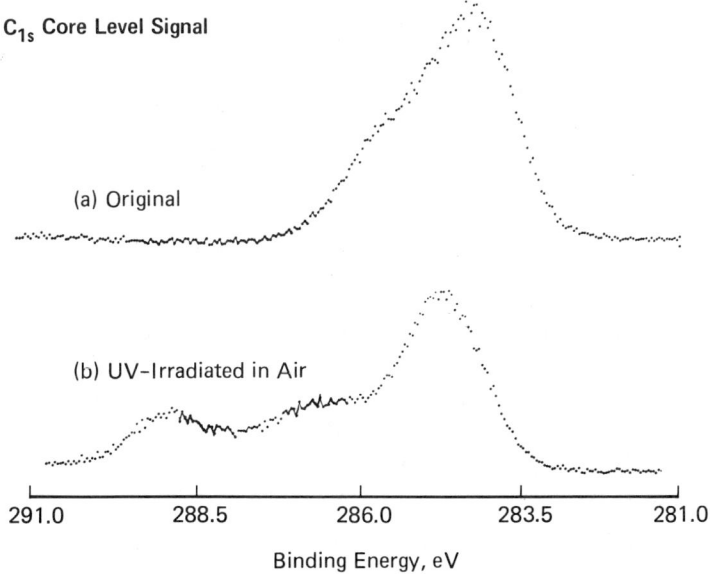

Figure 6. Change of carbon core level signals of PMPS in Varcum.

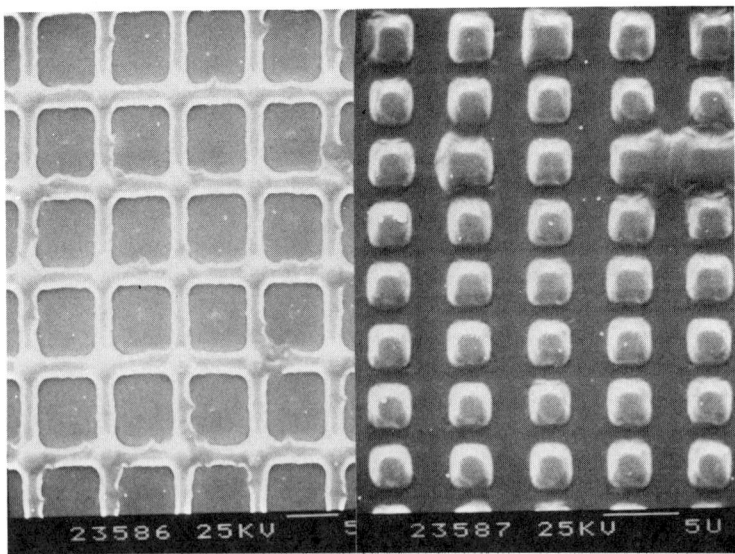

Figure 7. Deep UV polymer patterns obtained with a matrix resist of poly(2-methyl-1-pentene sulfone) and Varcum resin; positive tone images on the left and negative tone images on the right.

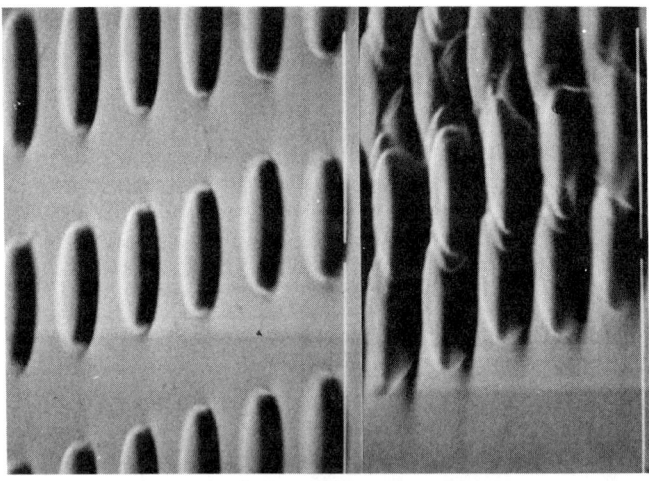

Figure 8. Negative tone resist patterns obtained with a matrix resist of poly(2-methyl-1-pentene sulfone) and Varcum resin.

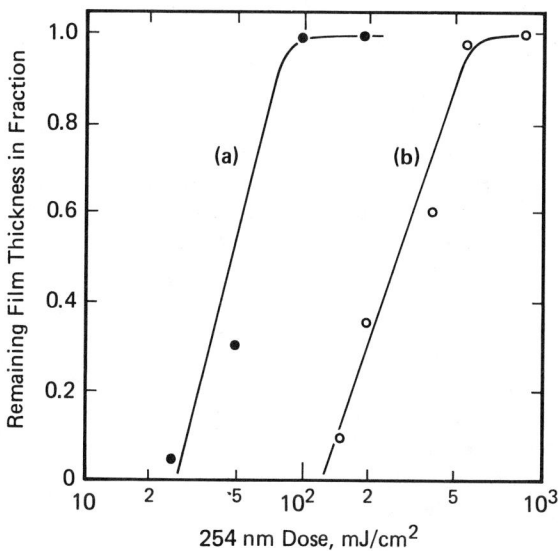

Figure 9. Relations of remaining film thickness *versus* 254 nm dose; (a) positive tone patterns, (b) negative tone patterns with a poly(2-methyl-1-pentene sulfone)/Varcum resist.

dry developable photoresist described earlier. The nature of the energy transfer is not clear; it is probably through complexes between the polysulfones and the resins described above, or through dissociative triplet states as in $Hg(^3P)$ photosensitized reactions of simple olefin sulfones.

Literature Cited

1. Hiraoka, H., Presented at the 167th ACS National Meeting at Los Angeles, California, May 1974; Abstract No. Phys. 137.
2. Hiraoka, H., Presented at the First Chemical Congress of the North American Continent, Mexico City, Mexico, November 1975; Abstract No. 60.
3. Recently a different view of electron beam induced degradation of poly(olefin sulfones) has been presented, particularly in a very low electron dose range, by J. Pacansky, R. Kroeker, E. Gipstein, and A. Gutierrez, The Electrochemical Society 1982 Spring Meeting, Montreal, Canada, May 1982; Abstract No. 273.
4. Bowden, M. J. and Thompson, L. F. J. Appl. Poly. Sci. 1973, 17, 3211; Gipstein, E.; Moreau, W.; Chiu, G.; Need, O. J. Appl. Poly. Sci. 1977, 21, 677.
5. Himics, R. J.; Kaplan, M.; Desai, N. V.; Poliniak, E. S. Tech. Papers, Soc. Plastics Eng. Mid-Hudson Sec., October 13-15, 1976.
6. Decout, J. L.; Lablanche-Combier, A.; Loucheux, C. Photographic Sci. and Eng. 1980, 24, 255.
7. Hiraoka, H. IBM Tech. Discl. Bull. 1982, 24, 5756.
8. Bowden, M. J.; Thompson, L. F.; Fahrenholts, S. R.; Doerries, E. M. J. Electrochem. Soc. 1981, 128, 1304

RECEIVED September 2, 1983

Method for a Comparative Study of Positive Photoresist Lithographic Performance

C. C. WALKER and J. N. HELBERT

Process Technology Lab/SRDL, Motorola SPS, Phoenix, AZ 85008

> A quantitative method has been developed to measure photoresist sensitivity and development/exposure latitude in order to compare positive photoresist systems under equivalent conditions. The method is based upon image size measurement by a Nanoline measurement system as a function of exposure and development. Using this method, an extensive series of first and second generation organic positive photoresists have been quantitatively compared for lithographic performance. The second generation resists generally exhibit greater sensitivity or photospeed. Unfortunately, the gains in photospeed are generally offset by slightly lower contrast values. Most importantly, however, the second generation resists are capable of exhibiting excellent processing latitudes, when the proper developer is employed.

Positive photoresists are composed mainly of three basic components (1) a base soluble resin usually of the novolac type, (2) a photoactive and dissolution-inhibiting component usually of the diazide type, and (3) a solvent system usually a cellosolve acetate (1). While commercial positive photoresists are generally of this generic formulation, their exact compositions are usually not known due to proprietary considerations. As a result, resist lithographic performance cannot be correlated with formulation composition, but it is still very necessary that their relative lithographic performance values be obtained and compared. This data, then, is of value to chemists functioning as photoresist engineers in the fabrication of electronic circuits. Really, this information is of greater importance than the former because most of these professionals will never formulate their own resist, but they are expected to develop and control the lithographic resist processes within 5-10% in device fabrication.

0097-6156/84/0242-0065$06.00/0
© 1984 American Chemical Society

In this work, a quantitative evaluation method has been developed to measure photoresist sensitivity or speed (2,3) and process latitude (2,4) in order to compare positive photoresist performance under equivalent conditions. Performance data using this method will be reported and compared for an extensive group of first and second generation commercial photoresists.

Nanoline Measurement Method

In photolithography, photoresist performance is determined ultimately by how well you can control the size of images being replicated into a resist coating on an electronic device substrate. This task is governed by exposure, development and processing effects which must be well understood before success is realized. With this goal in mind, a simple straight forward ratio method for following the resist image dimension as a function of exposure/development using a Nanometrics Nanoline critical dimension (CD) measuring system was developed. This method circumvents absolute calibration by being responsive to the condition of exact image size transfer from the chromium patterned photomask to the resist wafer coating. A method based upon image size measurements by scanning electron microscopy (SEM), another measurement alternative, would not be practical for this volume of work, because SEM sample preparation and measurement is a very cumbersome and time-consuming task. Furthermore, the method employed is non-destructive as opposed to that of specimen mounted SEM.

Relative photoresist image dimension measurements were made with the Nanoline system attached to a Zeiss Axiomat microscope. Line and space pattern measurements are first made on the chrome mask or reticle, then on the actual test wafers. The resist image dimension measurements are obtained for the islands (I) or lines and the windows (W) or spaces by computer analysis at 50% line edge profile threshold using the substrate-appropriate software programs provided with the Nanoline system computer. The empirically generated critical dimension parameter, Δ, is obtained by subtracting the Nanoline I/W dimension ratio on the photomask from that on the wafer, and is a relative measure of how well image transfer from the mask to the wafer is occuring. A delta of zero is the desired condition of CD transfer from the mask to wafer, and is a unique relative exposure/development equivalence point for photoresist performance comparison. $\Delta = 0$ conditions were verified by filar measurement on the Axiomat microscope at 1600X and with a Cambridge S4-10 SEM.

A representative photoresist Δ vs. exposure curve is found in Figure 1 for PC-129SF, a representative first generation positive photoresist. Each point in this Figure represents a different photoexposure and development time combination, which results in a cleared or fully developed image to the substrate. Development times that are too short for image clearing result in no data for these plots, while data points below $\Delta=0$ are representative of developed line images that are dimensionally oversized (or undersized for islands) vs. the mask dimensions. Data to the left of these plots represent an underexposed condition where undesirably long development times are required, thus, creating unexposed resist thickness loss and resist scumming problems. The solid line through the uppermost data is a least squares fit of the data where the minimum development time was employed to just achieve development to the stubstrate; a shorter development at the fixed exposure would have yielded a non-cleared image. The greatest exposure where $\Delta = 0$ is the unique equivalance condition in exposure/development where resists can be compared for relative sensitivity. For this purpose, all the sensitivity exposures are converted to mJ/cm^2 by multiplying the light intensity by the exposure time to take into account differences in light intensities necessitated by practical considerations. The lithographically useful resist working range is determined by following the $\Delta = 0$ line across the Figure; the shortest exposure time is dictated, in a practical sense by the length of the development time as discussed above. Finally, a measurement of the development/exposure latitude is obtained from the slope of the upper solid fit curve; a lower slope is indicative of greater overall process latitude and conversely.

Experimental

Wafer Processing. Testing substrates were either 2 or 3" silicon wafers with 5000Å of SiO_2. Following DI water scrub and rinse, wafers were dehydrated in a 200°C N_2-purged oven for 1-2 hours. An adhesion promoter, either of the amino or halogenated silane type, was always employed prior to resist coat and prebake--as specified by the resist vendors.

Exposure Conditions. Exposures were carried out using a Cobilt 2020H contact aligner controlled by an Optical Associates Model 760D power supply and controller. The emission output of the lamp was measured and is found in Figure 2. The emission wavelengths and intensities are characteristic of systems utilizing compact Hg arc lamps, such as Perkin-Elmer 200 series systems without band pass filters, Cobilt systems like the model 2020H, and Ultratech-900 projection stepper systems. The Ultratech-900 spectrum also contains a broad and lower intensity continuum of emission stretching from 3500-4500Å. We have found emperically, however, that the relative resist exposures or ratios required for the Cobilt system are maintained reasonably well for the Ultratech system due to the similarity in emission characteristics of their lamps. The smaller emission at 3300Å wavelength has been omitted from Figure 2 due to its relatively low effectiveness in resist exposure.

Results and Discussion

Cobilt Evaluations. The photoresist sensitivity, relative process latitude (RPL) and development information data is located in Tables I and II. The observed sensitivities at $\Delta = 0$ range from

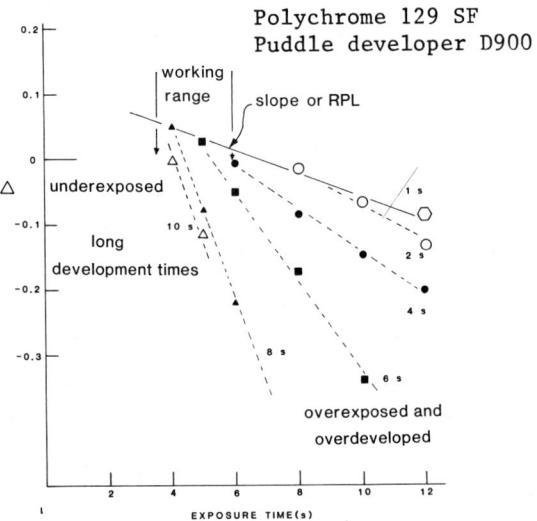

Figure 1. Δ vs exposure for PC-129.

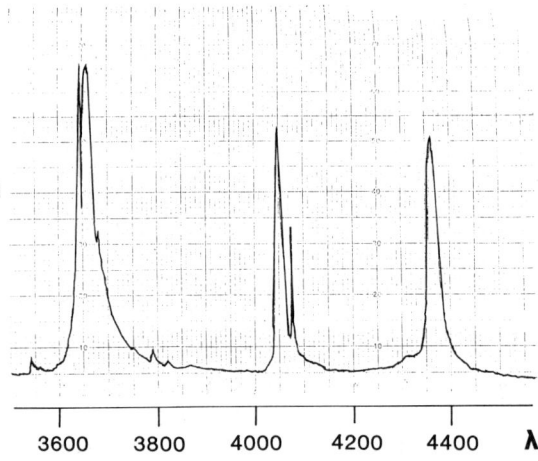

Figure 2. Cobilt Hg lamp emission characteristics.

12-130 mJ/cm^2, where the most sensitive systems are generally the newer second generation photoresists, those developed after the 1979-80 time-frame (see Table II). As a group they displayed sensitivities of 12-95 mJ/cm^2, depending upon development conditions. The first generation resists, such as PC-129, AZ-1350, KTI-II, K-809 and HPR-204, generally exhibit lower sensitivites (i.e., larger numbers) ranging from 60-130 mJ/cm^2. The developer, more specifically its concentration, is very important as Table I verifies. For example, PC-129 is greater than 2X more sensitive when a 4X more concentrated developer is employed. This gain, however, is accompanied by a severe loss in RPL, and this resist/developer combination would not be useful in a production environment. This effect is similar to that reported by Elliot in reference 2.

Table I: Photoresist Performance Data for First Generation Resists.

Resist	RPL	Q, mJ/cm^2	Developer/Method
PC-129 (Allied P2025)	1.0	110±20	D900/DIP
	1.6	102	D900/PUDDLE
	1.6	119	D910/DIP
	4-8[a]	50	CONC 900/DIP
AZ-1350	1.4	81	1:1 MF312/DIP
KTI-II	2.1	105	DE-3/SPIN-SPRAY
-II	1.0	132	DE-3/DIP
Kodak-809	0.4	100-160[b]	809 developer
HPR-204	2.7	91	1:1 LSI/DIP
-204	3.5	60	1:1 LSI/PUDDLE

a Hard to fit date exactly - large data scatter - (4X D900 concentration).
b Data from Kodak not measured in this work.

The observed RPL values range from 0.4 to approximately 8, and again as found for sensitivity, they are highly developer dependent (See PC-129 and other data of Tables I and II). OFPR-800, a second generation resist, displays a very low RPL value when developed by MF-312, a developer not specifically designed for that resist. Significantly, these second generation resists, such as AZ-1470, KMPR-820, MAC-9500's and HPR-100's, are capable of good (i.e., low) RPL when developed properly. Therefore, the sensitivity gains of up to a factor of 10 are not sacrificed in RPL as a trade-off in overall performance.

Table II. Photoresist Performance Data for Second Generation Resists.

Resist	RPL	Q, mJ/cm^2	Developer/Method
Allied P-5019[a]	1.0	13	2:1 D100/DIP
	3.9	72	D150/DIP
	3.0	95	LMI500/DIP
AZ-1470	2.3	40	1:1 MF312/DIP
-4110	1.5	43	1:1 MF312/DIP
OFPR-800-1	1.3	74	NMD3-1/DIP
-800-2	2.7	75	NMD3-2/DIP
-800-1	0.5	37	1:1 MF312/DIP
KMPR-820-AA[a]	1.6	18	AA-0980-52[a]/DIP
-820	2.2	17	AA-0980-52/DIP
Hunt-159[a]	0.9	73	1:3 LSI/PUDDLE
-118[a]	1.4	62	1:3 LSI/PUDDLE
MAC-9564	3.8	24	1:1 9562(MF)/DIP
	4.2	24	1:1 9564/DIP
-9574	1.9	12	1:1 9571/DIP

a Experimental or prototype system.

Unfortunately, the trade-off for sensitivity and RPL does come when the relative contrast (γ) values are considered (see Table III). In the Table, contrast is measured as described in reference 1. The photoresist contrast values are important, because they are a measure of the relative resolution of the photoresist system as has been pointed out by Blais (5). As suspected, the second generation systems do generally exhibit lower or poorer contrast, but the loss is only 20-40%. This points out, however, the question of whether further gains in positive photoresist sensitivity and RPL will be realized without causing significantly lower resolution capability? The notable exception to this question would be the Allied P-5019 system, which exhibited very high contrast. Unfortunately, this resist is an experimental resist only and was not commercially available at the time of manuscript completion. Table III also contains representative resist image edge wall angle (EWA) data taken for one micron images under $\Delta = 0$ conditions, and consistent with Blais (5), the edge wall angles correlate well with the measured contrast values obtained independently.

Table III. Photoresist Contrast[a] (Relative Resolution) Values.

Resist	γ	Resist	γ
PC-129 (P-2025)	2.0[b]	P-5019	2.5-3.8
AZ-1350	1.5[c]	AZ-1470	1.8
		AZ-4110	1.6
KTI-II	1.8	OFPR-800	1.4[d]
K-809	1.6	KMPR-820	1.2
HUNT-204	1.7	HUNT-100	1.5-1.6
		MAC-9500	1.1

a High contrast is best.

b EWA= $84°$
c EWA= $79-81°$
d EWA= $73-77°$

Tables I and II also contain photoresist data for three different development methods. They are (1) spin/spray, (2) temperature controlled dip, and (3) puddle development methods (see Figure 3).

The RPL values obtained by puddle development for PC-129 and HPR-204 are both larger than those obtained by dip development, thus indicating that the dip method is better with the developers employed. In addition, resist image clearing uniformity across the 3" test wafers was better for the dip developed wafers than for those puddle developed. Furthermore, the dip development method yielded statistically better wafer to wafer development reproducibility for PC-129. This is not to say that puddle development is not a useable entity but that dip is preferred. In fact, Leonard and coworkers (6) have developed a puddle development method with production capability. Similarly, the RPL data for KTI-II favors dip development over that of spin/spray, therefore, dip development is the favored development method overall for these three example resists.

Stepper Example. The resist evaluation method developed in this work for Cobilt contact aligners is a general method, and has also

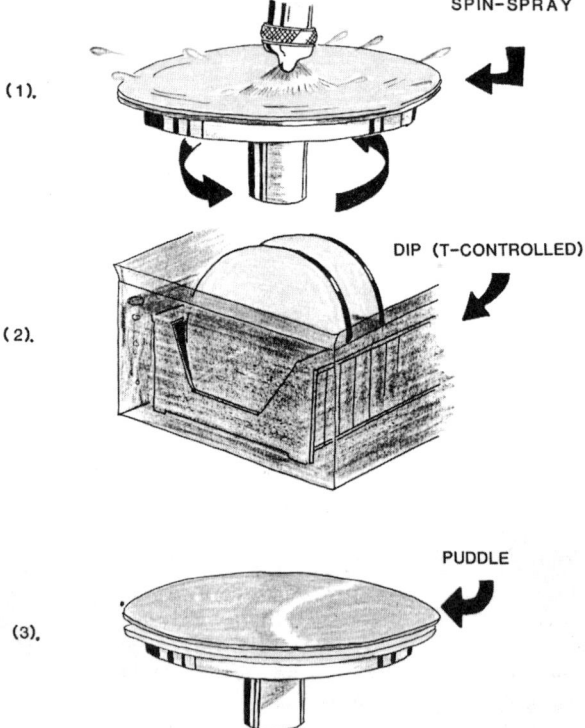

Figure 3. Photoresist development methods.

been employed to evaluate resist performance under Ultratech-900 projection stepper conditions. Results for one of the promising second generation resists from the Cobilt study, namely KMPR-820, are included in this section. The developers used were (1) 933, (2) 932 and (3) KMPR-experimental, all diluted 1:1 with water. Two of the Δ vs dose characteristic curves for KMPR-820 are found in Figures 4 and 5. The curve for 933 developer is similar in nature except the line slopes through $\Delta = 0$ are from 2-3X steeper, thus indicating lower or poorer RPL performance for 933 developed KMPR-820.

The Δ vs dose slopes of Figures 4 and 5 differ only slightly, thus indicating similar RPL values for 820 developed by these developers. Developer 932 increases the effective 820 resist sensitivity (see Figure 4) 50% over that observed for 933 and KMPR-experimental, while the $\Delta = 0$ condition was only achievable with developers 932 and 933. Figure 5 clearly illustrates that the KMPR-experimental developer is not able to achieve $\Delta = 0$ image transfer, even under the lowest clearing exposure (see circled data points on Figures 4 and 5). This behavior was not totally unexpected, because KMPR-experimental developer was formulated expressly for use with existing biased reticle sets which have been previously oversized for other resist systems.

Finally, best KMPR-820 edge wall angles were obtained by employing the 933 developer as can be seen in Figure 6. Note in the Figure that the KMPR-experimental developed resist image is visably smaller, consistent with the Δ vs. dose plots.

Summary

A quantitative method has been developed to comparatively measure photoresist sensitivity and development/exposure latitude. The second generation resists evaluated exhibit improved sensitivity, good to excellent RPL, but also exhibit slightly lower contrast. The older established resists were generally slower, but also exhibit good RPL and highest contrast/resolution. Of the resists evaluated, the two resists with the best overall combination of RPL, contrast and sensitivity were the AZ-4110 and Dynachem OFPR-800 systems. The newest resist, Allied P-5019, bears watching also due to its very high contrast performance.

This paper addresses photoresist lithographic performance only. For a more complete view of these systems, the accompanying paper should be read where thermal flow characteristics as well as plasma etch resistance data are presented and compared for many of the photoresists evaluated in this work.

Acknowledgments

The authors thank Andrea Kress and Liz Ramos for diligent sample preparation and measurement support. The art work of Ralph Alvarado is also much appreciated.

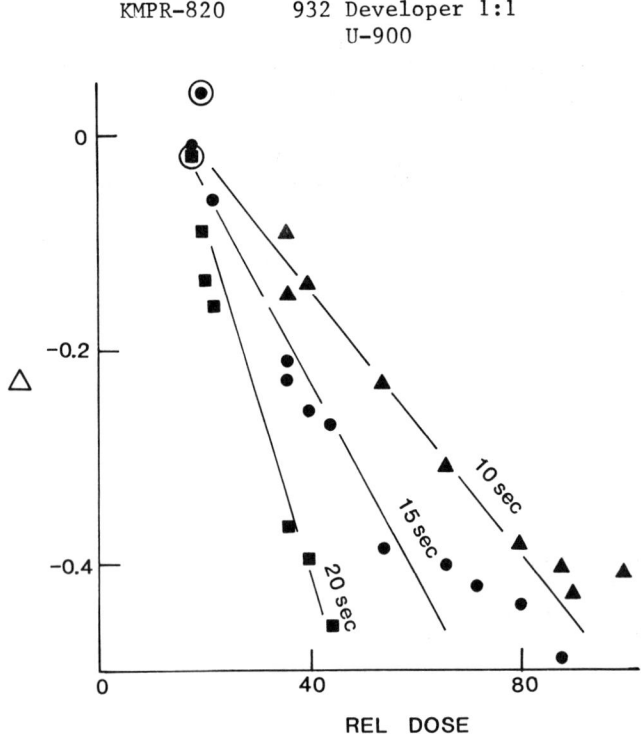

Figure 4. Δ vs relative dose for 932 developed KMPR-820 resist images

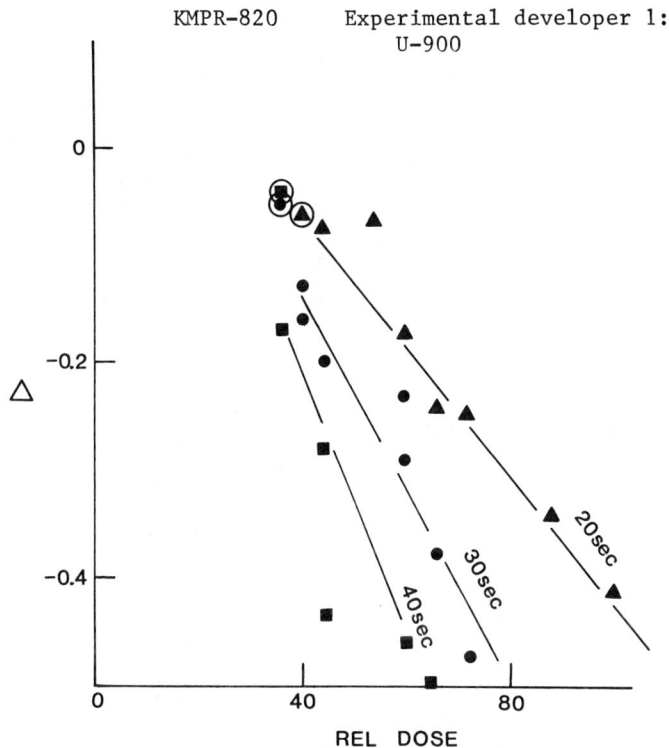

Figure 5. Δ vs relative dose for KMPR-experimental developer developed KMPR-820 resist images.

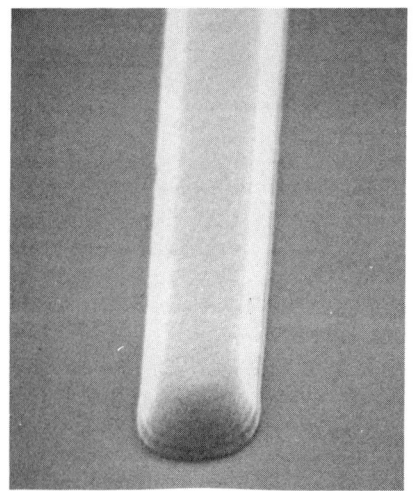 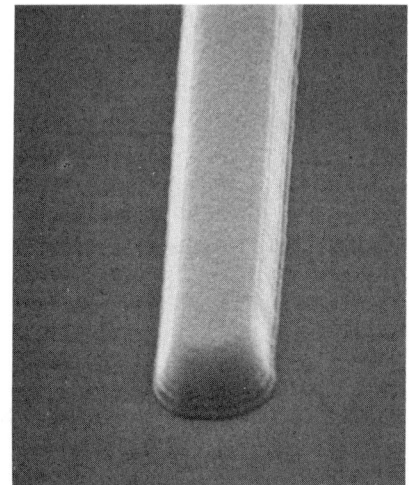

Developer 932   1 um   Developer 933

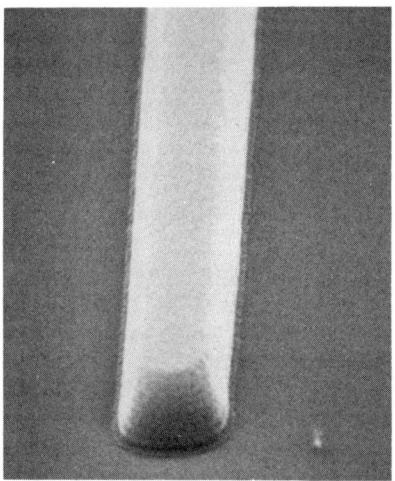

Experimental developer

Figure 6. KMPR-820 resist images following development under $\Delta = 0$ conditions (see text).

Literature Cited

1. Thompson, L.F.; Kirwin, L.E. Ann. Rev. of Materials Sci. 1976, 6, 267.
2. Elliot, D.J. Solid State Technology 1977, September.
3. Hatzakis, M; Shaw, J.M. Proceedings of Eighth International Conference on Electron and Ion Beam Science and Technology 1978, 78-5, p. 285.
4. Deckert, C.A.; Peters, D.A. Solid State Technology 1980, January.
5. Blais, P.D. Solid State Technology 1977, August.
6. Leonard, R.F.; Sim. G.; Weiss, R. Solid State Technology 1981, June.

RECEIVED September 2, 1983

Dependence of Dissolution Rate on Processing and Molecular Parameters of Resists

A. C. OUANO[1]

T. J. Watson Research Center, IBM Corporation, Yorktown Heights, NY 10598

In this paper we explore factors beyond the well known molecular weight and chemical effects on the dissolution kinetics of polymers. In our studies we show that factors controlling solvent mobility in polymeric matrices, such as stereo chemistry (tacticity), processing (heat treatment) and solvent molecular size may affect the dissolution kinetics of polymers far more strongly than factors affecting the equilibrium solubility, e.g. molecular weight of the polymer. It is shown that while the dissolution rate scales to approximately -0.7 power of the polymer molecular weight (P(MMA), 10K < Mw < 1000K) it can scale to -16 power of the solvent molecular size (propyl to hexyl acetate). Isotactic P(MMA) dissolves 300 times faster than the syndiotactic form of equivalent molecular weight (Mw ~ 100K). Heat treatment of P(MMA) is also shown to change the dissolution rate by several orders of magnitude. The results obtained in this study point to the prime importance of the molecular dynamics of both the polymer and the solvent on the resist dissolution kinetics.

Lithographic images develop because of differences in the dissolution kinetics between the exposed and unexposed areas of the resist. Resist are predominantly polymeric in nature of substantial molecular weight and may be a single component, e.g., poly(methyl methacrylate), P(MMA) or a 2 component (a Novolac resin and a diazo sensitizer) system. It is the high molecular weight (long chain) and the glassy nature of the resist material which make their dissolution dynamics complicated compared to small crystalline molecules such as NaCl and succrose where transport of solvent molecules deep into the crystalline lattice is not necessary for dissolution. The large pervaded volume and

[1]Current address: Dept. H36/025, General Products Division, IBM Corporation, 5600 Cottle Road, San Jose, CA 95193.

the high degree of chain interpenetration of polymeric material
requires considerable swelling of the glassy matrix (1-4) for
dissolution to proceed. Computer simulation has shown (1) that
in some cases the dissolution kinetics is controlled by solvent
diffusion in the glassy phase, where the diffusion coefficient
of the solvent could be less than $10^{-14} cm^2/sec.$ (5) It is also
well known (6) that solvent mobility in polymeric matrices is
strongly affected by its local environment, e.g., free volume
backbone flexibility and concentration of solvent. Thus solvent
transport is very strongly influenced by polymer architecture and
by the thermal and mechanical history of the polymer, and is only
weakly influenced by long range interaction characterized by
molecular weight of the polymer. This insight forms the core of
this investigation which covers tacticity, annealing, radiation
exposure and solvent molecular size effect on the dissolution
kinetics of polymers.

Processing Effects
The effect of heat treatment on the dissolution rate of spin
coated P(MMA) film on silicon has been studied by Greeneich (7&8).
The results he obtained showed very strong dependence of the
dissolution rate of P(MMA) on the prebake (hot plate) temperature
and time as shown in (figure 1). This result has been interpreted (7 & 8) as mostly due to a reduction in the solvent
content (drying) during heat treatment. More recent study on the
effect of prebaking on the dissolution rate of a photo resist
(a Novalac resin sensitized with 20 weight % diazo compound)
showed a non monotonic decrease in the dissolution rate (S)
(see figure 2). The photo resist showed a maximum at about
110°c., 30 min. prebaking time. However at temperature higher
than 120°c. both the photo resist and the Novalac resin showed
a steep decrease in their dissolution rate.

It is known that the diazo sensitizer is temperature sensitive
and decomposes almost completely on heating above 110°c. for 30
min. The decomposition products in the presence of moisture (in
the atmosphere) are nitrogen (N_2) and indene acid, a base
developer soluble material. On extended heating at higher
temperature there is a marked decrease in the dissolution rate
of both P(MMA)(see figure 1) and the photo resist (figure 2).
One asks: Is this continued decrease in the dissolution rate
simply a consequence of a continued loss of solvent in both P(MMA)
and the photo resist? In response to this question we conducted
a series of drying experiments on a 1.5μm P(MMA)(Elvacite 2041)
film spun on 1" silicon wafer from a 10% chlorobenzene (CB)
solution. Drying was carried on a hot plate equipped with a
thermo couple bonded to its surface to monitor prebaking
temperature. The loss of the CB solvent was measured using a
Cahn 1000 balance (sensitive to a 1μ gram change in weight). A
plot of the percent solvent remaining as a function of temperature

7. OUANO Dissolution Rate Dependence on Resist Parameters 81

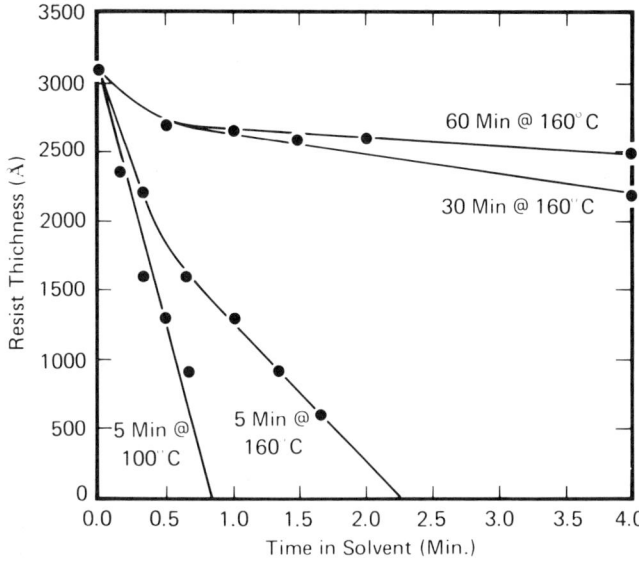

Figure 1. The effect of heat treatment on the dissolution rate of spin coated P(MMA). (Reproduced with permission from Ref. 8. Copyright 1953, Cornell University Press.)

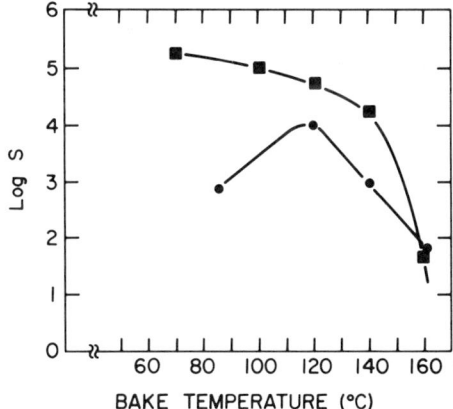

Figure 2. The prebake temperature vs. the solubility rate (°A/Min) of ■ - Novolac resin and ● - photo resist (Novolac + Diazo sensitizer). Prebaking time is 30 Min., 0.25N Acq. KOH at 25°c. developer.

and time is shown in the drying curve (figure 3). It is clear (figure 3) that above 137°c. the P(MMA) film is almost completely devoid of the CB solvent in 30 secs. of baking time. This is not surprising when one considers that the 137°c. prebaking temperature is very close to the CB boiling point and above the P(MMA) Tg. Yet, considering figure 1, it is clear that it takes more than 30 minutes of prebaking time to reduce markedly the dissolution rate of P(MMA) 2041 at 160°c. Since it takes less than a few seconds (20 secs.) to completely dry the P(MMA) film at 160°c., the extra prebaking time must be for annealing the film and to reduce the extra free volume created by the rapid loss of the solvent. This data appears to support the hypothesis that P(MMA) dissolution is markedly affected by free volume which facilitates solvent transport in the P(MMA) glassy matrix.

Molecular Weight Effects

It is well known (9) that the equilibrium solubility of polymers is dependent upon their molecular weight. This is mostly due to the decrease in the entropy of mixing and in the 2nd virial coefficient with increasing molecular weight of polymer in solution. The dissolution rate of P(MMA) in Amyl Acetate for example, scales with the molecular to the -0.7 power, (10) e.g.

$$S = KM^{-0.7} \qquad 20K < M < 800K \qquad (1)$$

Where S = solubility rate in A°/Min.

It is interesting to note that the 2nd virial coefficient (α) (10) and the diffusion coefficient (Dp)(1) of polymers in solution scales to about -0.3 power of its molecular weight, e.g.

$$\alpha = K_1 M^{-0.3} \qquad (2)$$
$$Dp = K_2 M^{-0.3} \qquad (3)$$

Radiation Exposure Effects

The use of P(MMA) as high energy radiation resist is due to its large increase in solubility rate on exposure to e-beam, deep UV and x-ray radiation. The cross section of a developed image on P(MMA) as function of radiation (shown in figure 4) is remarkable. The highest dosage shows a very sharp edge with a positive slope on the sidewall, such a profile can only be produced if very large (several orders of magnitude for example) dissolution rate difference exist between the unexposed and exposed region of the resist. If the increase in the dissolution rate on exposure is due only to a decrease in molecular weight then a decrease of several orders of magnitude in molecular weight must take place since the solubility rate increase is less than the inverse of molecular weight to the 1st power.

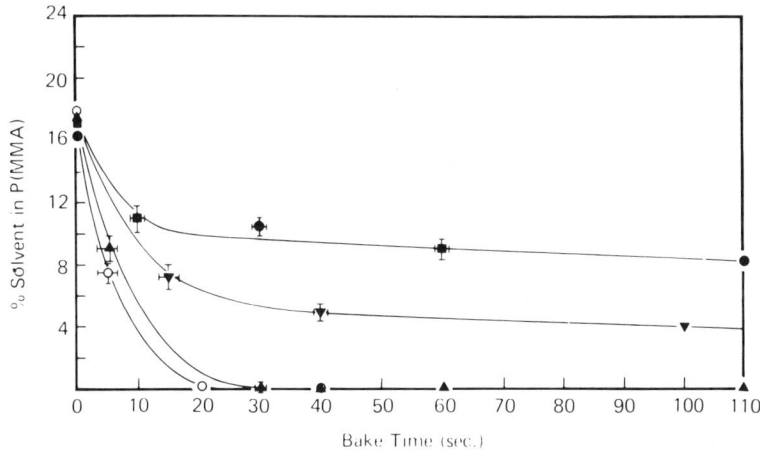

Figure 3. Residual solvent (chlorobenzene) in 1 μm P(MMA) film on silicon wafer vs. prebaking time ■ - 85°c., ▼ - 100°c., ▲ - 117°c., ● - 137°c. and ○ - 158°c.

Figure 4. SEM photo micrograph of developed P(MMA) profiles at exposures of 10^{-4} coul/cm^2 (top), 8×10^{-5} coul/cm^2 (middle) and 5×10^{-5} coul/cm^2 (bottom).

On exposure to high energy radiation, P(MMA) undergoes predomenantly chain scission and decomposition of its ester side groups resulting in Mw reduction and formation of volatil products, e.g., CO_2, CO, CH_4, CH_3OH, H_2 with G value for the formation of these products ranging from 0.5 for CH_3OH to 1.4 for H_2 (11). The chain scission G-value has been reported in literature to be about 1.3 (12). Thus the loss in mass due to evolution of volatil compounds (CO_2, CO, CH_4 . . .) on irradiation at temperatures much below Tg should result in a decrease in the density (increase in free volume), and increase stress in the irradiated P(MMA) film. This increase in free volume was verified by measuring the increase in the x-ray intensity emmitted from various thickness of P(MMA)(2500°A to 104°A) coated on silicon. Figure 5 shows the x-ray intensity emmitted from the silicon substrate as function of P(MMA) thickness and dosage. Measurement of thickness change due to volume relaxation caused by the loss of mass was negligible, thus most of the mass loss was reflected by a decrease in density.

This loss in density of P(MMA) on exposure to high energy radiation should result in a very large (several orders of magnitude) increase in the diffusion coefficient of the developer into the P(MMA) matrix. It is well known that relatively small increases in the free volume, e.g., polymers heated from below to above Tg is accompained with a large increase in the diffusion coefficient of solvents in the polymer matrix (5 & 6). For example, Kambour (13) reported very large increases in the diffusion coefficient of ethanol in crazed poly carbonate.

The effect of radiation on the solubility rate of P(MMA) in Amyl acetate is shown in figure 6. Figure 6 also shows the solubility as a function of molecular weight. Because of the exact relationship between exposure dosage and molecular weight through the G value for P(MMA) it is possible for us to plot the dissolution rate (S) due to exposure and the dissolution rate vs. molecular weight of unexposed P(MMA)(as in figure 6). Note that the rate of increase in S with decreasing Mw is much faster in the exposed than in the unexposed P(MMA). The excess change in S (above that due solely to the change in Mw) of the exposed P(MMA) must be the contribution of the increase in D_s (due to increase in free volume).

Stereo Chemistry Effects

P(MMA) can exist in 3 stereo forms with different Tgs. The isotactic, the atactic and the syndiotactic have Tg's of ∼ 40°c., ∼ 117°c, and 170°c. respectively. These differences in Tgs. is strongly reflected in the difference in the solubility rates among the 3 different tactic forms (as shown in figure 7). It

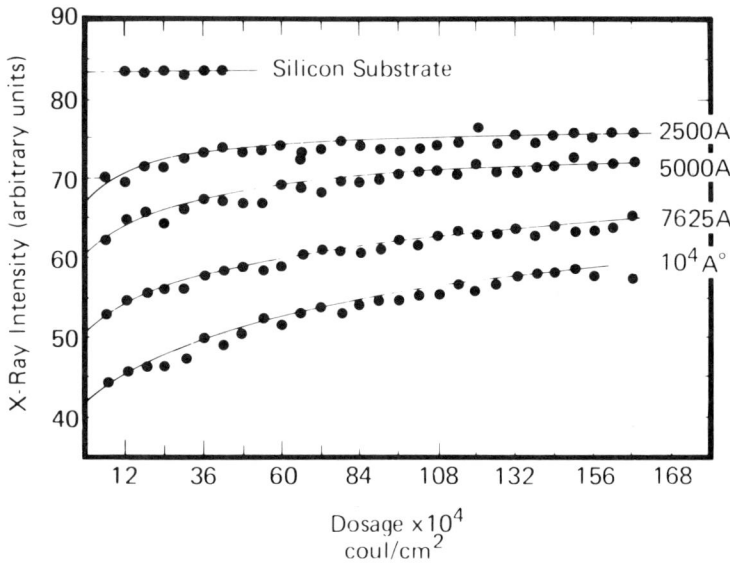

Figure 5. X-ray intensity emitted from a silicon substrate of various thickness of P(MMA) film vs. e-beam dosages.

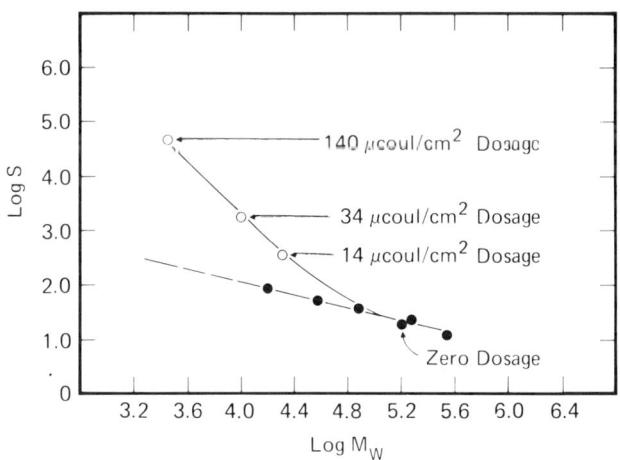

Figure 6. Solubility rate (°A/Min) of 1 μm P(MMA) on silicon in Amyl Acetate at 24.5°c. vs. molecular weight ○ - exposed and ● - unexposed.

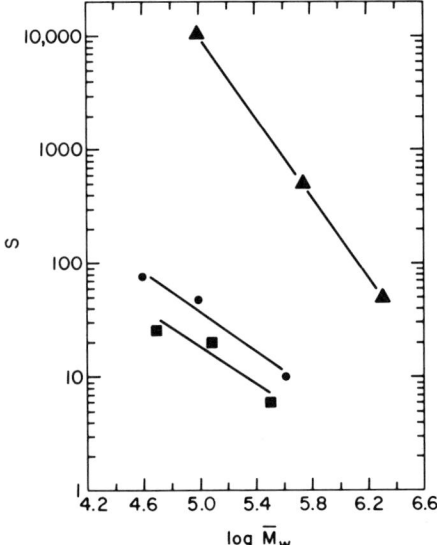

Figure 7. The solubility rate (°A/Min) of various tactic forms of P(MMA) in Amyl Acetate at 25°c. ▲ - isotactic, ● - atactic and ■ - syndiotactic 1μm film on silicon versus the molecular weight (Mw).

is also clear (figure 7) that the molecular weight dependence is stronger (steeper slope for the low Tg (high D_s) than for the higher Tg syndio and atactic forms. Thus for polymers which have Tgs at the dissolution temperature, the dissolution rate could be dictated by the molecular weight since the D_s is very large at temperatures above the glass temperature. It is useful at this point to note the relationship between the Tg and the solvent content in P(MMA). Figure 8 shows the rapid decrease in the Tg of P(MMA) with CB content. Again linking solubility rate to polymer segmental mobility.

Solvent Size Effects

It is amply clear at this point that the solvent diffusion rate is a gating factor in the dissolution kinetics of well annealed polymers with high glass temperature. Since solvent translational mobility is strongly affected by the relative size of the interstial distances (free volume) between the highly interpenetrating chain of the polymer and the size of the solvent molecule, it is therefore expected that the polymer solubility rate (S) will be greatly affected by the solvent molecule size. That this is in fact true is clearly demonstrated by figure 9. Figure 9 illustrates the effect of both radiation exposure (increase free volume and lower molecular weight) and solvent size. In the unexposed and annealed P(MMA) a sharp break in the log S/ Log Ms plot is clearly evident between the propyl and butyl acetate solvent. The power dependence of S on Ms between methyl and propyl acetate is relatively weak, but between butyl and the higher acetate homologs is very strong. It appears that there could be a transition of the diffusion of the solvent between propyl and butyl acetate, e.g., from a relatively free (correlated motion between polymer main chain and the solvent molecule is not necessary for solvent diffusion) to a highly hindered (correlated motion between the large scale motion of the polymer chain and the solvent molecule is required for solvent diffusion) diffusion. Note that the sharp bend in the log S/log Ms curve shift to the higher homologs and is also less sharp at higher radiation (higher free volume) dosages.

Concluding Remarks

It now appears that the dissolution dynamics of glassy polymers is at the very least a complicated process and its kinetics depends on many parameters of the solvent, the polymer and the processing condition. We have also found that one of the dynamical processes that strongly affect the dissolution kinetics of polymers with high Tg is the diffusion rate of the solvent into the polymer matrix which primarily depend on at least 4 factors: polymer free volume and segmental mobility, solvent size and the osmotic pressure (chemical potential difference between the polymer solution and pure solvent). We believe that the above finding has important implication in resist technology. Positive

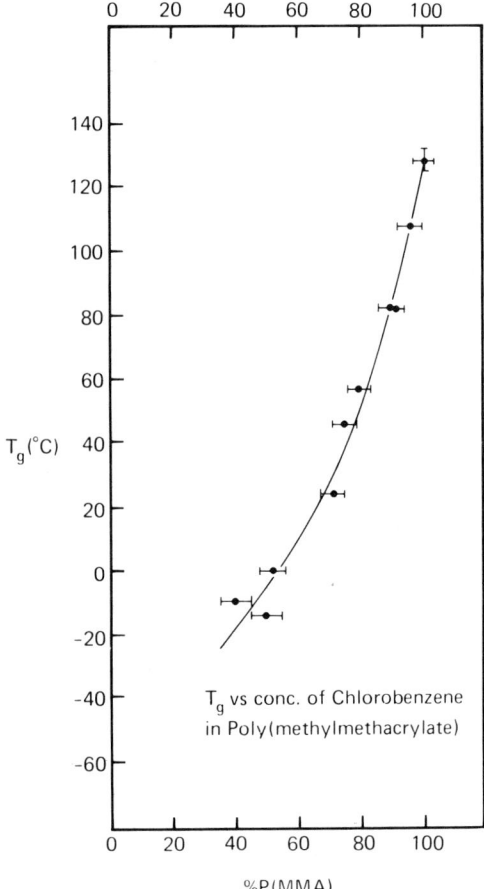

Figure 8. The glass transition temperature (Tg) vs. the solvent content of P(MMA). (Reproduced with permission from Ref. 10. Copyright 1978, Polym. Eng. Sci.)

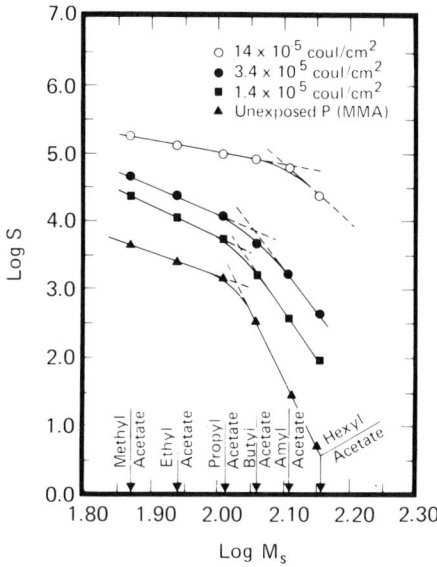

Figure 9. The solubility rates (°A/Min) of P(MMA)(Elvacite 2041) exposed to various e-beam dosages plotted against solvent molecular weight. The film thickness is 1 μm on silicon substrate.

polymer resist sensitivity should not only depend on main chain scission G value but also on the G value for small fragments (volatil or liquid like products), the G value for the formation of highly soluble products (acids for base developers lower molecular weight for polymers). Thus the search for higher sensitivity and high, contrast resist can now be broadened from materials that only undergo main chain scission to a much wider class of photo and high energy radiation sensitive material which undergo, chemical as well as physical changes on exposure.

Literature Cited

1. Yih-o Tu and A. C. Ouano, IBM Journal of Research and Development, (1977), 21, No. 2, 131.
2. F. Asmussen and K. Ueberrieter, J. Polymer Sci., (1962), 57, 199.
3. A. C. Ouano and J. A. Carrothers, Polymer Engineering and Sci., (1980), 21, 160.
4. Yih-o Tu, "Miltiphase Stefan Problem Describing the Swelling and Dissolution of Glassy Polymer", IBM Symposium on Mathematics and Computation, October 6-7, 1976. Yorktown Heights, N.Y.
5. K. Ueberrieter, "The Solution Process in Diffusion in Polymers", J. Crank and G. Park Eds., Academic Press, New York (1968).
6. C. E. Rogers, J. R. Sernancik and S. Kapur, "Transport Processes in Polymers" in "Structures and Properties of Polymer Films", R. Lenz and R. Stein, Eds., Plenum Press, N.Y. (1973).
7. J. S. Greeneich, J. Electrochem. Soc., (1974), 121, 1669.
8. J. S. Greeneich, J. Electorchem. Soc., (1975), 122, 920.
9. P. J. Flory, Principles of Polymer Chemistry, Cornell Univ. Press, Ithaca, N.Y. (1953).
10. A. C. Ouano, Polymer Engineering and Sci., (1978), 18, 42306
11. H. Hiroaka, IBM J. Res., (1977), 21, 121
12. A. C. Ouano, D. E. Johnson, B. Dawson and L. A. Pederson, J. Polymer Sci., Poly. Chem. Ed., (1976), 14, 701.
13. R. P. Kambour, Polymer, (1964), 5, 107.

RECEIVED December 19, 1983

Effect of Composition on Resist Dry-Etching Susceptibility
Novel Vinyl Polymers

J. N. HELBERT, M. A. SCHMIDT, and C. MALKIEWICZ

Process Technology Lab/SRDL, Motorola SPS, Phoenix, AZ 85008

E. WALLACE, JR., and C. U. PITTMAN, JR.

University of Alabama, University, AL 35486

> In an effort to devise a quick screening test for vinyl polymer resist dry-etch susceptibility as part of a total resist design criteria, we have adopted a simple and relatively easy CF_4/O_2 plasma etch test. This test is empirically found to be a good indicator for predicting etch resistance to even more harsh and anisotropic etch processes, such as reactive-ion etching and ion-milling. Using this test, the plasma etch rates of a large number of novel vinyl polymers have been measured and correlated with vinyl polymer composition. The effect of vinyl resist polymer composition upon plasma etch rate is found to range over a factor of 50. The polymer thermal stabilities have also been measured as a further part of the overall design criteria, because polymer thermal flow is also a vital issue to dry-etching technology. Although no correlation is found between pasma etch rate and polymer thermal stability, the polymers with lowest thermal stability were also most susceptible to detrimental resist image thermal flow.

Polymeric resists (1) are utilized widely in the electronics industry to lithographically delineate circuit patterns onto normally non-patternable inorganic substrates. As device dimensions shrink, an increasing number of the levels of these devices will have to be etched by dry-process (e.g., rf plasma) etching techniques (2). As a result, new polymeric resists must be able to withstand these harsh environments in order that the necessary dielectric, gate and metal material levels may be patterned during device fabrication. Thus, resist compatibility with these processes must be established before lithographic utilization.

It has been shown that plasma etching (PE), reactive-ion etching (RIE) and ion-milling (IM) equipment subject the resist in the etch masking application to reactive free radicals, atoms or ions which are capable of decomposing polymeric resists (3). Harada (3) has demonstrated that poly(methyl methacrylate) (PMMA) decomposes in a CF_4/O_2 plasma by a free radical-induced random chain scission mechanism. Furthermore, reference 2 establishes good tracking or correlation between relative etch rate ratios as the etch technique is made more harsh and anisotropic, and that these relative values are in fact dependent upon polymer composition. The dependence of etch rate ratio upon polymer composition can be as large as 100X (2). Etch rate ratios varying over a factor of 10 are more common, and results of other researchers (3,4,5) have recently appeared as focus upon polymer resist dry-process compatibility intensifies.

Strictly speaking, the dry-process compatibility for a resist is very process dependent, and must be measured for the specific process involved. A general idea of compatibility, however, can be obtained by doing a CF_4/O_2 plasma etch test vs SiO_2 and/or PMMA references, and this test has been adopted as a quick screen test for dry-process compatibility for new resists. The relative etch ratios vs these references usually, but not always, remain constant when the process requiring resist masking is changed (e.g., PE to RIE), thus, what is measured is resist compositionally-dependent.

In this work, we provide dry-process plasma etch rate ratios versus a standard for an expanded list of novel vinyl polymeric resists and commerical photoresists. The novel vinyl resists have been synthesized as part of a larger resist development program aimed towards the development of improved x-ray and e-beam lithographic resists (6).

The variety of vinyl homopolymers and copolymers, most containing quarternary centers and synthesized for plasma etch rate ratio evaluation, contain the following monomers to varying proportions:

$CH_2=C(CH_3)CO_2CH_2CH_2CN$ $CH_2=C(CH_3)CO_2CH_2CN$
    CEMA   CMMA

$CH_2=C(CN)_2$ $CH_2=C(CH_3)CN$
  VDCN   MCN

$CH_2=C(Cl)CN$ $CH_2=C(CF_3)CN$
  ACAN   TFMAN

CH$_2$=C(Cl)CO$_2$CH$_2$CF$_3$
ACTFEMA

CH$_2$=C(Cl)CO$_2$CH$_2$CCl$_3$
TCECA

CH$_2$=C(CH$_3$)CO$_2$CCl$_3$
TCEMA

CH$_2$=C(Br)CO$_2$CH$_3$
MBA

CH$_2$=C(CO$_2$CH$_2$CH$_3$)$_2$
BCEE

CH$_2$=C(CO$_2$CH$_3$)$_2$
BCME

CH$_2$=C(H)C$_6$H$_5$CH$_2$Cl
CMS

CH$_2$=C$\overset{CO-O}{\frown}$
AMBL

In addition, eight aromatic-like novolac resin containing first and second generation photoresist systems have been evaluated.

Experimental

Plasma etch measurements were carried out at 60°C in a Tegal Model 421 barrel plasma reactor (see ref 2 for conditions). To determine the etch rate of the polymer resist, a resist coated oxidized silicon substrate is exposed to the plasma for a specified time interval (10-20 minutes), then the original or pre-exposure resist step and a new post-exposure step are measured with a Tencor alpha-step profilometer. These steps are made by removing the resist by mechanically scratching or solvent dipping the substrate. The change in the resist is obtained by subtracting the post-exposure from the pre-exposure step height (see Figure 1). The SiO$_2$ substrate loss is measured by stepping down from the post-etch step to the original resist step, where the oxide had been previously exposed to the plasma. The oxide loss in these etch tests is usually 1000-2000Å, a value typically encountered in real device fabrication.

The etch rate ratio of the resist to the SiO$_2$ reference is referred to as the resist process selectivity, and coupled with resist thermal stability data is representative of the overall resist process compatibility. The lower the selectivity ratio (i.e., $\ll 1$), the better the resist polymer dry-process compatibility. PMMA, for example, has a marginal selectivity of 0.9-1.2, or etches at the undesirably same rate as SiO$_2$. In addition, PMMA is also very susceptible to thermally-induced image flow due to low tg and TGA parameters, and also undergoes surface "frying" phenomena (see Figure 2). Therefore, PMMA has very poor overall dry-process compatibility.

Results and Discussion

<u>Novel Vinyl Polymers.</u> Polymethacrylonitrile (PMCN) was found to possess a PE rate ratio vs SiO$_2$ of 0.3, and this relatively low ratio

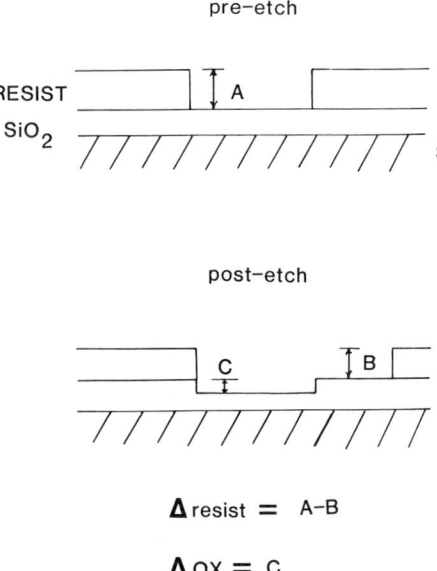

Figure 1. Etch rate ratio measurement method.

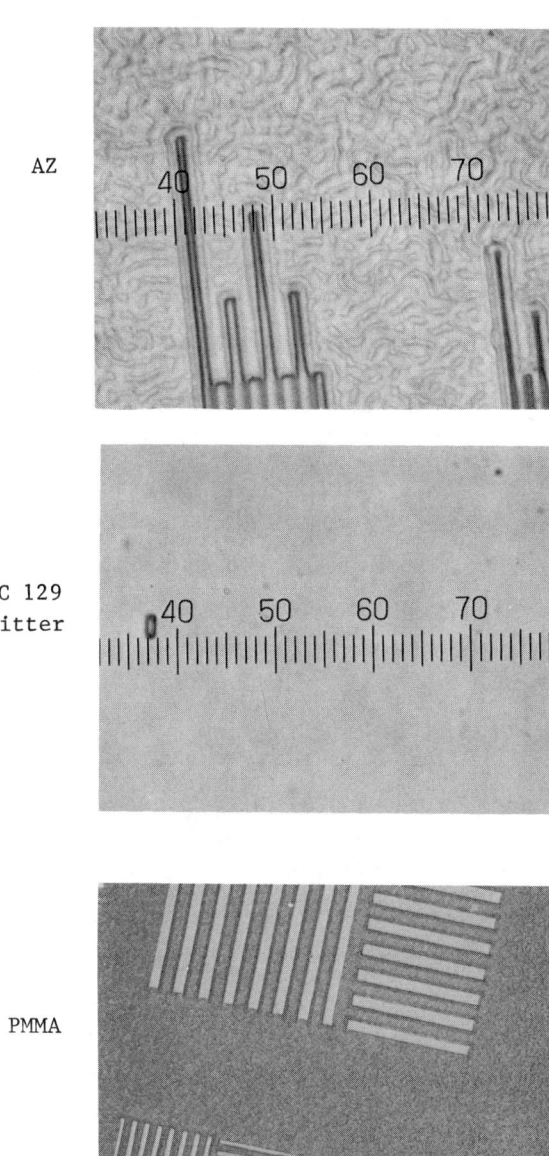

Figure 2. RIE resist "fry" examples for AZ-2400 (top) and PMMA (bottom) vs. non-frying PC-129.

was attributed to the alpha-cyano group which is a strongly bonded side group (2). To ascertain the effect of further CN incorporation (i.e., at the CH_3 site of PMCN) the homopolymer PVDCN was synthesized for etch rate measurement. Unfortunately, this polymer is insoluble in all known spinning solvents, therefore, measurement attention was focused upon the more soluble MMA/VDCN copolymers. Results for all the CN-containing systems are tabulated in Table I. A linear least squares fit of the VDCN/MCN copolymer data yields an extrapolation value at 100% of VDCN of 0.3, a value very close to that obtained for PMCN. Like PVDCN, PACAN is also sparingly soluble. As a result, the ACAN/MCN copolymers were focused upon for PE rate ratio measurement (see Table I). A linear least squares fit yields a PE rate ratio for PACAN of 1.3. For PACAN, the alpha-Cl appears to dominate the PE behavior over that of the alpha-CN stabilizing group.

The PTFMAN homopolymer, $(CH_2-C(CF_3)CN)$, is difficult to synthesize, therefore, attention was focused upon the more soluble MCN and MMA/TFMAN copolymers. As Table I shows, the TFMAN copolymers exhibit improved dry-process compatibility over that of the PMMA reference, again, due to the CN group stabilization effect.

The CMMA and CEMA polymers were included to determine if the CN protecting group could influence the PE rate at the alkyl ester site. Data for both systems is located in Table I. Surprisingly, the CN does influence the PE rate ratio for CMMA but not for the CN-ethyl system, which appears to be more influenced by the ethyl group effect. This effect was first observed when PEMA, $(CH_2-C(CH_3)CO_2CH_2CH_3)$, was found to etch 1.7X faster than PMMA(2). As in reference 2, the CN group is found to be a dry-process compatibility enhancer or found to decrease the PE rate ratio when it is alpha to the main chain or one carbon removed.

Earlier, (2) PMCA, $(CH_2-C(Cl)CO_2CH_3)$, was reported to plasma etch at a rate greater than 1.8X SiO_2, and this was attributed to the weak C-Cl bond. PTCEMA has a PE etch rate ratio of 2.3, thus, the presence of the weak C-Cl bond at either the alpha-position or on the ester alkyl group apparently enhances the PE decomposition of these resists. To further study this phenomena, the obvious system with both alpha-Cl and the trichloroethyl ester group, PTCECA, was synthesized and studied. As predicted, (see Table I) PTCECA and the 29/71 TCECA/MMA copolymer both etch faster than PMMA reference. These values are significantly larger than the desired less than one value. EBR-9, the Toray Japanese E-beam resist, is structurally similar to

Table I: Plasma etch rate ratios vs SiO_2 reference for vinyl polymers and copolymers.

Polymer or Copolymer	Mole % ratio	t_g °C [a]	TGA[a] °C	Plasma etch Rate Ratio
PMCN[2]	100	120	335	0.3
PVDCN	100	165	260/500[b]	insoluble
PVDCN-CO-MMA	52/48	155	330/500[b]	0.4
	38/62	135	270/400[b]	1.0
PMMA[2]	100	107	275	0.9-1.2
PACAN-CO-MCN	50/50	131	210/500[b]	1.0
	39/61	128	223/500[b]	0.5
	11/89	123	216/500[b]	0.4
PACAN	100	-	-	1.3[c]
PCMMA	100	115	250	0.4-1.1
PCEMA	100	105	250/400[b]	2.1
PCEMA-CO-MMA	50/50	102	293	1.4
PTFMAN-CO-MCN	12/88	123	357	0.3
PTFMAN-CO-MMA	32/68	98	295	0.3
PMCA[2]	100	130;151	-	1.8
PTCEMA[2]	100	123	305	2.3
PTCEMA-CO-MCA	66/34	137	286	3.5
PTCEMA-CO-MCN	50/50	130	310	0.4
PTCEMA-CO-TCECA	90/10	123	290	2.0
PTCECA	100	147	290	2.0
PTCECA-CO-MMA	29/71	130	325	1.6
PACTFEMA (EBR-9)	100	133	277	2.1
PMFA	100	131	404	0.4
PMBA	100	130	170	2.0
PBCEE	100	> 100	293	1.7
PBCEE-CO-MMA	25/75	78	275	1.0
	50/50	92	292	1.1
PBCME	100	72	320	≥ 2
PBCME-CO-MMA	50/50	100	310	0.9
PEMA[2]	100	65	-	1.7
PAMBL	100	83	363	0.3
PCMS (Toyo Soda)	100	105	350	≤0.06[d]
PS[2]	100	103	322	0.1

a Measured by DuPont Model 900 DSC/TGA. b TGA shows two distinct breaks: the first break is attributed to the ladder-like reaction that occurs between adjacent alpha-CN groups.
c Extrapolated value from the Figure.
d 0.06 is the lowest measureable value of this technique.

PTECA, $\{CH_2\text{-}C(Cl)CO_2CH_2CF_3\}$ (ACTFEMA), and etches at a ratio of 2.1. Again, the value is large but not as large as the 3.1 value obtained by adding the PMCA and PTFMA PE rate ratios. Curiously, the observed etch rate ratio for EBR-9 is very close to the sum of the PMCA and PTCEMA PE rate ratio values minus the ratio for PMMA. This additivity relationship holds true for both PTECA and EBR-9, and may be a valid way to predict PE rate ratios for doubly substituted polymers, where the respective mono-substituted PE rate ratio values are known.

When the chlorine of PMCA is replaced by F, the PE rate ratio decreases from 1.8 to 0.4 (2). This large difference was attributed to the difference in bond dissociation energies betweeen C-Cl and C-F; the C-F bond dissociation energy is 107 kcal/mole compared to that for C-Cl of 80 kcal/mole. To check this hypothesis further, the alpha-Br analog, $\{CH_2\text{-}C(Br)CO_2CH_3\}$, was synthesized, tested and found to etch at a ratio of 2.0. Thus replacement of the C-F bond with an even weaker C-Br bond (D=67 kcal/mole) has lead to a less dry-process compatible polymeric resist, consistent with the earlier work of reference 2.

PBCEE like PEMA etches 1.7 times that of PMMA reference. As discussed earlier and previously in reference 2, this effect is attributed to the alpha-ethyl ester group. As found for PVDCN, $\{CH_2\text{-}C(CN)_2\}$, the influence of the second substituent to the quarternary carbon is negligible. BCME, on the other hand, etches at ≈ 2, but this polymer has a t_g of 72°C, which is very close to the measurement temperature of the plasma etching tool employed. Heating above 60°C during etching, however, has been observed, and evidence of thermal flow was observed on the test wafer after etching. The more thermally stable 50/50 BCME/MMA copolymer with a t_g of 100°C etched at the same rate as PMMA, therefore, the PBCME value must be biased high due to thermal flow or instability. Therefore, the effect of a di-ester quarternary carbon is negligible upon the PE rate ratio, but does tend to reduce the polymer flow resistance significantly.

PAMBL, a special vinyl polymer system due to the five membered ester ringed main chain quaternary carbon, has a surprisingly low PE rate ratio of 0.3. At first glance, one would predict an etch rate ratio for PAMBL similar to that of PMMA, due to the fact both systems contain a alpha-ester group. This result may indicate that the alpha-methyl group (i.e., in PMMA) participation in the plasma etch degradation mechanism has been underestimated. Unfortunately, the good PE rate ratio is somewhat overshadowed by the low observed t_g value of 83°C, a value 24°C lower than the marginal value of 107°C observed for PMMA reference. Therefore, overall this resist is considered to possess only marginal overall dry-process compatibility.

The last vinyl system investigated is PCMS (7). This resist has H at the alpha-position, which is well known to establish the e-beam or x-ray resist tone as negative (1). Like other negative e-beam resists such as PS and PMFA, PCMS resist is also dry-process compatible, that is, has an etch rate ratio ≤ 1.0. The PCMS resist is even more compatible than the parent PS system. This data implies that the same type of reactions may occur in a gas plasma as occur during ionizing radiation exposure (i.e., abstraction of alpha-H atoms leading to the same crosslinking site radical precursors as are known to be created by radiation exposures).

For each vinyl polymer resist listed in Table I, a value for t_g, the glass transition temperature, and TGA temperature, which is the temperature for each system where significant thermal degradation and weight loss begins to occur, are listed. There is no apparent correlation between these thermal characteristics and the PE rate ratio data. Pederson has also reported a similar lack of correlation (4). Still, this thermal data is important because it aids the lithography and processing engineers in the selection of the resist prebake temperature ($>t_g$) and etching temperature ($<t_g$). It must be remembered, however, that the polymer resists which are most susceptible to flow will not maintain their relief patterns with integrity during the dry-etching processes where heating occurs at or above t_g.

Conventional Photoresists. PE rate ratio values for several positive photoresists are also included in this study (see Table II), because several of these novolac resin containing formulations also function as positive e-beam and x-ray resists. Generally speaking, these formulations are more dry-process compatible than most of the vinyl systems (see also ref.2). This is due primarily to the aromatic nature of the novolak resins in the photoresists. Thus, the photoresist PE rate ratio data is close in value to those of the aromatic vinyl and negative behaving polymers.

As for the vinyl polymers, there appears to be no obvious correlation between resist thermal stability, as measured by the TGA parameter and resist image flow data, and PE rate ratio. It is obvious at this point however, that both thermal stability and low PE rate ratio are very desirable to achieve good overall dry-process compatibility. Based on the data of Table II, OFFR-800 and Hunt 204 are clearly examples of dry-process compatible resists.

Table II: Positive Photoresist Thermal Flow and Plasma Etch Rate Ratio Characteristics.

Resist	TGA, °C	Image Flow Temperature, °C	PE
PC-129SF	125	140	0.2
KTI-II	133	160-170	0.4
OFPR-800	150	160-170	0.3
AZ-1350	140	140-160	0.5
AZ-2400*	145	140	0.5
Kodak-809*	-	120-130	0.3
Hunt-204	-	180	0.5

*AZ-2400, and K-809 "fry" (see Figure 1).

Conclusions

In summary, the PE rate ratio and thermal data provided must be considered invaluable information to the resist designer. If a polymer is not dry-process selective and reasonably resistant to thermal flow and degradation, it has very little application potential, regardless of how lithographically good it may be.

And finally, is there then a PE rate ratio guideline for predicting resistance to more harsh and anisotropic RIE and ion-milling etches? The answer to this question is empirically found to be yes. When the PE rate ratio is 0.3-0.5, the dry-process compatibility is usually very good. For resists with ratios ranging from 0.5 to 1, marginal to unsatisfactory behavior is usually observed, while polymers with ratios greater than one usually exhibit poor to totally unacceptable performance. For example, PS, PCMS and many positive photoresists exhibit etch rate ratios <0.5, and are proven dry-process resists.

Acknowledgments

The resist processing efforts of Charlie Walker and Liz Ramos are much appreciated. The editing efforts of Linda Marcy are also acknowledged.

Literature Cited

1. Thompson, L.F.; and Kerwin, L.E. Ann. Rev. of Materials Sci. 1976, 6, 267.
2. Helbert, J.N.; and Schmidt, M.A. ACS Symposium Series No. 184, Feit, E.D.; and Wilkens, C., eds., American Chemical Society, 1982, p. 60.
3. Harada, J. J. of Appl. Polym. Sci. 1981, 26 1961.
4. Pederson, L.A. J. Electrochem. Soc. 1982, 129, 205.
5. Harada, K.; Kogure, O.; and Murase, K. IEEE Transactions on Electron Devices 1982, ED-29, No. 4, 518.
6. Pittman, C.U.; Wallace, E.; Jayaraman, T.V.; Wang, P.; Ueda, M.; and Helbert, J.N. IUPAC-MACRO 82, U. of Mass., 1982, July.
7. Fukada, M., JEE, 1982, August.

RECEIVED December 19, 1983

ELECTRON BEAM RESISTS

9
Resists for Electron Beam Lithography

TOSHIAKI TAMAMURA, SABURO IMAMURA, and SHUNGO SUGAWARA

Polymer Section, Ibaraki Electrical Communication Laboratory, Nippon Telegraph and Telephone Public Corporation, Tokai, Ibaraki 319-11 Japan

>   Resist materials for electron beam lithography are
>   reviewed from the view point of application to
>   direct wafer-writing technology. In positive
>   resists the latest advances in improvement on
>   sensitivity, adhesion to substrates and dry-etching
>   durability are described. Resists based on
>   aromatic polymers are now important for dry-etching
>   technology. In negative resists, careful molecular
>   design of the polymer is necessary in order to
>   achieve high resolution.

The minimum feature of "very large scale integrated circuits" (VLSIs) will decrease to below 1 um in the near future. Conventional photolithography is already approaching its resolution limit and a new photolithographic technology using deep UV light (200 - 300 nm) is under development to achieve submicron resolution. Electron beam lithography has been in practical use for several years for the fabrication of photomasks for photolithography. Electron beam lithography is becoming increasingly important, because of its application to direct wafer-writing technology, which will be in competition with deep UV photolithography for submicron device fabrication. The latest progress from electron beam exposure machine side, particularly higher brightness electron sources ([1]) and higher data transfer speeds ([2]), make the future of this technology more promising.
   Another important change in VLSI technology is the application of dry-etching techniques rather than wet etching which does not have sufficient accuracy to transfer submicron resist patterns into substrates. The use of various halogenated hydrocarbon gases as etchant for reactive sputter etching has grown rapidly in the present VLSI fabrication processes.
   In spite of the availability of very high brightness

Table I Requirements for Electron Resists

Items	Photomask Fabrication	Direct Wafer-writing
Sensitivity	better than 1 $\mu C/cm^2$	1 - 2 $\mu C/cm^2$
Resolution	1 - 2 μm on a flat Cr-based surface	better than 1 μm on various uneven substrates
Adhesion	strong	stronger
Etching durability	high against etching of thin Cr film	high against various dry-etching processes
Contamination	-	no influence on devices

electron gun, electron resists with high sensitivity are still required in order to maximize the through-put and the chip-yield of the lithography. Table I compares the requirements of electron resists for photomask fabrication and direct wafer-writing technology. Obviously electron resists to be used in direct wafer-writing technology for submicron devices needs significant improvements in resolution, adhesion and dry-etching durability. Practical electron resists should be well balanced in all of these characteristics.

Both positive and negative resists depending on the nature of patterns to be written are required in the direct writing technology. These resists have been normally used not only as single layers, but recently, as top layers in multilevel resist systems. Multilevel resists allow fabrication of resist patterns with high aspect ratio, and will inevitably be used for the fabrication of submicron patterns on substrates with surface topologies, such as the metalization of electrode patterns in the last steps of VLSI fabrication processes. In the multilevel approach top resist has a significant advantage in resolution compared to the single level approach, because the top layer resist which defines the resolution of the system, can be thin and is exposed on a thick organic polymer layer, thereby reducing the influence of backscattered electrons.

It is instructive to review the resist materials reported so far, and to find the most practical resists. However, we encountered great difficulty in comparing the performances of different resists, because all resist characteristics strongly depend on exposure and process conditions, and because the details of some resists actually used are not reported. In this paper we attempt to give a brief review of the latest developments in electron resist materials from the view point of application to direct wafer-writing technology, and describe some results on new resist from our group. Though there are number of multilevel resist systems proposed, which should be referred in an excellent review (3), we concentrate on resist polymers themselves.

Positive electron resists

<u>High sensitivity resists</u>   Table II summarizes the properties of representative positive electron resists. The polymers are classified into four groups according to the chemical structure. Almost all positive electron resists operate by main chain scission of polymer, resulting in a molecular weight decrease in exposed areas. Resist patterns are produced by development in a suitable solvent in which degraded polymer dissolve much faster than unexposed polymer. The sensitivity is determined by the scission probability and the solubility rate ratio for the degraded polymers.

Methacrylic polymers have been extensively studied, because these polymers are readily degraded by high energy beam exposure. Poly(methyl methacrylate) (PMMA) which was the first material reported as electron resist (4), has rather poor sensitivity, but is still believed to show the highest resolution (18). Though a higher sensitivity can be achieved in polymers with a higher scission probability, the sensitivity of methacrylate based polymer is often enhanced by the development process. For instance, the G-value for main chain scission in PMMA is almost the same as in poly-n-butylmethacrylate (PnBMA), but the sensitivity of latter is about 200 times higher than that of the former (5). Despite its high sensitivity, PnBMA has not found practical use because of no repoducibility of the development process, leading to poor control of critical dimensions in the developed pattern. This low reproducibility is caused mainly by the low glass transition temperature of polymer.

We have succeeded in utilizing the high sensitivity of PnBMA via introduction of fluorine atoms into the n-butyl side group (8). Polyhexafluorobutylmethacrylate (FBM) for example shows rather higher sensitivity than PnBMA. This is achieved not by the increased electron scattering by heavier F atoms, but by the excellent development characteristics (19). The Tg of FBM is, however, significantly increased over PnBMA owing to the increased bulkiness of side chain and allows the resist to be practically used for photomask fabrication. Three drawbacks, viz., adhesion, dry-etching durability and developer-temperature dependence of the sensitivity, limit the application of FBM to submicron lithography.

Polydimethyltetrafluoropropylmethacrylate (FPM) was actually used for direct fabrication of a 256 kBit RAM with 1 um design rule (20). FPM shows good adhesion and better dry-etching durability, but the sensitivity was reduced by a factor of 10 compared with FBM (7). Adhesion problems in submicron FBM resist patterns were solved by the copolymerization with a small amount of glycidylmethacrylate, whose homopolymer is actually used as

Table II. Polymers for representative positive electron resist

Polymers	Abbre.*	Sensitivity (μC/cm^2)	Tg** (°C)	rf.
Group 1: Plain methacrylic polymers				
Polymethylmethacrylate	PMMA	100	104	4
Poly-n-butylmethacrylate	PnBMA	0.5	19	5
Polyphenylmethacrylate	PPhMA	150	110	6
Group 2: Halogenated methacrylic polymers				
Polydimethyltetrafluoropropylmethacrylate	FPM	5.0	93	7
Polyhexafluorobutylmethacrylate	FBM	0.4	50	8
FBM - glycidylmethacrylate copolymer	FBM-G	0.4	65	9
Trifluoroethyl-α-chloroacrylate	EBR-9	2.5	133	10
Group 3: Crosslinked methacrylic polymers				
MMA - methacrylic acid copolymer	PMMA-MAA	20	150	11
MMA - MAA - methacrylic chloride terpolymer	XXL	20	140	12
Polytrichloromethylmethacrylate	EBR-1	3.0	138	13
MMA -t-butylmethacrylate copolymer	CP-3	1.6	131	14
PhMA - MAA copolymer	φ-MAC	20	142	15
Group 4: Poly(olefine-sulfone)s				
Poly(butene-sulfone)	PBS	1.0	-	16
Poly(methylpentene-sulfone) + novolac resin	NPR	4.0	-	17

*Abbre.: abbreviation, ** Tg : glass-transition temperature of polymer

negative electron resist. The high sensitivity of FBM is not adversely affected by the crosslinkable epoxy group, and the new resist termed FBM-G shows much better adhesion (9). The dry-etching durability is also improved perhaps because of the higher Tg, and allows the application of FBM-G to submicron fabrication involving dry-etching processes. Submicron FBM-G resist patterns shown in Fig. 1-a were used as resist masks for etching of 0.6 μm thick silicon dioxide with an electron cyclotron resonance (ECR) type reactive ion-etching machine. Figure 1-b shows etched SiO_2 patterns using C_4F_8 gas as an etchant. Anisotropic dry-etchings with less damage to resist can be achieved in ECR type machine, because only ionic species drawn from a plasma chamber participate in the etching (21).

However, resists with a sensitivity higher than 1 $\mu C/cm^2$ seem to be rather difficult to handle for practical direct wafer writing technology and an optimum resist sensitivity for a well-designed exposure plant may be 1 - 2 $\mu C/cm^2$. Extremely fast resists are required for the x-ray lithography, where apart from synchrotron radiation the limited out-put of conventional x-ray sources causes a real limitation in the through-put (22). FBM-G will be the most important resist in x-ray lithography.

The high sensitivity of halogenated polymers seems to be attributed to their development characteristics, but in the case of EBR-9 (10) an enhanced probability in main chain scission due to the chlorine atom in α-position is also a contributing factor to the high sensitivity (23). Even in PMMA, stronger solvents can raise the sensitivity by about one order, but only at the expense of a serious thickness reduction in unexposed areas, or deterioration of pattern quality due to swelling.

Another method of enhancing sensitivity of methacrylic polymers utilizes a crosslinking technique during prebaking, as exemplified by the resists in Group 3 in Table II. Polymers containing methacrylic acid and/or methacrylic chloride can easily form inter- or intra-chain acid-anhydride bonds by heating the polymer films near 200°C (12), whereas crosslinking in EBR-1 (13) and CP-3 (14) involves a thermal decomposition of bulky trichloromethyl and t-butyl groups, respectively. Crosslinked resists are essentially insoluble in unexposed areas, and they also have an advantage in a higher Tg than that before prebaking. This class of resist is useful for double layer resist systems, where wide variations of spining solvent and developer are required. This is facilitated by the low solubility of resists after prebaking (11).

<u>Dry-etching durable resists</u> The real conflict requirement in positive electron resists exists in the relationship between dry-etching durability and sensitivity. Figure 2 depicts this situation. Clearly, aromatic polymers show high durability but low sensitivity (24). Poly(phenylmethacrylate) (PPhMA) withstands reactive sputter-etching more than 2 times longer than

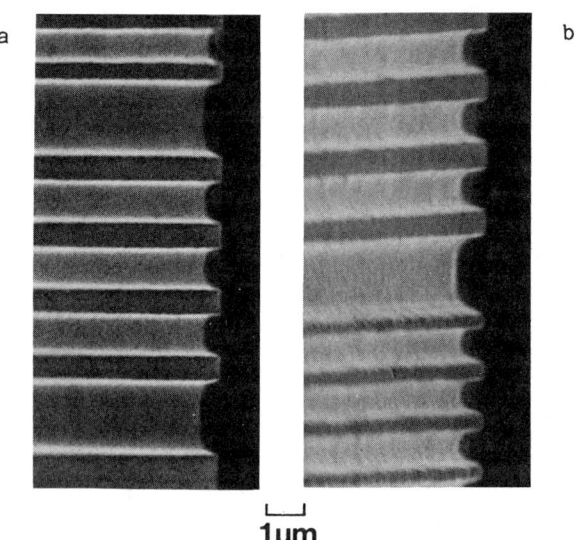

Figure 1. Submicron pattern fabrication with FBM-G resist. Key: a, 0.6-μm thick resist patterns; and b, after reactive ion beam etching of 0.6-μm thick SiO_2 using C_4F_8. The remaining resist thickness is about 0.15 μm.

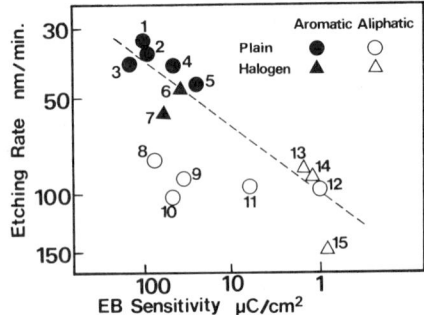

Figure 2. Conflict requirement between the sensitivity and dry-etching durability in positive electron resists based on methacrylate polymers. Sputter etching rates were measured under the same conditions using CF_4 gas. Key: 1, poly(p-methoxyphenyl methacrylate); 2, poly(phenyl methacrylate) (PPhMA); 3, poly(3-phenylpropyl meyhacrylate); 4, poly(benzyl methacrylate); 5, poly(p-methoxybenzyl methacrylate); 6, poly(p-fluorophenyl methacrylate; 7, poly(trichlorophenyl methacrylate); 8, poly(methyl methacrylate) (PMMA); 9, poly(tert-butyl methacrylate); 10, poly(ethyl methacrylate); 11, poly(isobutyl methacrylate); 12, poly(n-butyl methacrylate) (PnBMA); 13, poly(dimethyltetrafluoro methacrylate) (FPM); 14, poly(trichloromethyl methacrylate) (EBR-1); and 15, poly(hexafluorobutyl methacrylate) (FBM).

PMMA, but its sensitivity is decreased relative to that of PMMA. We have applied the crosslinking technique to improve the resist performance of PPhMA by introducing methacrylic acid into the chain (15). In the copolymer of PhMA and MAA (ϕ-MAC) a sensitivity of about 20 µC/cm^2 and excellent adhesion with substrates were obtained without losing high dry-etching durability of PPhMA. ϕ-MAC resist enables submicron patterns to be delineated and gives an unusually high aspect ratio, as shown in Fig. 3-a Lift-off processing is also possible with ϕ-MAC when double layer resist system is used. Very fine grating demonstrated in Fig. 3-b and other excellent resist performances indicate the utility as an alternative resist of PMMA in submicron lithography for the experimental studies of new devices.

As the high sensitivity of poly(olefin-sulfone)s originates from a high G-scission coupled with depolymerization during electron beam exposure (16), these polymers show little resistance to the dry-etching processes. To improve the dry-etching durability of poly(olefin-sulfone)s, a novolac resin was mixed with poly(methylpentene-sulfone) (17). In this resist called NPR, poly(methylpentene-sulfone) controls the dissolving rate of novolac resin like quinone-diazide in AZ type photoresist which has excellent dry-etching durability. If the adhesion of submicron resist patterns will not meet any problem, the sensitivity of several µC/cm^2 makes sure of the practical use of NPR resist.

Negative electron resists

Dry-etching durable resists Several representative negative electron resists are shown in Table III. As the sensitivity of negative electron resists is proportional to the weight average molecular weight Mw, the product of gel point dose, D_0, and Mw is used to compare the sensitivity. When Mw of the resist is unknown, a reported sensitivity is shown in the table. The mechanism of resist action of classical negative electron resists involves the crosslinking of polymer chains, and has been theoretically analyzed based on Charlesby's gel formation theory (39). The theory indicates that the type of crosslinking reaction and the polydispersity of the polymer significantly affect the contrast of the resist (40). Higher sensitivity is usually observed in polymers which crosslink by a chain reaction, although the contrast is usually low. Such lithographic properties are observed in polymers containing epoxy or vinyl group and with a low Tg. High reactivities of PGMA, COP and Sel-N are attributed to the fact that crosslinking occurs by chain reaction. Such resists often cause a lithographic trouble due to the so called "post-polymerization" effect (41). Stepwise crosslinking reaction is preferable to get higher contrast and to eliminate this "post-polymerization" effect. Polymers having a high Tg and suitable reaction groups such as halogens react mainly via step-wise reactions. Narrow molecular weight

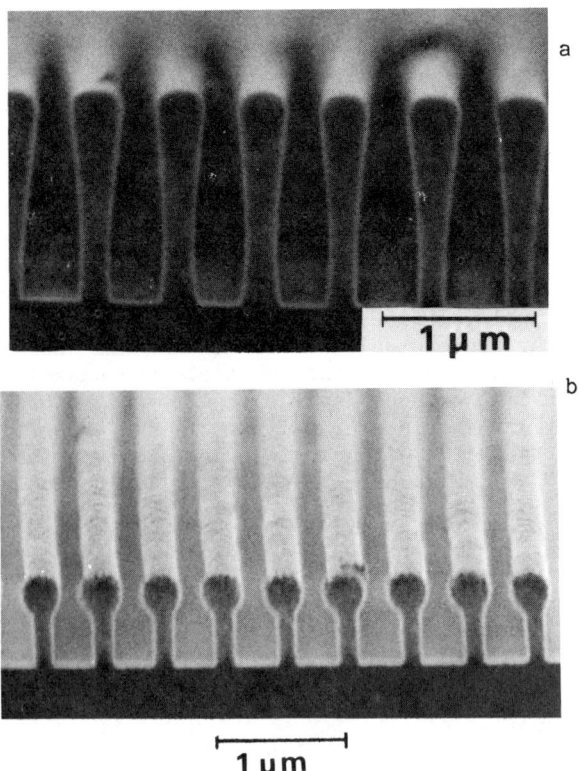

Figure 3. Submicron resist patterns with ϕ-MAC resist. Key: a, 0.6-μm pitch grating patterns in 1.5-μm thick single layer ϕ-MAC; and b, 0.45-μm pitch grating patterns with ϕ-MAC double layer resist. Top layer of 0.2-μm thick ϕ-MAC shows two-thirds of the sensitivity of the 0.5-μm thick bottom ϕ-MAC.

Table III. Polymers for representative negative electron resists

Polymers	Abbre.	Reactivity D_0 Mw*	Tg (°C)	rf.
Group 1: Epoxy or vinyl group polymers				
Polyglycidylmethacrylate	PGMA	0.028	78	25
GMA - ethylacrylate copolymer	COP	0.023	10	26
Polychlorohydroxypropylmethacrylate containing maleic acid methyl ester	SEL-N	0.026	–	27
Polydiallylorthophthalate	PDOP	0.036	160	28
Polyvinylmethylsiloxane	PVMS	2.0 μC/cm^2	-100	29
Group 2: Halogenated aromatic polymers				
Polytstyrene	PSt	5.5	85	30
Polychlorostyrene	PClSt	0.28	122	31
Chloromethylated PSt	CMS	0.12	110	32
Iodinated PSt	IPS	0.42	156	33
Chloromethylated poly-α-methylstyrene	αM-CMS	0.23	170	34
Chloromethylated polynaphthylmethacrylate	CMN	0.28	113	35
Vinylnaphthalene – vinylbenzylchloride copolymer	PVN-VB	0.41	–	36
GMA – chlorostyrene copolymer	GMC	0.09	77	37
GMA – vinylbenzylchloride	GMA-VB	0.067	78	37
PSt containing tetrathiafulvalene units	PSt-TTF	6μC/cm^2	–	38

*D_0: gel point dose, Mw: weight average molecular weight

dispersity in polymers also helps to improve not only the resist contrast but practical resolution in the fine pattern delineations.

Many negative electron resists recently developed for dry-etching processes are based on the polystyrene (PSt) structure because of the relatively high stability against dry-etching reactions and high Tg of PSt. Another advantage of PSt is the availability of nearly mono-dispersed polymer. Although mono-dispersed PSt acts as a negative resist with the highest contrast known, it exhibits low sensitivity (30). However, halogenation of PSt can substantially increase the sensitivity without losing the mono-dispersity of polymer. By using this technique, we have synthesized a high performance negative electron resist: chloromethylated polystyrene (CMS), which has been applied for the direct writing of 256 kBit RAM (19). As shown in Table III, the chloromethyl group confers the largest sensitization of PSt among various halogen groups (42). The interesting feature observed in halogenated aromatic polymers is the saturation of crosslinking reactivity with halogen group content. Unlike polymers containing epoxy or vinyl groups, whose reactivities are almost proportional to the functional group content, the reactivity of halogenated aromatic polymers shows a marked increase at low halogen contents but shows the saturation about 20 - 40 % of halogenation.

A higher reactivity relative to styrene homopolymers is observed in copolymers of GMA with halogenated styrenes, indicating substantial contribution of the epoxy group to the reactivity in these copolymers (37). GMC resist which is a copolymer of GMA and 3-chlorostyrene has been used for direct wafer-writing as a top layer resist for tri-level resist system (43). Polymers containing naphthalene ring show even higher dry-etching durability than PSt based polymers (36). When naphthalene rings in such polymers are substituted with chloromethyl groups, the polymers with a suitable sensitizer show high crosslinking reactivity to conventional UV light as well as electron beam and x-ray (44)

Reactive ion etching with oxygen is becoming widely used for the transfer of submicron patterns into a thick organic polymer layer in multilevel resist systems. Since it is difficult for organic polymers to possess sufficient resistivity against oxygen reactive ion etching, inorganic layers are inserted between the top layer of electron resist and thick bottom polymer layer (45). Silicone resins show virtually no thickness reduction in oxygen plasma etching, due to the formation of surface layer of silicon dioxide during etching. High molecular weight polyvinylmethylsiloxane (PVMS) has been reported to show reasonable resolution with a sensitivity of about 2 μC/cm , and can be applied to the top layer of unique bi-level resist system (29). In this system, the high resolution patterns are defined in thin PVMS films and are transferred to the thick organic polymer layer, producing a high aspect ratio. This type of resists will be a great interest because of the reduced number of processing steps relative to the tri-level systems.

High resolution resists   In the case of negative electron resists, the requirement of high sensitivity and high resolution, is mutually incompatible, because both are strong function of molecular weight of resist polymer but in opposite directions. The use of a low molecular weight polymer improves the resolution at the expense of sensitivity. The swelling of crosslinked polymer becomes less in a lower molecular weight polymer due to a higher crosslinking density. Even if a low molecular weight polymer is used, submicron pattern fabrications are not easy for negative resists with a lithographically useful sensitivity. Only two negative resists have been reported to show resolution enabling submicron pattern formation in a thick film.

We have refined the molecular design of CMS and synthesized high resolution negative resist, chloromethylated poly-α-methylstyrene (αM-CMS), which was derived from nearly monodispersed poly-α-methylstyrene as a starting polymer instead of polystyrene as in CMS (34). The lithographic performance of αM-CMS is compared with CMS, as illustrated in Fig.4. When the same molecular weight polymers are used, the sensitivity of αM-CMS is about half of that of CMS but still 20 times higher than that of PSt. On the other hand, the contrast is as high as that of mono-dispersed PSt. Consequently, higher sensitivity can be achieved in αM-CMS than in CMS, when we compare the sensitivity of resists having equal contrast.

The high contrast of αM-CMS may be due to the lack of hydrogen atom on the α-position of PSt. α-hydrogen subtraction by chlorine radicals formed by the dissociation of the chloromethyl group in CMS can proceed by a chain mechanism, thus leading to lower contrast. Such a chain mechanism would also explain the difference in reactivity between CMS and αM-CMS. The high Tg of αM-CMS (170°C) coupled with its high contrast makes high resolution lithography possible. Figure 5-a shows complex patterns with various sizes in a range of 0.4 - 4.0 μm, which have almost the same dimensions as designed. Very fine patterns as small as 0.1 μm were delineated in 0.4 μm thick resist as shown in Fig. 5-b. Figure 5-c shows 0.1 μm wide ion-milled gold lines fabricated with a mask of 0.1 μm wide αM-CMS patterns, and verifies the high dry-etching durability of αM-CMS. The ultimate resolution of αM-CMS is well below 0.1 μm, when a thin resist film is exposed on a thin substrate to minimize a resolution loss by electron scattering (46).

Our approach to reduce the effects of swelling in negative resists was to use a high Tg polymer. Other approaches have been reported (38). Tetrathiafulvalene (TTF) is a strong electron donor and forms its radical cation by forming a complex with an electron acceptor such as tetracyanoquinodimethane and halogen.

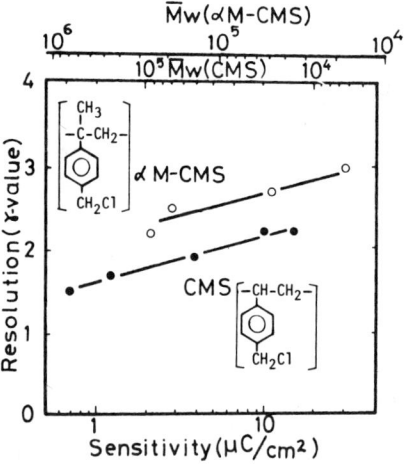

Figure 4. Molecular weight dependences of sensitivities and contrasts of CMS and *a*-M-CMS.

Figure 5. Submicron pattern fabrication with a-M-CMS. $D_{0.5} = 33$ μC/cm^2. Key: a, complex resist patterns with 0.75-μm initial thickness (minimum width is 0.4 μm); b, 0.1-μm wide lines on 0.4-μm pitch in 0.4-μm thick a-M-CMS (aspect ratio is 4); and c, 0.1-μm wide, 0.1 μm of thick gold lines Ar-ion milled with a mask of 0.3-μm thick a-M-CMS.

When TTF is introduced into the PSt structure and irradiated with an electron beam in the presence of CBr_4, some TTF units in the PSt side chain are easily converted to the radical cations via the electron transfer to CBr_4. The large difference in polarity between neutral species and cation radicals results in a large difference in solubility with little consequent swelling on development. 0.2 μm resolution patterns were delineated in this resist with about 10 μC/cm^2.

Conclusion

Our review is oriented toward the resist chemistry. This may be due to the fact that not a great deal of progress has been made in understanding the development, which still remains largely an art. A completely dry-fabrication process would be attractive, because many of the problems involved in wet processes such as small defects formed by the spin coating, the resolution loss due to resist swelling and the handling of large volumes of organic or inorganic solvents can be eliminated. While several studies indicate the possibility of dry-development of resist in electron beam lithography (47-49), the practicality of an all dry process is not clear at the present stage. As far as resolution is concerned, wet development of resist can resolve much finer dimension than that necessary for the future VLSI. However, dry-development processes may have more adaptability to future system, if VLSI fabrication process lines are to be completely automated.

The current technological competition for practical fabrication of 0.5 - 1.0 μm feature is in between photolithography and electron beam direct writing, and between single layer and multi-level resist. For less than 0.5 μm, the use of electron beam writing with multilevel resist will be inevitable. Further developments in electron resists from the standpoint of both resist chemistry and process development will be necessary to establish the electron beam lithography.

Acknowledgments

The authors wish to thank Messrs. K. Murase, O. Kogure, K. Harada, and H. Asakawa for their valuable discussions.

Literature Cited

1. Moor, H. D.; Caccoma, G. A.; Pfeiffer, H. C.; Wever, E.V.; Woodard, O.C. J. Vac. Sci. Technol. 1981, 19, 950.
2. Kelly J.; Groves, T; Kuo, H. P. J.Vac.Sci.Technol. 1981, 19, 936.
3. Bowden, M. J. Solid State Technol. 1981, June 73.
4. Hatzakis, M. J. Electrochem. Soc. 1969, 116, 1033.

5. Kitakohji, T.; Yoneda, Y.; Kitamura, K.; Okuyama, H.; Murakawa, K. Extended Abstracts Electrochem. Soc. Mtg. Pittsburg, 1978, p. 42.
6. Saeki, H. Proc. 39th Mtg. Japan Soc. Appl. Phys. 1978, 45, p. 983.
7. Murase, K.; Kakuchi, M.; Sugawara, S. Proc. Int. Conf. on Microlithography Paris, 1977, p. 261.
8. Kakuchi M.; Sugawara, S.; Murase, K.; Matsuyama, K. J. Electrochem. Soc. 1977, 124, 1648.
9. Asakawa, H.; Kogure, O.; 1982 Symp. VLSI Technology Kanagawa, 1982, 88.
10. Tada, T. J. Electrochem. Soc. 1979, 126, 1829.
11. Tada, T. J. Electrochem. Soc. 1979, 126, 1635.
12. Hazakis, M. J. Vac. Sci. Technol. 1979, 16, 1984.
13. Roberts, E. D. Appl. Polym. Symp. 1974, 23, 87.
14. Saeki, H.; Kohda, M. Proc. 17th Symp. Semiconductor & Integrated Circuit Technol. Tokyo, 1979, p. 48.
15. Harada, K.; Kogure, O.; Murase, K. IEEE Trans. 1982, ED-29, 781.
16. Bowden, M. J.; Thompson, L. F. J. Vac. Sci. Technol. 1975, 12, 72.
17. Bowden, M. J.; Thompson, L. F.; Fahlenholtz, S. R.; Doerries, E. M. J. Electrochem. Soc. 1981, 128, 1304.
18. Broers, A. N. J. Electrochem. Soc. 1981, 128, 166.
19. Tamamura, T.; Sugawara, S. in "Semiconductor Technology"; Nishizawa J. Ed.; OHM - North-Holland, Tokyo, 1982; Chap. 10.
20. Sakakibara, Y.; Ogawa, T.; Komatsu, K.; Moriya, S.; Kobayashi, M.; Kobayashi, T. IEEE Trans. 1981, ED-28, 1279.
21. Matsuo, S.; Adachi, Y.; Japan J. Appl. Phys. 1982, 21, L4.
22. Economou, N. P.; Flanders, D. C. J. Vac. Sci. Technol. 1981, 19, 868.
23. Helbert, J. N.; Wagner, B. E.; Caplan, P. J.; Poindexter, E. H. J. Appl. Polym. Sci. 1975, 19, 1201.
24. Harada, K. J. Appl. Polym. Sci. 1981, 26, 3395.
25. Taniguchi, Y.; Hatano, Y.; Shiraishi, H.; Horigome, S.; Nonogaki, S.; Naraoka, K. Japan J. Appl. Phys. 1979, 18, 1143.
26. Thompson, L. F.; Ballantyne, J. P.; Feit, E. D. J. Vac. Sci. Technol. 1975, 12, 1280.
27. Ohnishi, Y.; Ito, M.; Mizuno, K.; Nagasawa, K.; Ochi, H.; Sugawara, I.; Shibata, Y. Proc. 25th Autumn Mtg Japan Soc. Appl.Phys. 1978, 252.
28. Yoneda, Y.; Kitamura, K.; Naito, J.; Kitakohji, T.; Okuyama, H.; Murakawa, K. Polym. Eng. Sci. 1980, 20, 1110.
29. Hatzakis, M.; Paraszczak, J.; Shaw, J. Proc. Microcircuit Engineering, Laussanne, 1981, 386.
30. Lai, J. H.; Shepherd, L. T. J. Electrochem. Soc. 1979, 126, 696.
31. Feit, E.D.; Stillwagon, L. E. Polym. Eng. Sci. 1980, 20, 1059.

32. Imamura, S., J. Electrochem. Soc. 1979, 126, 1628.
33. Shiraishi, H.; Taniguchi, Y.; Horigome, S.; Nonogaki, S. Polym. Eng. Sci. 1980, 20, 1054.
34. Sukegawa, K.; Sugawara, S.; Japan J. Appl. Phys. 1981, 20, L583.
35. Imamura, S.; Tamamura, T; Sukegawa, K.; Sugawara, S. Polym. Preprints Japan 1982, 31, 1617.
36. Ohnishi, Y. J. Vac. Sci. Technol. 1981, 14, 1136.
37. Thompson, L.F.; Stillwagon, L. E.; Doerries, E. M. J. Vac. Sci. Technol. 1978, 15, 938.
38. Kaufman, F. B.; Shroeder, A.H.; Engler, E. M.; Patel, V. V. Appl. Phys. Lett. 1980, 37, 314.
39. Charlesby, A. Proc. Roy. Soc. 1954, A222, 60.
40. Atoda, N.; Kawakatsu, H.; J. Electrochem. Soc. 1976, 123, 1519.
41. Crosslinking reaction lasts after stopping irradiation in vacuum, causing a non-uniformity in the residual resist thickness in a wafer.
42. Imamura, S.; Tamamura, T.; Harada, K.; Sugawara, S. J. Appl. Polym. Sci. 1982, 27, 937.
43. Watts, R. K.; Fichtner, W.; Fuls, E. N.; Thibault, L. R., Johnston, R. L. IEEE Trans. 1981, ED-28, 1338.
44. Imamura, S.; Tamamura, T.; Kogure, O., unpublished data.
45. Moran, J. M.; Maydan, D. J. Vac. Sci. Technol. 1979, 16, 1620.
46. Tamamura, T.; Sukegawa, K.; Sugawara, S. J. Electrochem. Soc. 1982, 129, 1831.
47. Yoneda, Y.; Kitamura, K.; Miyagawa, M.; Narusawa, T.; Okuyama, H. Proc. 3rd Technical Conf. Photopolymer, Tokyo, 1982, p. 95.
48. Tsuda, M.; Oikawa, S.; Kanai, W.; Hashimoto, K.; Yokota, A. Nuino, K.; Hijikata, I.; Uehara, A.; Nakane, H. J. Vac. Sci. Technol. 1981, 19, 1351.
49. Hiraoka, H.; J. Electrochem. Soc. 1981, 128, 1065.

RECEIVED September 2, 1983

10
Chain-Scission Yields of Methacrylate Copolymers Under Electron Beam Radiation

C. C. ANDERSON, P. D. KRASICKY, and F. RODRIGUEZ

School of Chemical Engineering, Olin Hall, Cornell University, Ithaca, NY 14853

Y. NAMASTE and S. K. OBENDORF

Department of Design and Environmental Analysis, Van Rensselaer Hall, Cornell University, Ithaca, NY 14853

> Copolymers of methyl methacrylate, MMA, with itaconic acid and its esters were prepared for preliminary evaluation as positive-working electron-beam resists. Materials were polymerized using slow, free-radical initiation. Polymer films about 1 μm thick were spin-coated on 3-inch diameter silicon wafers. The films then were exposed to measured doses of 40 kv electrons in a modified transmission electron microscope. In order to calculate yields of chain scissions per 100 e.v., G(s), the molecular weights were estimated using a Waters HPLC calibrated with polystyrene using tetrahydrofuran as the solvent. Homopolymer of dimethyl itaconate gives G(s) twice that of PMMA. On the other hand, di-n-butyl itaconate gives copolymers with G(s) only slightly higher than PMMA. Itaconic acid definitely enhances the sensitivity of MMA. A copolymer with 35 mol% itaconic acid has a G(s) of about 5.

Poly(methyl methacrylate), PMMA, has remained the standard by which to judge positive-working electron-beam resists for over a dozen years (1,2,3,4). Hundreds of rival polymers have been disclosed. Most of them exceed PMMA in sensitivity. However, the combination of properties which include stability, sensitivity, contrast, adhesion, and solubility have kept PMMA in the limelight.

The ultimate test of usefulness is, of course, the lithographic performance of a resist. The often-quoted sensitivity of PMMA of 50 μC/cm^2 assumes conventional developing conditions and is measured by competitive dissolution rates of exposed and unexposed films. A secondary measure of sensitivity for positive-working resists is the G(s) value, the yield of chain scissions per 100 e.v. of absorbed energy. In this case, measurement of the number-average molecular weight, M_n, before and after a measured dose of energy suffices. Because G(s) represents only one aspect of resist performance, it is an incomplete, but rather convenient, screening technique. Unlike the lithographic test, it does not require detailed knowledge of solubility behavior.

Two directions have been taken towards improving the performance of PMMA without altering the processing characteristics drastically. If the original molecular weight is increased, the change in dissolution rate on exposure for the same G(s) value can be increased. Hess (5) claims a threefold enhancement in sensitivity for plasma-initiated PMMA ($M_w = 10 \times 10^7$) over that for the standard polymer ($M_w = 7 \times 10^5$). One can look at the methly methacrylate-methacrylic acid and methyl methacrylate-methacryoyl chloride copolymers which are crosslinked by intermolecular anhydride formation as being extremely high molecular weight molecules (6,7). The major advantage lies in the decreased dissolution rate of the unexposed polymer rather than in an inherent change in G(s).

The most popular way of changing sensitivity is to copolymerize methyl methacrylate. Copolymerization is presumed to preserve many of the desirable and well-established features of PMMA while introducing enough of a second (or third) component to judge its contribution to sensitivity. The second monomer is typically a 1,1-disubstituted vinyl type compound containing electron withdrawing substituents such as halogen, cyano, or carboxylic acid groups. Copolymers of methyl methacrylate with methacrylonitrile or α-halogen acrylates, for example, have been reported (4,8,10,11). The specific interest in the present work is in itaconic acid and its esters. The structure of the monomer bears a family resemblance to methacrylic acid and its esters (Table I).

Table I. Repeat Structures

		R-O-C=O
	CH$_3$	CH$_2$
	-CH$_2$-C-	-CH$_2$-C-
R = H, CH$_3$, or CH$_3$CH$_2$CH$_2$CH	C=O	C=O
	R-O	R-O
	Methacrylate	Itaconate

Polymer Preparation (Table II)

Poly(dimethyl itaconate), PDMeI, and poly(di-n-butyl itaconate), PDnBI, were prepared by slow bulk polymerization under vacuum at 50°C using benzoyl peroxide as the initiator. The bulk polymers

Table II. Synthesis and Molecular Weights of Polymer Samples

Polymer Copolymer $P(M_1-M_2)$	Initial Feed Composition M_1/M_2	Copolymer Composition from H+ NMR % M_2	Conversion, %	$M_n \times 10^{-3}$	$M_w \times 10^{-3}$
PMMA	-	-	-	340	600
PDMeI	-	-	70	405	580
PDnBI	-	-	60	127	170
P(MMA-DMeI)	12.7	4.4	22	120	185
P(MMA-DMeI)	5.25	10.9	15	114	160
P(MMA-DMeI)	2.6	16.9	23	118	173
P(MMA-DMeI)	1.0	41.0	9	100	135
P(MMA-DMeI)	0.45	53.8	6	70	100
P(MMA-DnBI)	19.4	3.7	13	137	195
P(MMA-DnBI)	8.1	12.9	14	121	179
P(MMA-DnBI)	4.0	18.7	13	140	205
P(MMA-DnBI)	1.6	40.1	8	105	140
P(MMA-ItaA)	13.0	5.7	35	290	545
P(MMA-ItaA)	5.2	12.1	15	108	180
P(MMA-ItaA)	2.6	19.5	27	105	190
P(MMA-ItaA)	2.6	19.5	22	245	370
P(MMA-ItaA)	0.67	35.0	17	37	53

All molecular weights listed are polystyrene equivalent

were dissolved in acetone and then precipitated using methanol as the nonsolvent. The precipitated PDnBI was also rinsed in 2-propanol to remove residual monomer.

The methyl methacrylate-itaconic acid copolymer, P(MMA-co-ItaA), was prepared by slow free-radical solution polymerization in methanol under nitrogen using 2,2'-azobis-(2,4-dimethyl valeronitrile)(du Pont Vazo 52) as initiator. The molar ratio of monomer to initiator was in the range of 5×10^3 to 10×10^3. Reaction at 50°C for 30 to 40 hrs gave conversions of 10 to 30%. The reaction mixture was added to cold, deionized water and the precipitated polymer obtained was rinsed with 2-propanol.

The MMA-DMeI and MMA-DnBI copolymers were made in acetone in the same manner except that they were precipitated in chilled methanol. The PMMA used as a control was obtained from Esschem Corp. All polymer samples were dried at 80° in a vacuum oven for 24 hrs prior to their use.

Copolymer compositions were determined by proton NMR. The relative proportions of the comonomers were estimated by comparing the areas of the MMA β-methylene peak and the pendant methylene peak of the itaconates. Compositions determined by NMR compare very well with those calculated from relative reactivity ratios given in the literature.

Evaluation

Resist films 0.5 to 1 μm thick were spin-coated on 3-inch diameter silicon wafers (oxide coated). The polymers were applied from 10% solutions in chlorobenzene, 2-ethoxyethanol, or 2-methoxyethyl acetate using a Headway Research Model EC-101D spinner. Prebaked (1 hr, 80°C, vac.) films were exposed to the e-beam and then extracted with about 4 ml of tetrahydrofuran, THF. About 4 to 8 hrs was allowed for dissolving to take place. Molecular weights were estimated using a Waters Model 201 HPLC equipped with 4 μ-Styragel columns of nominal sizes 500, 10^3, 10^4, 10^5 Å. The eluting solvent was THF. The molecular weights are reported on the basis of polystyrene equivalents since it was not feasible to calibrate the column systems for each polymer composition.

In order to have a uniform exposure to electrons over the entire wafer, a standard transmission microscope (RCA EMV-3) was used. The apertures were opened and the 40 kv beam spread out over a circle larger than the wafer itself. The coated wafer was introduced into the microscope via the film cassette drawer (below the fluorescent screen). Charge density was measured with a Faraday Cup. Each polymer was irradiated at doses of 2, 5, 10, and 20 $\mu C/cm^2$. The radiation chemical yield was measured from plots of $1/M_n$ versus the incident dose. A sample plot is shown in Figure 1, in this case for itaconic acid-MMA copolymers. The slope of such a plot is given by:

$$\text{slope} = \frac{V_a \int \Lambda(f) df \; \overline{G}(s)}{e \; z \; 100 \; N_a}$$

where V_a is the accelerating voltage, $\Lambda(f)$ is the normalized depth-dose function, e is the electronic charge, N_a is Avogadro's number, and z is the film thickness. Since there is no broadening of the molecular weight distribution beyond the "most probable" polydispersity of two, the assumption can be made that crosslinking is negligible. Figure 2 shows the molecular weight distribution for PMMA and poly(α-methyl styrene) after exposure in the RCA transmission electron microscope. The results indicate a uniform field of exposure over the entire resist coated wafer. In addition, values of G(s) calculated from plots of $1/M_n$ versus dose for these two polymers agree very well with those reported in the literature.

Results and Discussion

The slopes of the plots of $1/M_n$ versus dose give G(s) values which do not change much with composition (Figure 3). There is only a slight increase in G(s) with DMeI content until very high DMeI content is reached. The homopolymer was found to have G(s) about twice that of PMMA. One might indeed expect higher yields for DMeI since two ester groups are present for each quarternary carbon. Main chain scission of PMMA through ester side group elimination has been suggested in several studies (12).

Pittman and his co-workers reported values based on gamma radiation. The values of G(s) were derived in various ways, including membrane osmometry. The statement is made that the homopolymer of DMeI crosslinks measurably above 2 Mrad (10,11). No such crosslinking has been detected in the present work or in that reported by Harada (9). Pittman's values for G(s) were in the range of 3.2 to 3.5 for copolymers (60% DMeI with MMA) compared to 1.5 for PMMA in one study (10). However, a second study gave G(s) = 0.9 for a copolymer and 3.2 for PDMeI at low doses (11). The reason for the low values with copolymer in the second study are not clear. Harada (9) reported a G(s) of 1.7 when a homopolymer with a M_w of 54×10^3 was used. This is lower than his value for PMMA (G(s) = 2.2).

If an analogy with methacrylate is valid, one expects the n-butyl ester of itaconic acid to be less sensitive than the methyl ester. Reported values for poly(n-butyl methacrylate) are about half those for PMMA (9). In the present work, PDnBI indeed shows little advantage over PMMA (Figure 3). The ratio of G(s) for the n-butyl ester of itaconic acid to that for the methyl ester was found to be approximately the same as the ratio reported for the corresponding methacrylate esters.

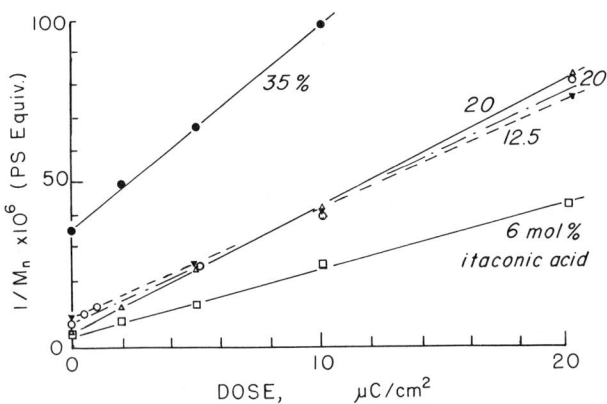

Figure 1. $1/M_n$ as a function of dose for MMA-IA copolymers at 40 kV.

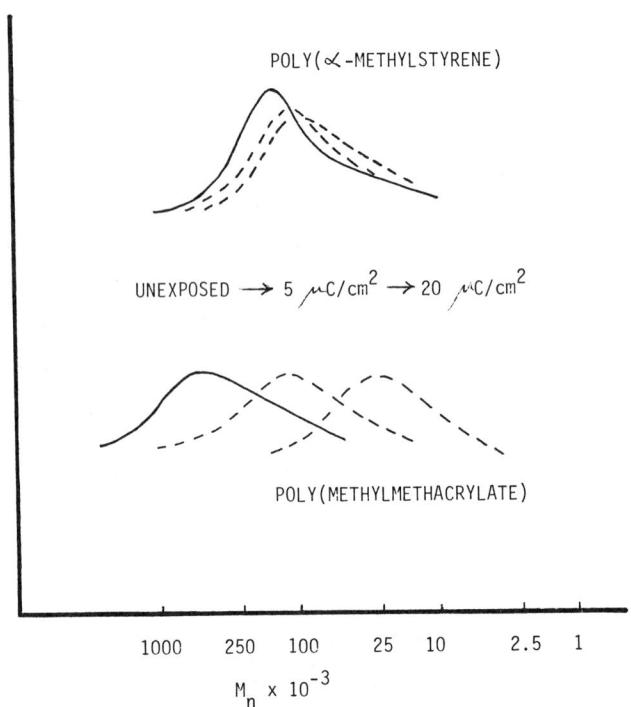

Figure 2. Typical molecular weight distributions for polymer samples exposed in the modified RCA transmission electron microscope.

The structure of itaconic acid bears some resemblance to
methacrylic acid. While it is know that methacrylic acid and
methacrylic anhydride do enhance sensitivity when copolymerized
with methyl methacrylate, some of the increase in speed may be
due to porosity (7,12). The gas formation may cause a higher
rate of dissolution then can be attributed to chain scission
alone. Pittman and co-workers have reported G(s) values for
poly(methacrylic anhydride) of 0.4 (10) and 1.8 and 2.9 (14)
based on gamma radiation experiments. Hiroaka (12) measured
G(s) by gas evolution on irradiation of films with electron
beams and established values in the ration 1:2:6 for the methyl
methacrylate. A terpolymer with the three components in the
molar ratio of 70:15:15 (same three monomers) was selected by
Moreau et al (13) on the basis of complete lithographic evaluation. The speed is 4 to 10x that of PMMA. As aforementioned,
an increase in G(s) may be only partially responsible.

In the absence of anhydride, one might expect itaconic acid
to fall in the same category as methacrylic acid. That is, an
increase of 2x that of PDMeI might be reasonable. However, the
results show a dramatic increase in G(s) with itaconic acid content (Figure 1 and 3). Looking at polymers in the 30-40% comonomer range, the acid stands quite apart from the esters. The
apparent sensitivity is probably not due to the pendant methylene group which is present in all three copolymers, but likely
to be associated with the free carboxyls.

One may suggest several mechanisms for the radiation degradation of the itaconate copolymers. These are depicted below
(for clarity, only reactions involving the itaconate repeat unit
are shown). Scheme I (Figure 4) involves β-scission induced by
ester group (or COOH) elimination from a main chain carbon analogous to that proposed by Hiroaka for PMMA or poly(methacrylic
acid) (12). An ester group may also be cleaved from the methylene carbon as shown in scheme II with subsequent main chain
scission following rearrangement. Scheme III shows chain scission resulting from direct hydrogen abstraction from the pendant
methylene. Chain scission induced by hydrogen abstraction from
the α-methyl group for PMMA under electron-beam radiolysis has
been proposed by Hiroaka (12).

Further investigations involving analytical techniques such
as epr and mass spectroscopy are needed to reveal the radiation
degradation mechanism of the itaconate copolymers.

Although chain scissioning yields are only a secondary
measure of sensitivity, the high G(s) values reported here for
MMA-itaconic acid copolymers have been supported by the good
electron-beam sensitivities determined from a complete lithographic evaluation of these copolymers. A full discussion of
these results is beyond the scope of this paper but they have
been reported elsewhere (15).

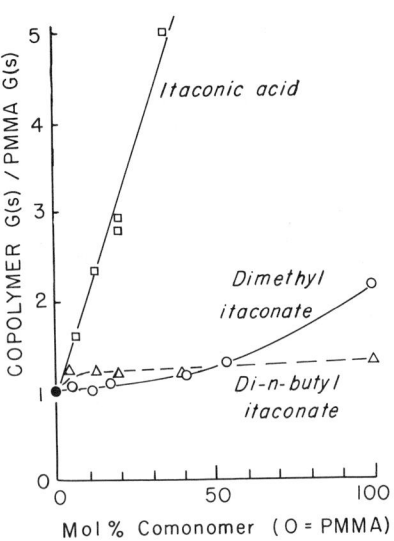

Figure 3. Effect of copolymer composition on chain scissioning yield, G(s)

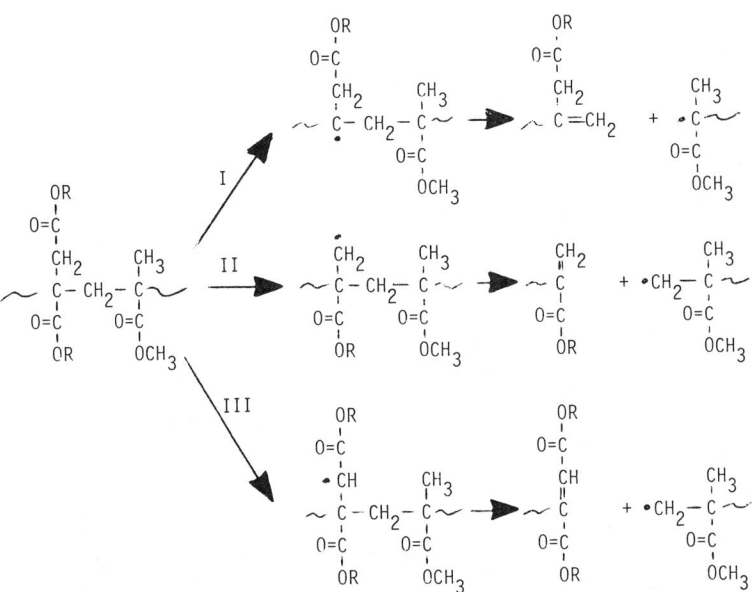

Figure 4. Possible schemes for radiation degradation of the itaconate copolymers.

Acknowledgment

The authors gratefully acknowledge the support of this work by the International Business Machines Corporation.

Literature Cited

1. Bowden, M. J. CRC Critical Revs. Solid State Sci., 1979, 8 223.
2. Broers, A. N.; Hatzakis, M. Sci. Am., 1972, 220 (11), 34.
3. Mitra, S. J. Appl. Photo. Eng. 1981, 7 (2), 37.
4. Hatzakis, M. J. Electrochem. Soc., 1969, 116 (7), 1033.
5. Hess, D. W. Air Force Report No. AFOSR-TR-81-0729, July 20, 1981.
6. Roberts, E. D. in "Scanning Electron Microscopy of Polymers and Coatings, II", Princen, L. H., Ed.; APPLIED POLYMER SYMP. No. 23, 87, Wiley: New York, 1974. Also in ACS Org. Coats. Plast. Div. Preprints, 33 (1) 359, 1973. Further described in ibid., 35 (2) 281, 1975.
7. Howard R. E.; Hu, E. L.; Jackel, L. D. Appl. Phys. Lett. 1980, 36 (2) 141.
8. Pittman, C. U., Jr.; Ueda, M.; Chen, C. Y.; Kwiatkowski, J. H.; Cook, C. F., Jr.; Helbert, J. N. J. Electrochem. Soc. 1981, 128 (8) 1758.
9. Harada, K. J. Appl. Polym. Sci. 1981, 26, 3395.
10. Helbert, J. N.; Poindexter, E. H.; Stahl, G. A.; Chen, R.C.; Pittman, C. U., Jr. ACS Polym. Preprints 1977, 18 (1) 772.
11. Chen, C.-Y.; Igbal, M.; Pittman, C. U.; Helbert, J. N. Makromol. Chem. 1978, 179, 2109.
12. Hiroaka, H IBM J. Res. Dev. 1977, 21, 121.
13. Moreau, W.; Merritt, D.; Moyer, W.; Hatzakis, M.; Johnson, D.; Pederson, L. J. Vac. Sci. Tech. 1979, 16, 1989.
14. Helbert, J. N.; Poindexter, E. H.; Stahl, G. A.; Chen, C-Y.; Pittman, C. U., Jr. J. Polym. Sci.-Chem., 1979, 17, 49.
15. Namaste, Y.; Obendorf, S. K.; Anderson, C. C.; Krasicky, P. D.; Rodriguez, F.; Tiberio, R. Intl. Symp. on Electron, Ion, and Photon Beams, Los Angeles, CA, June, 1983.

RECEIVED September 2, 1983

Poly(tetrafluorochloropropyl methacrylate) as Positive Electron Resist

B. BEDNÁŘ, J. DEVÁTÝ, J. KRÁLÍČEK, and J. ZACHOVAL

Department of Polymers, Prague Institute of Chemical Technology, 166 28 Prague 6, Czechoslovakia

Poly/tetrafluoro-chloropropyl methacrylate/ (PFCPM) prepared by solution free radical polymerization was tested as a potential candidate for electron positive resist. Thermal stability of the polymer was studied by means of TGA and DSC methods. The sensitivity of PFCPM to electron beam ranged from 7 $\mu C/cm^2$ to 2 $\mu C/cm^2$. The sensitivity is as high as more than an order of magnitude larger than that of PMMA with the same molecular weight. PFCPM can be employed as a positive electron resist at the dose less than 50 $\mu C/cm^2$. At the dose of more than 50 $\mu C/cm^2$ it can be employed as a negative electron resist because crosslinking predominates in this dose region. The influence of polymer chain structure on scission and crosslinking reactions is discussed.

The fabrication of LSI circuits, and of VLSI circuits in particular, requires patterns of micron and submicron dimensions, and consequently polymer resists with a high degree of resolution (1). So far the most frequently used positive electron resist has been poly(methyl methacrylate) (PMMA), which affords a high resolution power together with a relatively good thermal stability (2-4). A serious limitation of PMMA with respect to the efficiency of the electron lithography system is its low sensitivity to electron irradiation ($10^{-5} - 10^{-4} C/cm^2$). For the preparation of polymer electron resists it is desirable that the resist should possess such sensitivity that doses of the order of $\sim 1 \mu C/cm^2$ could be used during exposure. This requirement can be easily met in the preparation of negative electron resists (5-6). Of the positive electron resists used so far, only the PBS resists (copolymers of olefins with SO_2) meet this requirement (7), and recently some substituted methacrylate polymers, the most successful of which appears to be the halogen-substitued methacrylates (8-10). One of the above mentioned polymers is poly(tetrafluoro-chloropropyl methacrylate) (PFCPM), some lithographic properties of which will be presented in the present paper.

Experimental

Polymers used in the study were prepared by solution, free radical-initiated polymerization. TGA, DTGA and DSC measurements were performed using the DuPont 951 and DuPont DSC-910 devices connected to an electronic device DuPont TA-990. The device was not calibrated to precise values of heat effect. The determination of the lithographic properties was made using a modified scanning microscope JSM 35. To determine the sensitivity always 4 areas 1x1mm large were exposed with line scanning of the density of 200 lines per mm with a 1 μm electron beam diameter.

Results and Discussion

The polymerization conditions and some properties of the prepared PFCPM are presented in Table I. Polymerizations of samples of PFCPM-1, PFCPM-2 and PFCPM-3 were carried out in the presence of 10% tetrafluoro-chloropropanol used for the preparation of the monomer. As seen from the data presented in Table I, the presence of alcohol does not substantially affect the rate of polymerization, but the Mn values of the prepared polymers are lower than those obtained in the polymerization of a pure monomer. The decrease in molecular weight values is apparently due to the transfer of the tetrafluoro-chloropropanol during polymerization. TGA, DTGA and DSC measurements have shown that the thermal stability is not markedly affected by the molecular weight of the polymer. In contrast with PMMA, the TGA curve can be divided into two main regions (Fig. 1), the first decrease in weight occuring at 270°C, and the other change occurring at 355°C. The Tg values determined by DSC measurements are not the same for all the PFCPM, which cannot be attributed to the dependence on molecular weights, but more probably to different amounts of impurities present (e.g. solvent). Similarly as with other chlorosubstituted methacrylates the crosslinking of the polymer takes place during thermal treatment at about 200°C, but at ambient temperature the

Table I. Characteristics of poly(tetrafluoro-chloropropyl methacrylate)

Polymer	[I] x10³ mol/l	Polym. time hrs.	Conversion	$\bar{M}_n$x10⁻⁴ g/mol	Tg °C	Q uC/cm²
PFCPM-1	4,4	6	23,5	-	-	-
PFCPM-2	14,7	6	42,2	15,2	-	-
PFCPM-3	22,3	6	50,5	10,6	-	2,5
PFCPM-4	4,0	9	22,0	51,4	93	5,0
PFCPM-5	6,9	9	26,5	35,8	85	-
PFCPM-6	21,6	9	46,4	27,8	85	4,9

I - initiator ABIN, $\bar{M}_n$ - data from membrane osmometry

polymer is stable. The sensitivity of the PFCPM-3, PFCPM-4 and PFCPM-6 polymers to electron irradiation has shown a relatively small dependence on the molecular weight values (Table I) but, similarly as with some previous methacrylates (9), at higher irradiation doses the crosslinking of PFCPM takes place, the polymers thus changing from positive resists to negative ones (Fig. 2). On the basis of the present experimental results, however, it is impossible to decide whether the change in the ratio of the crosslinking and the degradation reactions with the increasing irradiation dose is not affected in a decisive way by the reactions induced by increased temperature during irradiation. The change in the ratio of the crosslinking and the degradation reactions was not observed during irradiation of the polymers with γ-radiation or with accelerated electrons (without any temperature change), not even at extremely high doses (11). The sensitivity values (Table I) indicate that PFCPM is one of the most sensitive positive electron resists. The sensitivity value for the positive electron resist, however, is always affected to a considerable degree by the developer employed (in our case it was a mixed solvent, 1,4-dioxane/80 vol.% n-heptane), so that by choosing a proper developer the sensitivity will most probably be increased. The region of doses in which the crosslinking of PFCPM takes place starts with the dose of 4.5×10^{-5} C/cm^2 (from this value onward the samples were developed in pure 1,4-dioxane) for PFCPM-3, increasing slightly with the increasing value of molecular weight, which is a course opposite to that to be expected. This can be attributed to the degradation reactions preceding the crosslinking and consequently changing the values of molecular weights and their distribution, and to the thermal degradation during the crosslinking.

The resolution power of PFCPM permits patterns of submicron dimensions to be prepared. Their resistance to wet etching is very good, in plasma etching being lower than that of PMMA but substantially better than that of PBS.

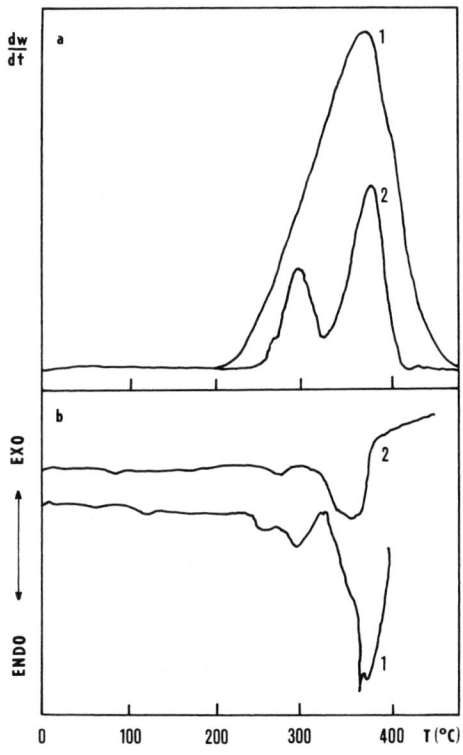

Figure 1. Results of the thermal treatment. (a) DTG curves. (b) DSC curves. Key: 1, PMMA; and 2, PFCPM-6.

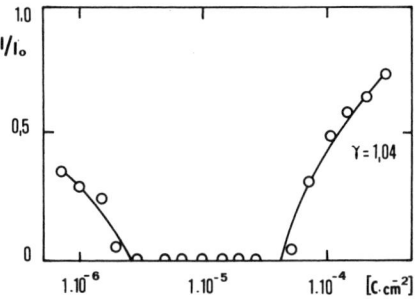

Figure 2. Dependence of reduced resist film thickness after development on the radiation dose for PFCPM-3. Development conditions:(up to 3×10^{-5} C/cm^2) 1 min, mixture 1,4-dioxane/ 80% vol. heptane; and (from 5×10^{-5} C/cm^2) 2 min, 1,4-dioxane.

Literature cited

1. Thompson, L.F.; Kerwin, R.E. Ann.Rew.Mater.Sci. 1976, 6, 267.
2. Hatzakis, M. J.Electrochem.Soc. 1969, 116, 1033.
3. Chang, H.P.; Hatzakis, M.; Wilson, A.D.; Speth, A.J.; Kern, A. Electron. 1977, 8, 51.
4. Zeitler, H.V.; Hieke, E.K. J.Electrochem.Soc. 1979, 126, 1430.
5. Thompson, L.F.; Stillwagon, L.E.; Joerries, E.M. J.Vac.Sci.Technol. 1978, 15, 938.
6. Thompson, L.F.; Van, J.; Doerrie, E.M. J.Electrochem.Soc., 1979, 126, 1703.
7. Bowden, M.J.; Thompson, L.F. J. Electrochem.Soc. 1974, 12, 1620.
8. Tada, T. J. Electrochem.Soc. 1979, 126, 1829.
9. Tada, T. J. Electrochem.Soc. 1979, 126, 1635.
10. Lai, J.H.; Helbert, J.N.; Cook, C.F.; Pittman, C.V. J.Vac.Sci.Technol. 1979, 16, 1992.
11. Dole, M. in "The radiation Chemistry of Macromolecules", (Dole, M. Ed.), Academic Press, New York 1973.

RECEIVED December 19, 1983

12

Radiation-Induced Degradation of Poly(2-methyl-1-pentene sulfone)
Kinetics and Mechanism

M. J. BOWDEN, D. L. ALLARA, W. I. VROOM, J. FRACKOVIAK, L. C. KELLEY, and D. R. FALCONE

Bell Laboratories, Murray Hill, NJ 07974

>PMPS degrades by a mechanism involving random chain scission followed by extensive depropagation from the chain ends. Both radical and cationic depropagation pathways may exist. The initial rate of loss of SO_2 is greater than the rate of loss of 2-methyl-1-pentene although at high doses the amount of SO_2 and olefin lost from the film are roughly comparable. The discrepancy between the amounts of SO_2 and olefin lost at low doses is attributed to oligomerization of the olefin (possibly by cationic reactions) as it diffuses through the film. These oligomers can be easily removed by post-exposure baking leaving "pure" PMPS behind. These results are consistent with degradation studies of other poly(olefin sulfones), but are contrary to the findings of Pacansky et al. whose results appear to be anomalous. We may also note that the kinetics of radiation-induced depolymerization in pure film or in mixture with a novolac resin appear to be qualitatively similar although there is a slight inhibition caused by the novolac. It appears though that the contrast of the vapor development process increases with dilution in spite of slight inhibition by the novolac resulting in efficient material removal at high doses.

Poly(2-methyl-1-pentene sulfone) PMPS, is an alternating copolymer of 2 methyl-1-pentene and SO_2 produced by free radical polymerization. The radiation-induced degradation of PMPS has been studied extensively using both 1.0 MeV electrons (1) (Van de Graaff electrostatic generator) and 10-20 keV electrons (2–4). In common with other poly(olefin sulfones), PMPS undergoes random scission followed by depropagation from the broken chain ends. In a previous paper, the degradation reaction occurring during irradiation with 1.0 MeV electrons was followed by monitoring changes in the infrared spectra of thin films as a function of dose. Since depropagation leads to material loss, the spectral absorbance decreases with increasing dose.

A priori, one might expect that depropagation of radical chain ends produced by random scission should lead to equal yields of SO_2 and the parent olefin. However, it was found that, in common with other poly(olefin sulfones) (5–8), the yield of SO_2 was greater than that of the olefin provided the irradiation temperature was <100°C. Above 100°C, the peaks in the infrared spectrum attributable to SO_2 and olefin moities decreased at equal rates. Bowden (1) suggested that the discrepancy in the rate of removal of SO_2 and 2-methyl-1-pentene at lower temperatures might be due to reaction of the olefin, e.g., cationic homopolymerization, as it diffused through the film. Evidence for cationic reactions during radiolysis of poly(olefin sulfones) was presented by Bowmer et al (7). in their studies on the effect of radical and ionic scavengers on the yields of volatile radiolysis products from several poly(olefin sulfones). In particular, they found that in the presence of cation scavengers, the overall product yield was reduced with concurrent reduction of the SO_2/olefin ratio towards unity. They also found that olefin isomerization reactions which frequently accompanied depropagation (8) were eliminated. They accounted for these results in terms of two pathways for depropagation, viz., free radical and cationic, with associated reactions for cationic homopolymerization of olefin and for isomerization in the formation of olefin. Cationic homopolymerization was believed to be initiated at cationic sites on the polymer chain.

This overall mechanistic picture has been questioned recently by Pacansky et al. (9–10) They reported virtually total elimination of SO_2 from PMPS films at an incident electron dose of 0.4 $\mu C/cm^2$ at 25 kV with little or no loss of olefin. Olefin loss took place at much higher doses. Pacansky proposed a two-step mechanism involving SO_2 elimination via a chain process leaving hydrocarbon polymer which subsequently decomposed to olefin. The results are totally contrary to those previously reported by Bowden who found that at doses corresponding to 0.4 $\mu C/cm^2$ only ~35% of the SO_2 was removed while 30% of the olefin was removed. The mechanism proposed by Pacansky et al. to explain their results is counter to the widely accepted mechanism of polysulfone degradation which is in accord with the thermodynamics of these polymers. It is difficult, for example, to conceive of a chain mechanism for SO_2 formation that does not include olefin formation. The "instantaneous" reformation of the polyolefin in the case of PMPS after multiple C-S bond scission as suggested by Pacansky would seem to be highly improbable.

The material loss occurring as a result of depolymerization is reflected as a thickness loss which can also be monitored as a function of dose (2-4). This technique was used by Bowden and coworkers to follow the degradation of thin PMPS films (<0.6 μm) on silicon substrates using the Bell Laboratories electron beam exposure system (EBES) (2,11). Measurement of the thickness loss as a function of dose showed a marked dependence on the manner in which the dose was delivered, (see Fig. 1), viz., whether it was delivered in a single pass (with varying beam current) or accumulated over several passes (at constant beam current). This phenomenon was attributed to dose rate effects coupled with the fact that the time for depolymerization (depropagation) was small relative to the dwell time in the beam. Studies on the effect of processing variables further showed that the extent of material loss could be greatly increased by post-exposure thermal treatment in the range 90-130°C. The optimum post-exposure condition was found to be 1-2 hr in air at 130°C which increased the amount of material removed for an incident dose of 3.5×10^{-6} C/cm^2 delivered in a single pass from 40% to 70% (see Fig. 1). This phenomenon was attributed to additional depolymerization initiated at "trapped" radical sites. It was difficult to obtain higher degrees of material loss. At higher doses, the curve obtained by post-exposure treatment appeared to approach a limiting value and a question might be raised concerning the chemical composition of the films after such treatment, particularly if extensive homopolymerization of the olefin has occurred.

PMPS is one of the components of NPR positive electron resist where it functions as a dissolution inhibitor of the matrix polymer (12). The latter is a novolac resin whose solubility in aqueous base is greatly reduced by the presence of PMPS dispersed in solid solution with the novolac. The mechanism of resist action depends upon the degradation behaviour of PMPS. As a result of the material loss stemming from radiation-induced depropagation, dissolution of the novolac resin in aqueous base is no longer inhibited in the irradiated area which can be subsequently developed to give a positive image. The sensitivity of this resist system is correlated with the rate of PMPS depropagation, i.e., removal, hence it is important to determine the depolymerization kinetics of PMPS and in particular to ascertain what effect, if any, the novolac matrix has on the depolymerization reaction. Since novolac resins are condensation products of formaldehyde and phenols, their presence might lead to inhibition of degradation due to the fact that phenols are known inhibitors of free radical processes.

Infrared spectroscopy provides a simple convenient means of following structural changes (in addition to material loss). This paper reports results obtained from IR measurements on PMPS both during exposure and following post-exposure treatment for pure films and also in composite mixtures with a novolac resin.

Experimental

PMPS was synthesized by UV initiation of monomer mixtures in methylene chloride solutions in the manner described previously (13). The intrinsic viscosity of the polymer in MEK was 250 cm^3/g.

Films of pure PMPS were cast from a 1% solution of the polymer in chloroform on the surface of sodium chloride plates. The solvent was allowed to evaporate at room temperature and the films finally dried for 1 hr at 110°C. Alternatively, films were spin coated from chlorobenzene solution onto silicon substrates which were undoped and polished on both sides. As such they are transparent in the IR down to 600-700 cm^{-1}.

The novolac resin used in NPR formulations was a proprietary resin manufactured by Polychrome Corp. to Bell Laboratories specifications. Samples of NPR containing 20% and 50% PMPS were prepared in Hunt Photoresist Thinner (mixture of ethoxyethyl acetate, butyl acetate and xylene), filtered through 1 μm Teflon filters and spin coated onto undoped silicon wafers to give films ~9,000 Å thick. All films were prebaked at 110°C for 2 hr. Film thickness was measured directly by a Nanometrics Nanospec AFT microarea thickness gauge. Alternatively, the samples were coated with a thin layer of aluminum and measured mechanically with a Dektak.

Samples to be irradiated in the electron accelerator were placed in a pyrex dish which was then covered and sealed with polyethylene film. The apparatus was continuously flushed with nitrogen during irradiation. Provision was made for monitoring the temperature rise during irradiation by drilling a small hole into the side of the sodium chloride plate and inserting a chromel alumel thermocouple which was connected to a digital voltmeter located outside the irradiation area. Temperature rise was monitored in this manner with the thermocouple in several different positions.

Irradiations were carried out using a Dynamitron 1.5 MeV electron accelerator, or EBES. In the former case, samples were placed on a conveyor tray and passed underneath the beam. At a velocity of 5.7 cm/sec and a beam current of 1 mA, an absorbed dose of 0.5 Mrads is delivered in a single pass. The dose is thus built up by multiple passes. Alternatively the sample could be held stationary underneath the beam for the requisite period of time necessary to attain a given dose. It will be shown later that the latter approach causes the temperature of the sample to rise significantly during irradiation.

Degradation was followed by measuring the infrared absorption intensities of the aliphatic and sulfone groups in the chain as a function of dose. Measurements were made on a Perkin Elmer Model 257 grating spectrophotometer and by Fourier Transform infrared spectrometry using a Nicolet 5DX FTIR spectrometer operating at 2 cm^{-1} resolution. Absorbance spectra of PMPS in novolac/PMPS blends were corrected for the contribution due to novolac absorption by subtraction of an appropriately scaled absorbance spectrum of pure novolac.

Powdered samples for cobalt-60 γ-irradiation were evacuated for 24 hr and sealed under vacuum. They were then irradiated at ambient temperature in a Co-60 Gamma Cell-220 radiation unit operating at a dose rate of 0.210 Mrad/hr.

Results and Discussion

IR Spectrum of PMPS

The infrared spectrum of PMPS is shown in Fig. 2. As discussed previously (1), it is characterized by two strong bands, one at 1298 cm^{-1} which is due to the symmetrical stretching vibration of the SO_2 group and the other, a split band with peaks at 1136 and 1119 cm^{-1}, which is due to the unsymmetrical vibration. The band at 2985 cm^{-1} is due to C-H stretching vibrations and therefore represents the saturated aliphatic component of the polymer, i.e., the polymerized 2-methyl-1-pentene units. In the virgin material, the ratio of the absorbance of the aliphatic peak to the sulfone peak at 1298 cm^{-1} is 0.46 ±0.01.

Temperature Considerations

The problem of temperature control at the sample site is particularly important at the high dose rates attained with electron accelerators. The Dynamitron, for example, operating at 1 mA spread over an area of ~300 cm^2 has a current density of 3 μA/cm^2 at the window (the actual current density at the irradiation position is considerably less due to scattering of the beam in air). This may be compared with a current density of ~0.25 μA/cm^2 for the Van de Graaff accelerator used in the previous work. Figure 3 shows temperature versus irradiation time profiles for several experimental configurations. In all cases, the temperature rose markedly with dose and reached a limiting value which was dependent on the particular configuration. In case A, the sample remained stationary under the beam which was set at 0.5 mA and 1.5 MeV. The resulting dose rate was 3.5 Mrad/min. After about 10 min (35 Mrad) the temperature had stabilized at 125°C. The temperature increase was not nearly as marked when the dose was accumulated by repeatedly passing the sample underneath the beam because of heat transfer during the non-irradiation portion of the cycle. The actual plateau reached (after about 10 min) was in the range of 50-70°C depending on the thermocouple configuration. Thus for PMPS samples irradiated to doses in excess of 5 Mrads (≥4 min irradiation time), the sample temperature increased throughout the run. For example, the irradiation temperature for the sample given 5 Mrads varied from 25 to about 45°C (curve D, Fig. 3) giving an average temperature of 35°C. At 10 Mrad, the sample reached 54°C with a proportionately greater time spent in the upper temperature region.

The dose rate in EBES is several orders of magnitude greater than that provided by the Dynamitron so that temperature control might be expected to be more of a problem. However, it should be pointed out that the films are very thin and are supported on silicon wafers which facilitate efficient heat dissipation. Further, since the diameter of the beam is typically ≤ 1.0 μm, the corresponding

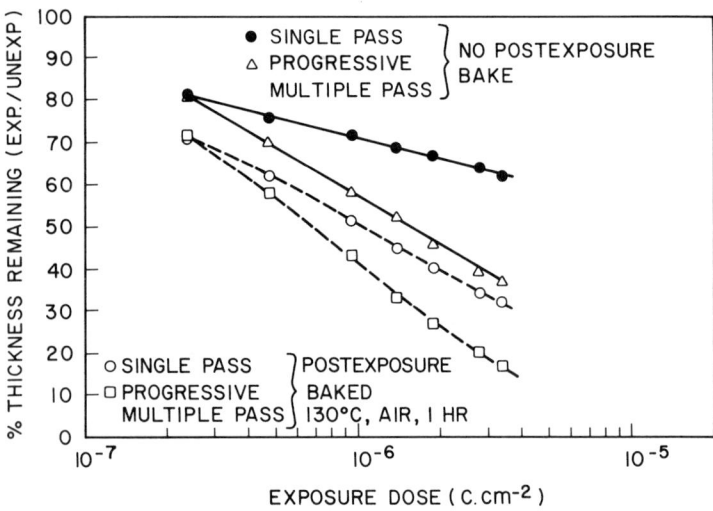

Figure 1. Effect of postexposure baking on vapor development rate for single pass and multiple pass exposures.

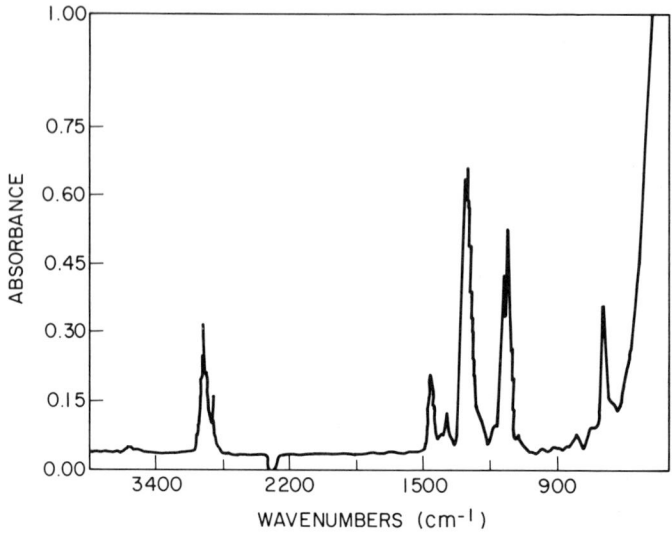

Figure 2. IR spectrum of PMPS.

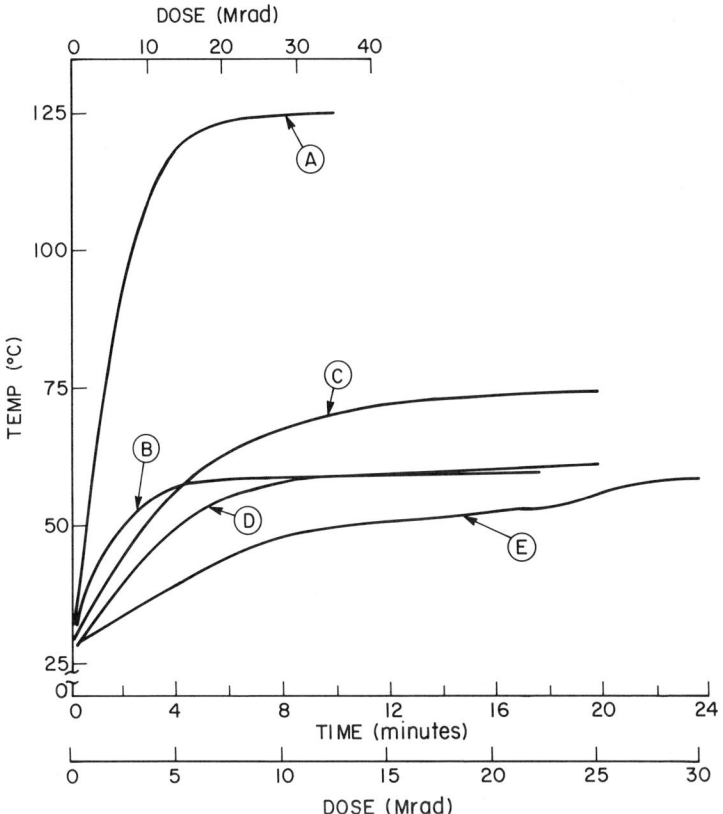

Figure 3. Temperature vs. irradiation time (dose) profile at the sample site for several experimental configurations. All sample doses exposed in multiple pass mode except A. Upper dose scale applies to A only. Key: A, exposed TC in stationary mode; B, exposed TC; C, TC at center of 1/4-in. thick plexiglass disk; D, TC at center of NaCl plate; and E, TC at center of molten paraffin.

irradiated area per beam address is very small relative to the area of the heat sink. Thus heat transfer should be sufficiently rapid so as not to be a problem. Resists such as poly(butene -1 sulfone) show no measurable vapor development during EBES exposure although at elevated temperatures, significant vapor development occurs (50% removed after a dose of 1×10^{-6}C/cm^2 at 70°C (4)). The absence of vapor development of PBS during EBES exposure at room temperature is thus indicative of no significant temperature increase under this condition.

Dose Dependence of Pure Film

The application of Beer's Law to quantitative analysis using infrared spectroscopy requires that

$$A = \epsilon\, C \ell \qquad (1)$$

where A is the absorbance, ϵ the absorption coefficient, C the concentration and ℓ the thickness. We may note that the product of concentration and thickness is a measure of the relative number of molecules in the path of the infrared beam. Thus provided the peak shape does not change, a linear relationship should be observed between absorbance and concentration (at constant thickness) and absorbance and film thickness (at constant concentration). This was checked by measuring the IR absorbance of the aliphatic and 1298 cm^{-1} sulfone peak as a function of sample thickness. As shown in Figure 4, a linear plot is obtained. Figure 5 shows a plot of absorbance of both the 1298 cm^{-1} and 1119 cm^{-1} bands as a function of film thickness. The varying thicknesses were obtained by exposing the original film to doses ranging from 0.2 to 3.4 μC/cm^2 in EBES and measuring the residual film thickness with a Dektak or Nanospec. It is immediately apparent from Fig. 5 that Beer's Law is obeyed to a first approximation (the dotted line has been drawn through the origin and the absorbance corresponding to the initial film thickness) although the intermediate points lie slightly to the right of this theoretical line, i.e., the films are slightly thicker than what one would estimate based on absorbance measurements. This could be caused by variation in density with dose as a result of depolymerization throughout the bulk of the film creating a porous matrix. Alternatively, it is possible that measurement of sulfone absorbance underestimates the actual film thickness by a small amount due to residual oligomers in the film. We can differentiate between these two possibilities by following the relative spectral absorbances of both aliphatic and sulfone peaks as a function of dose.

The effect of dose on relative absorbance (taking the initial absorbance as 1.0) is shown in Fig. 6 for both the aliphatic (open symbols) and sulfone peak (closed symbols). Also shown in Fig. 6 are data taken from our previous paper involving degradation of PMPS under electron irradiation from a Van de Graaff accelerator. The results obtained from the two different radiation sources show good agreement and confirm our previous observations that the peaks attributed

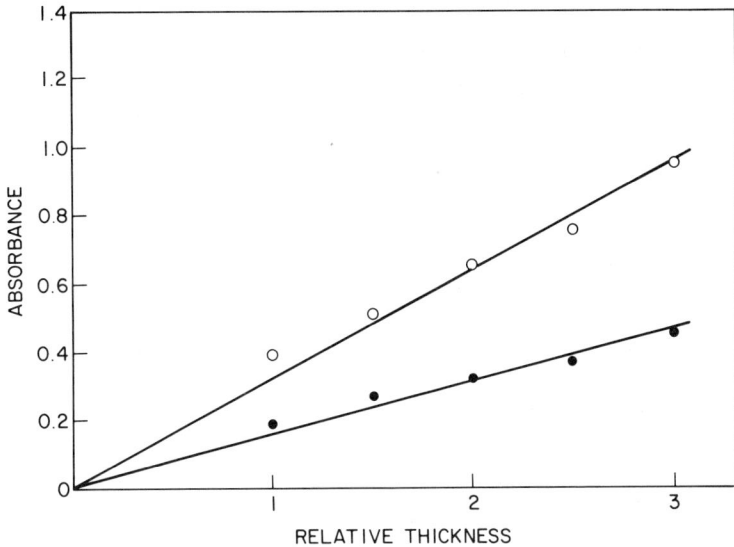

Figure 4. IR absorbance of aliphatic (●) and sulfone (○) peaks as a function of relative sample thickness.

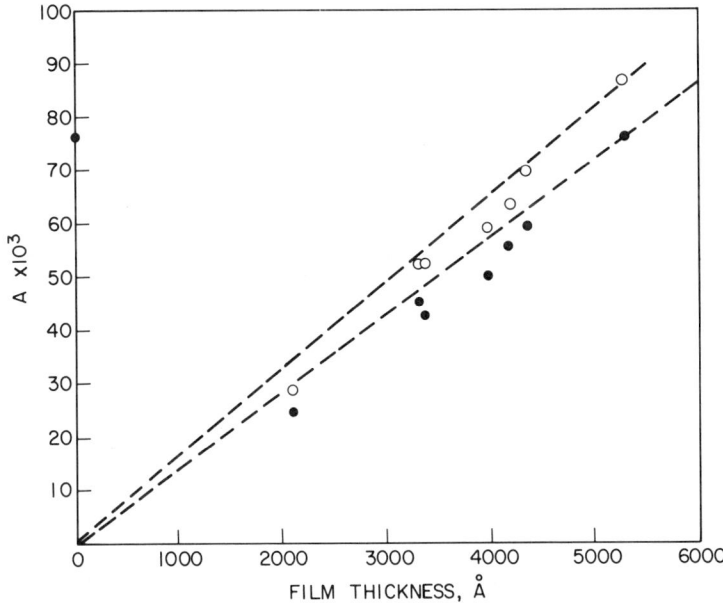

Figure 5. Plot of absorbance measured at 1298 (○) and 1190 cm^{-1} (●) vs. film thickness for pure PMPS.

to the SO_2 in the polymer decrease at a faster rate initially than the peak attributed to the olefin. Contrary to the results of Pacansky et al. which showed only SO_2 was liberated from the polymer at low doses ($<\sim 1$ Mrad), we found that $\sim 30\%$ of the olefin was removed along with $\sim 40\%$ of the SO_2 for a dose of 1 Mrad. These results are in line with the accepted mechanism of poly(olefin sulfone) degradation and agree with recent radiolysis product measurements of γ-irradiated PMPS (14). We also conclude that the film density remains constant with dose.

The results obtained on post-exposure baking the films for 1 hr at 110°C are also shown in Fig. 6. As found previously in our studies of the effect of processing variables on the vapor development of PMPS (2), the extent of degradation is further increased although only slightly. (Post-exposure heating at 130°C is required to effect considerable additional depropagation). It is further seen in Fig. 6 that the major change caused by post-exposure heating at 110°C is to decrease the intensity of the aliphatic peak, the intensity of the SO_2 peak changing only slightly. In fact, the intensity ratio changes back to that of the virgin material so that the plots of SO_2 and olefin absorbance as a function of dose coincide. Bowmer and O'Donnell (15) had previously shown that heating irradiated poly(olefin sulfones) above their ceiling temperatures initiated rapid depolymerization (in addition to that occuring during irradiation) generally yielding equal amounts of SO_2 and olefin. It is interesting to note that with PMPS, additional depropagation is not induced till ~ 130°C, some 160°C above the ceiling temperature suggesting a different mechanism is operating.

If the discrepancy between the extent of olefin and SO_2 loss produced by irradiation were due to cationic polymerization of the 2-methyl-1-pentene to *high* molecular weight polymer as previously suggested, this change in intensity ratio on heating should not be expected as the polyolefin would not be sufficiently volatile. It is possible some of the olefin remains in the film, and contributes to the absorbance at 2985 cm^{-1}. Alternatively, cationically initiated dimerization or trimerization might account for the discrepancy as these species might well be sufficiently volatile to be removed during post-exposure baking. The fact that we were unable to detect any peaks in the double bond region of the spectrum might be taken as evidence for the latter explanation. In any event, this "excess" material is removed by post-exposure baking leaving pure PMPS. These results imply that it is the SO_2 loss curve in Fig. 6 which reflects the true loss of PMPS by depolymerization at room temperature.

These conclusions were confirmed somewhat indirectly by exposing bulk samples (5-10g) of PMPS to 1.08 and 9.6 Mrad γ-irradiation, dissolving the irradiated material in acetone (a non-solvent for polyolefins) and precipitating into methanol. The irradiated polymer was soluble in acetone and the infrared spectrum of the precipitated polymer was identical to virgin PMPS ($I_{CH_2}/I_{SO_2} = 0.46$). Any oligomeric material formed from the olefin was presumably removed by the precipitation procedure. The data points corresponding to the percentage of the polymer remaining as a function of γ-ray dose are shown in Fig. 6. The fact that slightly more polymer was depolymerized at 10 Mrad in the case of electron exposure in the Dynamitron may reflect problems with temperature control during electron irradiation to such a high dose. At 1.08 Mrad, the results obtained from γ-irradiation and electron irradiation were identical. The equivalence of these results suggests that the unzipping reaction is dose rate independent under the Dynamitron exposure conditions where the unzip time is short relative to the total time of irradiation.

It is instructive to compare our results from infrared measurements with the vapor development curves obtained with EBES exposures. We have seen that the latter consist of two curves depending on the manner in which the dose is received. The vapor development curves obtained from film thickness measurements are plotted in Fig. 7 with the abscissa showing the incident electron dose ($\mu C/cm^2$) together with the equivalent absorbed dose (Mrads) calculated from the depth-dose curve for 20 kV electrons (16). Also plotted in Fig. 7 are the infrared data obtained from the Dynamitron-exposed samples. Only the loss curve for the olefin is shown as this should be equivalent to the thickness loss curve since it reflects the contribution of the oligomeric olefin - derived products to the total material content, i.e., thickness, of the residual film. A linear dependence on log (dose) is again observed with the slope (contrast) slightly greater than that found for the multiple passed EBES-exposed samples.

We suggest that the small difference in film removal rate can be explained in terms of dose rate effects. We have seen that the Dynamitron results are comparable to results obtained from γ-irradiation suggesting that at the dose rates used with the Dynamitron, the reaction is dose rate independent and little is gained by "compartmentalizing" the dose by multiple passes. At the much higher dose rates available to EBES, dose rate effects become important and important advantages can be gained by the multiple pass approach. The results of Fig. 7 suggest that the lowest available stable beam currents on EBES are still within the dose rate dependent range although not by much.

Effect of Novolac Resin on Depolymerization Kinetics

Fig. 8 shows a plot of absorbance as a function of PMPS concentration in novolac/PMPS composite mixtures for films of constant film thickness (~9,000 Å). Again a linear plot is obtained further demonstrating the applicability of Beer's Law.

The spectrum of the pure novolac is shown in Figure 9a and that of pure PMPS in Figure 9b (only a limited portion of the spectrum is shown). It is seen from Fig. 9a that the novolac absorbance at 1200 and 1485 cm^{-1} falls within a region of slight PMPS absorption whose strong peaks (due to SO_2 stretching vibrations) occur at 1298 and 1119 cm^{-1}. The spectrum of a 50% novolac/50% PMPS mixture is shown in Fig. 10a. By scaling the novolac absorption spectrum (Fig. 9a) to yield base line absorbance at 1200 and 1485 cm^{-1} and subtracting this spectrum from that of the mixture (Fig. 10a) we obtain the spectrum shown in Fig. 10b which is corresponds to that in Fig. 9b. Since the sulfone stretching peaks are much more intense than those due to the hydrocarbon position they were followed for all subsequent quantitative measurements.

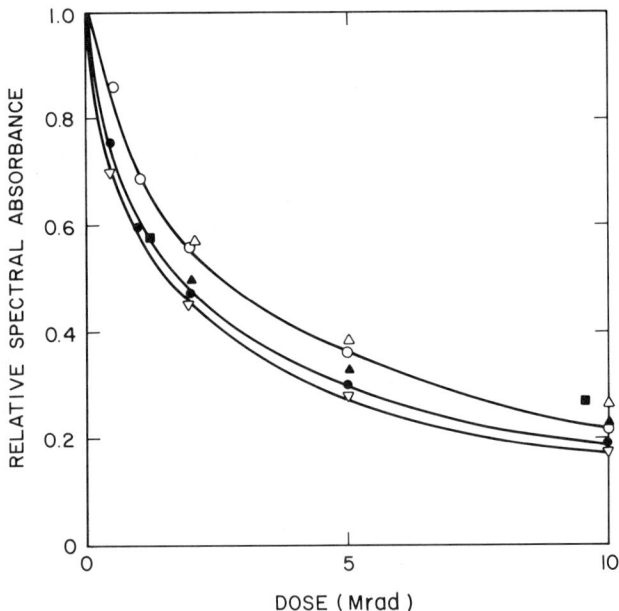

Figure 6. Fraction of film remaining (expressed as relative absorbance) vs. dose. Key: ○ and △, aliphatic peak; and ● and ▲, sulfone peak. Key(aliphatic and sulfone peak): ■, γ-irradiated sample; and ▽, after postexposure baking (110 °C).

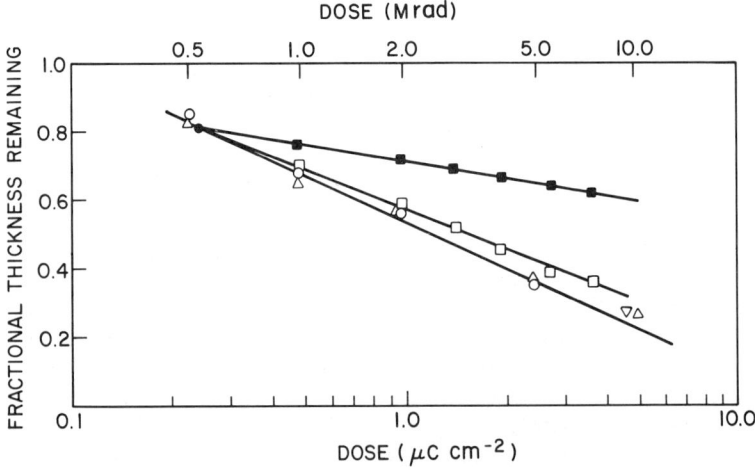

Figure 7. Plot of film thickness remaining vs. dose. Key: ■, EBES single pass; □, EBES multiple pass; and △, ○, Dynamitron exposed (calculated from olefin peak absorbance.

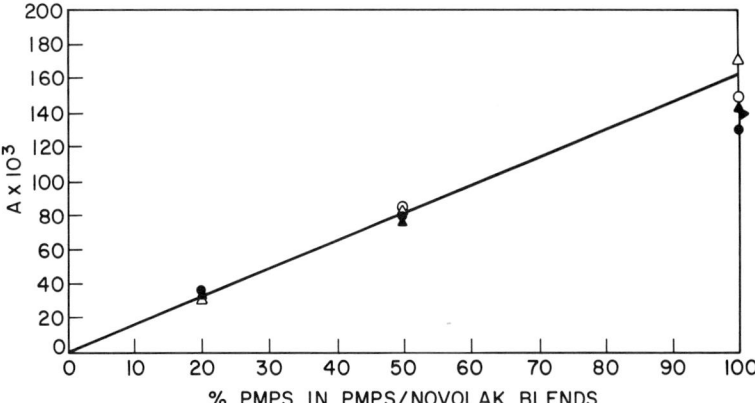

Figure 8. Plot of absorbance vs. percent PMPS in novolac-PMPS composite films.

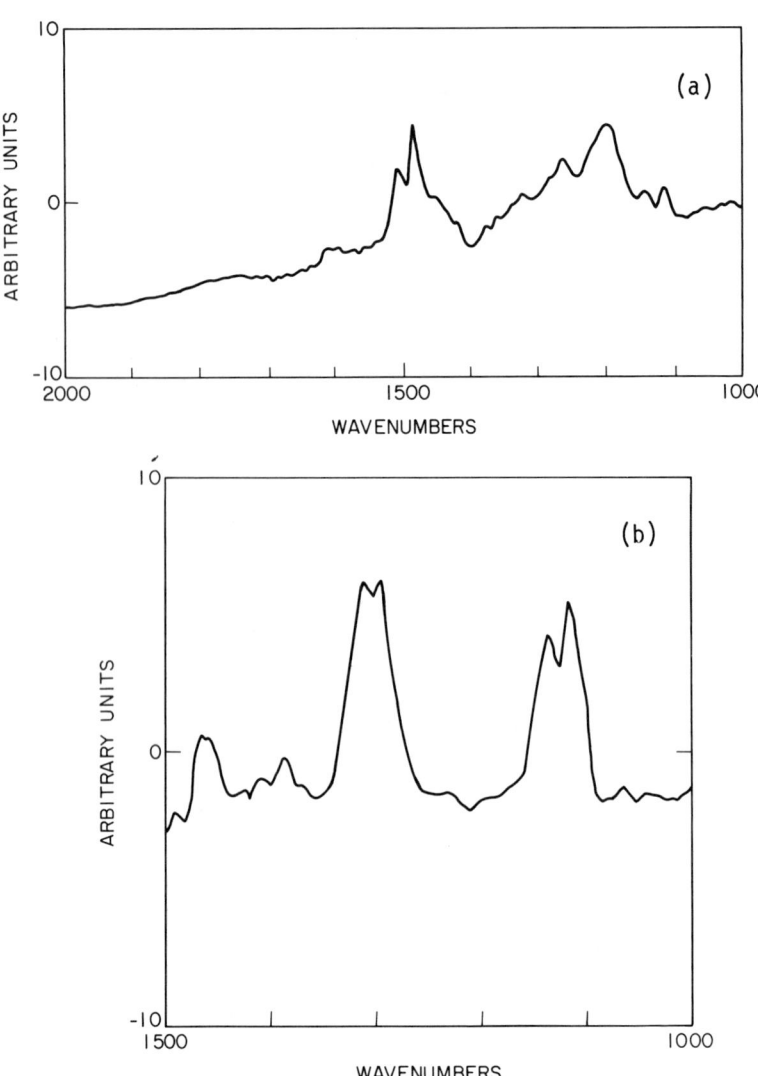

Figure 9. Absorbance spectra of (a) pure novolac film and (b) pure PMPS film.

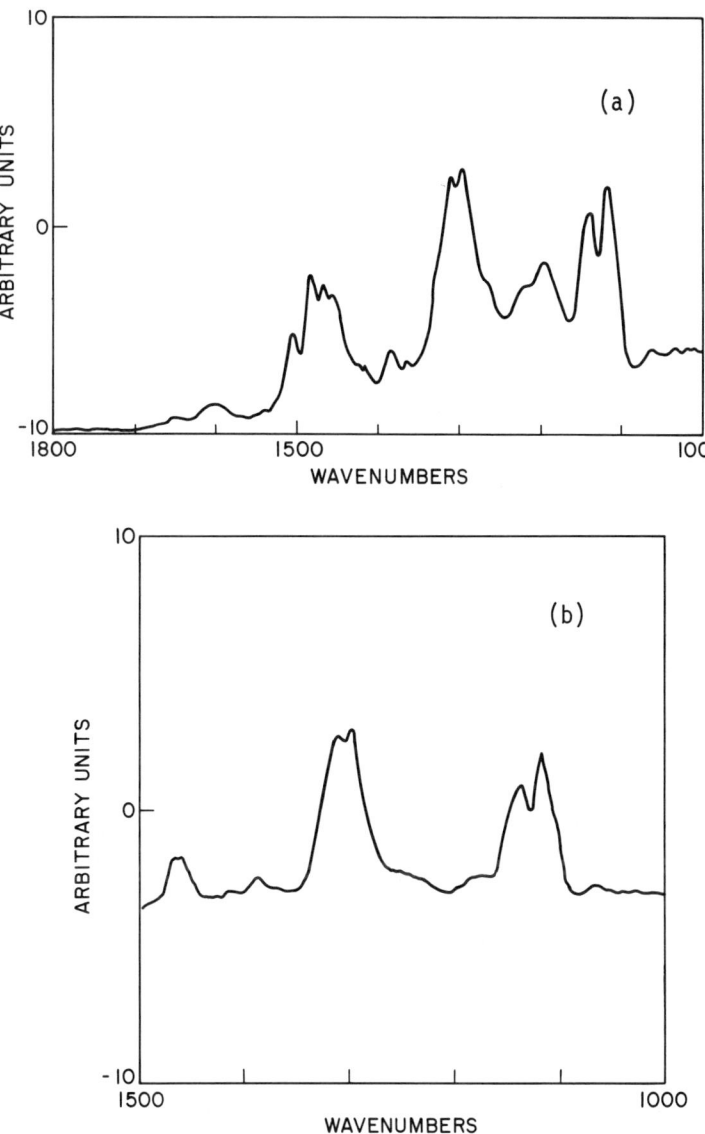

Figure 10. Spectra of (a) composite film containing 50% of each component and (b) "pure" PMPS obtained by subtraction of novolac component spectrum (See Figure 9) from composite spectrum (a).

The PMPS vapor development curve for the composite mixtures containing 20% and 50% PMPS are shown in Figs. 11 and 12. Qualitatively, the vapor development curves are similar to what is observed for pure PMPS although there appears to the some interesting inhibition and contrast effects. For samples which have not been post-exposure baked, the curve for 50% novolac lies above that of the pure film but with approximately the same contrast suggesting a general inhibiting effect of the novolac. With 80% novolac, there is also evidence of inhibition, particularly at low dose but the contrast of the vapor development curve is somewhat greater than that of the pure film leading to more efficient removal at high dose. We suggest that increasing the dilution may achieve the same result as "compartmentalizing" the dose by multiple passing, i.e. intermolecular effects are reduced leading to enhanced contrast.

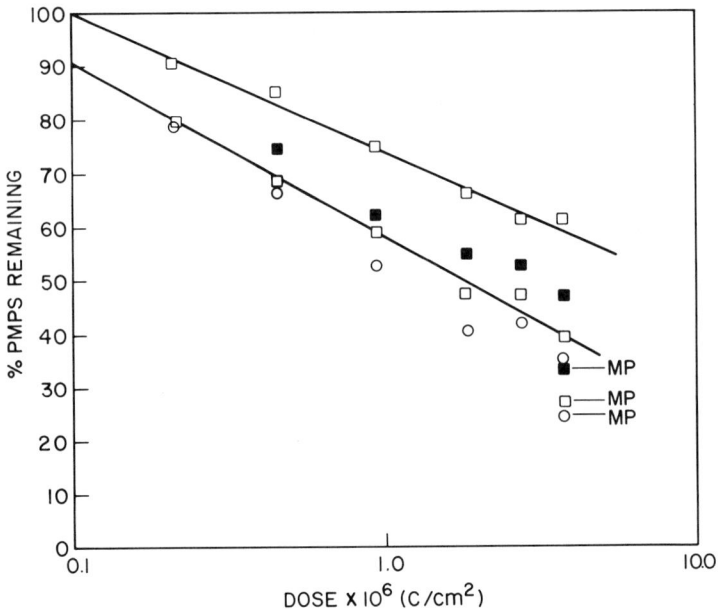

Figure 11. Vapor development curves for PMPS in 20 PMPS/80 novolac composite film (lower curve postexposure baked at 130 °C for 2 h). Key: □, 1298 cm^{-1}; ■, 1119 cm^{-1}; and ○, Dektak.

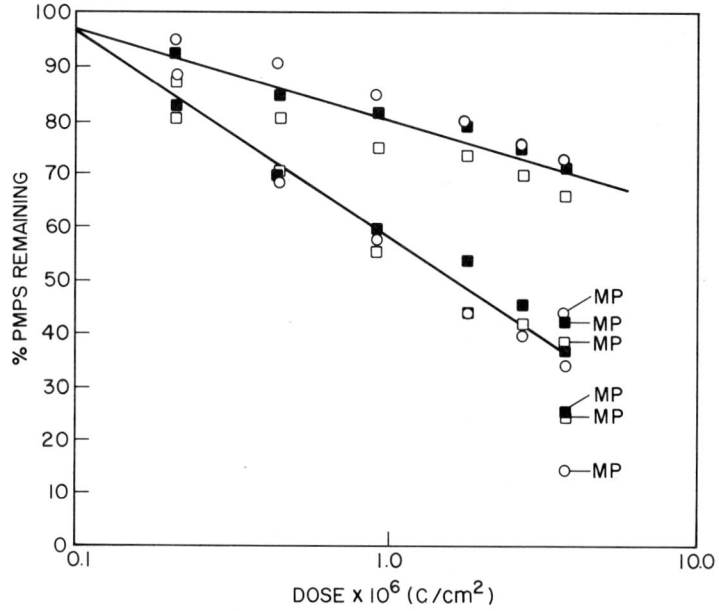

Figure 12. Vapor development curves for PMPS in 50 PMPS/50 novolac composite films (lower curve postexposure baked at 130 °C for 2 h). Key: □, 1298 cm^{-1}; ■, 1119 cm^{-1}; and ○, Dektak.

Literature Cited

[1] Bowden, M. J. J. Polymer Sci., 1974, 12, 499.
[2] Bowden, M. J.; Frackoviak, J. F. paper presented at Electronics Materials Conference, Fort Collins, Colorado June (1982).
[3] Bowden, M. J.; Thompson, L. F. Polymer Eng. and Sci., 1974, 14(7), 525,
[4] Bowden, M. J.; Thompson, L. F. Polymer Eng. and Sci., 1977, 17(4), 269.
[5] Brown, J. R.; O'Donnell, J. H. Macromolecules, 1970, 3 265.
[6] Brown, J. R.; O'Donnell J. H. Macromolecules, 1982, 5, 109.
[7] Bowmer, T. N.; O'Donnell, J. H.; Wells, P. R..nr 99 11 Makromol. Chem. Rapid Commun. 1980, 1, 1.
[8] Bowmer, T. N.; O'Donnell, J. H.; Well, P. R. Polymer Bulletin, 1980, 2, 103.
[9] Pacansky, J. paper presented at Electrochemical Society Meeting 10th Intl. Conf. on Electron and Ion Beam Sci. and Technol., Montreal May 9-14, (1982).
[10] Gutierrez, A.; Pacansky, J.; Kroeker, R. Org. Coatings and Appl. Polym. Sci. Proceedings, 1982, 46, 520.
[11] Bowden, M. J.; Frackoviak, J. ASTM Proceedings on Symposium on Silicon Processing San Jose, Jan. 19-22, (1982) in press.
[12] Bowden, M. J.; Thompson, L. F.; Fahrenholtz, S. R.; Doerries, E. M. J. Electrochem. Soc., 1981, 128(6), 1304.
[13] Bowden, M. J.; Novembre, T. Org. Coatings and Appl. Polym. Sci. Proceedings 1982, 46, 681.
[14] Bowmer, T. N.; Bowden, M. J. Org. Coatings and Appl. Polym. Sci, 1983, 48, 169.
[15] Bowmer, T. N.; O'Donnell, J. H. Polymer, 1981, 22, 71.
[16] Bowden, M. J. CRC Critical Rev. in Solid State Sci, 1979, 8, 223.

RECEIVED September 2, 1983

The Radiation Degradation of Poly(2-methyl-1-pentene sulfone)
Radiolysis Products

T. N. BOWMER and M. J. BOWDEN

Bell Laboratories, Murray Hill, NJ 07974

> Poly(2-methyl-1-pentene sulfone), PMPS, is a 1:1 alternating copolymer of 2-methyl-1-pentene with sulfur dioxide. Upon irradiation extensive main chain scission occurs via C-S bond cleavage, followed by depolymerization to monomers. These properties make it attractive for use as an electron beam resist. In this paper, the radiolysis products are reported as a function of dose. At low doses the SO_2/olefin ratio is ~2:1. However, the ratio of SO_2/hydrocarbon moieties in the initial copolymer has a value of unity. The distribution of product yields and species from PMPS is explained in terms of a general mechanism for degradation of poly(olefin sulfone)s involving free radical and cationic depropagation pathways as well as oligomerization of free olefin by cationic intermediates. A post-exposure isomerization of the product olefin is described and is attributed to rearrangement in a SO_2/olefin charge transfer complex.

Copolymerization of sulfur dioxide and olefins generally yields 1:1 alternating copolymers (1). Upon exposure to ionizing radiation, the weak C-S bond in the main chain is cleaved and rapid depropagation occurs from the fractured chain ends (2-6). Values for G(scission) of 7-10 have been reported for poly(olefin sulfone)s (4-6). G(scission) is the number of main chain scissions per 100 eV of energy absorbed per gram of material. These G(scission) values are among the highest reported for polymer radiolysis (7) and have resulted in a number of poly(olefin sulfone)s being used as electron beam resists in the manufacture of integrated circuits (8).

Bowmer and O'Donnell studied the degradation of a variety of poly(olefin sulfone)s with different olefin structures (2,3,9-11). The predominant process was C-S bond scission producing free radical and cationic fragments followed by depropagation. A general mechanism was presented which involved (1) depropagation via both radical and cationic species, (2) oligomerization of free olefin initiated by cationic species on the residual polymer and (3) isomerization of olefin via the cationic species (10,11) The extent of depropagation was correlated with the ceiling temperature of the copolymer (3). The ceiling temperature (T_c) is the temperature above which liquid phase copolymerization does not occur (1) and may be taken as an indicator of the thermodynamic stability of the polymer. Thus the lower the value of T_c, the more extensive was the depropagation.

Poly(1-butene sulfone), PBS, and poly(2-methyl-1-pentene sulfone), PMPS, are used widely as electron beam resists (8,12—14). PBS degrades according to the mechanism of Bowmer and O'Donnell (10,11). The degradation of PMPS using 1.0 MeV and 10-20 keV electrons has been studied by Bowden and Thompson (12—14) and found to undergo random chain scission followed by depropagation. However, Pacansky et al. (15,16) degraded PMPS with electron beams and from infrared studies suggested a different mechanism. They proposed that SO_2 was exclusively produced at low doses with no concomitant formation of olefin. The residual polymer was considered to be essentially pure poly(2-methyl-1-pentene) and this polyolefin underwent depolymerization after further irradiation. Generally a chain reaction is postulated to account for high yields of SO_2. However a chain reaction that produces only SO_2 and no olefin is difficult to envisage. The rapid depolymerization of poly(2-methyl-1-pentene) to monomer is unexpected, since G(total gas) for poly(isobutylene), a related polymer, is only 2.0 (7).

The volatile products from the radiolysis of PMPS are reported in this paper. The product distribution and its dose dependence are used to decide whether PMPS degrades like other poly(olefin sulfone)s or whether a novel mechanism is required to explain its degradation.

EXPERIMENTAL

PMPS was prepared at -50°C in methylene chloride using UV initiation (17). The structure of PMPS is:

$$\mathrm{\{SO_2-CH_2-\underset{\underset{CH_2-CH_2-CH_3}{|}}{\overset{\overset{CH_3}{|}}{C}}\}_n}$$

The polymer was dissolved in acetone, precipitated into methanol, and then dried in a vacuum oven at 35°C for 24 hr. The powdered polymers (20-40 mg) were placed in glass ampoules, degassed at less than 10^{-4} mm Hg for 40 hr at ambient temperature and then sealed under vacuum. The samples were irradiated in a ^{60}Co Gamma-cell at a dose rate of 0.29 Mrad/hr at ambient temperatures (25-30°C). Absorbed doses up to 10 Mrad were used.

The volatile products formed by the radiolysis were analyzed by gas chromatography (18). A sealed ampoule was placed in an injection system attached to a Hewlett Packard-5734A Gas Chromatograph (GC). The ampoule was broken in line by depression of a plunger and the products were injected into the GC. A thermal conductivity detector and a flame ionization detector were used in series to measure the product yields. Helium was used as the carrier gas. A Porapak Q column (80/100 mesh 6' length, 1/8" diam) was used with the oven temperature programmed from 80 to 190°C to optimize resolution of chromatographic peaks. The detectors were calibrated for quantitative measurement of yields with pure gases (Matheson) and reagent grade chemicals (Aldrich). Products with large yields (e.g., 2-methyl-1-pentene) were collected for ^{1}H NMR analysis. The effluent gas from the thermal conductivity detector was bubbled through a cold CCl_4 solution (-5°C) and the proton spectra of the solutions recorded on a Varian T-60A NMR spectrometer.

PMPS samples, 30-40 mg, were degassed and sealed in NMR tubes under vacuum. After irradiating up to doses of 10 Mrad, some samples were opened immediately, dissolved in CCl_3D and the ^{1}H NMR spectra recorded. Spectra were recorded for up to ~300 hours after irradiation. Other samples were kept at room temperature for 100-200 hours and then opened, dissolved and their spectra recorded. During these various processes, viz., opening the tube, adding solvent and capping the tube, the vessel was maintained at -78°C to minimize loss of the products from the tube.

A modified ampoule preparation vessel was used to study the effect of gaseous products trapped in the polymer. A side arm with glass break-seal was added through which a solvent was introduced after irradiation (1.0 Mrad) of the polymer. Chlorobenzene (~0.3 cm^3) was used as the solvent since its chromatographic retention time was significantly longer than the retention times for the radiolysis products. After dissolution, the irradiated polymer, volatile products and solvent were sealed into an ampoule and analysed as outlined above. The addition of solvent, dissolution and ampoule preparation were all performed in a closed system, i.e., the irradiated sample was never exposed to air.

Mixtures of sulfur dioxide (SO_2) and 2-methyl-1-pentene (2M1P) were prepared by (1) bubbling sulfur dioxide into a NMR tube containing 2-methyl-1-pentene at -78°C, and (2) condensing sulfur dioxide and 2-methyl-1-pentene into a NMR tube which was sealed under vacuum. Mole ratios for the 2M1P/SO_2 mixtures of 20:1 (method (1) above) and 1:1 (method (2)) were prepared and the samples irradiated in the ^{60}Co gamma-cell.

RESULTS

The major products upon irradiation were sulfur dioxide and olefin with small yields of hydrogen, C_1 to C_4 hydrocarbons and C_{10} hydrocarbons. The SO_2 and the olefin were produced by depropagation after C-S bond scission and their yields as a function of dose are shown in Figure 1. All other products have linear dependence on radiation dose. The G-values are listed in Table 1.

TABLE 1. Initial G-values for Products

PRODUCT	INITIAL G-VALUE
SO_2	3000 ± 300
OLEFIN	1500 ± 300
CH_4	0.02
C_2H_4	0.06
C_2H_6	0.005
C_3H_8	~0.1
isobutene	~0.4
dimers	~0.5

The initial G-value is the slope of the yield/dose plots at low doses. At low doses, the SO_2/olefin ratio was ~2 and this ratio decreased expotentially towards unity as the radiation dose increased. In the unirradiated PMPS the ratio of $-SO_2-$ to $-C_6H_{12}-$ moieties was unity.

Methane and propane were produced by fracture of the side chains from the hydrocarbon moiety, followed by hydrogen abstraction. Isobutene, ethylene and ethane were formed by the secondary reactions of the methyl fragment ($CH_3\cdot$) with itself and the propyl fragment ($C_3H_7\cdot$). A multiplet peak on the GC trace at high retention time was assigned to dimers of the parent olefin, i.e., branched C_{12} hydrocarbons. Impurities of acetone (0.2%) and methanol (0.005%) were detected in the PMPS from its preparation, and were retained in the glassy polymer even after 40 hr degassing at less than 10^{-4} mm Hg.

Figure 2 shows the proton NMR spectra obtained after a dose of 1.47 Mrad. Spectrum (a) in Figure 2 was recorded immediately after irradiation. This spectrum was attributed to 2-methyl-1-pentene (2M1P) plus residual PMPS. The following resonances were assigned to 2-methyl-1-pentene:

(A) 4.70 ppm, vinylic protons;

(B) 2.2-1.8 ppm, allylic methylene protons;

(C) 1.7 ppm, allylic methyl protons;

(D) 1.3-2.0 ppm, multiplet due to $-\underline{CH_2}-CH_3$ protons;

(E) 0.9 ppm, triplet due to $-CH_2-\underline{CH_3}$ protons.

Resonances at the same chemical shift as peaks (B)-(E) were present in the PMPS spectrum and were assigned to the corresponding protons. For example, the 1.7 ppm peak had contributions from the allylic methyl protons of 2M1P and from the protons on the methyl side branch in PMPS. The 3.68 ppm resonance

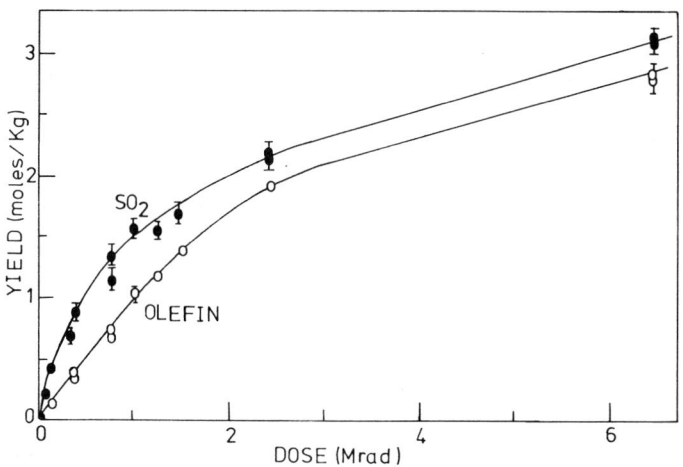

Figure 1. SO_2 (●) and olefin (○) yields in moles/Kg versus radiation dose in Mrad.

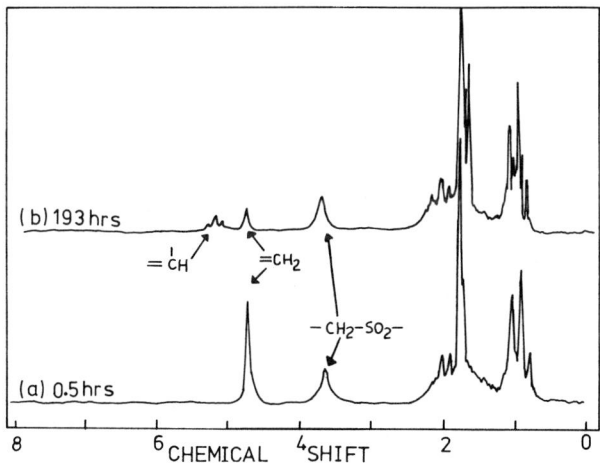

Figure 2. ^{1}H NMR spectra of PMPS irradiated to 1.47 Mrad. Spectrum recorded (a) 0.5 hrs and (b) 193 hrs after irradiation. The solvent was CCl_3D.

was assigned to the methylene protons adjacent to the $-SO_2-$ moiety in PMPS. 2-Methyl-1-pentene from depropagation of PMPS and residual PMPS were the only species present in spectrum (a) in Figure 2.

If the NMR spectrum was recorded ~190 hr after irradiation, spectrum (b) in Figure 2 was obtained. New resonances at 5.15 ppm (triplet) and 1.65 ppm were found. In addition, the 4.7 ppm peak due to the vinylic protons of 2M1P decreased in magnitude. These changes are consistent with the formation of 2-methyl-2-pentene (2M2P) which has the following resonances:

(F) triplet centered at 5.15 ppm, vinylic proton;

(G) multiplet centred at 2.1 ppm, allylic methylene protons;

(H) doublet at 1.7 and 1.65 ppm, allylic methyl protons;

(I) triplet centred at 1.0 ppm, due to $-CH_2-\underline{CH_3}$ protons.

The resonances at 3.68, 4.7 and 5.15 ppm are due to PMPS, 2M1P and 2M2P respectively.

The kinetics of the 2-methyl-2-pentene formation was quantitatively studied using these resonances. The magnitude of the residual polymer resonance (3.68 ppm) remained constant, i.e., no further depolymerization occurred. The total concentration of olefin remained constant as the 2-methyl-1-pentene was isomerized to 2-methyl-2-pentene. Figure 3 shows the kinetic data for the isomerization reaction after 1.47 Mrad. The post-exposure isomerization reaction took place both in the solid state (symbols ■, □ in Figure 3) and in solution (symbols ○, ● in Figure 3). Similar results were found for samples after a dose of ~10 Mrad. However the 10 Mrad sample had significant amounts of 2M2P present immediately after irradiation, see Figure 4.

DISCUSSION

In a previous study, Bowmer and O'Donnell (3) had found that radiation product yields were correlated to the ceiling temperature for a number of poly(olefin sulfone)s. Their results are summarized in Figure 5 which shows that the lower the T_c for a particular poly(olefin sulfone), the higher is the gas yield from irradiation, i.e., the lower the T_c the more strongly the right-hand side of the equilibrium:

$$(R-SO_2)_n \rightleftarrows nR + n\,SO_2$$

is favored. In Figure 5 the correlation reported for $G(SO_2)$ versus T_c is shown. In Figure 6 this plot is expanded to include the present results for PMPS. T_c for PMPS is -34°C and as seen in Figure 6, irradiation well above T_c (25°C) highly favors the depropagation reaction as expected.

The degradation scheme suggested by Bowmer and O'Donnell for poly(olefin sulfone)s is summarized in Figure 7. It includes (1) depropagation via free radical and cationic intermediates, (2) oligomerization of the free olefin initiated by

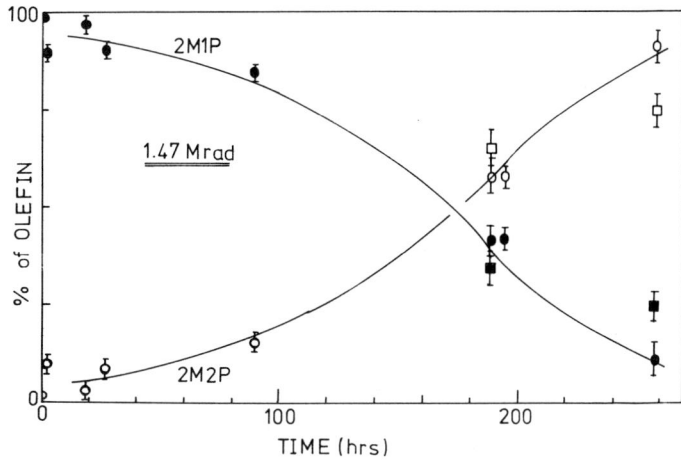

Figure 3. Kinetic curve for post-exposure isomerization of product olefin after dose of 1.47 Mrad. 2-methyl-1-pentene, ■, ●; 2-methyl-2-pentene, □, ○. Circles, tube opened immediately after irradiation. Squares, tube opened 193 hrs after irradiation.

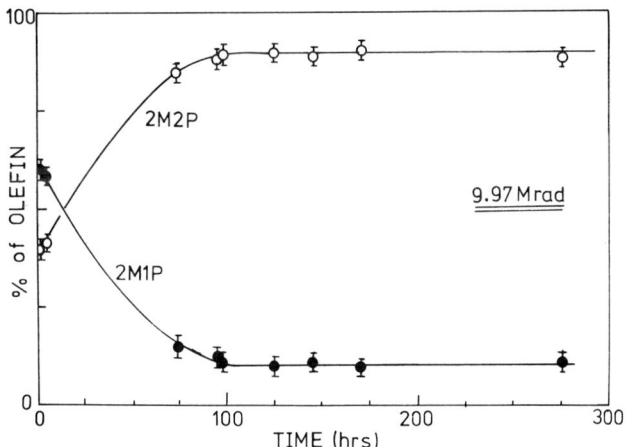

Figure 4. Kinetic curve for post exposure isomerization reaction of olefin after 9.97 Mrad dose. 2-methyl-1-pentene, ●; 2-methyl-2-pentene, ○.

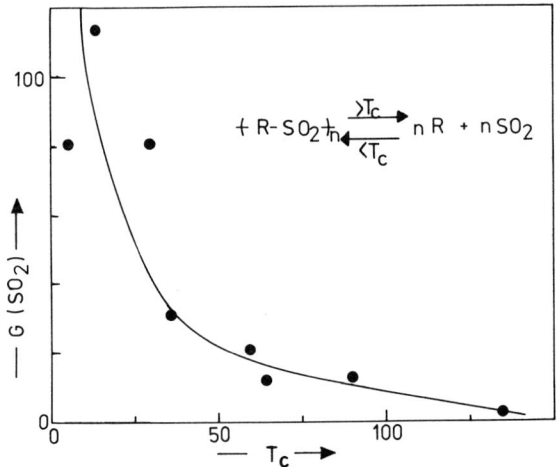

Figure 5. $G(SO_2)$ versus ceiling temperature (T_c) for a variety of poly(olefin sulfone)s. Figure 3 in reference 3.

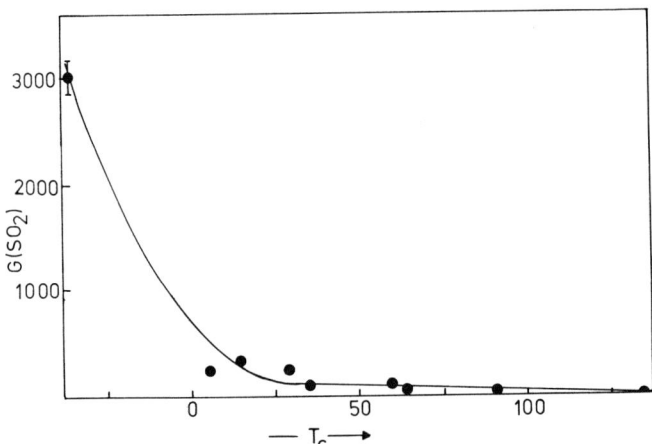

Figure 6. $G(SO_2)$ versus ceiling temperature, PMPS data from this work $(T_c = -34°C)$. Other radiolysis yields from reference 3.

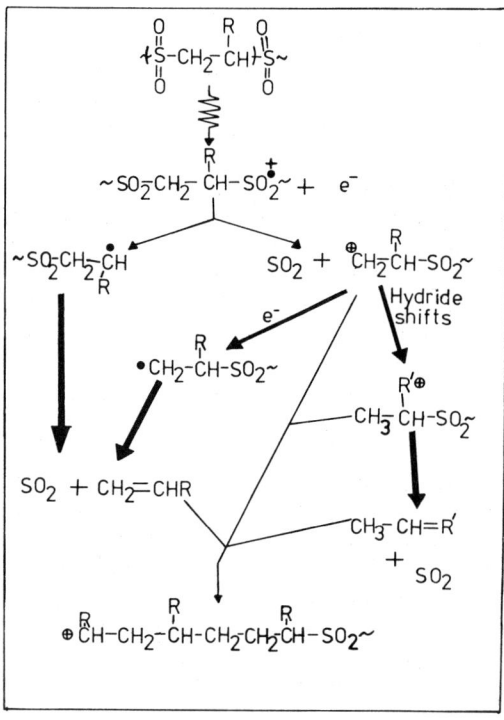

Figure 7. Reaction scheme for poly(olefin sulfone) degradation.

cationic sites and (3) isomerization of the olefin via the cationic intermediates (10,11). We attribute the discrepancy between G(SO$_2$) and G(olefin) in PMPS to the fact that oligomerization of the free olefin occurs at low doses. The oligomerization appears to be reversible as seen from the reduction of the SO$_2$/olefin ratio from two at low doses towards unity at high doses (see Figure 1). It is possible the free olefin is physically trapped in the polymer and is not chemically retained as the oligomer. Such trapped gases would be released by dissolution of the polymer. Upon dissolving PMPS in chlorobenzene prior to analysis, we found no significant change in the SO$_2$/olefin ratio. After a dose of 1.0 Mrad the ratio was found to be 1.6 ± 0.2 with dissolution and 1.4 ± 0.1 without dissolution. Therefore observation of a SO$_2$/olefin ratio greater than unity can not be attributed to trapped olefin.

Infrared studies of the radiation-induced degradation of PMPS by Bowden et al. (19) supports the oligomerization process and also shows that the oligomers can be removed by post-exposure baking. These effects have not been seen for other poly(olefin sulfone)s (2,3). Figure 8 and Figure 9 show the yield versus dose curves for irradiation of poly(1-butene sulfone) and poly(cyclohexene sulfone) respectively (20). No comparable shift of the SO$_2$/olefin ratio towards unity is observed in the radiolysis of these polymers.

In contrast to other poly(olefin sulfone)s, little isomerization of the olefin occurs during cationic depropagation. The degradation of PMPS is extremely rapid and it is possible that the lifetime of the cationic intermediates may be too short for appreciable isomerization to occur. However, isomerization of the 2-methyl-1-pentene product to 2-methyl-2-pentene did occur during the post-exposure period. The isomerization proceeded both in solution and in the solid state. It seems unlikely that the reactive intermediates (e.g. cations and free radicals) produced by irradiation catalyse the isomerization reaction since they would be expected to be destroyed during dissolution. However sulfur dioxide is well known to form charge transfer complexes (CTC) with olefins (1) and we propose that the isomerization occurs via such an intermediate:

$$2M1P + SO_2 \rightleftharpoons [olefin - SO_2]_{CTC} \rightleftharpoons 2M2P + SO_2$$

Figure 10 shows the isomerization kinetics for a ~20/1 mixture of 2-methyl-1-pentene with SO$_2$ after a dose of 0.8 Mrad. No significant isomerization occurred upon irradiation of pure 2-methyl-1-pentene, even after 40 Mrad of absorbed dose. However, after the small dose of 0.8 Mrad 2M1P (in the presence of SO$_2$) underwent isomerization to 2M2P as seen in Figure 10. The kinetics are similar to that observed following PMPS radiolysis (compare Figure 3). The equi-molar mixtures prepared under vacuum also showed accelerated isomerization after irradiation, see Figure 11. During preparation of the 1:1 mixtures upto 20% isomerization occurred prior to any irradiation. This effect was not observed for the 20/1 mixture (Figure 10) and its cause is not known at present.

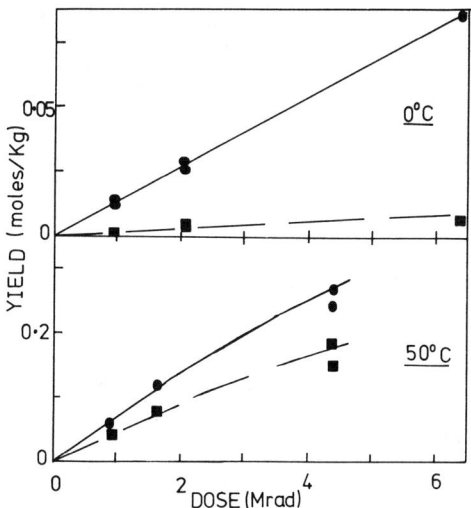

Figure 8. Yield (moles/kg) versus radiation dose (Mrad) for poly(1-butene sulfone). SO_2, ● and 1-butene, ■; for irradiations at 0° and 50°C.

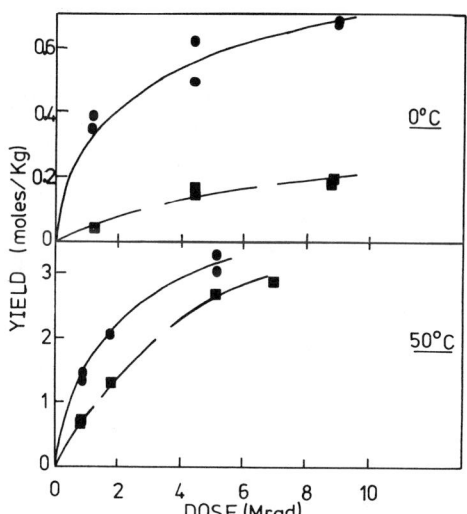

Figure 9. Yield (moles/kg) versus radiation dose (Mrad) for poly(cyclohexene sulfone). SO_2, ● and cyclohexene, ■ irradiated at 0 and 50°C.

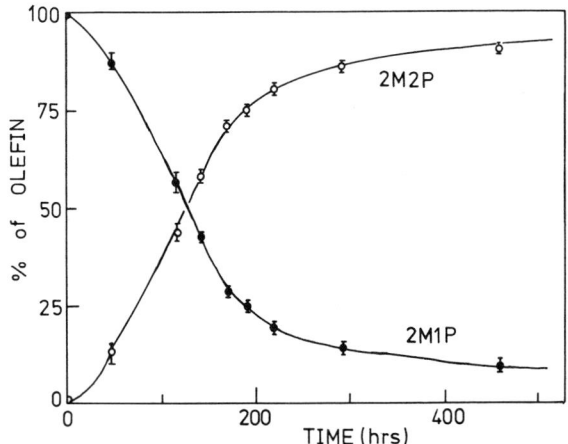

Figure 10. Kinetic curve for isomerization of 2-methyl-1-pentene/SO_2 (20/1) mixture after 0.8 Mrad. 2-methyl-1-pentene, ●; 2-methyl-2-pentene, ○.

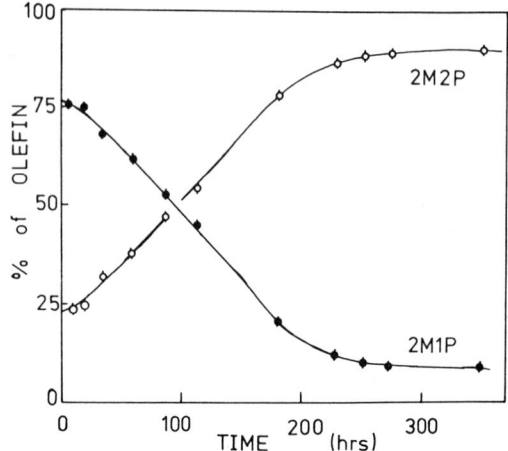

Figure 11. Kinetic curve for isomerization of equi-molar mixture of 2-methyl-1-pentene and sulfur dioxide. 2-methyl-1-pentene, ●; 2-methyl-2-pentene, ○.

CONCLUSION

PMPS has been found to depropagate rapidly upon exposure to ionizing irradiation producing SO_2 and 2-methyl-1-pentene. Oligomerization of the free olefin was observed probably occurring via cationic intermediates. This degradation scheme is consistent with that proposed by Bowmer and O'Donnell (10,11). No evidence was found for the mechanism suggested by Pacansky et al. (15,16).

A post-exposure isomerization reaction was observed in which 2-methyl-1-pentene isomerized to 2-methyl-2-pentene. We propose that this reaction proceeds via SO_2-olefin charge transfer complexes.

ACKNOWLEDGMENTS

The author wishes to thank A. Novembre for preparation of the polymer and L. Shepherd for discussions of the degradation mechanism.

REFERENCES

1. Ivin, K. J. and Rose, J. B. "Advances in Macromolecular Chemistry" (Editor, W. M. Pasika) Vol. 1., p. 335 (1968).
2. Bowmer, T. N. and O'Donnell; J. H.; Jour. Polym. Sci., Polym. Chem. Ed., 19, 45 (1981).
3. Bowmer, T. N. and O'Donnell, J. H.; Jour. Macromol. Sci.-Chem., A17, 243 (1982).
4. Stillwagon, L. ACS Symposium Series No. 184, "Polymer Materials for Electronic Applications" (Editors, E. D. Feit and C. W. Wilkins, Jr.) p. 19 (1982).
5. Brown, J. R. and O'Donnell, J. H.; Macromolecules, 3, 265 (1970).
6. Brown, J. R. and O'Donnell, J. H.; Macromolecules, 5, 109 (1972).
7. Dole, M. "The Radiation Chemistry of Macromolecules", Academic Press (1973).
8. Bowden, M. J. CRC Critical Reviews in Solid-State Sciences, 8, 223 (1979).
9. Bowmer, T. N., O'Donnell, J. H. and Wells, P. R. Makromol. Chem., Rapid Communication, 1, 1 (1980).
10. Bowmer, T. N., O'Donnell and Wells, P. R.; Polymer Bulletin 2, 103 (1980).
11. Bowmer, T. N. and O'Donnell, J. H., Polymer, 22, 71, (1981).
12. Bowden, M. J. Polym. Sci., 12, 499 (1974).
13. Bowden, M. J. and Thompson, L. F., Polym. Eng. and Sci., 14 (7), 525 (1974).

14. Bowden, M. J. and Thompson, L. F., Polym. Eng. and Sci., 17(4), 269 (1977).
15. Gutierrez, A., Pacansky, J. and Kroeker, R. Org. Coatings and Appl. Polym. Sci. Proceedings, *46*, 520 (1982).
16. Pacansky, J. paper presented at Electrochemical Society Meeting 10th Intl. Conf. on Electron and Ion Beam Sci. and Technol., Montreal, May 9-14, (1982).
17. Bowden, M. J. and Novembre, T., Org. Coatings and Appl. Polym. Sci. Proceedings, *46* 681 (1982).
18. Bowmer, T. N. and O'Donnell, J. H., Polymer, *18*, 1032 (1977).
19. Bowden, M. J., Allara, D. L., Vroom, W. I., Frackoviak, J., Kelley, L. C. and Falcone D. R.; preceding paper.
20. Bowmer, T. N., Ph.D. Thesis, University of Queensland, Australia (1980).

RECEIVED September 14, 1983

14

Novolac Based Positive Electron Beam Resist Containing a Polymeric Dissolution Inhibitor
Preparation and Exposure Characteristics

HIROSHI SHIRAISHI, ASAO ISOBE[I], FUMIO MURAI, and SABURO NONOGAKI

Central Research Laboratory, Hitachi Ltd., Kokubunji, Tokyo 185, Japan

A novolac-based positive electron beam resist has been investigated for use in direct device fabrication. The resist is a composite system consisting of an alkaline soluble novolac resin, and poly(2-methylpentene-1 sulfone)(PMPS) which serves as a dissolution inhibitor. Upon exposure to an electron beam, PMPS decomposes and loses its dissolution inhibiting ability. One difficulty with the use of PMPS as a dissolution inhibitor is that film uniformity is not always sufficient, because of phase separation in the spin coating. It was however found that most of novolac or phenolic resins containing PMPS can form homogeneous films when isoamyl acetate is used as a coating solvent. A cresol novolac resin has been synthesized which exhibits a much greater dissolution-inhibiting effect than in various commercially available novolac or phenolic resins. Using this resin, a positive electron beam resist was prepared and its exposure characteristics were examined. A tetramethylammonium hydroxide aqueous solution was used as the developer. The sensitivity reaches 3×10^{-6} C/cm^2 without post-exposure baking. It was found that the sensitivity to double exposure was much higher than that to single exposure with the same total dose. A similar phenomenon was also found for PMPS film. The mechanisms for these phenomena are discussed.

Electron beam lithography for direct device fabrication requires high performance electron beam resists having high sensitivity, resolution and especially dry etching resistance. Various kinds of negative electron beam resists showing excellent dry etching resistance have been developed for the purpose of direct device fabrication (1) (2) (3).

[I]Current address: Yamazaki Works, Hitachi Chemical Co., Ltd., Hitachi, Ibaraki 317, Japan

0097-6156/84/0242-0167$06.00/0
© 1984 American Chemical Society

However conventional positive electron beam resists like PMMA(4) or PBS(5) do not have excellent dry etching resistance. The electron beam sensitivities of these positive resists primarily result from radiation-induced degradation of polymer main chains. If the main chain bonding force of these polymers is weakened in order to improve sensitivities, the dry etching resistances of these polymers will decrease. In such cases, sensitivity to electron beam exposure and dry etching resistance are in a trade-off relationship.

Rai and Shepherd(6) have showed that polystyrene has good sputtering resistance. It is well known that the incorporation of phenyl groups into resist polymers or the use of aromatic polymers such as polystyrene or novolacs enhances the dry etching resistance of the polymer.

Conventional photoresists which consist of alkali-soluble novolac resins and photoactive dissolution inhibitors(7) have excellent dry etching resistance. Hatzakis and his colleagues(8) have investigated such positive photoresists as positive electron beam resists. However, sensitivity to electron beam exposure was not so good(2×10^{-5} C/cm^2 at 20 kV).

Bowden and his coworkers(9) proposed a new type of positive electron beam resist which consists of an alkali-soluble novolac and polymeric dissolution inhibitor. The positive working mechanism of this new type positive resist(NPR) is similar to that for the conventional positive photoresist(10). It was also found that poly(2-methylpentene-1 sulfone)(PMPS) is good as a polymeric dissolution inhibitor for NPR(10). In addition, it was clarified that one of the difficulties with NPR is phase separation in the resist films(10)(11).

In this paper we report on the synthesis of novolac resins suitable to NPR. A look is also taken at the anomalous exposure characteristics of NPR.

Experimental

PMPS Preparation: PMPS was prepared using the same procedure described by Bowden and his coworkers(12). 2-Methylpentene-1 was refluxed over lithium aluminum hydoride for two hours, and then distilled. 50ml of distilled 2-methylpentene-1 was transferred into a 300ml reaction vessel attached to a vacuum line containing a magnetic stirrer tip, and then degassed. The vessel was cooled to -90 °C and then charged with 100ml of liquid sulfur dioxide dried through a phosphor pentoxide column. 3g of t-butyl hydroperoxide was transferred into the vessel by means of vacuum distillation. The mixture was warmed slowly to -50°C through magnetic stirring, and kept at that temperature for 5 hours. After evacuation of excess sulfur dioxide, the contents were dissolved in 1000ml of acetone and the polymer was then precipitated into methanol. The polymer was finally dried for 48 hours at 30 °C.

Novolac Preparation: Novolac resins were prepared by an acid catalyzed condensation of m- and p-cresols with formaldehyde. A three necked flask with a distillation reflux condenser, thermometer and mechanical stirrer was charged with m- and p-cresols, formaldehyde aqueous solution, and conc. hydrochloric acid as a catalyst. The flask was immersed in an oil bath and heated to 90 °C and kept for 2 hours while undergoing stirring. After the flask was cooled to room temperature on standing, the supernatant layer of the contents was removed by decantation. Then the volatile components were eliminated by distillation under a nitrogen gas flow and slow heating to 175°C. The molten content was poured into a stainless steel tray to cool.

Characteristics: The structure of the obtained PMPS was confirmed by measurements of infrared spectra, and nuclear magnetic resonance spectra using a Digilab FTS-20C/D Fourier transform infrared spectrometer, and a Hitachi R-24 NMR spectrometer, respectively. The molecular weight and molecular weight distribution of the novolac or phenolic resins were determined by gel permeation chromatography using a Hitachi 635 liquid chromatography system.

Dissolution Rate Measurements: Sample films were spun onto silicon wafers from isoamyl acetate solutions of sample resins or sample resin and PMPS. The films were immersed in an aqueous solution of tetramethylammonium hydroxide. Dissolution rates were obtained from a plot of the measured residual film thickness against the immersion time. The film thickness was measured with an interferometer.

Resist Preparation: Resist solutions were prepared by dissolving novolac resins and PMPS in isoamyl acetate. The solutions were then filtered through an 0.2 um Teflon filter.

Sensitivity Measurements: Resist films were spun onto silicon wafers. The films were then prebaked and exposed in a modified Hitachi electron microscope to an undeflected, nearly collimated electron beam at an acceleration voltage of 15kV. After exposure the resist films were developed in an aqueous solution of tetramethylammonium hydroxide. Sensitivity curves were obtained by plotting film thickness against incident dose.

Patterning of the Resist: Resists were delineated with a vector scanning type, shaped electron-beam drafting machine, specially designed by the Hitachi Central Research Laboratory. The acceleration voltage was 30kV.

Results and Discussion

Coating solvents were investigated for composite systems of PMPS and various commercially available phenolic or novolac resins. It was found that most of novolac or phenolic resins containing PMPS can form homogeneous films when isoamyl acetate is used as a coating solvent.

PMPS has been shown elsewhere to be vapor developable, which means that the polymer is vaporized and removed during electron beam exposure(13). The vapor development characteristics of PMPS are shown here in Figure 1. PMPS was not completely removed by a dose of 2×10^{-5} C/cm^2 at 15kV. Therefore, a resin suitable to NPR should show a large solubility increase with a decrease in its concentration in NPR.

To select a resin suitable for NPR, we defined parameter R_{10} as

$$R_{10} = T_0/T_{10}$$

where T_0 and T_{10} are respectively the dissolution rates of the resin without and with 10 wt.% of PMPS to the resin. A large R_{10} value indicates the large dissolution inhibiting effect of PMPS. We measured R_{10} values for various commercially available phenolic and novolac resins, together with those for synthesized novolac resins. The results are summarized in Table I.

Table I. Investigated resins and their R_{10} values.

Resin	Origin	R_{10}
Resin M	Maruzen Oil Co., Ltd.	3.6
Resin MB	Maruzen Oil Co., Ltd.	18.0
Alnovol PN430	Hoechst Japan Ltd.	11.1
Resitop PSF2803	Gunei Chemical Industry Co., Ltd.	9.5
Resitop PSF2807	Gunei Chemical Industry Co., Ltd.	21.5
Resitop XPS4800B	Gunei Chemical Industry Co., Ltd.	3.4
Penol Novolak	Shinko Tech. Research Co., Ltd.	9.6
m-High Cresol Novolak	Shinko Tech. Research Co., Ltd.	14.5
m-Cresol Novolak	Shinko Tech. Research Co., Ltd.	16.8
Hitanol HP-607N	Hitachi Chemical Co., Ltd.	6.7
Sample Resin-1	This work	28.3
Sample Resin-3	This work	>45

As can be seen in Table I. one synthesized novolac resin (Sample Resin-3) showed a very large R_{10} value compared with other resins. It was found that R_{10} is strongly dependent on the ratio of m- to p-cresol concentrations in the resins. Dissolution rates are shown as a function of PMPS content for this resin in Figure 2.

Exposure characteristics for NPR with use of the cresol novolac are shown in Figure 3. This NPR contains 12% PMPS agaist the resin weight. Bowden and his coworkers showed that post-exposure baking increases the sensitivity of NPR(10). However, with the present NPR, the post-exposure baking does not remarkably affect sensitivity. An example of the fine pattern for this NPR as obtained by electron beam delineation is shown in Figure 4.

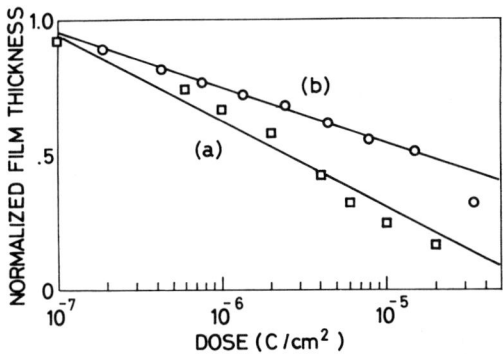

Figure 1. Exposure characteristics of PMPS prepared by vapor development: (a) at 15kV; (b) at 30kV.

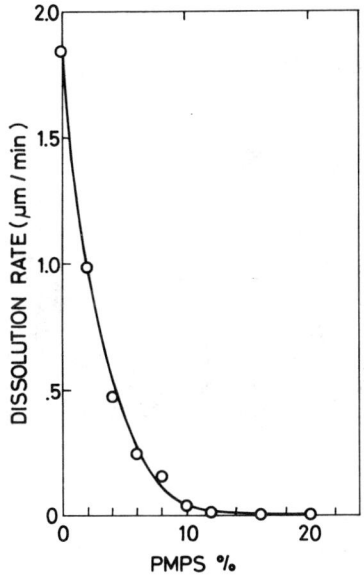

Figure 2. Dissolution rate of resin film, shown as function of PMPS content.

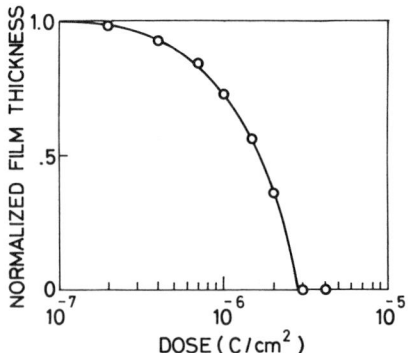

Figure 3. Exposure characteristics of NPR. Acceleration voltage: 15kV.

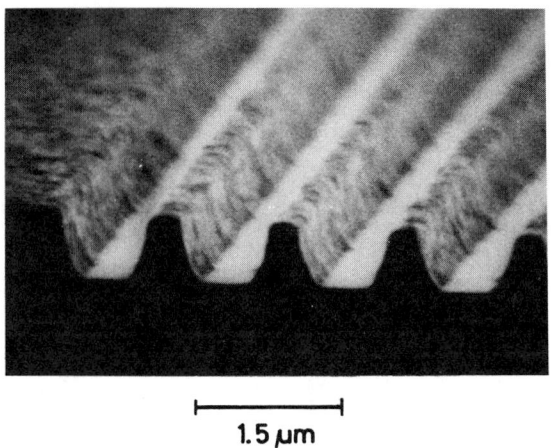

Figure 4. SEM photomicrograph of NPR pattern. Acceleration voltage: 30kV; Dose: 8×10^{-6} C/cm^2.

It was also found that the sensitivity to double exposure is much higher than that to single exposure with the same total dose. This double exposure effect is shown in Figure 5. The degree of sensitivity enhancement is represented by the equivalent single dose with which the exposed area would show the same solubility as for the doubly exposed area. Sensitivity enhancement depends on the time interval between the first and second exposure, and also the dose fraction ratio, as can be seen in Figure 5. The sensitivity enhancement peak is reached at a first dose fraction of 0.25.

A similar double exposure effect was found for PMPS film, as Figure 6 shows. In this figure, the enhancement is expressed by the equivalent single dose with which the exposed film thickness would become the same as for a doubly exposed area. In this case, sensitivity enhancement does not depend on the time interval between the first and second exposure, but the sensitivity enhancement rapidly reaches a maximum in the region of the low first dose fraction, whereby it slowly falls.

It was found that both sensitivity and double exposure effect are reduced with mixing of a small amount of radical scavenger, 1,1-diphenyl-2-picrylhydrazyl (DPPH), into the PMPS. This indicates that radical reactions are involved in the mechanism of double exposure sensitivity enhancement.

The double exposure effect can be explained on the basis of a few assumptions. We have assumed that pulsed electron beam irradiations produce PMPS radicals in PMPS or NPR films. We also assumed that the initial amount of PMPS radicals is proportional to the exposure dose, and that the concentration, C, decreases with both first-order and second-order reactions. The first-order reaction may be the end of zipping or stabilization by chain transfer, and the second-order reaction may be a radical-radical recombination reaction. Equation 1 can be obtained such that

$$-\frac{dC}{dt} = k_1 C + k_2 C^2 \qquad (1)$$

where t is the time after pulsed exposure, and k_1 and k_2 are constants.

If we assume that a decrease in PMPS results from the zipping reaction, the total amount of decomposed PMPS, M, can be expressed as

$$M = \int_{t=0}^{\infty} kC\, dt \qquad (2)$$

where k is a constant. This integration can then be calculated using Equation 1 as

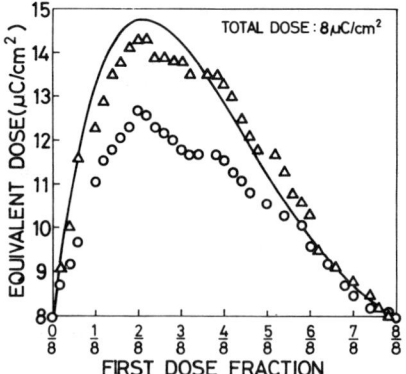

Figure 5. Double exposure effect in NPR. Exposure time interval: 30s (circle); 26min (triangle). Solid curve represents simulation.

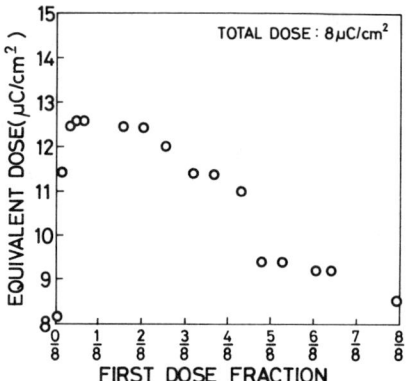

Figure 6. Double exposure effect in PMPS. Exposure time interval: 30s.

$$M = \int_{C=C_0}^{0} kC \frac{dt}{dC} dC$$

$$= \int_{C=C_0}^{0} kC \left(-\frac{1}{k_1 C + k_2 C^2}\right) dC$$

$$= k \int_{0}^{C_0} \frac{dC}{k_1 + k_2 C}$$

and finally, we obtain

$$M = \frac{k}{k_2} \log\left(1 + \frac{k_2}{k_1} C_0\right) \qquad (3)$$

where C_0 is the initial PMPS radical concentration.

Since we have assumed that the initial radical concentration, C_0, is proportional to the dose, Equation 3 indicates that the total amount of decomposed PMPS, M, is almost linearly dependent on log(C_0) and therefore log(Dose), provided (k_2/k_1)C_0 is much larger than unity. This relationship was also obtained experimentally, as Figure 1 shows.

The doubled exposure effect follows directly from the assumption that PMPS decreases in accordance with Equation 3 in every exposure. Furthermore, asymmetry in the enhanced sensitivity with respect to the dose fraction ratio can be explained by assuming that the efficiency of PMPS radical production in the second exposure is lower than in the first exposure. The solid curve in Figure 5 illustrates an approximation resulting from the assumption that (k_2/k_1)C_0 = 10, the initial concentration of PMPS radicals at first exposure = pC_0 ($0 \leq p \leq 1$), and that at second exposure = $(1-p)C_0\exp(-2p)$. The curve is in good agreement with the observed data.

Acknowledgments

The authors are grateful to Kozo Mochiji and Kiyotake Naraoka for their assistance in the electron beam fabrication experiments. They also thank Takumi Ueno for useful discussions regarding the sensitivity enhancement mechanism.

Literature Cited

1. S. Imamura, J. Electrochem. Soc. 1979, 126, 1628.
2. H. Shiraishi, Y. Taniguchi, S. Horigome, and S. Nonogaki, Polymer Eng. Sci. 1980, 20, 1054.

3. E. D. Feit and L. E. Stillwagon, Polymer Eng. Sci. 1980, 20, 1058.
4. I. Haller, M. Hatzakis, and R. Srinivasan, IBM J. Research and Development 1968, 12, 250.
5. M. J. Bowden, L. F. Thompson, and J. P. Ballantyne, J. Vacuum Sci. Tech. 1975, 12, 1294.
6. J. H. Rai and L. T. Sheperd, Preprints ACS Organic Coatings and Plastics Chemistry 1975, 35, 252.
7. J. Pacansky, Polymer Eng. Sci. 1980, 20, 1049.
8. M. Hatzakis and J. M. Shaw, Proceedings 8th Int. Conf. on Electron and Ion Beam Sci. and Tech., Electrochem. Sci. 1978, 78-5, 285.
9. L. F. Thompson, M. J. Bowden, E. M. Doerries, and S. R. Fahrenholtz, paper presented at Electrochem. Soc. Meeting, Seattle, Wash. 1978, May 21-26.
10. M. J. Bowden, L. F. Thompson, S. R. Fahrenholtz, and E. M. Doerries, J. Electrochem. Soc. 1981, 128, 1304.
11. M. J. Bowden, J. Applid Polymer Sci. 1981, 26, 1421.
12. M. J. Bowden and L. F. Thompson, J. Applied Polymer Sci. 1973, 17, 3211.
13. L. F. Thompson and M. J. Bowden, J. Electrochem. Soc. 1973, 120, 1722.

RECEIVED September 2, 1983

15

Molecular Design for Cross-linking Negative Resists
Relationship Between Sensitivity and Component Ratio in Copolymer Resists

KATSUMI TANIGAKI, YOSHITAKE OHNISHI, and SHOZO FUJIWARA

Fundamental Research Laboratories, NEC Corporation, Miyamae-Ku, Kawasaki-City 213, Japan

> The relationship between sensitivity and component ratio in copolymer negative resists was studied theoretically on the basis of Charlesby's gel formation theory. The formulas for sensitivity as a function of component weight ratio are derived for a nonchain reaction and for a chain reaction, respectively. Copolymer sensitivities for any component ratio can be estimated numerically using the derived formulas, from the data on sensitivities for homopolymers composed of individual constituent monomers in the copolymer. The calculated results are in good agreement with the experimental data reported. The derived formulas make it possible to predict copolymer sensitivities and are useful in designing copolymer negative resists.

Several characteristics, such as high sensitivity, high resolution and high dry-etching durability, are required for electron beam resists to realize integration to a great extent of microelectronic devices. Because homopolymer resists could not often satisfy all these requirements at the same time, various kinds of copolymer resists have been developed to improve the homopolymer resist properties. For example, though polystyrene is highly resistant to dry-etching, its sensitivity is low. Thus, glycidyl methacrylate or chloromethylstyrene is copolymerized with styrene to increase sensitivity in a negativie resist system(1,2).

In the course of the research on copolymer negative resists, numerous experimental data on sensitivities have been reported as a function of component ratio in order to optimize each resist system. Summarizing these data, it is found that the relationship between sensitivity and component ratio can be divided into two groups. In one group, sensitivity increases steeply at first as the mole fraction of highly sensitive monomer in the copolymer increases, and then it tends to saturate. In the other group, sensitivity saturation does not occur. It has been presumed that these different dependences on component ratio result from the difference in crosslinking reaction scheme, i.e., a nonchain reaction or a chain reaction.

In this paper, the relationship between sensitivity and component ratio was studied theoretically with the intention to confirm the above stated presumption. The formulas for sensitivity as a function of component ratio are derived in a nonchain reaction case and in a chain reaction case, respectively, on the basis of Charlesby's gel formation theory. Copolymer sensitivities for any component ratio can be estimated numerically using the derived formulas, from the data on sensitivities for homopolymers composed of individual constituent monomers in the copolymer. Discussion will be given on the validity of the derived formulas.

Theoretical basis

According to Charlesby's gel formation theory, one crosslinking unit per weight-average molecule is required to gel a polymer(3). This means that sensitivity for a crosslinking negative resist is determined by the weight-average molecular weight and the relationship between the radiation dose and the number of crosslinking units formed.

In a negative resist composed of several kinds of constituent monomers, the total number of crosslinking units n_T per gram can be calculated by summing up the number of crosslinking units on an individual monomer.

$$n_T = (x_I/m_I)q_I N_0 = (x_M/m_M)q_M N_0 + (x_N/m_N)q_N N_0 + \cdots \quad (1)$$

where $I(I=M,N,\ldots)$ is the constituent monomer, q_I denotes the probability of a given monomer I being crosslinked, x_I is the weight ratio for I in a polymer, m_I is the molecular weight of a monomer I, and N_0 is Avogadro's number. Introducing n_I, Equation 1 is transformed as

$$n_T = x_M n_M + x_N n_N + \cdots \quad (2)$$

where n_I is defined by

$$n_I = (1/m_I)q_I N_0 \quad (3)$$

n_I denotes the number of crosslinking units formed on I per gram.

The following assumptions are made in order to estimate n_I as a function of the radiation dose. They are:

(I)   The number of active bonds induced by irradiation is proportional to the electron dose.
(II) All radiation-induced active bonds lose their activity only through a crosslinking reaction.
(III) An intramolecular crosslinking reaction (cyclization) is neglected.

Based on assumption (I), the number of induced active bonds on I per gram is expressed as

$$k_I = K_I D \quad (4)$$

where K_I denotes the number of induced active bonds on I per gram, D is the electron dose and K_I denotes the active bond production efficiency per gram and per unit dose for I.

Based on assumptions (II) and (III), the relationship between n_I and k_I can be obtained. It varies with whether induced active bonds react through a nonchain reaction or through a chain reaction.

In the case of a nonchain reaction, an active bond reacts once, through the active bond coupling reaction or the abstraction reaction. Figure 1 illustrates an example of the two kinds of reaction paths, which possibly occur in a nonchain reaction. Therefore, one active bond makes one crosslinking unit through the active bond coupling reaction and two crosslinking units through the abstraction reaction, respectively.

$$n_I = k_I \quad \text{; for the active bond coupling reaction}$$
$$n_I = 2k_I \quad \text{; for the abstraction reaction} \tag{5}$$

On the other hand, in the case of a chain reaction, an induced active bond continues to react until it loses its activity through the active bond coupling reaction($\underline{4}$). In this case, Charlesby showed that an active bond makes j/(1-j) crosslinking units, where j is termed the chain reaction activity (i=1-j, where i is termed the inhibition activity by Charlesby) and it depends on the individual monomer concentration($\underline{5}$). Thus, the relationship is expressed as

$$n_I = [\ j_I(x_I)/(\ 1-j_I(x_I)\ )\ ]\ k_I \tag{6}$$

Using Equations 4-6, n_I can be written as a function of the radiation dose both for a nonchain reaction and for a chain reaction. Accordingly, sensitivity can be expressed as a function of component weight ratio using Equation 2.

<u>Application to a copolymer resist of monomer M and monomer N</u>

Let us consider a copolymer of monomer M and monomer N. From Equation 2, the total number of crosslinking units per gram is

$$n_T = x_M n_M + x_N n_N$$

Three cases must be considered.

(i) Both kinds of monomers are nonchain reaction monomers.
(ii) One kind of monomer is a chain reaction monomer and the other is a nonchain reaction monomer.
(iii) Both kinds of monomers are chain reaction monomers.

Equations 5 and/or 6 are applied to each case.

(i) When both kinds of monomers are nonchain reaction monomers, n_T is expressed as

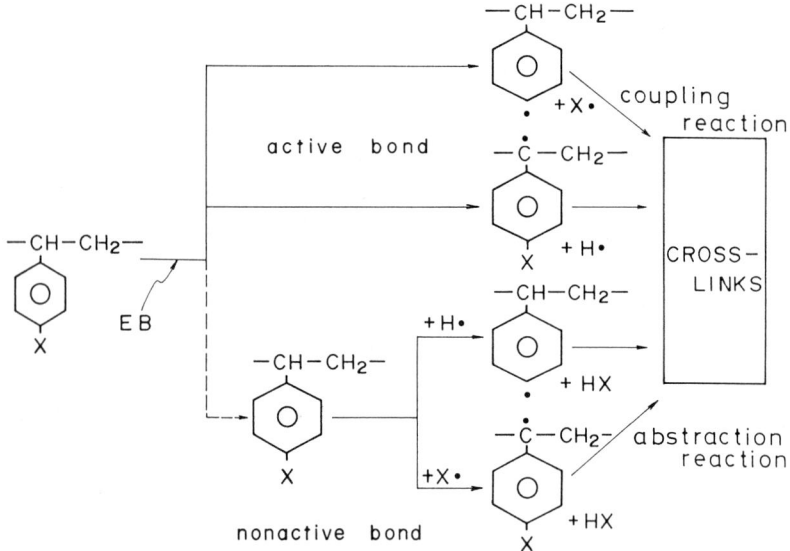

Figure 1. Reaction paths for ring-substituted PS in a non-chain reaction.

$$n_T = h_M(x_M)x_M k_M + h_N(x_N)x_N k_N$$
$$+ 2(1-h_M(x_M))x_M k_M + 2(1-h_N(x_N))x_N k_N \quad (7)$$

where $h_M(x_M)$ and $h_N(x_N)$ denote the active bond coupling reaction ratio for the induced active bond on M and N, respectively, and they depend on the individual monomer concentration. Thus, the first two terms represent the number of crosslinking units formed through the active bond coupling reaction, and the last two terms represent the number of crosslinking units formed through the abstraction reaction. Substituting Equation 4 into Equation 7, the following equation is obtained.

$$n_T = (2-h_M(x_M))K_M D x_M + (2-h_N(x_N))K_N D x_N \quad (8)$$

According to Charlesby's gel formation theory, $N_0/M_w(M-N)$ crosslinking units per gram are needed at a gel point, where $M_w(M-N)$ is the weight-average molecular weight of a polymer and N_0 is Avogadro's number (3). Therefore, copolymer sensitivity $D_g^i(M-N)M_w(M-N)$ (this product was introduced by Ku and Scala to characterize sensitivity for negative resists (6)) is expressed as a function of component weight ratio.

$$1/D_g^i(M-N)M_w(M-N) = K_M(2-h_M(x_M))x_M/N_0 + K_N(2-h_N(x_N))x_N/N_0 \quad (9)$$

where $D_g^i(M-N)$ is the copolymer gel dose. Constants K_M and K_N can be estimated from sensitivities for homopolymers ($D_g^i(M)$ and $D_g^i(N)$) composed of individual constituent monomers in the copolymer. Then,

$$K_M = N_0 / (2-h_M(1))D_g^i(M)M_w(M) \quad (x_M=1) \quad (10)$$

$$K_N = N_0 / (2-h_N(1))D_g^i(N)M_w(N) \quad (x_N=1)$$

Substituting K_M and K_N into Equation 9, sensitivity is expressed as

$$1/D_g^i(M-N)M_w(M-N) = (2-h_M(x_M))x_M / (2-h_M(1))D_g^i(M)M_w(M)$$
$$+ (2-h_N(x_N))x_N / (2-h_N(1))D_g^i(N)M_w(N) \quad (11)$$

For simplicity, the individual monomer concentration dependences of $h_M(x_M)$ and $h_N(x_N)$ are neglected, i.e., $h_M(x_M)$ and $h_N(x_N)$ are approximated to be constant. Then, Equation 11 can be reduced to

$$1/D_g^i(M-N)M_w(M-N) = x_M / D_g^i(M)M_w(M) + x_N/D_g^i(N)M_w(N) \quad (12)$$

(ii) When one kind of monomer is a chain reaction monomer and the other is a nonchain reaction monomer, the total number of crosslinking units is given as

$$n_T = (2-h_M(x_M))K_M D x_M + [j_N(x_N)/(1-j_N(x_N))]K_N D x_N \quad (13)$$

Thus, the formula for sensitivity can be written as follows:

$$1/D_g^i(M-N)M_w(M-N) = (2-h_M(x_M))x_M / (2-h_M(1)) D_g^i(M)M_w(M)$$
$$+ (1-j_N(1))j_N(x_N)x_N / j_N(1)(1-h_N(x_N))D_g^i(N)M_w(N) \quad (14)$$

(iii) When both kinds of monomers are chain reaction monomers, the formula for sensitivity as a function of component weight ratio becomes

$$1/D_g^i(M-N)M_w(M-N) = (1-j_M(1))j_M(x_M)x_M / j_M(1)(1-j_M(x_M))D_g^i(M)M_w(M)$$
$$+ (1-j_N(1))j_N(x_N)x_N / j_N(1)(1-j_N(x_N))D_g^i(N)M_w(N) \quad (15)$$

Atoda et al. showed experimentally that the chain reaction activity depends on the chain reaction monomer concentration(7). Thus, it is approximated that the chain reaction activities are proportional to the individual monomer concentration, i.e., $j_M(x_M) = \xi_M x_M$ and $j_N(x_N) = \xi_N x_N$, where ξ_M and ξ_N are the proportionality constants. Therefore, Equations 14 and 15 can be reduced to

$$1/D_g^i(M-N)M_w(M-N) = x_M/D_g^i(M)M_w(M) + (1-\xi_N)x_N^2/(1-\xi_N x_N)D_g^i(N)M_w(N) \quad (16)$$

and

$$1/D_g^i(M-N)M_w(M-N) = (1-\xi_M)x_M^2 / (1-\xi_M x_M)D_g^i(M)M_w(M)$$
$$+ (1-\xi_N)x_N^2 / (1-\xi_N x_N)D_g^i(N)M_w(N) \quad (17)$$

As a result, sensitivities for copolymers composed of M and N can be estimated for any component ratio using Equation 12, 16 or 17, from the experimental data on the constituent homopolymer sensitivities $D_g^i(M)$ and $D_g^i(N)$. However, in the case of (ii) and (iii), in addition to $D_g^i(M)$ and $D_g^i(N)$, ξ_M and ξ_N must be determined from one or two copolymer gel dose values in a P(M-N) system.

Results and discussion

Reported experimental data on sensitivities for various copolymer resists are summarized in Figure 2 (sensitivity is expressed by $D_g^i(M-N)M_w(M-N)$). This figure shows that sensitivity vs. component ratio curves are divided into two groups.

One group consists of chloromethylated polystyrene (P(ST-CMS))(8) chlorinated polystyrene (P(ST-CS)) (9) iodinated polystyrene (P(ST-IS) (10) etc.. For copolymer resists in this group, their sensitivities increase steeply as the mole fraction of highly sensitive monomers N increases, and then tend to saturate. In this case, constituent monomers are considered to be nonchain reaction monomers. Thus, sensitivities can be calculated from Equation 12.

In calculation, the following experimental data are used:

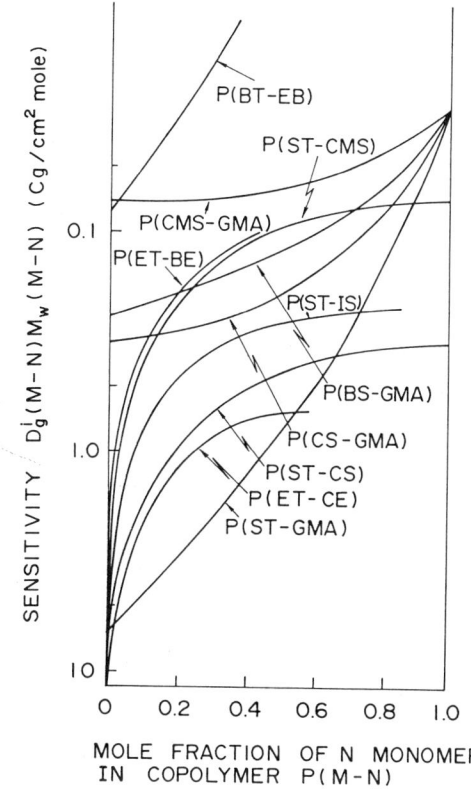

Figure 2. Experimental data on copolymer resist sensitivities in relation to constituent monomer mole fraction. The following references and abbreviations are used: P(ST-CMS), chloromethylated polystyrene (8); P(ST-CS), chlorinated polystyrene (9); P(ST-IS), iodinated polystyrene (10); P(ET-CE), chlorinated polyethylene (16)[a]; P(ET-BE), brominated polyethylene (16)[a]; P(ST-GMA), poly(styrene glycidyl methacrylate) (1); P(CS-GMA), poly(chlorostyrene glycidyl methacrylate) (2,14,15); P(BS-GMA), poly(bromostyrene glycidyl methacrylate) (14); P(CMS-GMA), poly(chloromethylstyrene glycidyl methacrylate) (2); and P(BT-EB), epoxidized polybutadiene (17)[b].
Note: In P(M-N), which denotes a copolymer of monomer M and monomer N, sensitivity for N is higher than that for M.
[a] $M_n = 34,460$; M_w is not reported
[b] 10kV acceleration voltage

$D_g^i(CMS)M_w(CMS)=0.07$ Cg/cm^{2}mole,(10) $D_g^i(IS)M_w(IS)=0.2$ Cg/cm^{2}mole (extrapolated from the experimental data) (10), $D_g^i(CS)M_w(CS)=0.32$ Cg/cm^2 mole (2), and $D_g^i(ST)M_w(ST)=6.5$ Cg/cm^{2}mole (8).

Calculated sensitivities are shown as a function of mole fraction for highly sensitive monomer N in Figure 3. The results are in good agreement with the experimental data. It must be noted that sensitivity saturation occurs in the high mole fraction region of N, as clearly demonstrated above.

The crosslinking reaction mechanisms for substituted polystyrenes were reported and it was shown experimentally that the abstraction by halogen and hydrogen radicals exists in a crosslinking reaction(12). However, the individual monomer concentration dependence of the ratio for the abstraction reaction has not been clear yet. In calculation, $h_M(x_M)$ and $h_N(x_N)$ were approximated to be independent of the individual monomer concentration.

Several papers have been published to explain the sensitivity saturation shown in the high mole fraction region of N. In those reports, the saturation is interpreted by the steric hindrance effect (8) or the recombination effect(10). Assumption (II) made in this paper means that these two effects are neglected. However, the calculated results explain the experimental data on sensitivities well. Therefore, the contribution of these effects seems to be small.

The other group consists of poly(styrene-glycidyl methacrylate) P(ST-GMA), poly(chlorostyrene-glycidyl methacrylate) P(CS-GMA), etc.. In this case, the sensitivity saturation does not occur. GMA is a chain reaction monomer containing an epoxy group. Thus, sensitivities can be calculated from Equation 16.

In calculation, the following data on sensitivities are used:

$D_g^i(GMA)M_w(GMA)=0.025$ Cg/cm^{2}mole,(13) $D_g^i(ST)M_w(ST)=$ 6.5Cg/cm^{2}mole(8) and $D_g^i(CS)M_w(CS)=0.32$ Cg/cm^{2}mole(2)

From the copolymer P(ST$_{35}$-GMA$_{65}$) gel dose(1), $\xi_N=0.96$ (this value is used in a P(ST-GMA) copolymer resist system) is obtained, and $\xi_N=0.92$ (this value is used in a P(CS-GMA) copolymer resist system) is also obtained from the copolymer P(CS$_{50}$-GMA$_{50}$) gel dose(2). The calculated results are in good agreement with experimental data, as shown in Figure 4. It is noted that, in contrast to the case of a nonchain reaction, the shape of sensitivity vs. component ratio curves shows no sensitivity saturation in the high mole fraction region of GMA. This is due to the component ratio dependence of the chain reaction activity.

In calculation, it was approximated that the chain reaction activity

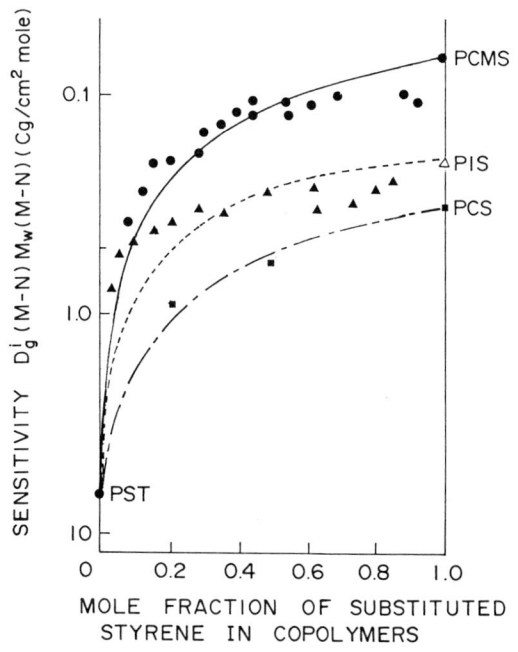

Figure 3. Comparison between calculated and experimental results in a nonchain reaction. Key(experimental): ●, P(ST-CMS); ▲, P(ST-IS); and ☐, P(ST-CS). Key(theoretical): ——, P(ST-CMS); ---, P(ST-IS); and — - —, P(ST-CS).

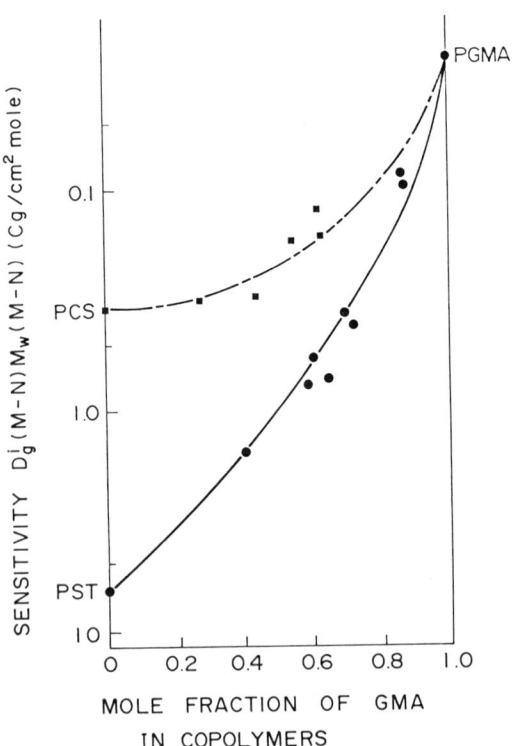

Figure 4. Comparison between calculated and experimental results in a chain reaction. Key(experimental): ●, P(ST-GMA); and ■, P(CS-GMA). Key(theoretical): ——, P(ST-GMA); and — - —, P(CS-GMA).

was proportional to the chain reaction monomer concentration. However, it is reported that reactivity is dependent on the mobility of the radicals formed within the films, and mobility of the radicals is affected by glass transition temperature T_g ($\underline{18}$). Therefore, this approximation will be valid when T_g of a copolymer is nearly the same throughout the constituent monomer ratio.

Ogata et al. reported the relationship between sensitivity and component ratio in a 3-aminoperhydroazepine (APA)-3-hexenedionyl-chloride(HC)-aziponylchloride(AC) polymer system ($\underline{19}$). In this case, HC is the chain reaction monomer containing a carbon-carbon double bond. Their result shows a complicated relationship, due to T_g varying with component ratio.

Molecular design method concerning sensitivity

The information on designing sensitivity for crosslinking copolymer negative resists is obtained from the above described results. The general relationship between senitivity ($D_g^1(M-N)M_w(M-N)$) and component ratio is shown in Figure 5. The shape of the sensitivity vs. component ratio curves depends on which type of monomer is copolymerized to enhance the sensitivity for a homopolymer P(M). The highly sensitive monomer N affects the shape mainly. Therefore, when nonchain reaction monomer N is copolymerized to enhance the sensitivity, low mole fraction region of N (lower than 0.5) is desirable. On the other hand, when chain reaction monomer N is copolymerized, high mole fraction region of N (higher than 0.5) is desirable.

Summary

The formulas for sensitivity as a function of component weight ratio in crosslinking copolymer negative resists are derived for a nonchain reaction and a chain reaction, respectively. It is found that the shapes of the sensitivity vs. component ratio curves differ for the two reaction categories, due to the individual monomer concentration dependence of the chain reaction activity. The derived formulas not only explain the experimental features well, but also interpret the experimental data quantitatively. Therefore, the derived formulas make it possible to predict sensitivity for copolymers in advance to synthesizing them and measuring their sensitivities. These formulas are useful in designing crosslinking copolymer negative resists.

Acknowledgments

The authors thank M. Suzuki and K. Saigo for their meaningful discussions, and also thank D. Shinoda for his encouragement in this work.

Symbols

P(M-N)   copolymer of monomer M and monomer N

continued on page 189

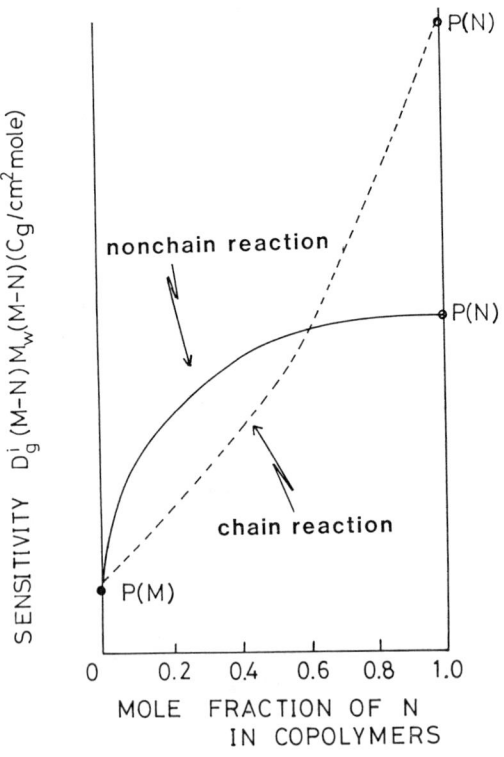

Figure 5. General relationship between sensitivity and component ratio.

$I(I=M,N,...)$	constituent monomer
$M_w(M-N)$	weight-average molecular weight of a copolymer $P(M-N)$ (g/mole)
D_i	radiation dose (C/cm^2)
$D_g^i(M-N)$	gel dose for a copolymer $P(M-N)$ (C/cm^2)
N_0	Avogadro's number
X_I	weight ratio for I in a polymer
m_I	molecular weight of a monomer I (g/mole)
q_I	probability of a given monomer I being crosslinked
n_T	total number of crosslinking units per gram
k_I	number of radiation-induced active bonds on I per gram
K_I	active bond production efficiency per gram and per unit dose for I
$j_I(x_I)$	chain reaction activity for induced active bonds on I
$h_I(x_I)$	active bond coupling reaction ratio for induced active bonds on I
I	proportionality constant for I

Literature Cited

1. Thompson, L.F.; Stillwagon, L.E.; Doerries, E.M. J.Vac.Sci. Technol. 1978, 15, 938.
2. Feit, E.D.; Thompson, L.F.; Wilkins, C.W.Jr.; Doerries, E.M.; Stillwagon, L.E. J. Vac. Sci. Technol. 1979, 16, 1997.
3. Charlesby, A. Proc. R. Soc. 1954, A222, 542.
4. Ohnishi, Y.; Itoh, M.; Mizuno, K.; Gokan, H.; Fujiwara, S. J. Vac. Sci. Technol. 1981, 19, 1141.
5. Charlesby, A. Proc. R. Soc. 1957, A241, 495.
6. Ku, H.Y.; Scala, L.C. J. Electrochem. Soc. 1969, 116, 980.
7. Atoda, N.; Kawakatsu, H. J.Electrochem. Soc. 1976, 123, 1519.
8. Imamura, S.; Tamamura, T.; Harad, K.; Sugawara, S. J.Appl. Polym. Sci. 1982, 27, 937.
9. Feit, E.D.; Stillwagon, L.E. Polym. Eng. Sci. 1980, 20, 1058.
10. Shiraishi, H.; Taniguchi, Y.; Horigome, S.; Nonogaki, S. Polym. Eng. Sci. 1980, 20, 1054.
11. Choong, H.S.; Kahn, F.J. J. Vac. Sci. Technol. 1981, 19, 1121.
12. Weir, N.A.; Mikei, T.H. J.Polym. sci., Polym. Chem. Ed. 1979, 17, 3735.
13. data mesured in our laboratories using OEBR 100 (trade name of Tokyo Ohka Kogyo Company).
14. Thompson, L.F.; Doerries, E.M. J. Electrochem. Soc. 1979, 125, 1699.
15. Thompson, L.F.; Yau, L.; Doerries, E.M. J.Electrochem. Soc. 1979, 126, 1703.
16. Sikorski, R.T.; Gabrys, R. J. Appl. Polym. Sci, 1980, 25, 1131.
17. Nonogaki, S.; Morishita, H.; Saitou, N. Appl. Polym. Symp. 1974, 23, 117.
18. Mitani, K.; Ogata, T.; Awaya, H.; Tomari, Y. J. Polym. Sci. Polymer Chemistry 1975, 13, 2813.
19. Ogata, N.; Sanui, K.; Azuma, C.; Tanaka, H.; Oguchi, K.; Nakata, T. J. Appl. Polym. Sci. in press.

RECEIVED September 2, 1983

16

Molecular Design for Cross-linking Negative Resists
Optimum Design for Poly(chloromethylstyrene-*co*-2-vinylnaphthalene)

YOSHITAKE OHNISHI, KATSUMI TANIGAKI, and AKIHIRO FURUTA[1]

Fundamental Research Laboratories, NEC Corporation, Miyamae-ku, Kawasaki-City 213, Japan

> Poly (chloromethylstyrene-co-2-vinylnapthalene) has been developed as a negative electron resist.
> Optimum design method for this material is presented, which is based on theoretical analysis for copolymer sensitivity and on dry etch rate dependence on polymer structure obtained by a series of experiments. A resist with high sensitivity (D_g^i=1 μC /cm^2), high resolution (<0.5 μm) and high dry etch resistance was obtained.

Dry etch resistance, as well as sensitivity and resolution capability, has become one of the most important factors for resist materials for microfabrication. Much effort has been made to develop resist materials which meet severe requirements for lithography and processes needed in VLSI fabrication, and many resist materials have been reported so far.

Nevertheless, to the authors' knowledge, there have been few reports on theoretical predictions of resist material characteristics. Theoretical predictions on resist material characteristics, if they can be made prior to polymer synthesis and/or evaluation, can save much time and elaborate work.

Relationship between dry etch resistance under ion bombardment and polymer structure has recently clarified (1). As for sensitivity predictions, theoretical formulas for sensitivities of crosslinked positive resists have been proposed (2). Formulas for sensitivities of copolymer negative resists have also been derived and reported (3). Unfortunately, however, no valid schemes are yet available for predicting resolution capabilities of resist materials.

In this paper, optimum design method for poly(chloromethylstyrene-co-2-vinylnaphthalene) is presented, based on theoretical analysis for copolymer sensitivity and on the dry etch rate dependence on polymer structure obtained by a series of experiments. 2-vinylnaphthalene was selected as a main constituent of the polymer, as poly(2-vinylnaphthalene) was found to be a negative resist with high contrast and high dry etch resistance (4).

[1]Current address: Central Research Laboratories, Sumitomo Chemical Co., Ltd., Tsukahara, Takatsuki 569, Japan

Design Method

A feasibility of designing a dry etch resistant negative electron resist was discussed by Thompson et al. (5). They discussed parameters affecting negative resist characteristics in a qualititative way.

This section describes a quantitative and theoretical method to design negative eletron resist for both high dry etch resistance and high sensitivity.

Dry Etch Resistance. Polyvinylnaphthalene and its derivatives were reported to be highly resistant to dry etching (4). Under Ar ion milling or CCl_4 sputter etching conditions, etch rates for poly(2-vinylnaphthalene) were found to be about two-thirds of those for polystyrene or novolak resin resists. To find relationships between etch resistance and chemical structure, etch rate measurements for various metal-free polymers were made under Ar or O_2 ion beam incidence. It was found that the etch rate under ion bombardment depends linearly on $N/(N_C - N_O)$, where N denotes the total number of atoms in a monomer unit, N_C and N_O are the number of carbon atoms and the number of oxygen atoms in a monomer unit, respectively (1). The dependence on "$N/(N_C-N_O)$ factor" indicates that the dry etch resistance under ion bombardment is determined by the effective carbon content in the material, irrespective to chemical bonds. Since ion bombardment is an indispensable technology for anisotropic or precise etching in microfabrication, this relationship would provide a design method on metal-free polymer resists to dry etching. The dependence on "$N/(N_C-N_O)$ factor" also implies that carbon itself is the most highly resistive material among metal-free organic polymers. Polyvinylnaphthalene has the smallest $N/(N_C-N_O)$ value among known metal-free organic resists. Although polymer materials with smaller $N/(N_C-N_O)$ value than polyvinylnaphthalene can be designed, only a slight improvement can be expected in dry etch resistance, as shown in Figure 1.

Sensitivity. Poly(2-vinylnaphthalene) is a low sensitive negative electron resist, and $D_g^i \cdot M_W$ value for the polymer was estimated as 10 C·g /cm^2·mole (4). Product $D_g^i \cdot M_W$ is a constant that characterizes the sensitivity for negative resists, where D_g^i and M_W denote gel dose and weight average molecular weight, respectively (6). To attain higher sensitivity, modification or copolymerization with other sensitive monomers is needed.

The relationship between copolymer sensitivity and component ratio was derived theoretically (3). According to the theory, copolymer sensitivity can be predicted for any component ratio and molecular weight, utilizing data on each homopolymer sensitivity.

Theoretical sensitivity calculations were carried out for two copolymer systems, poly (chloromethylstyrene-co-2-vinylnaphthalene) (P(CMS-2VN)) and poly(glycidyl methacrylate-co-2-vinylnaphthalene) (P(GMA-2VN)). Polychloromethlstyrene is known as a high sensitivity negative resist (7,8) ($D_g^i M_w$=0.07 C·g /cm^2· mole), which shows no post-irradiation polymerization. The crosslinking reaction in P(CMS-2VN)

TABLE I List of organic polymers examined in the present experiment

	NAME	REMARKS
SEL-N		SOMAR MANUFACTURING
PMMA	Poly(methyl methacryylate)	
COP	Poly(glycidyl methacrylate-co-ethyl acrylate)	MEAD ASSOCIATES
CP-3	Poly(methacrylate-co-t-butyl methacrylate)	t-BMA 30 %
EBR-9	Poly (α-chloro-trifluoro ethylacrylate)	
PBZMA	Poly(benzyl methacrylate)	
FBM	Poly(hexa fluorobutyl methacrylate)	
FPM	Poly(fluoro propyl methacrylate)	
PMIPK	Poly(methyl isopropenyl ketone)	
PS	Poly styrene	
CMS	Chloromethylated polystyrene	
PαMS	Poly(α-methyl styrene)	
PVN	Poly(vinyl naphthalene)	
PVB	Poly(vinyl biphenyl)	
AZ1350J		SHIPLEY CO., INC.
CPB	Cyclized polybutadiene	

Reproduced with permission from Ref. 1. Copyright 1983, J. Electrochem. Soc.

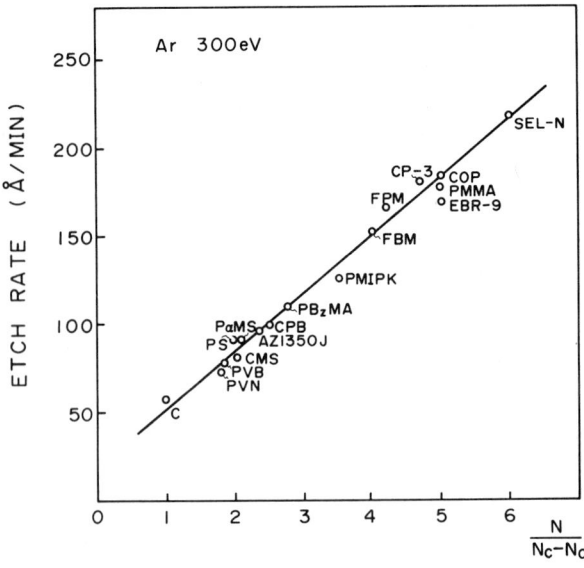

Figure 1. Etch rates under 300 eV Ar ion bean etching condition vs. $N/(N_C - N_O)$. N, N_C and N_O denote the total number of atoms, the number of carbon atoms and the number of oxygen atoms in a monomer unit, respectively. Organic polymers examined are listed in Table 1. (Reproduced with permission from Ref. 1. Copyright 1983, J. Electrochem. Soc.)

system is assumed as a nonchain reaction. Sensitivity for P(CMS-2VN) is expressed in clude approximation as

$$\frac{1}{(D_g^i \cdot M_w)_{P(CMS-2VN)}} = \frac{x}{(D_g^i \cdot M_w)_{PCMS}} + \frac{1-x}{(D_g^i \cdot M_w)_{P2VN}}$$

where x denotes molar fraction of CMS in the polymer (3). Poly(glycidyl methacrylate) is also known as a highly sensitive negative resist (9) ($D_g^1 M_w$=0.025 C·g/cm^2·mole) which shows post-irradiation polymerization (10). The crosslinking reaction in P(GMA-2VN) system is assumed as a chain reaction. Sensitivity for P(GMA-2VN) is expressed in brief approximation as

$$\frac{1}{(D_g^i \cdot M_w)_{P(GMA-2VN)}} = \frac{(1-\xi_{GMA}) \cdot x^2}{(1-\xi_{GMA} \cdot x) \cdot (D_g^i \cdot M_w)_{PGMA}} + \frac{1-x}{(D_g^i \cdot M_w)_{P2VN}}$$

where x denotes molar fraction of GMA in the polymer (3), ξ denotes proportional constant for the chain reaction activity, and assumed the same value ξ_{GMA} =0.96 as in the polymer system poly(glycidyl methacrylate-co-styrene) (3,5).

The results of these theoretical calculations are graphically shown in Figure 2. It is obvious from these results which copolymer is more desirable. Though sensitivity of poly(glycidyl methacrylate) is higher than that of polychloromethylstyrene in equal molecular weight, P(GMA-2VN) requires more than 80 % of GMA content in the polymer to compete P(CMS-2VN) in sensitivity. P(GMA-2VN) is not expected to attain both high sensitivity and high dry etch resistivity. Thus, P(CMS-2VN) was chosen as a promising candidate for high sensitive dry etch resistance negative resist. Futhermore, chloromethylstyrene is a more preferable monomer than glycidyl methacrylate from resolution capability standpoint. As we discuss later, a nonchain reaction monomer (chloromethylstyrene) can give higher contrast than a chain reaction monomer (glycidyl methacrylate) does (11).

So far sensitivity calculations were carried out to estimate $D_g^i \cdot M_w$, which is a parameter that characterizes the sensitivity for negative resist materials. Theoretically, gel dose (D_g^1) can be decreased without limit as molecular weight (M_w) increases. Figure 3 shows predicted sensitivity for P(CMS-2VN) with M_w as a parameter.

In the case of negative resists, molecular weights greater than one million are unfavorable because of swelling problems during development processes.

As for dry etch resistance, the etch rate decreases as the fraction of 2VN in the polymer increases.

Taking these requisites into account, favorable combinations can be found in the region of $M_w < 10^6$ and CMS fraction < 20 mole %. Setting desired sensitivity as D_g^i =1 μ C / cm^2, possible combinations are easily

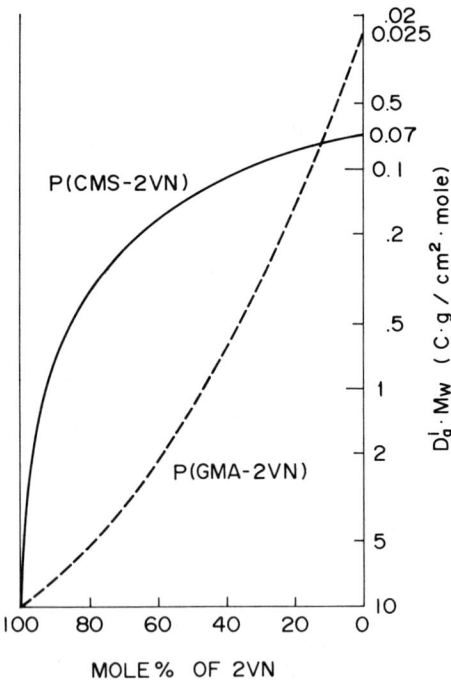

Figure 2. Theoretical sensitivity predictions for the polymer systems P(CMS-2VN) and P(GMA-2VN).

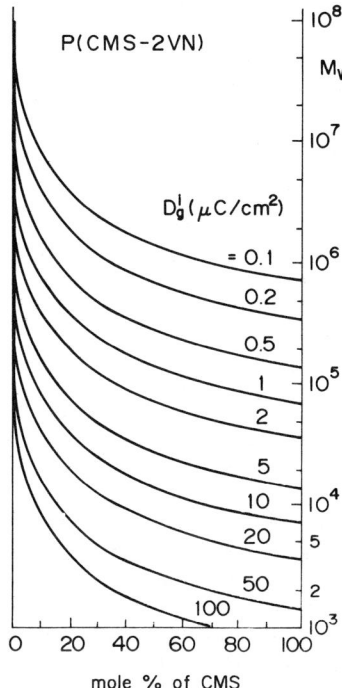

Figure 3. Theoretical prediction for the system P(CMS-2VN), showing relationship among component ratio, molecular weight and sensitivity.

selected from the line $D_g^i = 1 \, \mu C / cm^2$, indicated in Figure 3, for example: $P(CMS_{20} - 2VN_{80})$ $M_w = 3.4 \times 10^5$, $P(CMS_{15} - 2VN_{85})$ $M_w = 4.5 \times 10^5$, $P(CMS_{10} - 2VN_{90})$ $M_w = 6.6 \times 10^5$ and so on.

Experimental Results and Discussion

Synthesis of P(CMS-2VN), aiming at $D_g^i = 1 \, \mu C / cm^2$ sensitivity, was carried out by ordinary radical polymerization in an inert-gas-filled flask with stirring (60 °C, 5 hrs.). The polymer after fractionation was $M_w = 7.8 \times 10^5$, and polydispersivity $M_w / M_n = 1.3$ (as measured by gel permeation chromatography). The mole fractions in the polymer were found to be 9 mole % of CMS and 91 mole % 2VN (as determined by elemental analysis). These values are very close to the mole fractions in the feed. The polymer was dissolved in xylene (5 wt.%), spin-coated (2000 rpm) and prebaked (100 °C for 30 min) to get uniform $0.5 \, \mu$ m-thick-films on Si substrate. Pattern delineation was made by a JEOL JBX-5A (20 kV acceleration) electron beam exposure system without proximity corrections. Images were developed by dipping in 1,1,2,2-tetrachloroethylene for 60 sec, then rinsing in isopropanol for 30 sec.

Sensitivity characteristics are shown in Figure 4. D_g^i was found to be $0.9 \, \mu C / cm^2$, which agreed well with the expected value. Examples of pattern profiles are shown in Figure 5. Dose was $2.15 \, \mu C / cm^2$ for 0.5 μm lines and $0.5 \, \mu$ m spaces pattern. As shown in these SEM photos, these sub-micron resist prifiles are rectangular in shape. The resolution is good, considering its high molecular weight ($M_w = 7.8 \times 10^5$).

As for quantitative prediction concerning crosslinking negative resist resolution, no theories are available except relationships between polydispersivity (M_w / M_n) and contrast (γ). Charlesby showed relationship between crosslinking density and gel fraction (12). Crosslinking density can be considered proportional to dose, and gel fraction is, in principle, normalized thickness after development. Atoda et al. utilized these results to describe exposure characteristics of negative resists (11). According to their results, a nonchain reaction polymer gives higher contrast than a chain reaction polymer does, and smaller M_w / M_n gives higher contrast.

These results are applicable to all crosslinking negative resists. However, it is not useful in explaining the difference in resolution power among various polymer resists.

Sub-micron fine resist patterns could not be realized with polystyrene or styrene based resists of high molecular weight ($M_w > 5 \times 10^5$), although their polydispersivity is nearly unity. The difference in resolution capability between styrene based resists and vinylnaphthalene based resists possibly comes from difference in their developing characteristics. Thompson et al. have pointed out that polymer swelling can be minimized by designing a polymer which has a strong backbone and is rheologically "stiff" during development cycles, and polymers possessing this rigidity generally have high glass transition temperature (T_g) (5). Poly(2-vinylnaphthalene) has a higher glass transition temperature T_g (424 K) than that of polystyrene ($T_g = 373$ K) (13), and vinylnaphthalene based resists are more stiff and more resistant to swelling than styrene based resists.

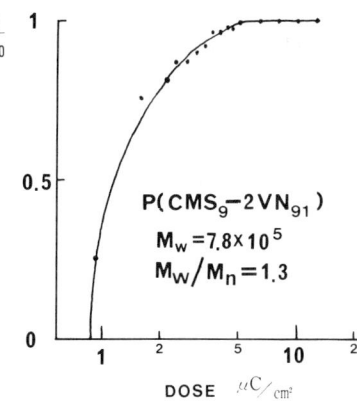

Figure 4. Sensitivity characteristics for P(CMS$_9$-2VN$_{91}$). Acceleration voltage is 20 kV.

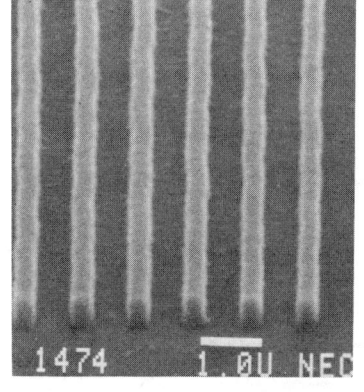
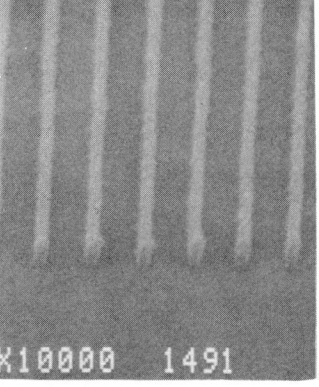

(a) 0.5 μm lines and 0.5 μm spaces, (b) 0.3125 μm lines and 0.5625 μm spaces,

(c) 0.1875 μm lines and 0.4375 μm spaces.

Figure 5. Developed resist profiles of P(CMS$_9$-2VN$_{91}$). Remaining thickness d = 0.41 μm.

Resolution has become a crucial requirement for microlithography. It depends not only on material itself, but also on lithographic apparatus limitations, development processes and so many other complex factors. In designing a high resolution resist, therefore, starting material, or basic structure has to be chosen by an evidence of resolution capability. Low molecular weight (M_w = 3 x 10^4) poly(2-vinylnaphthalene) showed high resolution less than 0.1 μm line on Si substrate by ordinary procedure.

Conclusion

Optimum design was made theoretically on P(CMS-2VN) system. Design method was developed based on theoretical analysis for copolymer resist sensitivity and on dry etch rate dependence on polymer structure obtained by a series of experiments. P(CMS$_9$-2VN$_{91}$) (M_w =7.8 x 10^5, M_w/M_n = 1.3) proved to have high sensitivity (D_g^i =0.9 μC/cm^2) and high resolution (<0.5 μm) as well as high dry etch resistance. This resist is suitable for microfabrication especially for dry etching processes.

Literature Cited

1. Gokan, H.; Esho, S.; Ohnishi, Y. J. Electrochem. Soc. 1983, 130, 143.
2. Suzuki, M.; Ohnishi, Y. J. Electrochem. Soc. 1982, 129, 402.
3. Tanigaki, K.; Ohnishi, Y.; Fujiwara, S. Proc. 185th Am. Chem. Soc. Mtg. 1983,
4. Ohnishi, Y. J. Vac. Sci. Technol. 1981, 19, 1136.
5. Thompson, L. F.; Stillwagon, L. E.; Doerries, E. M. J. Vac. Sci. Technol. 1978, 15, 938.
6. Ku, H. Y.; Scala, L. C. J. Electrochem. Soc. 1969, 116, 980.
7. Feit, E. D.; Thompson, L. F.; Wilkins, C. W., Jr.,; Wurtz, M. E.; Doerries, E.M.; Stillwagon, L. E. J. Vac. Sci. Technol. 1979, 16, 1997.
8. Choong, H. S.; Kahn, F.J. J. Vac. Sci. Technol. 1981, 19, 1121.
9. Hirai, T.; Hatano, Y.; Nonogaki, S. J. Electrochem. Soc. 1971, 118, 669. (The value D_g^i in the text was measured in our laboratory.)
10. Ohnishi, Y.; Itoh, M.; Mizuno, K.; Gokan, H.; Fujiwara, S. J. Vac. Sci. Technol. 1981, 19, 1141.
11. Atoda, N.; Kawakatsu, H. J. Electrochem. Soc. 1976, 123, 1519.
12. Charlesby, A. "Atomic Radiation and Polymers"; Pergamon: London, 1960.
13. Brandrup, J.; Immergut, E. H., Eds. "Polymer Handbook" 2nd Ed., John Wiley and Sons: N. Y., 1975.

RECEIVED September 2, 1983

17

Poly(butadiene-*co*-glycidyl methacrylate) as Negative Electron Resist

B. BEDNÁŘ, J. DEVÁTÝ, J. KRÁLÍČEK, and J. ZACHOVAL

Department of Polymers, Prague Institute of Chemical Technology, 166 28 Prague 6, Czechoslovakia

> Copolymers of 2,3-epoxypropyl methacrylate and butadiene prepared by solution free radical polymerization were tested as potential candidates for electrom negative resists. The sensitivity of the copolymers to ionizing radiation was determined with a linear accelerator. The experimental data were treated according to the Saito-Inokuti statistical theory. The calculated crosslinking yield values, $G(X)$, were dependent an both initial molecular weight and butadiene content in the copolymers. A contrast were found to have $\mu \simeq 1$ and the sensitivity $D_g^i = 0.04 - 0.4$ uC/cm^2 at 10 kV. The resist exhibited resolution of 1.0 μm and good adhesion for wet chemical etching of thermal SiO_2 and aluminium.

The choice of polymers as potential candidates for negative electron resists to be used the fabrication of LSI and VLSI circuits depends on a number of parameters, the most significant of which are the sensitivity to ionizing radiation, resolution, wet or dry etching resistance, thermal stability, and many other technological parameters such as adhesion etc. In this respect poly(2,3-epoxypropyl methacrylate) (PEPM) and its copolymers (1-3) prove to be the most appropriate candidates. The prevailing degradation processes induced in the methacrylate polymers by ionizing radiation (4) are in this case negligible in comparison with the crosslinking processes in which the reaction of epoxy groups plays a significant role (5). By a proper choice of monomers for the copolymerization of EPM some properties of PEM, undesirable from the technological point of view, can be repressed. The use of the copolymer EPM-butadiene (BD) in the preparation of radiation-crosslinked hydrophilic coating elastomers for medical applications (6) has suggested the use of this copolymer as a negative electron resist (7).

Experimental

Polymers used in the study were prepared using a solution, radical copolymerization technique (8). The polymerization conditions and some molecular charasteristics of the copolymers used for studying thermal stability and sensitivity to ionizing radiation are given in Table I. TG and DTG measurements were performed using the DuPont 951, heating rate 10 °C/min. In the irradiation by fast electrons a linear accelerator served as the source of ionizing radiation (4-5 MeV, output power of γ-radiation 1.2 kW). Samples were irradiated in the form of pellets on aluminium discs at a rate 10 kGy/s by individual doses of 20 kGy with pauses in between in order to prevent heating of the samples. The determination of the lithographic properties was made using a modified scanning microscope JSM 35. To determine the sensitivity always 4 areas 1x1mm large were exposed with line scanning of the density of 200 lines per mm with a 1 μm electron beam diameter.

Results and Discussion

Thermal Analysis

The TG and DTG measurements have shown that the weight changes resulting in the copolymers by heat treatment depend to some extent on the chemical composition of the copolymers, and can be divided into three main regions. In the temperature range of 70°--340 °C a 2.5-3% weight loss occurs, in the range of 340°-420 °C the sample weight decreases by 17-17.5%, and in the temperature range of 420°-500 °C gasification of the remaining part of the sample takes place. The thermal degradation of the sample is qualitatively similar to that of polybutadiene (9), but completely different from that of PEPM. For the thermal degradation of PBD it has been proved that in the temperature range corresponding to the first degradation region (70°-340 °C for the copolymer) the following reactions which result in polymer crosslinking proceed (9).

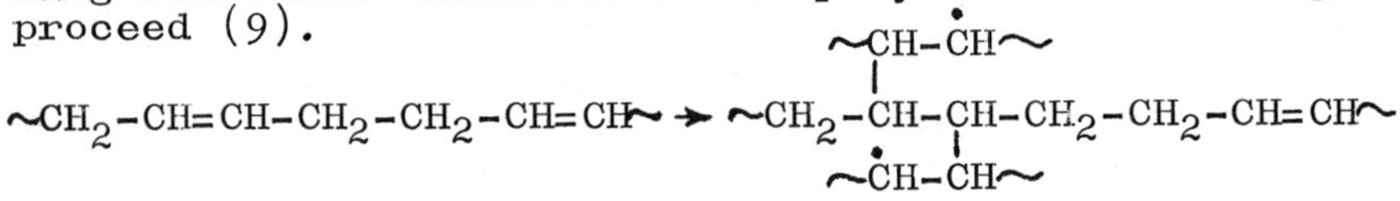

$$\sim CH_2-CH=CH-CH_2-CH_2-CH=CH\sim \rightarrow \sim CH_2-\overset{\sim CH-\overset{\bullet}{C}H\sim}{\underset{\sim CH-\overset{\bullet}{C}H\sim}{\underset{|}{C}H}}-CH-CH_2-CH_2-CH=CH\sim$$

Table I. Composition, microstructure of BD units and molecular parameters for copolymers EPM-BD used for determination of sensitivity on ionizing radiation

Polymers	EPM in co-polymer mol.%	$\bar{M}_w \times 10^{-4}$ g.mol^{-1}	$\bar{M}_n \times 10^{-4}$ g.mol^{-1}	Microstructure mol.% 1,2-	mol.% 1,4-
EBD-1	15,66	2,56	0,81	18,82	81,18
EBD-2	25,78	1,40	0,46	12,37	87,63
EBD-3	31,33	1,10	0,69	20,00	80,00
EBD-4	36,58	3,50	0,63	14,97	85,03
EBD-5	55,79	7,70	2,02	8,79	91,21
EBD-6	35,75	2,38	1,00	12,15	87,85

Owing to the fact that the copolymers up to a high EPM content contain in the polymer chain mainly higher number-average sequence lenght of butadiene units (8), it is probable that the behaviour of the copolymers during heat treatment will be similar to that of PBD. As seen from the dependence of the weight fraction of the crosslinked polymer (w_g) on the heating time (t), obtained by the sol-gel analysis of heat treatment samples, the content of the crosslinked polymer increases with increasing BD content in the copolymer (Figure 1). This we can assume that in the temperature range of $70°-340$ °C the reactions occuring in the copolymers are similar to those in polybutadiene.

Ionizing Radiation Sensitivity

The sensitivity of the copolymers to ionizing radiation was determined on the basis of a sol-gel analysis of samples after irradiation on a linear accelerator. The experimental dependences of the weight fraction of the crosslinked polymer on the relative value of the irradiation dose D/D_g (D_g being the irradiation dose to the gel point) were treated numerically using the Saito-Inokuti statistical theory (10,11) for the simultaneously proceeding crosslinking and degradation reactions of polymers. Figure 2 shows the experimental points of dependence of wg versus D/D_g for the EBD-3 copolymer, and a theoretically calculated curve which is a good representation of the experimental dependence. A similar agreement of the experiments with the theory has been obtained for all the copolymers, so that we can assume that the values obtained for the yields of crosslinking $G(x)$ from the parameters of theoretical dependences give a good description of the sensitivity of the copolymers to ionizing radiation. As seen from Figure 3 the dependence of $G(x)$ versus molar content of EPM in the copolymer has its maximum in the region of 25% mol EPM in the copolymer. The value for the yield of main chain scission $G(s)$ was virtually zero for all the copolymers, the share of the degradation reactions compared with that of the crosslinking reactions thus being negligible.

Lithographic Properties

To determine the lithographic properties 2 copolymers of different composition (Table II) were prepared. The parameter $D_g^{0,7}$ is defined as a dose to give a resist thickness after development, corresponding to a thickness 0.7 times the initial resist

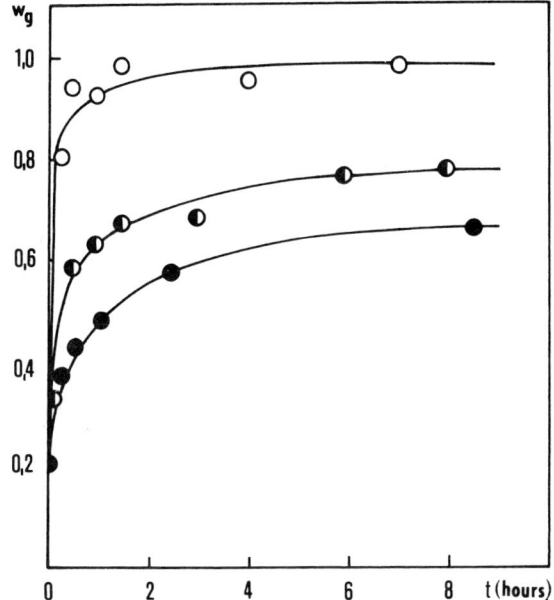

Figure 1. Dependence of w_g on the time of thermal treatment at 180 °C

● - EBD-5
○ - EBD-1
◐ - EBD-3

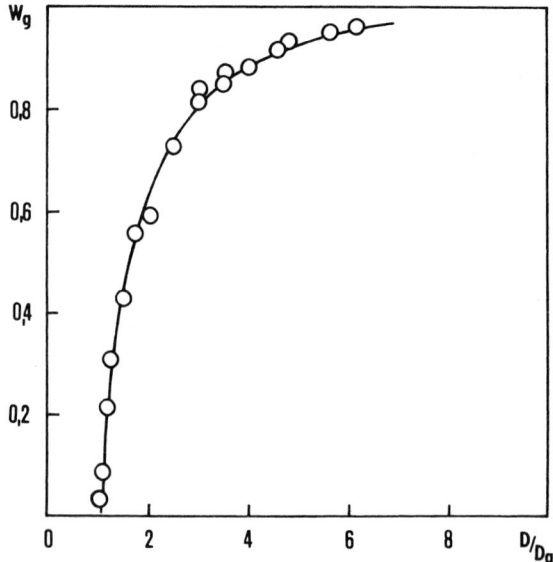

Figure 2. Dependence of w_g on D/D_g
O - experimental points, copolymer EBD-3, full line theoretical curve

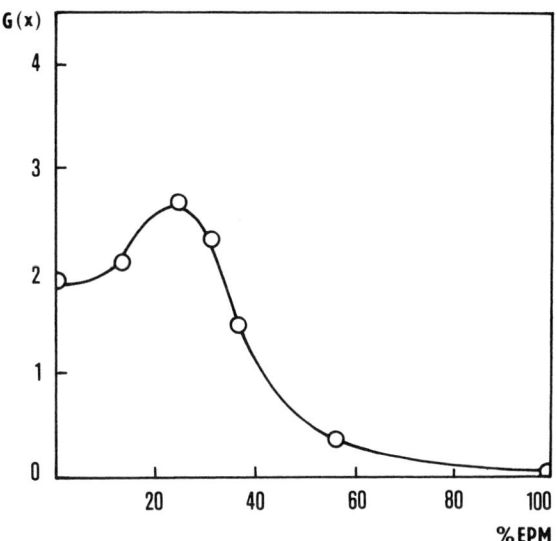

Figure 3. Dependence of w_g on the time of thermal treatment at $140°-250\ °C$ for copolymer EDB-A

◐ - 250 °C
○ - 180 °C
● - 140 °C

Table II Polymeric and lithographic parameters for EBD resist evaluated

Polymers	EPM mol.%	$\bar{M}_w \times 10^{-4}$ g.mol^{-1}	$\bar{M}_n \times 10^{-4}$ g.mol^{-1}	Microstructure mol.%			$D_g^{0.7}$ uC/cm^2		γ	
				1,2–	1,4–		10kV	15kV	10kV	15kV
EBD-A	33,94	6,57	2,76	15,4	84,6		1,5	1,8	1,0	1,1
EBD-B	51,68	8,98	3,99	20,0	80,0		0,3	1,2	0,9	1,1

$D_g^{0.7}$ – is defined as the exposure required to give a develeped film thickness 0.7 times the initial film thickness

thickness. The initial thickness of the resist was 0.7 μm in determining the values $Dg^{0.7}$ and γ listed in Table II. The values of $Dg^{0.7}$ indicate that it is necessary to apply a higher dose at higher voltage, for the same resist thickness after development, which has been observed previously for other negative electron resists (3). The contrast value defined as

$$\gamma = \log \left(Dg^1 / Dg^i \right)^{-1}$$

where Dg^i is the extrapolated dose for zero film remaining, and Dg^1 is the dose required for extrapolated 100% film remaining, showed a slight increase with increasing value of the accelerating voltage. Both the sensitivity and the contrast were virtually independent of the development time and of the thermodynamic quality of the solvent, but similarly as in the previous cases (2,3) there was a strong dependence of the developed pattern quality on the thermodynamic quality of the solvent, and consequently also on the development temperature. The solvents of a good thermodynamical quality (benzene, THF, 1,4-dioxane) caused substantial swelling of the structures, which resulted in the destruction of the pattern structures or even in the resist ripping off the substrate. A mixture of a good solvent (toluene) with a bad one in such a ratio as to cause a sufficiently fast dissolution of the non-irradiated polymer with the minimum swelling of the crosslinked polymer taking place proved to be the optimum one.

Prebaking at temperatures higher than 70 °C leads to crosslinking of the copolymer, especially if carried out in presence of oxygen. If prebake proves necessary because of adhesion, it can be carried out at the temperature 50 °C for 30min.max. Postbaking can be carried out at the temperatures of 70°-250 °C (Figure 4) without any detectable change in the resist thickness taking place; the only potential danger is that of the potential flow causing pattern distortion.

The resolution power of the resist depends on the initial thickness of the resist, the resolution of 1 μm being attainable in the thickness range of 0.5 - - 0.7 μm. The resist offers a very good resistance (adhesion) to wet etching of SiO_2 and Al, and after additional postbaking also satisfactory resistance to plasma etching of Si_3N_4 and Al.

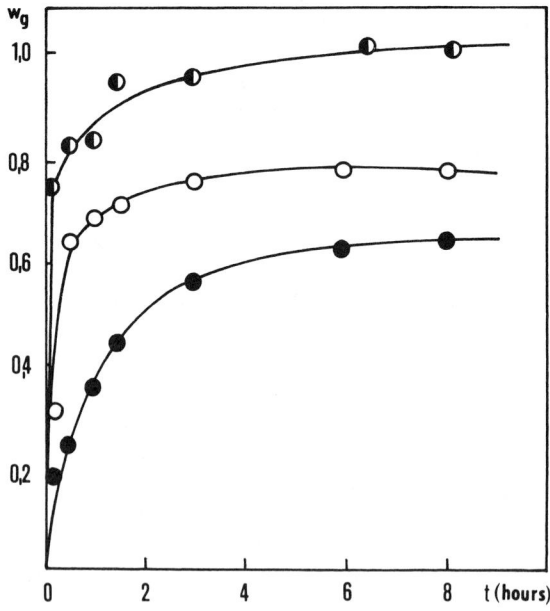

Figure 4. Dependence of $G(x)$ on copolymer composition

Literature Cited

1. Thompson,L.F. Solid State Technol. 1974, 17, 27.
2. Thompson,L.F.; Stillwagon,L.E.; Dobrries,E.M. J.Vac.Sci.Technol. 1978, 15, 938.
3. Thompson,L.F.; Dobrries E.M. J.Electrochem.Soc. 1979, 126, 1703.
4. Dole,M. in "The Radiation Chemistry of Macromolecules", Vol.II, (M.Dole,Ed.), Academic Press, New York 1973.
5. Bednář,B.; Doležal,J.; Králíček,J. to be published.
6. Kálal,J.; Bednář,B.; Houska,M. ACS Polymer Preprints 1975, 16, 363.
7. Bednář,B.; Zachoval,J.; Kálal,J.; Petr,J.; Pelzbauer,Z.; Švec,F. US pat. 4 269 962 (26.5.1981).
8. Bednář,B.; Chuchma,F.; Zajíc,S.; Kálal,J.; Králíček,J. Scientific Papers of Prague Institute of Chemical Technology 1982, S7, 255.
9. McGreedy,K.M.; Keskkula,H.J. Polymer 1979, 20, 1155.
10. Saito,O. in "The Radiation Chemistry of Macromolecules", Vol.I, (M.Dole,Ed.), Academic Press, New York 1972.
11. Inokuly,M. J.Chem.Phys. 1963, 38, 299.

RECEIVED September 2, 1983

Plasma Developable Electron Resists

JUEY H. LAI

Honeywell Corporate Technology Center, Bloomington, MN 55420

Negative electron resists are generally more sensitive than
positive resists but the resolution of negative resists is lower
(1). Negative resists are normally applied to the substrate by
a spin coating process (2), exposed using a high energy electron
beam and spray developed with a developer solvent. The dominant
effect of e-beam exposure on the organic polymers used as negative resists is a crosslinking process that increases the molecular weight of the polymers, making the exposed area insoluble.
During the development process in which the unexposed area in
the negative resist is removed, the solvent molecules invariably
penetrate into the exposed crosslinked area resulting in swelling
of the polymer which seriously degrades the resist resolution.
The swelling of negative resist patterns can be minimized, but
cannot be totally eliminated. This is because the solvent which
can dissolve the unexposed polymer has a natural tendency to
diffuse and penetrate into the exposed crosslinked polymer area.
The poor edge acuity and low resolution of the conventional
negative polymer resist is a well known problem.

Current work on negative electron resists has been concentrated on halogenated polystyrene (3). Although considerable
progress in improving the sensitivity and thermal stability
(glass transition temperature) of the polystyrene has been made,
the resists are still subject to solvent-induced swelling. Use
of a conventional negative resist to fabricate submicron devices
has been very difficult if not impossible.

The swelling problem in the negative resist can be avoided
if the differential solubility between exposed and unexposed
area does not totally rely on the crosslinking of the exposed
resist (4), or a dry development process, such as plasma development, is used.

In recent years, there have been some reports on plasma
developable photo (5) x-ray (6) and electron beam resists (7).
Although the basic principle behind the three types of resists
is similar, a plasma developable electron resist (PDE) is more
difficult to formulate due to the following reasons: (a) the

monomer contained in the electron resist could vaporize or sublimate under high vacuum resulting in inconsistent performance, and (b) the line broadening effect of scattering of electrons by the resist and substrate molecules.

In this paper, we will discuss some results of our work on PDE. The general approach we have taken for PDE is similar to that for plasma-developed photo and x-ray resists. The approach has been to formulate and evaluate organic resists that have three components, a monomer, a low plasma-resistant base polymer and a barrier polymer which is used to prevent the sublimation or vaporization of the monomer. The monomer, when exposed to high energy electrons, is spontaneously grafted to the base polymer or polymerized to a homopolymer. Since the graft copolymer and the homopolymer have a lower plasma etch rate than that of the base polymer by virtue of the formulation, this will result in a negative-acting PDE. Further, if the monomer can be sublimated or vaporized from the unexposed area by baking prior to plasma development, a thickness difference between exposed and unexposed area would be obtained. In this formulation, a thin barrier polymer is coated on the top of the electron resist film to prevent monomer sublimation and vaporization under high vacuum. The barrier polymer is removed before baking and plasma development.

Experimental

The monomers, N-vinyl carbazole (NVC), methacrylamide (MAM), octadecyl methacrylate (ODMA) and trichloroethyl methacrylate (TCEM) were obtained from Polyscience, Inc. Diphenyl acetylene (DPAE), poly(4-chlorostyrene) (PClS), poly(butadiene) (PBD), poly(styrene) (PS) were obtained from Aldrich Chemical Co. Poly (vinyl alcohol) (PVA), 99-100% hydrolyzed, was obtained from J. T. Baker Chemical Co. Poly(vinyl pyrrolidone) was obtained from Sigma Chemical Co.

Poly(trichloroethyl methacrylate) (PTCEM) was polymerized by free-radical solution polymerization in chlorobenzene using AIBN (2,2-azo-bis isobutylonitrile) as an initiator. The number-average molecular weight $\bar{M}_n$, weight-average molecular weight $\bar{M}_w$, and molecular weight distribution $\bar{M}_w/\bar{M}_n$ of PTCEM determined by GPC (gel permeation chromotograph) are 30,000, 77,000 and 2.60 respectively.

For e-beam exposure, the resist films were spin coated on SiO_2 (2000 Angstroms), prebaked and exposed at 20 kV in a Cambridge e-beam microfabricator. For our initial studies, an IPC 4005 barrel plasma etcher was used to develop the resist patterns. In our recent studies, a Plasma-Therm PK-1250 planar plasma etcher was used to develop the resist patterns.

Results and Discussions

In formulating a plasma developable electron resist (PDE), NVC, DPAE, MAM and ODMA have been evaluated as monomers along with PC1S, PS, PBD and PTCEM as base polymers. Aside from lithographic performance, the main issues concerning the formulation of PDE are the sublimation or vaporization of monomers under vacuum and the compatibility of the monomer with the base polymer.

Vaporization and/or Sublimation Rate of Monomers

A major concern for any candidate monomer is its vaporization or sublimation rate under vacuum. Ideally, the monomer should have a low vaporization or sublimation rate in vacuum at room temperature and a relatively high vaporization or sublimation rate at a convenient bake temperature for ease in removing the monomer from the unexposed area. Further, the monomer should have a significantly lower vaporization rate than that of the solvent used in spin coating of the monomer/polymer mixtures. The vaporization rate of some monomers (powders) was determined by using a microbalance which measured the weight loss as a function of both temperature and pressure. A schematic diagram of the apparatus is shown in Figure 1. The vaporization rate data for some monomers in low vacuum (1×10^{-3} Torr) are shown in Table I. The weight of each monomer before exposure to vacuum was approximately 100 mg. For comparison, the sublimation rate of iodine was also determined and included in Table I.

Monomer/Polymer Compatibility

In general, both NVC and DPAE are quite compatible with base polymers which have aromatic groups, e. g. PC1S and PS. Up to 50 weight % of NVC can be embedded in PC1S or PS without significant crystallization. However, as is well known, both PC1S and PS have low plasma etching rates. When NVC or DPAE is used as the monomer along with PS or PC1S as a base polymer, the plasma etch rate difference between exposed and unexposed area was not sufficiently high for lithographic applications.

The compatibility of MAM with the base polymers PBD, PC1S, PS and PTCEM, is generally poor. Although crystallization of MAM was not obvious in the spin coated composite films, the films generally appeared to be cloudy and exhibited an orange peel structure under microscopic examination.

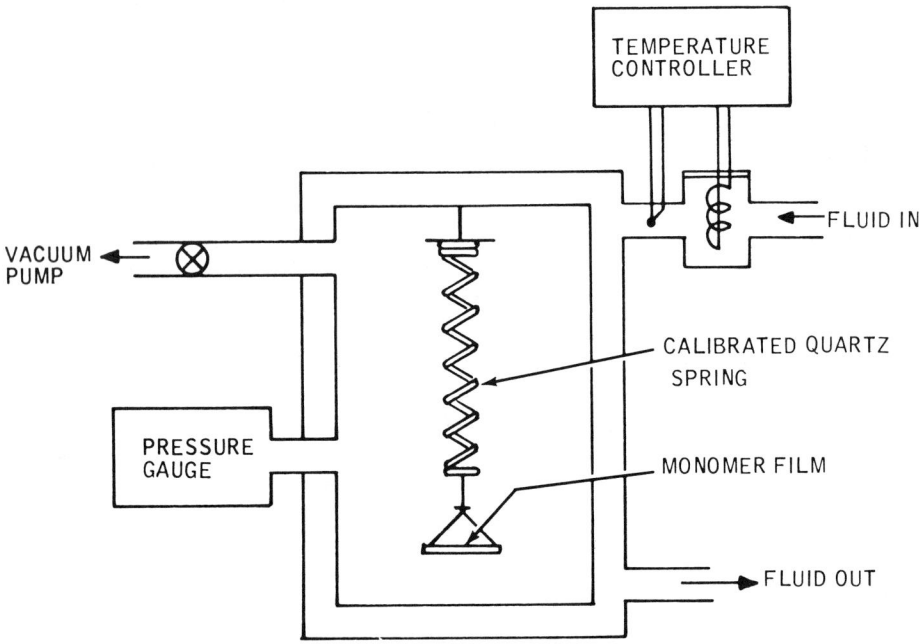

Figure 1. An apparatus for determining the sublimation rate of monomers.

Table I. Vaporization Rate of Some Monomers in Vacuum

Monomer	Temperature °C	Vaporization Rate (mg./hour)
NVC	25	0.35
	40	7.0
MAM	25	2.5
	40	30.0
DPAE	25	0.20
	40	1.51
NTBM*	25	1.05
	40	14.0
ODMA	25	< 0.1
	40	0.30
Iodine	25	35.5

*NTBM: N-t-butylacrylamide

The compatibility of ODMA with most base polymers is generally excellent. This may be, in part, due to its low T_m, i.e. it exists as liquid at room temperature.

Sublimation Barrier

Since the sublimation or vaporization rate of most monomers under vacuum is not insignificant, a thin polymer film was coated on top of the resist as a monomer sublimation barrier. Two water soluble polymers, poly(vinyl alcohol) (PVA) and poly(vinyl pyrroidone) (PVP), were used. It has been found that PVA solution yields more uniform films, and after e-beam exposure, can be removed easily by dipping in water for 30 seconds. There was no evidence of crosslinking in PVA after e-beam exposure. However, PVP acts as a negative resist with sensitivity ($D_g^{0.7}$) approximately equal to $60\mu C\ cm^{-2}$. PVA, therefore, was used as the monomer barrier for most experiments.

PDE Formulation and Lithographic Performance

Poly(trichloroethyl methacrylate) (PTCEM) is a polymer electron resist with a high glass transition temperature, $T_g = 126°$ C. It is soluble in many common organic solvents. The plasma etch rate of the polymer is relatively high (8). In our PDE work, PTCEM has been used as the base polymer. Several monomer/PTCEM mixtures have been evaluated as a PDE.

ODMA/PTCEM Resist

Octadecyl methacrylate (ODMA) can be polymerized by electron beam irradiation (9). When ODMA is used as the monomer with PTCEM as the base polymer, the resolution and line profile of the PDE is poor. This poor performance may be in part due to the low glass transition temperature of poly(octadecyl methacrylate) (PODMA) and the small plasma etch rate difference between PODMA and PTCEM.

NVC/PTCEM Resist

In this formulation, the ratio of NVC to PTCEM is 1:1 in the mixture. One micron thick film was spin coated on a silicon wafer using chlorobenzene as the coating solvent. The film was vacuum dried for an hour at room temperature and coated on the top with PVA film. The film was vacuum baked for an hour and exposed with an e-beam. After e-beam exposure, the PDE film was dipped in deionized water for 30 seconds to remove the PVA layer, and baked at 120° C for 30 minutes. The patterns were visible after the bake. The development of patterns was conducted in a barrel etcher using O_2 plasma. The temperature inside the barrel etcher during the development was maintained at $\leq 90°$ C. SEM micrographs of some plasma developed patterns are shown in Figures 2 and 3.

Figure 2 is a micrograph of a portion of 4 micron-wide line exposed at 1600 $\mu C\ cm^{-2}$. The line is 1.0 micron thick which is virtually the same thickness as before development which indicates that the monomer was completely grafted or polymerized at this high dose. SEM micrographs of plasma developed submicron resist lines are shown in Figure 3. The exposure dose was 60 $\mu C\ cm^{-2}$ and the line width as narrow as 0.3 microns was obtained. The thickness of the resist film after development is approximately 0.3 microns. It should be noted here that NVC has also been used successfully in plasma developed X-ray resists (6).

DPAE/PTCEM Resist

The maximum amount of DPAE which could be embedded in the base polymer PTCEM without crystallization is less than that of NVC, approximately 30 weight %. A 1:2 DPAE/PTCEM mixture in chlorobenzene was prepared and spin coated on silicon wafer at 800 rpm for obtaining 1.0 micron thick resist film. The resist film was

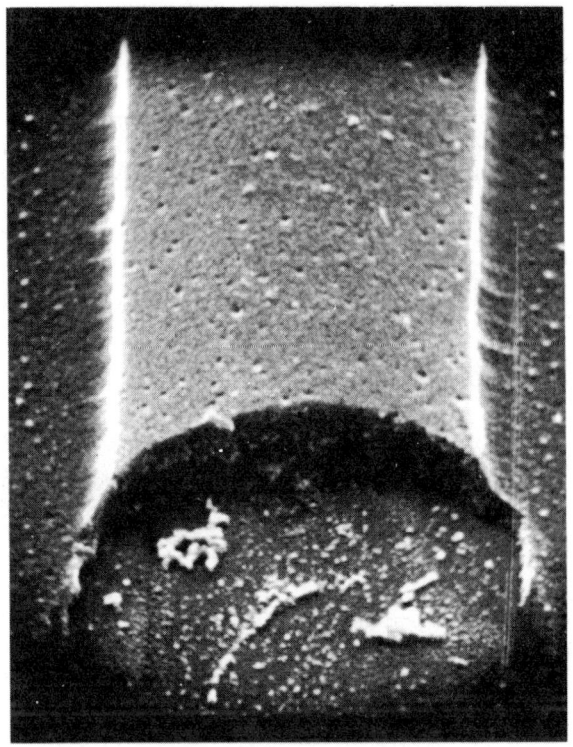

Figure 2. SEM micrograph (14,000X) of plasma developed resist (NVC/PTCEM) line profile. The resist film thickness is 1.0 μm and the line width is 4.0 μm.

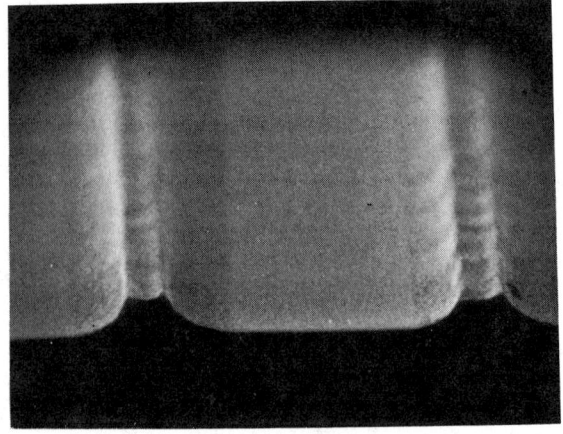

Figure 3. SEM micrographs (10,000X) of plasma developed resist (NVC/PTCEM) patterns. The resist film thickness after development is 0.3 μm and the exposure dose was 60 μC cm^{-2}.

vacuum baked at room temperature for one and one half hour and was spin coated with a monomer sublimation barrier, either PVA or PVP film. After coating, the resist film was again vacuum baked at room temperature for an hour. After e-beam exposure, the monomer sublimation barrier was removed and heat baked at 60-100° C for 30 minutes in a vacuum oven. Plasma development was carried out either in a barrel etcher or a planar etcher. For the barrel etcher, O_2 (oxygen) was used as the etch gas. In Figure 4, the normalized film thickness remained after development vs exposure dose is plotted for two different bake temperatures. It appears from the figure that 60° bake has produced a more sensitive resist but 100° bake has produced a thicker resist at high dose. The exposed patterns were 10 micron by 500 micron rectangles.

Although the PDE resist does have high resolution, as demonstrated by the fact that 0.3 micron line has been obtained, the contrast of the resist does not appear to be significantly better than conventional negative resists. In Figure 5, line profiles of two 1 micron lines separated by 5 microns exposed at 200 and 240 μC cm^{-2} and plasma developed in a planar etcher are shown. The resist film thickness after plasma development are 0.56 and 0.60 microns respectively. The vertical wall was not obtained and the wall angle is 45°. The line profiles are similar to those of wet-developed conventional, high sensitivity negative resists. However, significant improvement in wall angle was not obtained at lower exposure dose. The high resolution of the resist apparently faithfully replicated the electron scattering profile which resulted in non-vertical walls. Significant improvement in line profile, therefore, requires minimization of electron scattering effect.

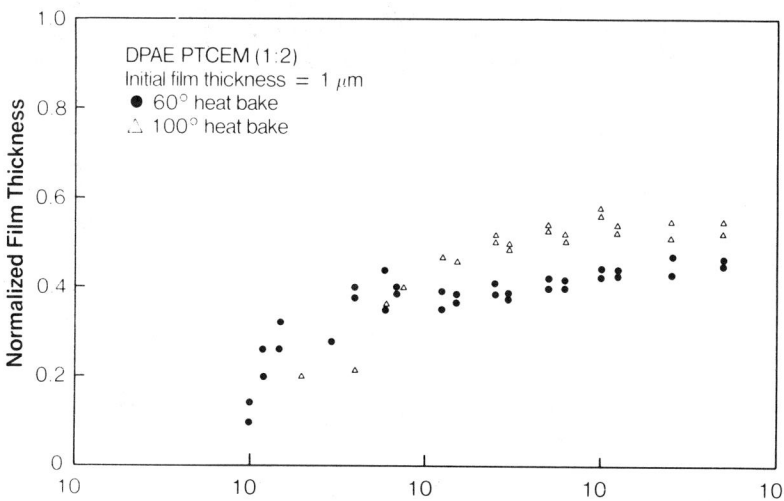

Figure 4. Normalized film thickness after development vs exposure dose (Coulombs cm^{-2}).

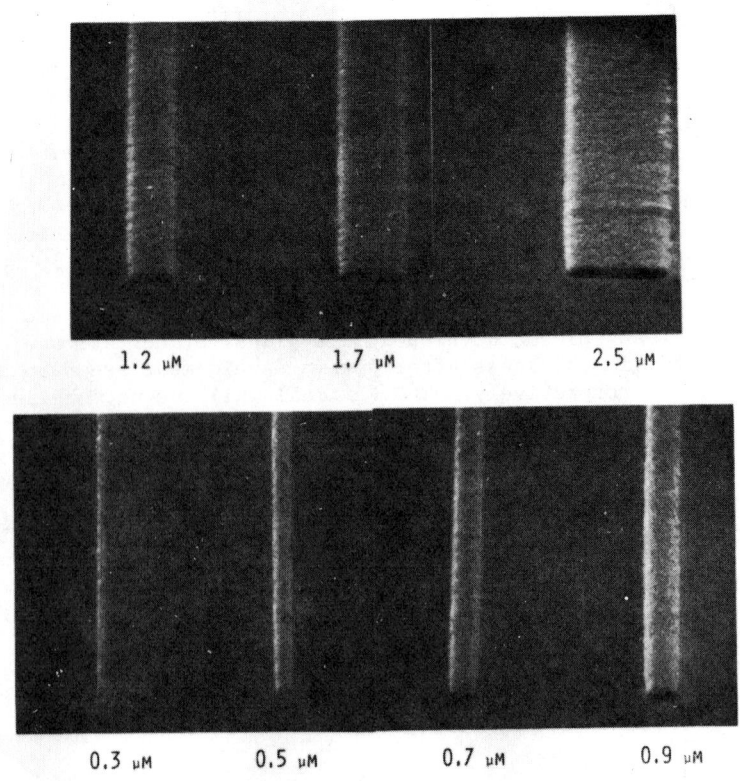

Figure 5. SEM micrograph (10,000X) of plasma developed resist (DPAE/PTCEM) line profiles. The exposure doses are from left to right, 240 and 200 µC cm^{-2} respectively.

It is expected that PDE with improved contrast and line profiles can be obtained in a multi-level resist structure when PDE is used as the top resist. Similar conclusion has been reached in plasma developed X-ray resists (6).

Acknowledgments

The author thanks L. T. Shepherd, Joe Sandstrom and Richard Douglas for their technical assistance.

Literature Cited

1. Thompson, L. F., Solid State Technology, 1974, 44, August.
2. Lai, J. H., Polym. Eng. & Sci., 1979, 19, 117.
3. (a) Lai, J. H. and Shepherd, L. T., J. Electrochem. Soc., 1979, 126, 696.
   (b) Feit, E. D. and Stillwagon, L. E., Polym. Eng. & Sci., 1980, 20, 1058.
   (c) Shiraish, H., Taniguich, Y., Horigome, S., and Nonogaki, S., Polym. Eng. & Sci., 1980, 20, 1054.
   (d) Tamauro, T., Sukegawa, K., and Sugawara, S., J. Electrochem. Soc., 1982, 129, 1831.
4. (a) Hofer, D. C., Kaufman, F. B., Kramar, S. R., and Aviram, Ari, Appl. Phys. Lett., 1980, 37(3), 315.
   (b) Tomkiewica, Y., Engler, E. M., Kustsis, J. D., Schad, R. G., Patel, V. V., and Hatzakis, M., Appl. Phys. Lett., 1982, 40, 1.
5. Hughes, H. G., Goodner, W. R., Wood, T. E., Smith, J. N., and Keller, J. V., Polym. Eng. & Sci., 1980, 20, 1093.
6. Taylor, G. N., Wolf, T. M., J. Electrochem. Soc., 1980, 127, 2665.
7. (a) Tsuda, M., Oikawa, S, Kanai, W., Yokota, A., Hijikata, I., Vehara, A., and Nakane, H., J. Vac. Sci. Technol., 1981, 19(2), 259.
   (b) Whipps, P. W., Extended Abstract, 16th Symposium on Electron, Ion, and Photon Beam Technology, Dallas, May 26-29, 1981.
8. Helbert, J. N., Schmits, M. A., and Lai, J. H., Organic Coating and Plastics Chemistry preprints, American Chemical Society Meeting, San Francisco, August 24-29, 1980.
9. Hatada, M., and Nishii, M., J. Polym. Sci., Polym. Chem., 1977, 15, 927.

RECEIVED September 2, 1983

POLYIMIDES

19

Solution Characterization of Polyamic Acids and Polyimides

P. METZGER COTTS and W. VOLKSEN

IBM Research Laboratory, San Jose, CA 95193

> The low dielectric constant, ease of processing, and high temperature stability of polyimide materials have led to the widespread use of these materials as insulating layers in electronic devices. Several characteristics of these materials are not yet well understood, such as the instability of solutions with time, structural changes on curing, and polyelectrolyte effects observed in dilute and concentrated solutions. In this study we have investigated these and other characteristics of the polyamic acid and polyimide formed from the condensation of pyromellitic dianhydride and 4,4'-diaminodiphenylether. The polyamic acids were studied in solution using various solvents, and the cured polyimide was studied in concentrated sulfuric acid. Techniques used included low-angle light scattering, osmometry and viscometry. Behavior characteristic of dilute polyelectrolyte solutions was only observed in NMP which had not been purified by distillation. In all other solvents studied, normal behavior characteristic of neutral, flexible chain polymers dissolved in good solvents was observed. Comparison of data for samples before and after cure showed that structural changes during cure were limited essentially to imidization with little degradation in molecular weight and no evidence of cross-linking, for samples cured thermally to 300 degrees C.

The condensation of pyromellitic dianhydride (PMDA) with 4,4'-diaminodiphenylether (DAPE) to a precursor polyamic acid followed by cyclodehydration (cure), produces a polyimide (Figure 1) which is widely used in the electronics industry as a high temperature insulating material. This particular polyimide is available in several forms commercially and has been the subject of

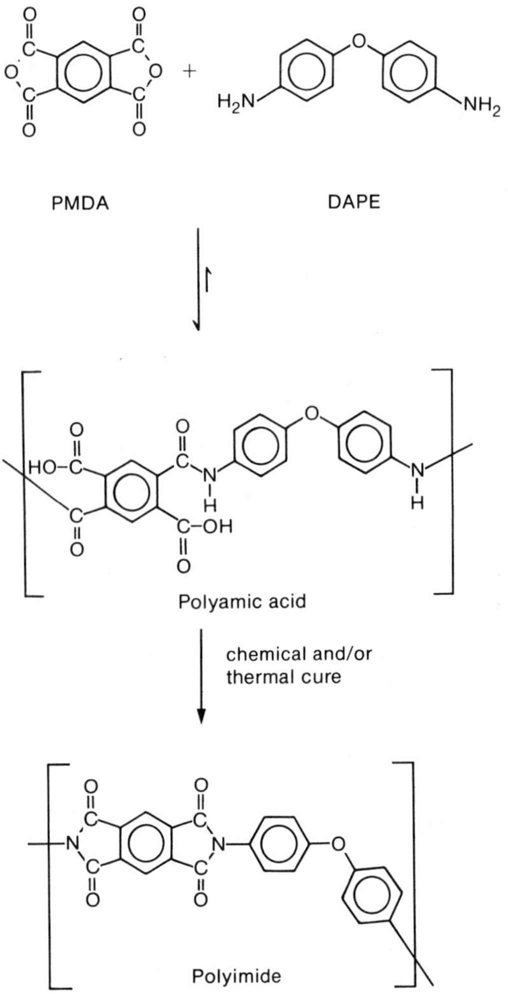

Figure 1. Synthesis and cure of the polyimide and precursor polyamic acid from the condensation of pyromellitic dianhydride (PMDA) and 4,4'-diaminodiphenylether (DAPE).

many investigations (1-5). However, some aspects of its synthesis, physical properties and processing requirements are not yet completely understood; such as control of molecular weight and polydispersity, solution stability, polyelectrolyte effects and changes in molecular structure with cure. We have investgated the properties of solutions of both the polyamic acid and the polyimide to increase our understanding of this particular polymer, and the complete class of polyimide materials. The use of dilute solutions for our studies permitted observation of molecular parameters essentially independent of interactions between polymer chains. Techniques used included low angle light scattering (LALS), viscometry (η) and membrane osmometry (OS). Synthesis of PMDA/DAPE samples in our laboratory (7) permitted conditions of the synthesis to be carefully controlled, eliminating many of the uncertainties inherent in a commercial synthesis. We were able to directly measure weight average molecular weights by LALS on the cured polyimide in concentrated sulfuric acid, and the precursor polyamic acid in organic solvents. Until recently reported from this laboratory (6), the molecular weights of cured PMDA/DAPE samples could only be compared by relative viscosity or final mechanical properties, where the interpretation can be ambiguous.

Experimental

Samples of polyamic acid were obtained commercially (DuPont) as concentrated solutions or were synthesized in this laboratory in N-methylpyrrolidone (NMP), with the polymerization solutions stored under argon until use (7). All dilute solutions were prepared by dilution from the concentrated solution with distilled NMP. Cured polyimide samples were either commercially available films (Kapton), or were cured in this laboratory from either commercial or laboratory synthesized polyamic acids, using a thermal or a combination chemical/thermal cure. Solvents used were all reagent grade, and at times were redistilled before use.

Low angle light scattering (LALS) was measured using a Chromatix KMX-6 light scattering photometer. Weight average molecular weights (M_w) were obtained by determining the Rayleigh factor, R_θ, of several different concentrations and extrapolating to infinite dilution to obtain M_w, using Equation 1 below:

$$[Kc/R_\theta]^{1/2} = (M_w)^{-1/2} + A_2 c (M_w)^{-1/2} \tag{1}$$

which is strictly valid only at $\theta=0°$ and is used here without correction for the low angles measured with this instrument (~4°). This square root equation is often found to be linear over a larger concentration range than the usual equation especially for polymers in very good solvents ($A_2 \gg 0$) as is found here.

The intrinsic viscosity ($[\eta]$) for each sample was obtained by measuring the viscosities of several dilute solutions of the polymer and extrapolating to infinite dilution using the usual relations:

$$\eta_{sp}/c = [\eta] + k'[\eta]^2 c + \ldots \qquad (2)$$

$$\ln(\eta_{rel})/c = [\eta] + (k'-1/2)[\eta]^2 c + \ldots \qquad (3)$$

where η_{sp}/c is the reduced viscosity and $\ln(\eta_{rel})/c$ is the inherent viscosity. Linear plots were obtained except where polyelectrolyte effects were observed, which will be discussed below. Concentrated solution viscometry was measured using a Brookfield viscometer equipped with a pressure transducer and temperature controller to allow simultaneous recording of temperature and viscosity.

Membrane osmometry was measured on a few samples to determine the number average molecular weight (M_n). Regenerated cellulose membranes were used, and were found to be highly swollen in NMP, leading to both very long equilibration times and excellent retention of low molecular weights. Table I below shows the results of these and the above measurements for the samples reported here.

Studies on solutions over a period of months, to be discussed below, were made on solutions stored under argon at room temperature until just before dilution and/or measurement. All dilutions were made with freshly distilled NMP, unless specific effects of the non-distilled NMP were being investigated.

Results and Discussion

Equilibrium Molecular Parameters. Polyelectrolyte effects, evidenced by increased solution viscosity and reduced scattering intensity at very low concentrations, were observed for polyamic acid solutions in NMP which had not been redistilled over P_2O_5. We attribute this effect to the abstraction of protons from the amic acid by amine impurities in the NMP. This is discussed in more detail in an earlier study (6). Measurements reported by other workers in other amide solvents such as dimethylacetamide, have been made with LiBr added to suppress the polyelectrolyte affect (3). Measurements described here were made on solutions in NMP which had been redistilled over P_2O_5, and no polyelectrolyte effects were observed.

Polymerizations such as the condensation of PMDA and DAPE are expected to follow step-growth kinetics and lead to a most probable distribution. Control of molecular weight should be possible via a stoichiometric imbalance, if the purity of the monomers is well controlled. The control of molecular weight by stoichiometric imbalance has been shown in this laboratory for the samples used here (7). These samples span a range of weight average molecular weights from 2000 to 250,000, where the highest molecular weight sample was obtained with equal stoichiometry. These molecular weights were all measured using LALS on solutions in distilled NMP. More details on the use of this technique to determine molecular weights for PMDA/DAPE polymers can be found in an earlier paper (6). Membrane osmometry results on two of the samples confirmed that M_w/M_n was in fact two as expected for a most probable distribution.

In some cases, notably when polymerizations were carried out at higher concentrations, initial molecular weights were higher than expected and decreased slowly over a period of days to the expected M_w. This was accompanied by a corresponding decrease in the concentrated solution viscosity. Results obtained for two samples are shown in Figure 2. Once the equilibrium molecular weight had been obtained, the viscosity and M_w measured by LALS remained constant for several days. Results reported in Table I were measured on these stable equilibrated samples. In all cases, intrinsic viscosities were measured within 24 hours of the LALS measurement so that relations between these parameters would be independent of any changes in the sample with time. We attribute this observed decrease in M with time to a very slow approach to the equilibrium distribution in samples polymerized at high concentration. This is caused by the combined effects of the very high reaction rate of the anhydride-amine condensation and the relative insolubility of the PMDA monomer in NMP. The initial local stoichiometry differences equilibrate later to the expected distribution through the reversibility of the condensation reaction. Details of the synthesis have been discussed in an earlier report (7). Reports from other laboratories indicating degradation in molecular weight with time (2), or synthesis of molecular weights increasing with concentration of the polymerization solution (5) may also be evidence of this observed slow equilibration.

Measurements on Cured Polyimide. Table II below shows results obtained for the cured polyimide samples dissolved in concentrated sulfuric acid. For the samples cured in this laboratory, where the precursor polyamic acid M is known, the observed molecular weight of the cured polyimide is comparable. This suggests that the final physical properties of the cured polyimide should be determined by the molecular weight of the precursor polyamic acid formed.

The samples which were cured chemically with acetic anhydride and pyridine, and then heated to remove solvent, appeared to be incompletely imidized, and yielded lower molecular weights than thermally cured samples from the same polyamic acid. In addition, these samples produced a bright red solution in sulfuric acid, in contrast to the orange-gold color observed in solutions from thermally cured samples. The red color disappeared within 24 hours, when the molecular weights were determined. This color was also observed by Wallach (4), who carried out viscosity measurements on sulfuric acid solutions of chemically cured PMDA/DAPE polyimides. Wallach observed a slow decrease in the dilute solution viscosity with time over a period of hours from the initial preparation of the solution. We have not observed any decrease in viscosity for 24 hours for solutions prepared from thermally cured samples. Polyimide samples which have been cured chemically have been shown to contain a small percentage of isoimide, which is then converted to the more stable imide at higher temperatures (8-9). The observed red color in sulfuric acid solutions may be because protonation of

Table I. Molecular Parameters for Polyamic Acids

Sample	$10^{-3} M_w^{LS}$ daltons	$10^{-3} M_n^{OS}$ daltons	$10^4 A_2^{LS}$ ml/g·dalton 25°C	$10^4 A_2^{OS}$ ml/g·dalton 40°C	$[\eta]$ in NMP ml/g	$[\eta]$ in NMP/dioxone ml/g	η (cp) at ~10 wt %
PAA-1	4.5	—	43	—	28	—	20
PAA-2	9.0	—	40	—	50	—	70
PAA-3	37.0	—	29	—	132	—	700
PAA-4	77.0	—	21	—	245	—	4000
PAA-5	250.0	—	16	—	585	—	200(2%)
PAA-6	2.0	—	45	—	—	14	12
PAA-7	4.0	—	42	—	—	23	—
PAA-8	6.0	3.0	35	45	—	—	7500(32%)
PAA-9	9.1	—	38	—	48	—	—
PAA-10	9.8	—	37	—	49	—	—
PAA-11	10.4	—	48	—	50	44	—
PAA-12	16.0	—	35	—	72	61	—
PAA-13	16.0	—	30	—	74	—	—
PAA-14	22.0	—	31	—	87	—	—
PAA-15	29.0	15.0	25	32	100	—	10000(18%)
DuPont A	28.0	—	—	—	120	—	500
DuPont B	18.0	—	—	—	70	—	—

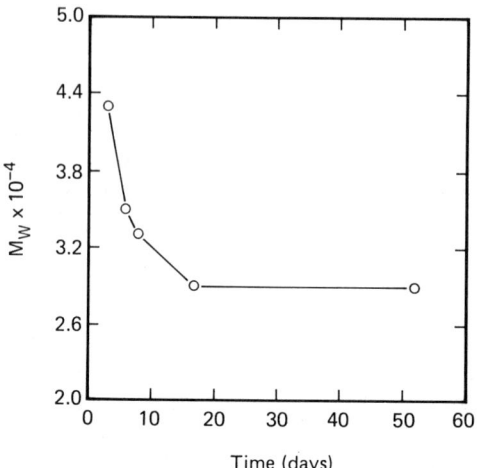

Figure 2. Equilibration of the weight average molecular weight to that expected for a most probable distribution.

the isoimide moiety by the acid, as has been observed for other heterocyclic polymers dissolved in strong acids (10). Slow degradation of the less stable isoimide in the sufuric acid would explain the observed loss of the red color, the lower molecular weight, and the slow decrease in solution viscosity observed by Wallach. The molecular weight of ~7,000 would suggest 1-2 scissions per chain, or 1-2% residual isoimide after the 200° thermal treatment.

Table II shows that a sample which was post-baked at 400°C for one hour after being thermally cured to 300°C was no longer soluble in concentrated sulfuric acid. We attribute this insolubility to a morphological change in the arrangement of the polyimide chains, leading to insolubility in the kinetic sense. Similar behavior has been observed for another heterocyclic polymer and was attributed to formation of aggregates which acted as crosslinks (11). In the latter study, significant partial solubility and swelling of the annealed polymer was observed. Partial cross-linking of the polyimide chains would be expected to produce a highly swollen sample in sulfuric acid, and some degree of solubility, which was not observed here. Thus, if crosslinks arise from an aggregation process, then it is apparently a relatively rapid process, since these samples are only annealed for 1 hour, in comparison to the 15 hour period used in reference 11. The molecular weight of the cured polyimide is essentially determined by the molecular weight of the precursor polyamic acid (see Table II and Figure 3, below). This was observed for both commercial and laboratory synthesized samples, and over a molecular weight range of 10,000 to 30,000. Thus physical properties of the final cured polyimide can be expected to be a direct function of the equilibrium molecular weight obtained for the precursor poyamic acid. Additional enhancement of these properties may arise from specific cure conditions, such as the chemical cure or post-bake above 300°C mentioned above, but these processes probably do not affect the actual molecular weight.

Comparison of Chain Dimensions. Knowledge of the intrinsic viscosities and molecular weights of a series of PMDA/DAPE polyamic acids allows estimation of the unperturbed chain dimensions. Comparison may then be made with the cured polyimide, with results obtained by other workers, and with calculated values. The expressions in the literature using values for $[\eta]$ obtained in good solvents generally involve extrapolation of $[\eta]/M^{½}$ to $M^{½} = 0$, where excluded volume effects are presumed to be minimal (12-14). Although this method is not strictly valid, it is useful for comparison purposes when direct measurement of the dimensions is impractical. The data in Table I for the polyamic acid in distilled NMP and in the poorer mixed solvent NMP/dioxane yield (see Figure 4):

$$(<r^2>_0/M_w)^{\frac{1}{2}} = 0.95 \text{ A} \tag{4}$$

where $<r^2>$ is the mean square end to end distance. These dimensions

Table II. Measurements of Cured PI in 97% H_2SO_4

Sample	$10^{-3} M_w^{LS}$ PAA (daltons)	Cure Conditions	Film (μm) Thickness	Age of Solution	$10^{-3} M_w^{LS}$ PI (daltons)	$[\eta]$ in H_2SO_4 ml/g
Kapton®	—	—	7	5 hours	18	100
Kapton®	—	—	7	4 days	16	—
Kapton®	—	—	12	3 days	20	—
Kapton®	—	—	25	3 days	18	—
Kapton®	—	—	50	3 days	23	—
DuPont A	28	thermal, 150°C	8-10	degraded	—	—
DuPont A	28	thermal, 200°C	8-10	1 day	20	—
DuPont A	28	thermal, 300°C	8-10	1 day	22	—
DuPont A	28	thermal, 300°C	8-10	14 days	25	—
DuPont A	28	thermal, 300°C, 400°C	8-10	insoluble	—	—
DuPont A	28	chemical, 150°C	8-10	4 days	7	—
DuPont A	28	chemical, 200°C	8-10	4 days	9	—
PAA-11	10.4	thermal, 300°C	8-10	7 hours	9	50
PAA-11	10.4	thermal, 300°C	25	4 days	11	—
PAA-3	37.0	thermal, 300°C	8-10	1 day	30	—

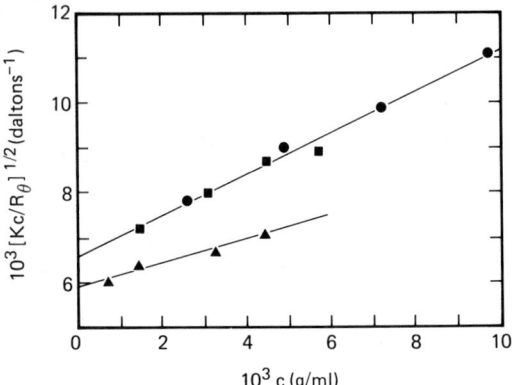

Figure 3. Light scattering measurement of weight average molecular weights of a polyimide (●, ■) and the precursor polyamic acid (▲), showing nearly equivalent molecular weight before and after cure.

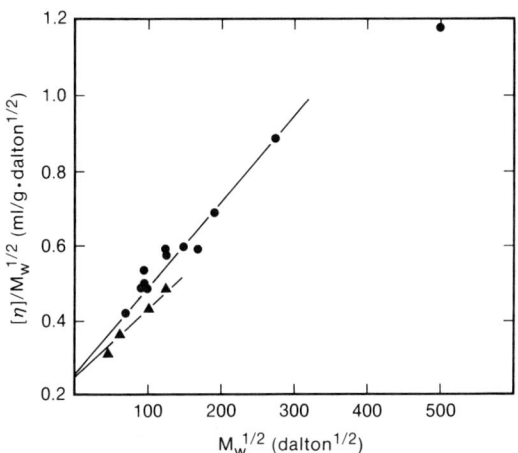

Figure 4. Estimation of the unperturbed dimensions of the polyamic acid from intrinsic viscosity data (●): distilled NMP, (▲): NMP/dioxane.

are larger than those reported by Wallach (3) for the same polymer in dimethylacetamide (DMAc) with 0.1 N LiBr added to suppress the polyelectrolyte effect. As mentioned above, the data reported here are in the absence of any polyelectrolyte effect or added salt, and thus the chain dimensions are larger. We have observed precipitation of the polymer when the concentration of LiBr is 0.2-0.3 N, indicating that the solvent quality is substantially reduced by addition of salt, as in Wallach's study. If the extrapolation to $M^{\frac{1}{2}} = 0$ is not sufficient to eliminate all excluded volume contribution, then the estimated unperturbed dimensions could be affected. The samples used in reference 3 were also of varying polydispersity, and the intrinsic viscosity molecular weight relation showed more uncertainty than we have observed in this study, which is also sufficient to explain the difference in estimated unperturbed chain dimensions.

The limited viscosity data obtained for the cured polyimide in concentrated sulfuric acid (Table II), show that the chain dimensions of the cured material, at least in the protonated form in solution, are essentially the same as in the amic acid form. The amide linkage in the uncured polyamic acid, rather than contributing substantially to the flexibility of the chain, appears rather rigid. This is probably because of the strong resonance stabilization achieved when the amide linkage is in the planar trans configuration, and is co-planar with the aromatic rings. For example, this is observed to be the case for solutions of polyphenyleneterephthalamide, which is highly extended in solution despite the amide linkages in the chain backbone (15,16). Thus the cyclodehydration to the imide ring would not be expected to increase the chain unperturbed dimensions. It should be noted here that this measurement of chain dimensions is an equilibrium average of single molecule size in solution, and is not sensitive to hindrances to segment movement in the bulk state that contribute to the very large glass transition temperature observed for the cured polyimide in comparison to the polyamic acid.

We have been able to measure both the weight average molecular weight and the intrinsic viscosity for equilibrated polyamic acid precursors in NMP, and for stable solutions of the cured polyimide in concentrated sulfuric acid. These measurements were not complicated by either polyelectrolyte effects in the amide solvent or degradation in the acid, as have been reported by other workers for these systems. The results show that molecular weight achieved in the condensation to the polyamic acid is retained in the final polyimide, at least for moderate molecular weights. The condensation reaction appears to follow the expected kinetics and leads to a most probable distribution at equilibrium.

Literature Cited

1. Bower, G. M.; Frost, L. W. J. Pol. Sci.:A 1963, 1, 3135.
2. Frost, L. W.; Kesse, I. J. Appl. Pol. Sci. 1964, 8, 1039.
3. Wallach, M. L. J. Pol. Sci.:A-2 1967, 5, 653.

4. Wallach, M. L. J. Pol. Sci.:A-2 1969, 7, 1995.
5. Birshtein, T. M.; et. al. European Pol. J. 1977, 13, 375.
6. Cotts, P. M. First Technical Conference on Polyimides November 10-12, 1982, Ellenville, N. Y.
7. Volksen, W. First Technical Conference on Polyimides November 10-12, 1982, Ellenville, N. Y.
8. Gay, F. P.; Berr, C. E. J. Pol. Sci.:A-1 1968, 6, 1935.
9. Baise, A. I.; Buchwalter P. L. First Technical Conference on Polyimides November 10-12, 1982, Ellenville, N. Y.
10. Berry, G. C.; Fox, T. G J. Macrom. Sci.:Chem. 1969, A3, 1125.
11. Berry, G. C. J. Pol. Sci., Pol. Phys. Ed. 1976, 14, 451.
12. Stockmayer, W. H.; Fixman, M. J. Pol. Sci.:C 1963, 1, 137.
13. Kurata, M.; Stockmayer, W. H. Adv. in Pol. Sci. 1963, 3, 196.
14. Berry, G. C. J. Pol. Sci.:B 1966, 4, 161.
15. Wong, C.-P.; Ohnuma, H.; Berry, G. C. J. Pol Sci.: Pol. Symp. Ed., 1978, 65, 173.
16. Erman, B.; Flory, P. J.; Hummel, J. P. Macromolecules 1980, 13, 484.

RECEIVED September 21, 1983

Ultrapure Polyimides: Syntheses and Applications

J. DURAN and N. S. VISWANATHAN

IBM Research Laboratory, San Jose, CA 95193

> Some of the problems seen with the commercially available polyimides such as limited shelf life, gelation and high ionic contamination are traceable to the raw materials themselves. A zone refining technique has been perfected for use with organic materials and these precursors have been used to synthesize ultrapure polyamic acids for IC device applications. The key feature of the synthesis is the use of solid ingots of the dianhydrides*. Materials prepared by this technique show low metallic impurities and have been shown to be excellent film formers for a variety of applications. In particular a polyimide derived from PMDA-ODA has been used to passivate magnetic bubble devices. IR techniques coupled with electrical measurements have been used to optimize the cure conditions and a simple resist process has been defined to passivate these devices. Device performance compares well with conventional inorganic insulators.

The major objective of this work was to synthesize a polyamic acid/imide material for bubble memory fabrication applications. Important qualities sought in this material are:

1. Improve adhesion over commercially available polyimides without the use of adhesion promoters.

2. Storage and Shelf life of several months.

3. Low level of ionic impurities and absence of organic particulate matter which would require filtration.

0097-6156/84/0242-0239$06.00/0
© 1984 American Chemical Society

4. Good mechanical properties to form low stress passivation films on devices. 5. Obtain batch to batch consistency in the synthesis for device reliability considerations.

The best approach to attempt to synthesize a suitable polymer was to study the effect of high purification of traditional monomers to higher levels of purity using zone refining; use of solid ingots in the synthesis to reduce thermal effects and charge-transfer complexing ; reactor design to avoid surface effects but to ensure smooth synthesis and ensuring completion of synthesis to avoid undissolved particulates in the system.

The increasing complexity of electronic devices require several materials as dielectric passivation layers. The ability to form uniform coatings and provide thermal and electrical insulations for IC devices has been usually associated with inorganic systems such as SiO, SiO_2 or Si_3N_4 layers.[1].These are usually deposited in the vapor phase using expensive high vacuum equipment and sputtering or other deposition technologies. Recently,extensive use of organic layers have been made. Table I lists some the electrical and thermal properties of polyimides as compared to the inorganic insulators.

TABLE I PROPERTIES OF TYPICAL INSULATORS

MATERIAL	THERM. EXP. /° C	HEAT COND. cal/cm sec°	DIELEC. STR.	DIELEC. CONST. V/cm
Polyimide	$2-4\times10^{-5}$	5×10^{-4}	10^6	3.5
SiO_2	$3-5\times10^{-6}$	5×10^{-4}	10^7	4.2
Si_3N_4	$2-3\times10^{-6}$	3×10^{-2}	10^6	9.4
Al_2O_3	9×10^{-6}	8×10^{-2}	10^5	9-10

Polyimides are a class of thermally stable,electrically insulating polymers and offer the simplicity of forming low stress insulating layers on IC devices [2,3].Major advantages are in the ease of film formation (the precursors are soluble in organic solvents and can be cast from a solution in a process analogous to photoresist coatings); ability to build up layers of varying thicknesses by modifications of the coating compositions and coating processes. In addition to possessing excellent adhesion to a variety of substrates, the polyimides offer coatings of less stress than the inorganic analogs and in many instances offer cost advantages. Also, the imide coatings have been shown to be excellent barriers for α particles in memory devices [4].

A variety of devices have been passivated with polyimide coatings and in general the process involves the use of a polyamic acid (or similar) precursors to form the film, pattern the desired areas and form the final pattern by a thermal imidization process. The final cured polyimide coatings are extremely thermally resilient and also good electrical insulators. Figure 1 schematically shows the processing sequence, and Figure 2 shows a composition used in fabrication. The final cure is a strong function of temperature and the extent of cure is highly dependent on the cure conditions. From the reliability point of view, the devices should be stable at extremes of humidity ambients and a complete cure is a must in the fabrication process. Cures around 200° C have been used successfully for a variety of silicon device applications [5,6]. In this paper, the application of polyimide passivation to a magnetic bubble device fabrication will be addressed.

Magnetic bubble devices are high density memory devices and use stored magnetic domains as storage areas [7]. The key features of these devices are their simplicity of fabrication and use of Ni/Fe as device elements. The passivation of these devices is similar to corresponding IC devices except in the following:
   1. The Ni/Fe permalloy elements are extremely sensitive to temperatures and the devices cannot be cycled above 250° C (due to shifts in coercivities of the magnetic films)[8].
   2. The polyimide passivation, by device design should be compatible with a variety of interfaces, in the present case, the Ni/Fe active elements, Al conductors, SiO_2 interlevel insulation and the terminal metallurgy using Cu and Cr.
For this end application, a variety of commercially available polyamic acid compositions are available and were examined. An in-house synthesized polyamic acid system appeared to have the best combination of all the desired features. The method of synthesis and the fabrication processes involved in the passivation application will be described in this paper.

MATERIALS PURIFICATIONS

Polyamic acids were prepared from a variety of dianhydrides aromatic amine precursors. Some compounds (Aldrich Chemical pany, Milwaukee Wisconsin) used were the following.
   1. PMDA 1:2:4:5 benzene tetracarboxylic dianhydride

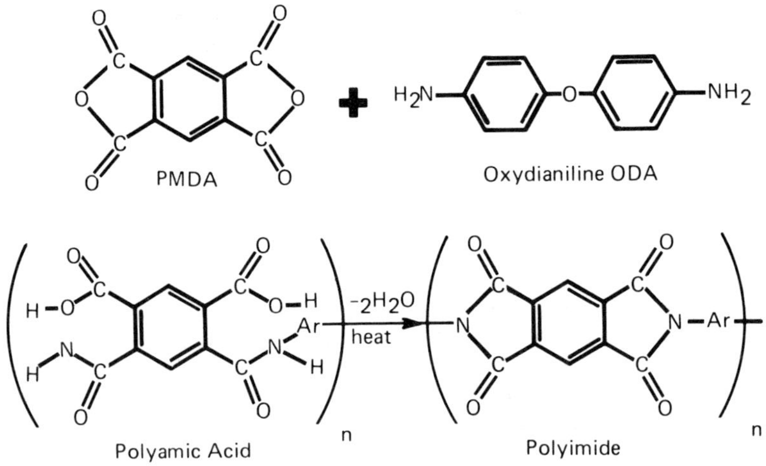

Figure 1. Polyamic acids - Polyamide reactions.

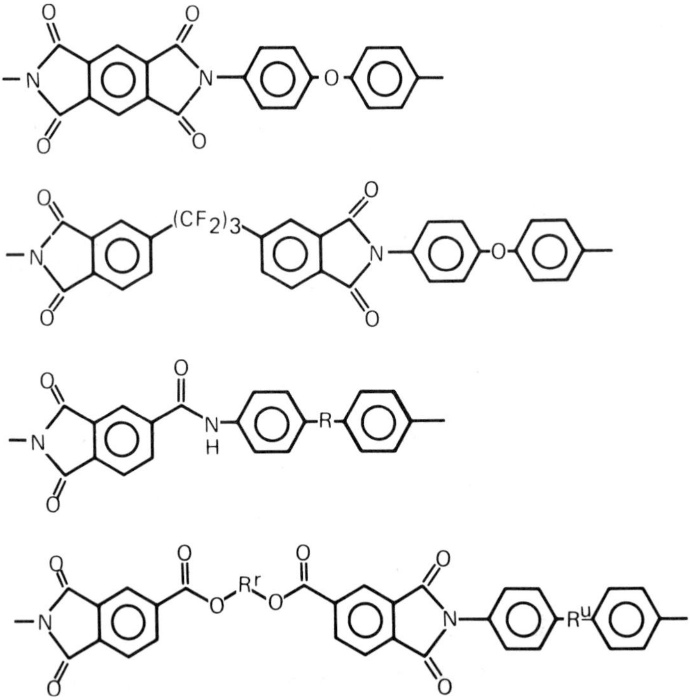

Figure 2. Typical imide structures.

2. ODA 4:4' oxydianiline

3. TMB 3:3':5:5' tetramethyl benzidine

4. BTDA 3:3':4:4' Benzophenone tetracarboxylic dianhydride
The purification steps are described in the following section.

In the case of the solids, initial experiments showed that the conventional purification techniques such as recrystallization from solutions do not offer sufficiently pure materials. For example, PMDA samples, after several recrystallizations and vacuum sublimations appear colorless, but show high levels of ionic contaminations when analyzed for inorganic ions. To eliminate this problem, the technique of zone refining as used in semiconductor materials purifications has been used to purify the starting materials[9]. This technique has been applied for a number of organic materials in our laboratory. PMDA for example, when subjected to a simple zone refining process (as few as 25 zones), shows removal of impurities as high as 3% even if preceded by recrystallization and sublimation. In syntheses of polyamic acids, a variation as small as 3% in stoichiometry can cause considerable variation in the final batch-to-batch synthesis.

PMDA was recrystallized from dry MEK solvent. The anhydride crystals were then mortar crushed to remove traces of the MEK solvent adsorbed in the material. A vacuum sublimation step (240° C, 10^{-6} torr) was followed by zone refining at a traverse rate of 3.0 cm/hour, through 25 zones.

ODA was purified by high vacuum sublimation followed by zone refining through 50 zones.

TMB was also vacuum sublimed and zone refined (100 zones) for use.

BTDA could not be zone refined due to its tendency to glassify (supercooling). The commercial BTDA was washed in MEK, vacuum dried, and sublimed at 230° C, at 1×10^{-6} torr.

The zone refined, purified precursors were colorless and in some cases clear after the zone refining operations. The materials remained stored in the zone refining tubes (glass containers) prior to their use. The major precaution was to avoid exposure to moisture.

All the solvents used in the synthesis of the polyamic acids were obtained from Burdick and Jackson laboratories. All were "Distilled in Glass" quality, usually with a specified upper limit for the water content. Most of the solvents contained less than 0.009% water. No further purification of the solvents were attempted.

The materials were preserved in the Solid Ingot* form for syntheses(cf section below). To achieve this consistency, the PMDA was carefully removed for use from the zone refining tubes for use. The ingots were 11 mm OD and varied from 6 to 25mm in length. The BTDA material was melted in a nitrogen purged furnace, then quenched to produce the solid ingot form. In the case of syntheses involving mixed anhydrides, co-melting was employed to achieve a homogeneous distribution of the precursors for polymerizations. As will be seen later this is an inherent feature of the solid ingot syntheses, not easily attainable in the conventional synthetic routes.

POLYAMIC ACID SYNTHESIS

Several vessel types were examined for use in the synthesis of the polyamic acid from the purified precursors. Several materials such as quartz, pyrex, sodalime glass, teflon, polycarbonates etc were tried. Of these only quartz and pyrex showed longer room temperature shelf lives for polyamide acid samples synthesized. Pyrex was preferred because of its ease to be modified by conventional glass blowing, and availability. The reactors used were modified pyrex bottles with built in vanes about 120° apart from each other (cf.Figure 3). Bottles with capacities from 1 liter to 4 liters were used as reactors for varying applications. Standard ground glass joints were used with the bottles. Teflon sleeves and Parfilm tapes were used to seal the reactors against moisture during the reactions.

Typical synthesis can be generalized as follows:

A stoichiometric amount of the diamine (to 0.1% accuracy in weight) was placed in the glass vaned reactor. A measured amount of solvent ($\simeq$ 80% by weight) was added to the amine. The bottle was then sealed and rotated in a conventional ball mill apparatus to ensure complete dissolution, usually 30 minutes. The reactor was then opened and a stoichiometric amount of solid ingots of the anhydride were placed into the diamine solution all at once. The bottle was then sealed again completely, placed in the ball mill and rotated at a speed $\simeq$ 30 RPM for 60 minutes. After this time, the rotation speed was reduced to about 10 RPM. The reaction is complete when no more anhydride is seen in the reactor, and this usually takes about 24 hours (Maximum 28 hours). This indicates the completion of the synthesis, and the vessel is stored at room temperature (20-23° C) for about 12 days. The container is kept sealed during this time to ensure normalization of the viscosity. Periods of storage less than 12 days usually results in lower viscosity value.

* US Patent 4,269,968; J. Duran et al.; assigned to IBM Corporation.

Synthesis of some polyamic acid compositions are given below:

a. PMDA/TMB <u>colorless</u> polyamic acid: To a well dissolved solution of 4.4 g of TMB in DMSO (30 ml),4.0 g solid ingots of zone refined PMDA were added. The PMDA ingots immediately appeared light yellow (as opposed to the expected orange color). Soon the solution became bright yellow,and with overnight stirring resulted in a colorless clear polyamic acid solution. A viscosity of 101 poise was obtained after dilution of this solution to 7.3% solids. Following precipitation from a little amount of water the tough sponge like solid appeared pure white. In all the cases where the solids content was above 20% the polyamic acid was colorless.The resultant PMDA/TDB films could be imidized and showed excellent thermal stability up to 450° C and optically were superior to other polyimides.

It should be mentioned that colorless PMDA/TMB acids were never obtained when the PMDA in the powder form was added to the amine solution. Slow addition also is not useful and the solution following the completion of the synthesis is yellow in color. Also when using powdered PMDA there were clear indications of exothermic reactions in the mixture,resulting in a rise in temperature.

Solid ingots of the anhydride serve to lower the surface area during the synthetic reaction. This in turn slows the reaction to produce only negligible complexing,thermal reactions and produces conditions for higher molecular weight polymers.

b. PMDA/ODA: This material has been the major material used in the bubbles passivation work and this syntheses has been repeated several times and has been extremely reproducible. Typical reactor charges were 120.0 g of PMDA in the ingot form used with a solution of ODA (110.163 g) in 1.05 liters of DMSO solvent. In this case both the precursors were zone refined (cf previous sections) and a pyrex vessel was used for the synthesis. After the introduction of the PMDA,the ingots usually turn orange and the solution turns pale yellow in color. A temperature rise of 1-2° C is seen when the solids content was higher than 20%. Normally very little temperature rise is seen in the reactor. The final polymer is light yellow in color and and with about 20% solids the final viscosity is about 850± 50 poise. The molecular weight is around 40,000. Higher Mol.Weight samples have been made using higher percentage solids at the start of the reaction.

The synthesized materials were used after the 12 day waiting time described earlier to obtain consistent products. The material is exceptionally stable even in the diluted form and materials have been successfully stored for several months at room temperature,and for more than four years when kept in sealed containers at 4° C. Long term storage usually results in a drop in viscosity as opposed to gelation of the polymer.

The material could be easily selectively diluted with DMSO and in many applications, a variety of solutions could be made for different applications using the viscous material as a source. Some discussions on the viscosity and related film thicknesses are dealt with in the subsequent sections.

CHARACTERIZATION

The polyamic acid synthesized by the technique described was extremely consistent from batch to batch. Table II describes the typical analytical data from a typical sample of the ultrapure material.

TABLE II TYPICAL ANALYSIS DATA FOR ULTRAPURE POLYIMIDE FILMS

PROPERTY	TYPICAL DATA
Metals(Na,K.Li)	< 5ppm
Metals(other)	< 30ppm
Chloride	< 20ppm
Acidity(KOH/g)	≈ 40
Residue upon Ignition	none detectable

From the data, it is clearly seen that the material has little ionic contamination and little non volatile residue, thus meeting the objectives set forth. Another important property was the excellent stability of the material when stored at room temperature. Figure 4 shows the viscosity of the material plotted as a function of the weight of polyamic acid in solution. It is clearly seen that the linearity is maintained well into the low viscosity region.

The control of viscosity of the polyamic acid solutions is useful in casting films of various thicknesses, Figure 5 shows the results of these. The films were cast using conventional spin techniques and the thicknesses measured with an IBM 7840 film thickness analyzer ^{10}, the optical constants were independently estimated by ellipsometry. In Figure 5 the thickness at a particular spin speed 4000 RPM is plotted against viscosity, and it is seen that the linearity is maintained throughout the range. In fact, the polyamic acid solutions could be modified to coat layers as thin as 20 nm by a single application of a dilute polymer film to several μ m (24μ m) by repeated applications (8) of a viscous film. This broad extendibility proved very valuable in customizing the coatings for a variety of applications.

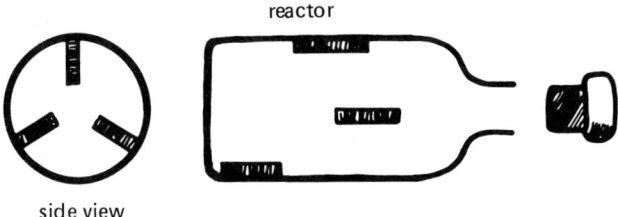

Figure 3. Reactor for solid ingot synthesis.

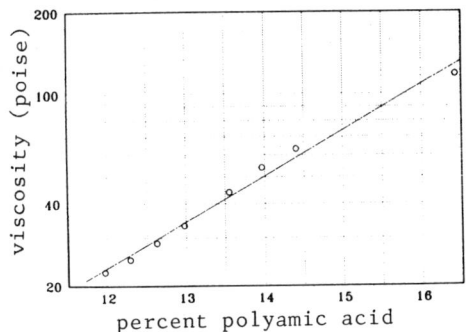

Figure 4. Viscosity vs. weight percent polyamic acid.

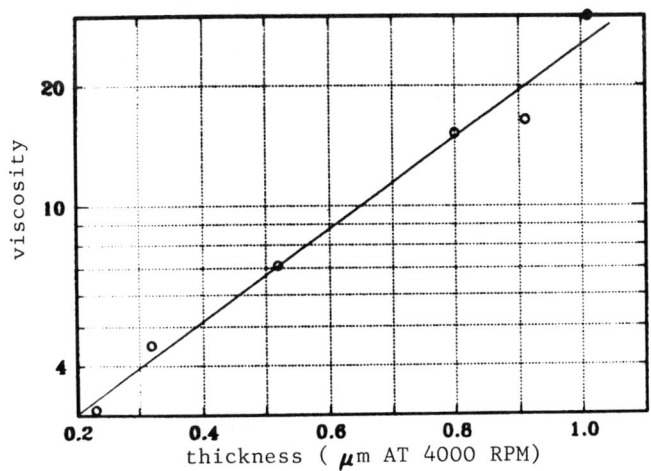

Figure 5. Film thickness vs. viscosity at 4000rpm.

Filtration was not required owing to the high purity of the precursors and to the control of the synthesis with the solid ingot method.

PROPERTIES OF CURED FILMS

Since the polyamic acid is imidized thermally, it was important to determine the thickness loss upon imidization. This was done for a number of films and the data is plotted in Figure 6. It was seen that in almost all the cases, the thickness change was about 30%. This was useful in altering initial polyamic acid thicknesses to suit the device needs, and in a few cases to form multiple layers of imides on devices.

The key step in the application of the polyamic acid for devices is the ability to form fully cured (imidized)films. For initial experimentations, the ultrapure material was compared with other commercially available polyimides and sputtered SiO_2. Some electrical data available is discussed in this section. For electrical testing, the films were cured at 350° C, and were totally imidized, as shown by conventional loss tangent and IR methods. The electrical tests were performed using a conventional Al-Imide-Al dot capacitors. In Figure 7 the leakage current at an applied field of 10^5 V/cm (film thickness = 700 nm) at various temperatures. Two different materials are compared. It is seen that the material synthesized in DMSO/DMF shows a lower leakage current than the material made in DMF solvent. Commercial imides showed a slightly higher leakage current. Figure 8 shows the effect of applied field on the leakage current at 170° C. In this case also, the trends noted in the previous data were clearly seen. Figure 9 shows the Na ion mobility through the polyimide films, and all the materials behave similarly. The test structures were thermal SiO_2-Imide-Al dots. These results indicate that the materials behave as good insulators and are useful for device passivations. The adhesion of the films on a variety of substrates was also well within expectations. Table III shows the data obtained with a number of substrates.

The important point to be stressed is that in all cases no surface adhesion promoters were used. The only precaution taken was to ensure a good cleaning of the substrates. The final step prior to application was a vapor degreasing operation with hot freon vapors.

Figure 10 shows the TGA data for polyimide films synthesized in two different solvents. It is clearly seen that the properties of the final imides are almost independent of the solvents used. In this case, a polyimide made in DMSO is compared with a sample

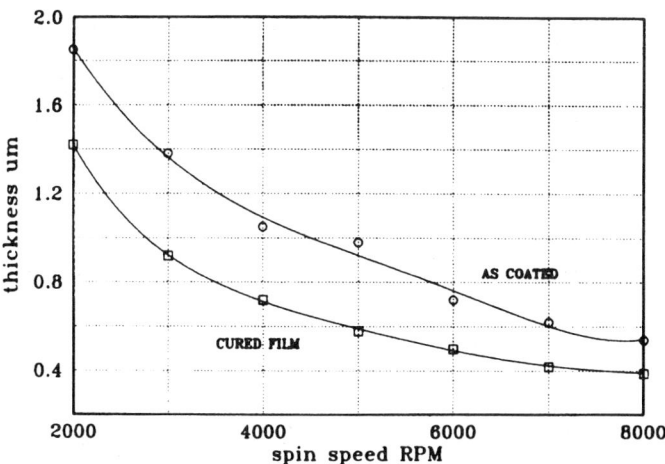

Figure 6. Thickness changes upon imidization.

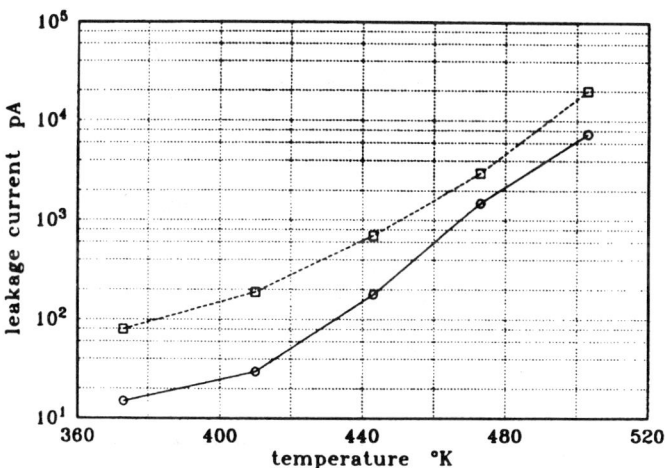

Figure 7. Leakage current at various temperatures.

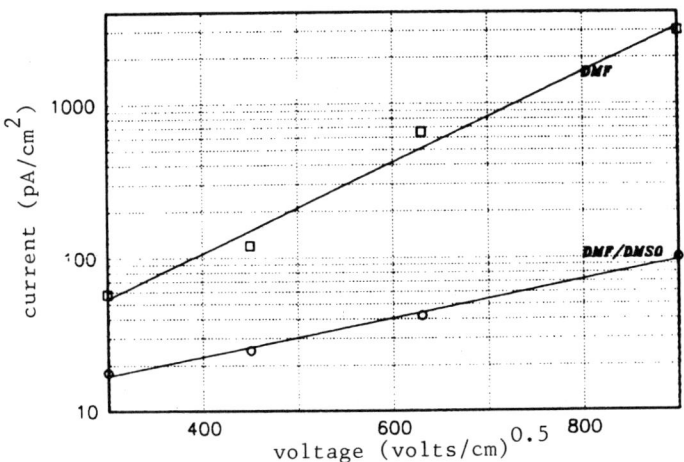

Figure 8. Effects of applied field on leakage current.

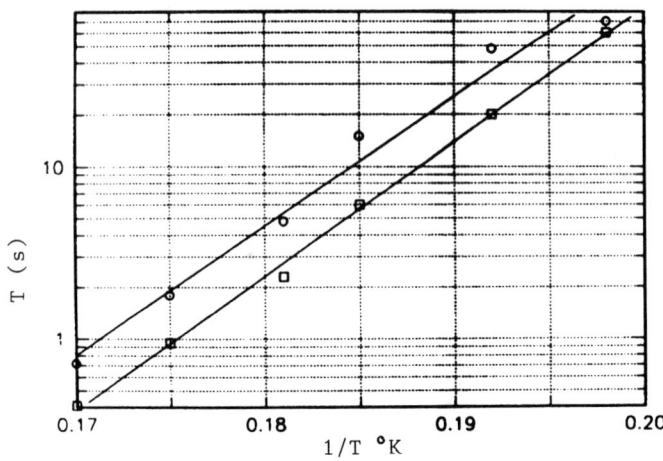

Figure 9. Sodium ion mobility through polyimide films.

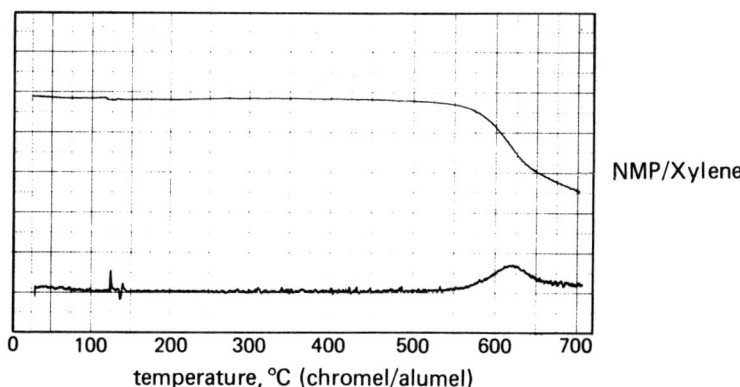

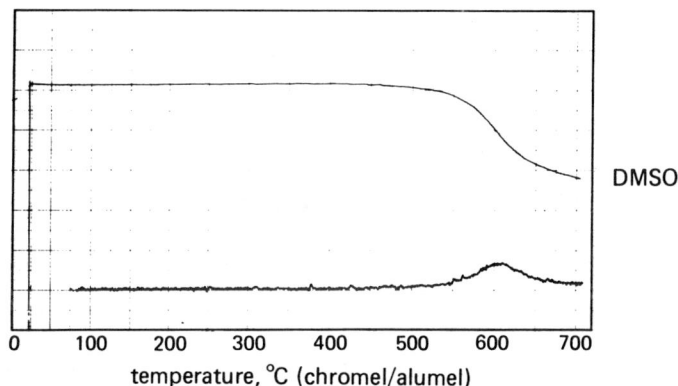

Figure 10. TGA data on polyimides in DMSO and NMP/xylene.

made in NMP/Xylene. DMSO was preferred in our applications since polyamic acids in DMSO showed better wettability to metal substrates when compared to NMP or NMP/Xylene polyamic acids.

TABLE III ADHESION TEST DATA : SEBASTIAN QUAD I TESTER

UNDERCOAT	TOPCOAT	ADHESION**	FAILURE MODE
Gold	Polyimide	5.3-6.4	Epoxy Conn. failure
Polyimide	Gold	1.7-2.0	Gold off from imide
Polyimide	Ni/Fe	8.4	Epoxy Conn. Failure
Polyimide	Tantalum	9.3	Epoxy Conn. Failure
Aluminum	Polyimide	5.2	Epoxy Conn. Failure

** in kpsi units.

CURE OPTIMIZATION

Initial experiments showed that the PMDA/ODA material can be successfully cured by a hot plate bake at 280° C for 3 hours. Electrical,Loss tangent[11] data showed that imidization was complete under these conditions. As noted before,the passivation process for bubble devices had to be optimized by using a lower temperature (to preserve the magnetic properties of the permalloy structures),without losing the electrical insulation characteristics of the final polyimide. IR technique was used as the method of detecting complete cure. A Perkin Elmer 283 spectrometer was used for the IR analyses routinely,and in a few cases the results were checked with a Digilab FTIR instrument.Routine resolutions were 2cm^{-1} with the Perkin Elmer instrument. The studies of imide formation by IR is a well known technique[12,13,14]. The procedure is to follow the disappearance of the amide/acid bands and simultaneously follow the appearance of the imide bands in the films. The former peaks are,3300, 3080,1410 and 1312 cm^{-1}.The imide formation can be followed by the bands at 1780,1720 (Carbonyl vibrations of the imide,imidal vibrations),1375 (the axial imide Imide II C-N vibrations),1120 (the Cyclic imides Imide III vibrations). Additionally the bands at 720 cm^{-1} out of plane imide vibrations proved to be of great value in our work). To quantify these peaks, the aromatic vibrations at 1503 cm^{-1} were used as the internal standards. By repeated experiments, it was seen that these peaks remain unaltered by temperature cycles. Significantly, the appearance of the peaks at 1720 and 720 appeared to be the best tests for imidization. Typical final values normalized to the 1503 cm^{-1} peak were, 0.21 for the 720 cm^{-1}, and 1.05 for the 1720 intensities. The IR technique was particularly useful in arriving at a bake condition usable with bubble memory devices. Figure 11 shows typical cure data as a function of cure conditions.

A series of device wafers were processed along with the KBr IR plates and magnetic and IR measurements were performed using varying cure ambients. It was found that a vacuum bake at 220° C was sufficient to cure the films completely and preserve the magnetic properties of the Ni/Fe films. Figure 12 shows typical IR comparisons of the air cured and vacuum cured films. For all device applications, a vacuum bake at 220° C was chosen as the standard processing condition.

RESIST PROCESS OPTIMIZATION

For passivation applications, the use of the ultrapure polyamic acid involves some changes in the routine processing. The polyimide passivation process consisted of the following steps:

a. Application of the polyamic acid solution

b. Resist application

c. Exposure and development of the resist

d. Resist stripping and

e. Cure of the polyamic acid to the final imide

Of these steps a, and e have been independently examined and conditions optimized by techniques without using the resists. Step b had to be optimized to prevent intermixing of the resist and polyamic acid layers. This was accomplished by control of the prebake of the polyamic acid films prior to resist applications. GC/MS data indicated that most of the DMSO was removable from the films by a bake of 180° C. However, at this temperature too much imidization of the films affect the development of the polyamic acid in the resist developer. A suitable compromise had to be arrived at to preserve the image integrity and allow a simple processing of the resist and the imide in a single development step. To achieve this goal, a variety of exposures were made with the resist films and the final images were measured (bias measurements were done). The data given in Figure 13 indicates

254   POLYMERS IN ELECTRONICS

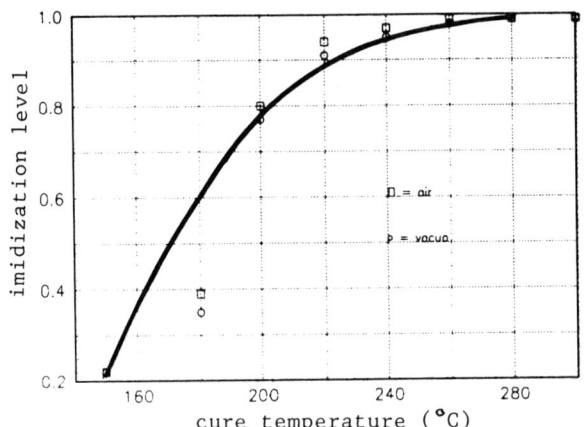

Figure 11. Imidization levels vs. temperature.

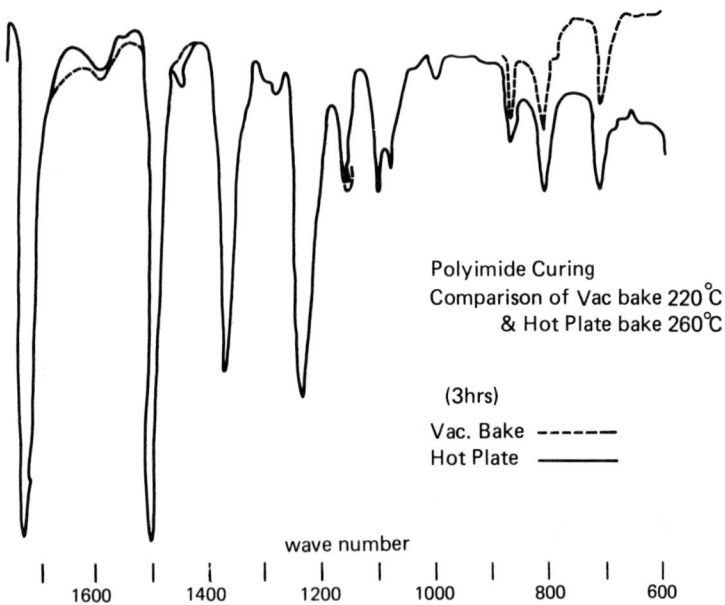

Figure 12. Cure ambients: Air vs. vacuum cure.

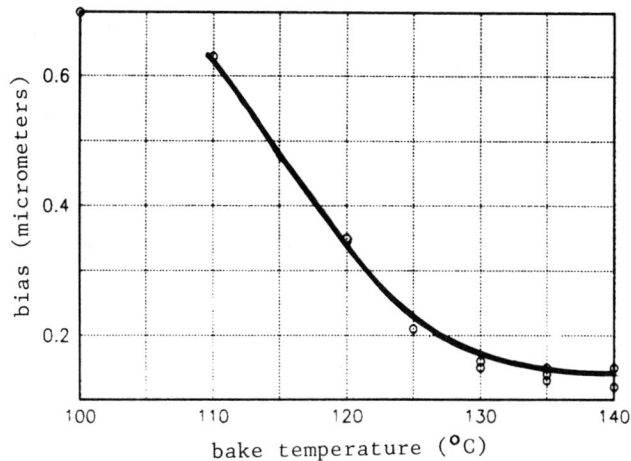

Figure 13. Image bias vs. prebake temperatures.

that a suitable compromise can be reached in defining the prebake conditions. The imidization level at the 130-140° prebake is estimated to be within 20%.

The other step, viz, resist removal without attacking the practically cured polyamic acid also proved to be tricky. By a series of experiments, a solution of aliphatic esters appeared satisfactory. Alternately, plasma etching in CF_4 or CF_4-O_2 mixtures also could be used. In the present system, these were avoided based on device reliability considerations.

The cure conditions used for passivation of the Ni/Fe devices were a vacuum oven bake at 220° C for 3 hours. The 220° C was reached in a programmed mode to avoid thermal stress on the substrates.

PASSIVATION PERFORMANCE

The devices passivated by the polyimide process were compared for performance with similar devices using SiO_2 passivations. In all the cases, it was found that an 0.8µ cured film of polyimide derived from PMDA/ODA performed equally well as the SiO_2 passivated devices. Bias margins (indicative of magnetic device performance) were within the specified limits. The polyimide passivated devices were compatible with standard terminal metallurgies and also chip mounting metallurgies. Device performance at the chip and module levels were excellent. References 15,16 show some of the advantages of the imide devices over similar SiO_2 devices. In particular, the thin film stresses of the polyimide films were at least a factor of 10 lower than the corresponding SiO_2 films used for passivations. Also, the bias margins of the devices showed a higher sensitivity to small variations in film thicknesses when SiO_2 was used as passivating films.

CONCLUSIONS

A variety of low impurity polyamic acid compositions have been prepared by using precursors purified by a zone refining process developed for organic materials. The levels of impurities in these precursors are in the ppm range and are very useful for high reliability microelectronic applications. Using these precursors, a simple solid ingot synthesis method has been developed and successfully employed to produce consistent high viscosity polyamic acids. The key objective of this technique was to reduce the surface area of the <u>anhydride</u> during synthesis to prevent charge-transfer complexing and reduce exothermal temperature changes during the synthesis.

The ability of these to be diluted to a variety of lower viscosities allows these to be used in thin film applications (such as interlevel spacers) as well as thicker film applications such as passivation layers for terminal and conductor metallurgies. The polyamic acids synthesized could be used with little or no filtration. Polyamic acid solutions could be stored easily at room temperature for several months. Additionally, these solutions never demonstrated a "gel-effect" when stored in pyrex containers.

Specifically for the passivation of temperature sensitive bubble memory devices, these ultrapure materials proved to be of great value. A cure process was optimized to obtain a reliable low temperature cure without affecting the magnetic coercivities of the bubble memory devices. A positive resist process, using a simple development step to pattern via holes in devices has been optimized and successfully used to fabricate devices. The devices fabricated using the the polyimide process have been compared with conventional SiO_2 offers reliable passivations with thinner stress free films for passivations. The fabrications involve simple inexpensive process steps and are compatible with conventional resist processes. The reliability of the imide passivated devices can be considerably enhanced by the use of ultrapure starting materials to preclude harmful ionic mobilities through passivated layers.

Acknowledgments

We would like to thank Laura Rothman for some of the electrical data used in this paper. T.Montelbano provided many helpful suggestions in optimizing the permalloy passivation processes.

Literature Cited

1. "Handbook of Thin film Technology " (L.Maissel and R.Glang, Editors)McGraw-Hill,(1970)

2. M.L.White, Proc. IEEE, 57, 1610 (1969)

3. A.M.Wilson,in "Polymeric materials for Electronic Applications " ACS symposium series, 184, (E.Feit,Editor) 1982. Chapter 11

Also,A.M.Wilson. Thin Solid Films 83, 145,(1981). (An excellent review on the subject)

4. Y.K.Lee and J.D.Craig in "Polymeric Materials for Electronic Applications" ACS Symposium Series 184, (E.Feit,Editor) 1982.Chapter 9.

5. S.Harada,T.Okubo,K.Mukai and T.Kimura,A.Saiki J.Electrochem.Soc, 124, 1619,(1977)

6. G.Samuleson,ACS Organic Coatings and and Plastics Chem.Preprints 43, 446,(1980). See also Ref.4 Chapter 8.

7. A.Eschenfelder,"Magnetic Bubble Technology " Springer-Verlag (1980)

8. R.L.Andersen,A.Gangulee and L.Romankiw,J.Electron.Materials, 2, 161,(1973)

9. J.Duran and D.Harrer,IBM Techn.Discl.Bulletin 19, 4741,(1977)

10. W.J.Daughton and W.J.Daughton,J.Electrochem.Soc, 1, 126, 269,(1979)

11. A.J.Gregoritsch,14th Annual Reliability Physics Conference (IEEE) 14, 228,(1976)

12. A.A.Berlin,Russ.Chem.Reviews, 40, 284 (1971)

13. I.K.Verma et al Die.Angew.Makromol.Chemie, 64, 10,(1977)

14. R.A.Dine-Hart,D.B.V.Parker and W.W.Wright,Brit.Polym.J. 3, 222,(1971)

15. N.S.Viswanathan Paper presented at the 2nd Int.Bubble Memory Conference, Santa Barbara,CA,December 1982.

16. C.T.Horng and R.O.Schwenker,J.Appl.Phys, 52, 2383,(1981)

RECEIVED September 2, 1983

21
Photosensitive Polyimide Siloxane

GARY C. DAVIS

Corporate Research & Development Center, General Electric Company, Schenectady, NY 12301

A photosensitive silicone polyamic acid is described. This material is easily prepared directly from the precursor silicone polyamic acid. Properties and characteristics of this polymer system are discussed including handling, use, and practical photochemistry.

Thermal tempering of the photosensitive or cross-linked polymer gives the polyimide siloxane which has been previously shown to be an excellent candidate as an insulating polymer in electronics. The use of such a directly patternable polyimide for dielectric and passivation applications, particularly in microelectronics, should become increasingly important as polyimides become more widely accepted in the industry.

The insulating properties of polyimides for microelectronics applications are well known (1). However, patterning of polyimides using photoresist technology can be cumbersome since inorganic protective layers are often required to protect polyimide films during dry etching. Even the use of wet solution development of polyamic acid films still requires the use of a photoresist to form the initial pattern. A photosensitive polyamic acid can simplify processing by acting essentially as its own photoresist. If designed properly, the photosensitive polyamic acid when thermally cured is converted to the parent polyimide with its inherent excellent thermal and electrical properties. Scheme I demonstrates a savings of six steps in device fabrication by the use of a photosensitive polyimide.

This paper will report on the preliminary work on the preparation of a photosensitive polyamic acid which has advantages over previously reported materials (2) in the ease of synthesis and in some properties of final imide film.

260　　　　　　　　　　　　　　　　　　　　POLYMERS IN ELECTRONICS

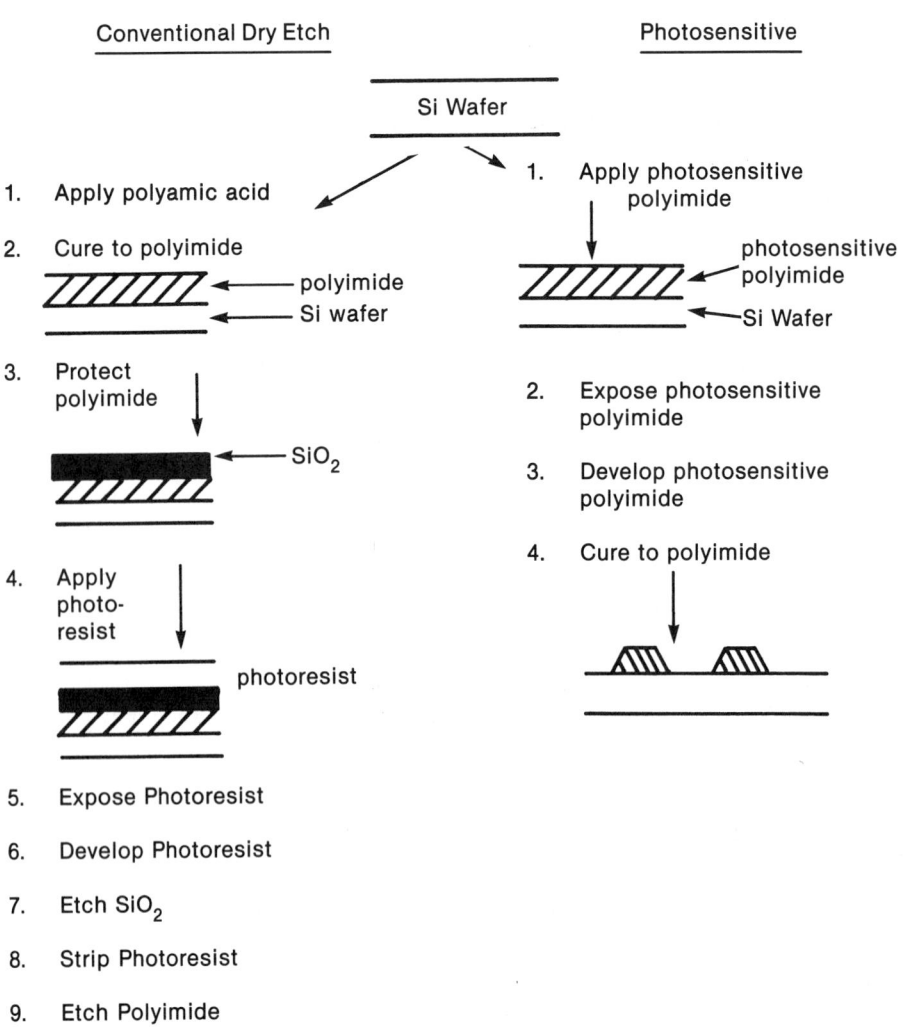

Scheme I.　Comparison of Patterning Process for Conventional and Photosensitive Polyimides.

Experimental

General. Infrared measurements were performed using a Nicolet 7199C FTIR on thin films ($\sim 1\mu$) coated on silicon wafers. These measurements were used to follow the UV curing and thermal tempering of polyamic acid films. Bulk IR's were performed on a Perkin Elmer 598 Infrared Spectrophotometer. Glass transition temperature (Tg) measurements were performed on a Perkin Elmer DSC-2 Differential Scanning Calorimeter. UV exposures were performed on a PPG Model QC 1202 Ultraviolet Processor run at 10 to 20 ft/min, total exposure time 15 to 30 sec, or under 4RS sunlamps located 10 inches from the sample on a turntable. RS sunlamp intensity was measured to be 4.3 mW/cm^2 at 365 nm. Film development was done in batch fashion. All solvents and reagents were used as received. Isocyanotoethyl methacrylate is an experimental monomer from Dow. N-phenyl phthalamic acid was prepared from a phthalic anhydride and aniline in methylene chloride.

Polyimide Siloxane (SiPI). Into a 500 cc 3 neck round bottomed flask equipped with a mechanical stirrer and nitrogen bypass is placed 54.00 gms (0.17 moles) of 3,3',4,4'-benzophenonetetracarboxylic dianhydride dissolved in 250 cc of sieve dried N-methyl pyrrolidone (B&J). To the stirred solution is added 23.22 gms (0.12 moles) of methylene dianiline followed by 12.6 gms (0.05 moles) of bis-1,3-gamma aminopropyltetramethyl disiloxane. The viscous solution is allowed to stir at room temperature for 24 hours.

Isocyanatoethyl Methacrylate Modified Polyimide Siloxane (PSiPI). Into a 100 cc 1 neck round bottomed flask protected from light and equipped with a magnetic stirring bar and a nitrogen bypass, is placed 35 gms of 28% solids in N-methylpyrrolidione (NMP) polyimide siloxane (SiPI) and 5.5 gms of isocyanotoethyl methacrylate. The reaction mixture is stirred at room temperature for 24 hrs. The evolution of CO_2 begins immediately, yet some isocyanate remains for about 24 hrs. In some reactions toluene is added to reduce the initial viscosity of the solution. Toluene is also found to improve spincoating capabilities of the polymer solution. Prior to coating, a sensitizer package is added. A typical package consists of 4% Michler's Ketone and 4% N-methyldiethanolamine. The final solution is stored cold ($\sim 4°C$); however, it is warmed to room temperature prior to use.

PSiPI Film Preparation. A cleaned (acid washed or oxygen plasma descummed) silicon wafer is spincoated with PSiPI and dried at 100°C for 1 hour. The wafer is then masked and exposed. Typical times are 30 sec on the PPG Ultraviolet Processor and 5 min under RS Sunlamps. The unexposed area is then developed typically using 0.5N NaOH. The patterned film is then thermally tempered at 200°C for 1 hr followed by 300°C for 1 hr to convert it to the imide.

Results and Discussion

The photosensitive polyamic acid discussed here is made directly from the precursor polyamic acid. Of the few other photosensitive polyamic acids reported in the literature (2), the workers initially photo-functionalize the monomers. This latter approach requires more steps and requires more care as the photosensitive moiety is isolated and carried through the entire synthesis process. A comparison of the two processes is shown in Scheme II.

Our polymer requires a reaction between a carboxylic acid or amide and a molecule containing a photolabile group. Candidates considered for that molecule included glycidyl methacrylate 1, acryloyl chloride 2, and isocyanotoethyl methacrylate (IEM) 3.

Model system studies with these materials using N-phenyl-phthalamic acid 4, indicated that 3 was the most promising candidate. As in typical reactions of epoxides with carboxylic acids (3), reaction of 4 with 1 required temperatures in excess of 100°C. Extended time at 100°C, however, causes undesired imidization. Reaction of 4 with 2 was shown to be feasible, but was not further pursued. The anhydride linkage formed would likely

be too hydrolytically unstable and prone to preimidization during film drying. Reaction of 4 with 3 was found to be slow at room temperature but like reported reactions between carboxylic acids and isocyanates could be accelerated by amines and tin compounds (4). Similar rate increases were found in the catalyzed reaction of IEM with the acid amide polymer. The use of a catalyst is not necessary to prepare PSIPI; however, the non-catalyzed reaction requires 24 hours compared to less than 1 hour for a catalyzed reaction. In the model system study, even after an equivalent of 3 has reacted with 4, ^{1}H and ^{13}C nmr as well as FTIR analysis of the product indicates that some carboxylic acid remains. It is conceivable that some of the 3 isocyanate reacts with the amide group of 4 (5). However, even if this reaction occurs, the

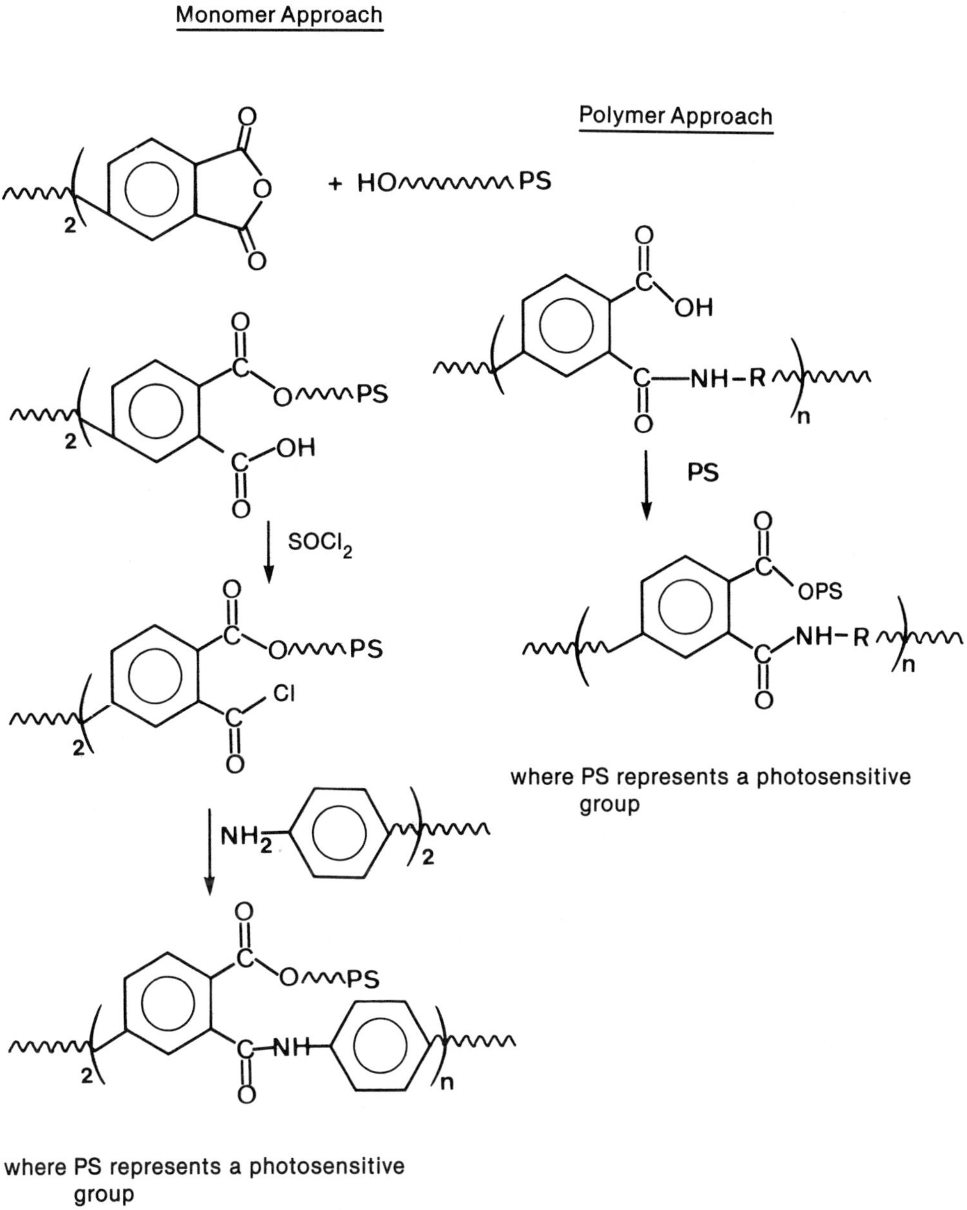

Monomer Approach

Polymer Approach

where PS represents a photosensitive group

where PS represents a photosensitive group

Scheme II. A Comparison of Processes for Preparing Photosensitive Polyimides.

original goal of acid amide functionalization has been accomplished to give a photosensitive material which would ultimately yield an imide on thermal treatment. The reaction product of 3 with 4 was not further characterized. The rest of this paper deals exclusively with the IEM modified SiPI.

The amount of IEM required to give a polymer which could be successfully photocrosslinked is important (Table I). As expected, the more polymer groups functionalized, the more photoreactive the polymer, yet it is important not to add too much IEM as this leads to gell formation prior to UV exposure. A 1:1 ratio of IEM to polymeric carboxylic acid group was found to be optimal.

Table I. Determination of Optimum Ratio of IEM to Carboxylic Acid for SiPI Polymer Modification

Equivalent IEM/COOH[a]	Solution After Reaction (24 hrs)	Exposure and Development Results[b]
0	No gell	No Pattern, Film Dissolved
0.2	No gell	Pattern, Film Thinned
0.5	No gell	Pattern, Some Thinning
1.0	No gell	Pattern, Minimal Thinning
1.2	No gell	Pattern, Minimal Thinning
1.5	Gell	---

[a] Reaction run at room temperature using 40% SiPI solution in NMP
[b] 1.8μ films (dried 60°C/20 min) prior to 30 sec exposure (PPG UV Processor) and 1 min developer (0.5N NaOH)

Shelf-life of photosensitive polyamic acids is important. We have found that the PSiPI polymer solutions (containing sensitizers) begin to gell after about 1 month at room temperature. They remain photoactive during the entire period prior to gell formation. If the polymer solutions are kept at 4°C, they are stable up to at least 6 months. We have found that inhibitors such as hydroquinone are not effective in increasing polymer solution stability. Unlike the precursor SiPI polyamic acid, the photosensitive analogue can be precipitated into methanol to give a solid which has shown increased photopolymer shelf stability.

Sensitizers were found to be necessary to effect a photocure of the PSiPI. Sensitizer effectiveness was measured by comparing final polymer film thicknesses after exposure and development. Michler's ketone was found to be most effective (Table II).

Films prepared using the Michler's ketone/N-methyldiethanolamine sensitizer system at 4% level showed identical development behavior when exposed in either air or nitrogen demonstrating the oxygen insensitivity of this sensitizer package. Two micron films (after equivalent exposure times under RS sunlamps) maintained 71% of their original thickness after 6 minutes of developing in 0.5N NaOH. In separate experiments the optimum amount of Michler's ketone for a 5 min RS sunlamp exposure was determined to be about 4 to 6% to polymer weight (Table III).

Table II. Sensitizer Effectiveness on PSiPI Curing

Sensitizer	% Sensitizer in Polymer[a]	Developer[b] Time[c] (min)	% Of Initial Film[d] Thickness After Exposure[e]
None	-	0.1	0
2,2-Dimethoxy-2-phenylacetophenone	3	0.5	32
Benzophenone	3	1.5	24
Michler's Ketone	3	1.5	80
2-Chlorothioxanthone	4	2	60

[a] An equal percentage of N-methyldiethanolamine added to each sample
[b] 0.5N NaOH
[c] Development time was determined by the time it took to open the pattern.
[d] Initial films dried 100°C/1 hr except 2-chlorothioxanthone film (100°C/2 hr).
[e] Exposed 5 minutes 10 inches under 4 RS sunlamps.

Table III. Effect of Sensitizer Concentration on Photocure of PSiPI

% Michler's Ketone[a]	Developer[b] Time[c] (min)	% of Initial Film[d] Thickness After Exposure And Development
2	2	63
4	2	85
6	2	91
8	3	92
10	3	92

[a] Same as Table II
[b] Same as Table II
[c] Same as Table II
[d] Initial films dried 100°C/1hr

Unlike other photosensitive polyimides (2), the use of N-phenylmaleimide (10% to polymer weight) as crosslinking agent did not improve the photopolymerization of PSiPI.

As expected, the more intense the light source, the faster the photocure. Cures in as short as 30 seconds have been obtained on a PPG Ultraviolet Processor. However, with this polymer system, UV radiation alone is not sufficient for fast cure rates. Both heat and light are required. This is demonstrated in Table IV.

Table IV. Effect of Heat and Light On Photocure of PSiPI

PSiPI Sample[a] Exposure	% of Initial Film Remaining After Exposure Development
4 RS sunlamps 5 min/80°C	84
4 RS sunlamps 5 min/10°C	36[b]
Heat lamp 5 min/90°C	0

[a] Film dried 100°C/1 hr and contain 8% Michler's ketone and 8% N-methyldiethanol amine.
[b] Film completely lost in spots.

Further work is being done in this area to understand this phenomenon.

Since PSiPI is a negative acting material, dissolution of the unexposed areas is required for patterning. Unlike some photo-acting polyimides which require organic solvents to dissolve the unexposed area (2), PSiPI is developed in dilute hydroxide solution which makes it compatible with positive Shipley development equipment and should minimize polymer swelling problems associated with organic developers. 0.5N NaOH has been found to be a convenient solvent for unexposed PSiPI. Some film thickness loss in exposed areas is observed during development. We have previously shown that a 1 hr 100°C dried SiPI film contains about 22% NMP (6). The film thickness loss upon development may partially reflect the leaching of the NMP from the film.

Once the PSiPI has been patterned, thermal tempering converts the crosslinked modified polyamic acid to the SiPI imide. The crosslinks are expelled during thermal ring closure. The resulting SiPI film has been shown to be more than adequate for electronic applications and has some properties, particularly adhesion, which are better than commercial polyimides (6). Figure 1 shows a set of IR spectra demonstrating various stages of PSiPI film life. IR spectrum A is the uncured, untempered film. Note the methacrylate double bond at 1640 cm^{-1}. IR spectrum B shows the same film after UV irradiation; the absence of the carbon carbon double bond is apparent. IR spectrum C shows the PSiPI film after thermal tempering at 300°C which converts it to SiPI imide. For comparison, a pure sample of imidized SiPI is shown in IR spectrum D. Also, when a sample of PSiPI was ring closed at 300°C, the Tg of the resultant polymer was 190°C which is the Tg of unmodified fully imidized polyimide siloxane.

During the final thermal tempering of PSiPI, about 40% weight loss of the polymer is expected and this is observed in a 40% reduction of the z (thickness) dimension of the polymer. Other

21. DAVIS *Photosensitive Polyimide Siloxane* 267

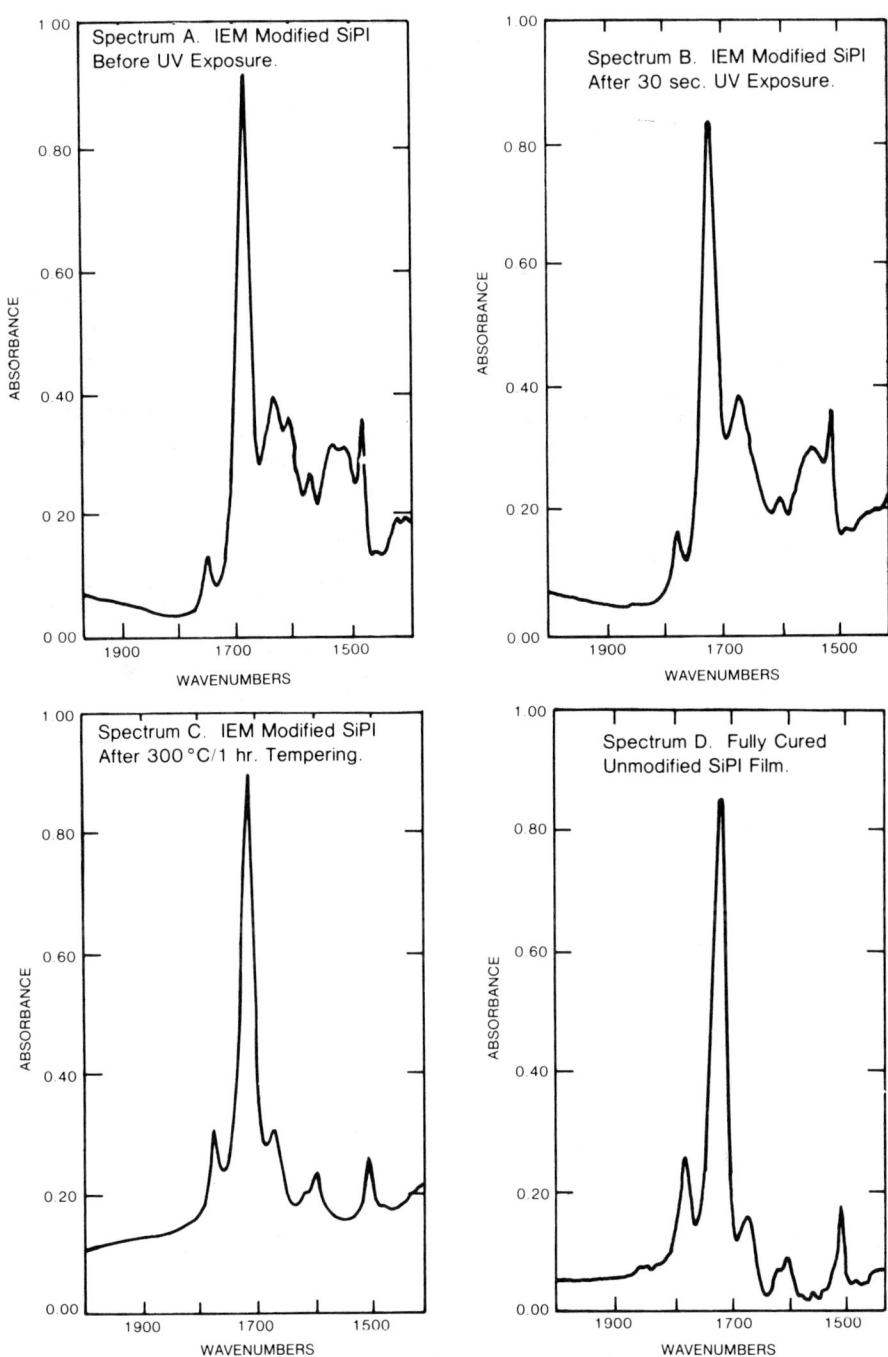

Figure 1. IR Spectra of SiPI After Various Treatments

pattern dimensions are not effected. Figure 2 demonstrates this phenomenon. A similar effect has been previously observed (7).

PSiPI films as thick as 12μ have been patterned and resolutions of 40μ have been accomplished. Work is continuing to improve the resolution, sensitivity, and speed capabilities of PSiPI.

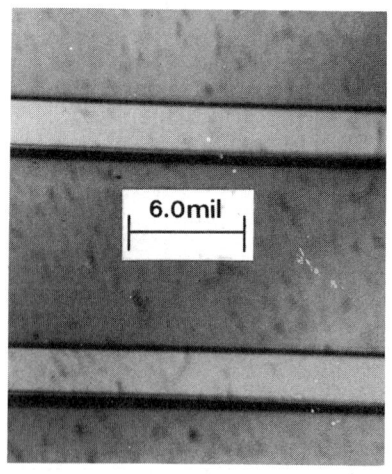 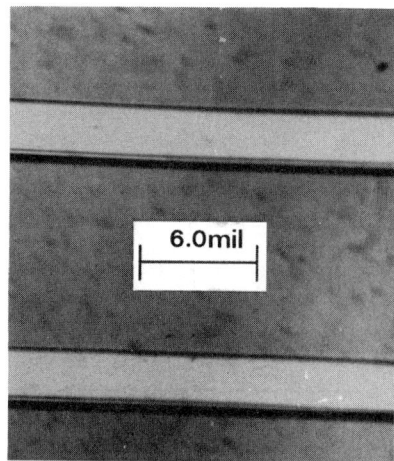

A – Exposed and developed polymer film before tempering (11.6). B – Same film after 300 °C/2h bake (8.1).

Figure 2. Photographs of PSiPI.

Summary

A photosensitive silicone polyamic acid has been described. This material is easily prepared directly from the precursor silicone polyamic acid. Properties and characteristics of this polymer system have been discussed including handling, use, and practical photochemistry.

Thermal tempering of the photosensitive or crosslinked polymer gives the polyimide siloxane which has been previously shown to be an excellent candidate as an insulating polymer in electronics. The use of such a directly patternable polyimide for dielectric and passivation applications, particularly in microelectronics, should become increasingly important as polyimides become more widely accepted in the industry.

Acknowledgments

The author wishes to thank Carol Fasoldt for her excellent technical support and Dr. Stephen Valenty for obtaining FTIR data.

Literature Cited

1. Wilson, A. M., Thin Solid Films, 1981, 83, 145.
2. Rubner, R., Siemens Forsch.-u. Entwickl.-Ber., 1976, 5, 235.
3. Sherter, L.; Wynstra, J., Ind. Eng. Chem., 1956, 48 (1), 86.
4. Ozaki, S.; Hashino, T., Nippon Kagaku Nasshi, 1959, 80, 434; Chem. Abstracts, 1961, 55, 4396i.
5. The author is indebted to the referee for pointing out this possibility.
6. Davis, G. C.; Heath, B. A.; Gildenblat, G., Proceedings of the First Technical Conference on Polyimides, in press.
7. Rubner, R.; Ahne, A.; Kühn, E.; Kololodziej, G., Photogr. Sci. Eng., 1979, 23, 303.

RECEIVED September 2, 1983

ENCAPSULANTS

22

Catalysts for Epoxy Molding Compounds in Microelectronic Encapsulation

WINSTON C. MIH

California State University, Chico, CA 95929

> Epoxy molding compounds encapsulate various microelectronics and represent a major part of polymer packages for semiconductors. The key to the development of the proper epoxy molding compounds for microelectronic encapsulation is the catalyst in the formulation. The many new developments in catalysts during the last few years have enabled tremendous improvements in microelectronic encapsulation. However, the exact curing mechanisms of various catalysts in epoxy molding compounds are still not fully understood. This paper intends to review the following catalysts and their curing mecahnisms in phenolic and anhydride cured epoxy molding compounds for microelectronic encapsulation: 1) Lewis acids, e.g., stannous or zinc octoate; 2) Lewis bases, e.g., tertiary amines, phosphorous compounds, imidazoles and their salts; 3) Organometallics, e.g., metal acetyl acetonates.

The concept of polymer encapsulation dates back almost to the invention of the transistor (1). To eliminate the costly hermetic seals, it was necessary to find low cost techniques that would also yield handling capability and environmental protection. Early systems had poor reliability performance and failed under the least severe environmental stress conditions (2).
    In 1962, polymer encapsulated transistors were produced and directed primarily at the consumer market. The major advantage of polymers was the economics of production where many parts could be packaged simultaneously with a relatively low cost material. Polymer packages were used on semiconductor devices when the cost of semiconductor packages became significant

0097-6156/84/0242-0273$06.00/0
© 1984 American Chemical Society

relative to the overall device cost. Despite several drawbacks
such as, moisture permeability, aluminum corrosion, thermomechanical stress, poor moldability and flow characteristics,
current leakage, and charge effects at the semiconductor/polymer
interface, the use of polymer packages caught on rapidly and
performance parameters were improved over the years (3).

In the period since 1962, the use of polymer packaged devices
has increased from purely consumer applications to industrial and
other high reliability applications. The pressures exerted by
commercial and industrial users on semiconductor suppliers have
significantly improved the reliability of the polymer encapsulated
semiconductors (2). Today, it is estimated that nearly 60% of
all semiconductor devices are encapsulated in polymers (3).

Electronic packages are sealed to prevent gross contamination, handling damage, and the entry of detrimental gases. The
technical solution would be to use a "true" hermetic package,
i.e., impermeable glass, ceramic, or metal housing. However,
this is not frequently desirable due to processing costs, rework
difficulty, temperature limitations of enclosed electronics, and
potential damage to expensive enclosed electronics. Polymers are
therefore considered to provide the needed protection.

Polymers used for semiconductor encapsulation must protect
against the environment to which the device is likely to be
exposed, such as moisture, chemical agents, wide temperature
fluctuations and mechanical shock. The polymeric material must
be able to do this with a minimum of effect on device parameters
over an extended period of time, and be relatively inexpensive
and easy to process (4). Many thermosetting polymers have been
used, but only epoxies, silicones, and phenolics have been used
extensively for large scale production (1, 4). Joint research
programs in the early 1970s between semiconductor manufacturers
and epoxy molding compound suppliers led to the use of epoxy
molding compounds as the major encapsulants for semiconductor
parts (3). Today, epoxy molding compounds represent a major part
of polymer packages for semiconductors. Epoxy encapsulants
dominate the market with the highest volume growth, about 15-20%,
and production is estimated at several billion units per year.
The use of silicones is declining and phenolics are essentially
out of the market (3).

Epoxy Curing Mechanism

The key to the development of the proper epoxy molding compounds
for microelectronic encapsulation is the catalyst in the
formulation. In spite of serious limitations in epoxy molding
compound performance in sensitive microelectronic devices about
ten years ago, the many new developments in catalysts during the
last few years have enabled tremendous improvements. However,
the exact curing mechanisms of various catalysts in epoxy molding
compounds are still not fully understood today.

The role of catalysts in epoxy molding compounds for microelectronics is in initiating ionic polymerization, and subsequently, crosslinking of phenolic or anhydride hardeners with epoxy prepolymers at the proper molding conditions. This is achieved either through homopolymerization or heteropolymerization, depending upon the type of catalyst and curing mechanism. The ideal crosslinked three-dimensional network can then be reached to provide the epoxy molding compound with the proper moisture resistance, physical, electrical, and electronic properties, and shelf stability for transportation and storage.

<u>Anhydride Cured Epoxy Reaction Mechanism.</u> In the case of anhydride cured epoxy reaction, catalysts will promote ring opening of the anhydride to provide carboxylic group for reaction with epoxide. Without catalysts, the reactions are slow and accompanied by extensive epoxide homopolymerization at elevated temperatures.

In the non-catalyzed anhydride cured epoxy reaction, Tanaka and Kakiuchi ($\underline{5}$, $\underline{6}$) found that the reaction rate was proportional to the concentration of epoxide, acid anhydride, and hydroxyl compound, suggesting a termolecular transition state for the rate determining step.

The reaction could be promoted by electron-withdrawing substituents on the hydroxyl groups, indicating hydrogen bonding with epoxide.

For Lewis base and Lewis base salt catalyzed anhydride cured epoxy reactions, Fischer ($\underline{7}$) proposed that the initial step of an anhydride, epoxide, and tertiary amine system was the activation of the anhydride (reaction 1):

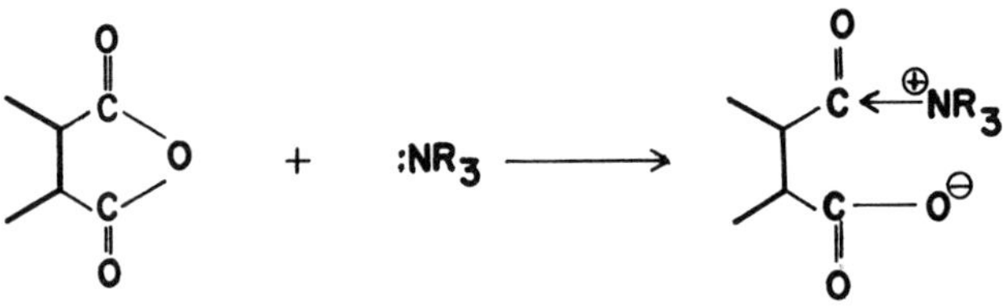

[Reaction scheme (1): anhydride + :NR₃ → zwitterionic intermediate]

The generation of the alkoxide ion was concluded to be the rate determining step (reaction 2):

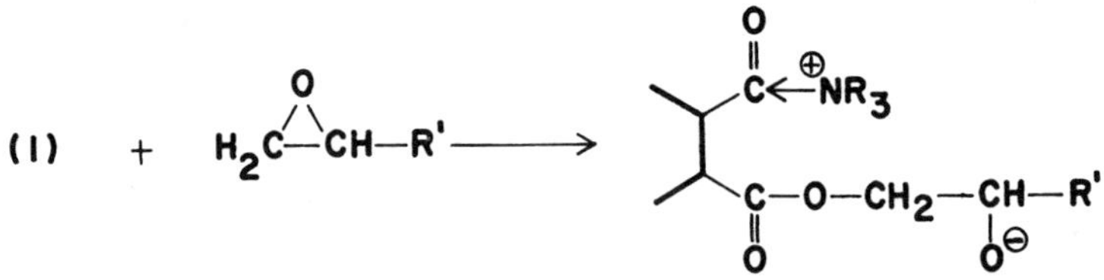

[Reaction scheme (2): (I) + H₂C—CH—R' (epoxide) → ring-opened alkoxide product]

Tertiary amine catalyzed reactions were also studied by Tanaka and Kakiuchi (8), who essentially supported the Fischer mechanism, but disagreed on the kinetic order. The copolymerization kinetic scheme proposed by both Fischer and Tanaka postulate three rates, R_1, R_2, and R_3, as follows:

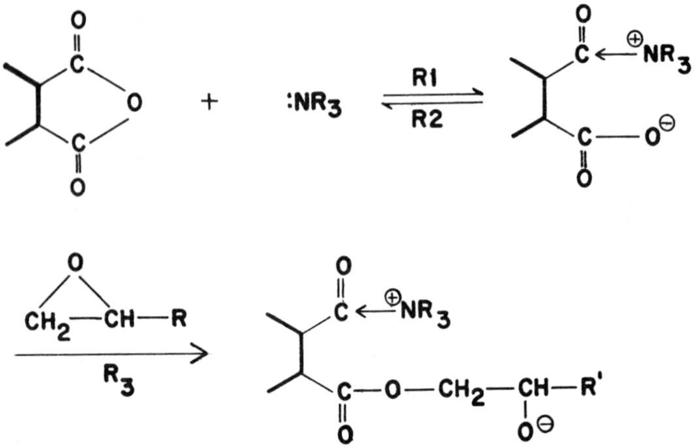

[Reaction scheme showing anhydride + :NR₃ ⇌ (rates R_1/R_2) intermediate, then + epoxide (rate R_3) → alkoxide product]

The rates R_1/R_2 and R_3 are different. Electron-withdrawing substituents on the epoxide group accelerate the heteropolymerization rate and suggested that R_3 is a propagation reaction rate.

Lewis base salts (imidazole versus imidazolium salts or tertiary amine versus quaternary amine salts) do not change the mechanism. They follow the same order as Lewis bases. Metal salt complexes with tertiary amines may be more latent than tertiary amines, probably due to the absence of solvolysis of the anhydride.

Sorokin, et. al. (9) proposed that the initial step is either solvolysis of the anhydride (reaction 3) or reaction of the tertiary amine with a co-catalyst (water, alcohol, or carboxylic acid) (reaction 4), followed by catalytic reaction of the monoester with the epoxide group (reaction 5).

$$\text{anhydride} + ROH \longrightarrow \text{monoester(OR, OH)} \quad (3)$$

$$R_3N: + ROH \longrightarrow R_3\overset{\oplus}{N}{:}H + {}^{\ominus}OR \quad (4)$$

$$\text{monoester} + CH_2\!\!-\!\!\overset{O}{\overset{\triangle}{}}\!\!-\!\!CH\!\!-\!\!CH_2R' \longrightarrow$$

$$\begin{array}{c} \text{diester} \\ C\text{-OR} \\ C\text{-OCH}_2\text{-CH-CH}_2\text{-R'} \\ |\\ OH \end{array} \quad (5)$$

(ROH can be free acid present as an impurity in the anhydride.)

Sorokin concluded that the formation of the amine-anhydride complex (as proposed by Fischer) did not have a decisive influence on the rate determining step, although it was an intermediate in the catalytic solvolysis of the anhydride.

Feltzin, et. al. (10) indicated that the initial step in the reaction was catalyst activation by the reaction of the tertiary amine with a co-catalyst to form a quaternary salt. Fischer supports the Feltzin mechanism because he reports that quaternary salts catalyze the reaction in much the same way as tertiary amines. Tanaka and Kakiuchi (8) also favor a catalyst activation. It can be concluded that the experts essentially agree that the mechanism consists of:

(a) Catalyst (or activated catalyst) reacting (or solvating the anhydride) to generate a carboxylate.

(b) The generated carboxylate reacting with the epoxide to generate an alkoxide.

(c) The generated alkoxide reacting with the anhydride, etc. However, Mika (11) found that in mixing highly purified anhydride and tertiary amine, there is no evidence for the formation of the complex proposed by Fischer or Tanaka. This was based on nuclear magnetic resonance spectra comparing imidazole with an equimolar mixture of imidazole and nadic methyl anhydride.

Recently, Crandall and Mih (12) studied the anhydride curing mechanism using near and middle IR and DSC. Their findings supported the anhydride-epoxide curing mechanism that most experts essentially agree upon.

Phenolic Cured Epoxy Reaction Mechanism. An important factor in the epoxy-phenolic reaction is the degree of reaction of the unreacted epoxide group with the generated secondary hydroxyl groups of the hydroxyl-alkyl products (13). In order to have optimum mechanical and electrical properties, the phenolic cured epoxy should be formulated to have stoichiometric quantities of the epoxy and phenolic. Epoxy-phenolic reactions which proceed without the reaction of the secondary hydroxyl group are selective reactions. The degree of selectivity of reaction will depend on the phenolic used, catalysts, temperature and other reaction variables. Nonselective reaction of the generated secondary hydroxyl group can upset stoichiometry in the epoxy-phenolic reaction, and thus reduce properties. Alvey (13) studied a series of nitrogen containing Lewis base catalysts in the phenolic cured epoxy system to show selectivity and reactivity.

Shechter and Wynstra (14) studied the uncatalyzed phenolic cured epoxy reaction and proposed the following mechanism, (with reaction 1 predominating):

$$\text{—C}_6\text{H}_4\text{—OH} + \text{H}_2\text{C}\underset{\text{O}}{\overset{}{\text{—}}}\text{CH—CH}_2\text{—OR} \longrightarrow$$

$$\text{—C}_6\text{H}_4\text{—O—CH}_2\text{—CH(OH)—CH}_2\text{—OR} \quad (1)$$

$$\text{—C}_6\text{H}_4\text{—O—CH}_2\text{—CH(OH)—CH}_2\text{—OR} + \text{CH}_2\underset{\text{O}}{\overset{}{\text{—}}}\text{CH—CH}_2\text{—OR}$$

$$\longrightarrow \text{—C}_6\text{H}_4\text{—O—CH}_2\text{—CH(O—CH}_2\text{—CH(OH)—CH}_2\text{—OR)—CH}_2\text{—OR} \quad (2)$$

The strong base catalyzed phenolic cured epoxy reaction proceeds via the phenoxide ion which reacts with epoxide, generating the alkoxide ion. The high basicity of the alkoxide ion reacts with phenolic to regenerate the phenoxide ion and the reaction repeats itself.

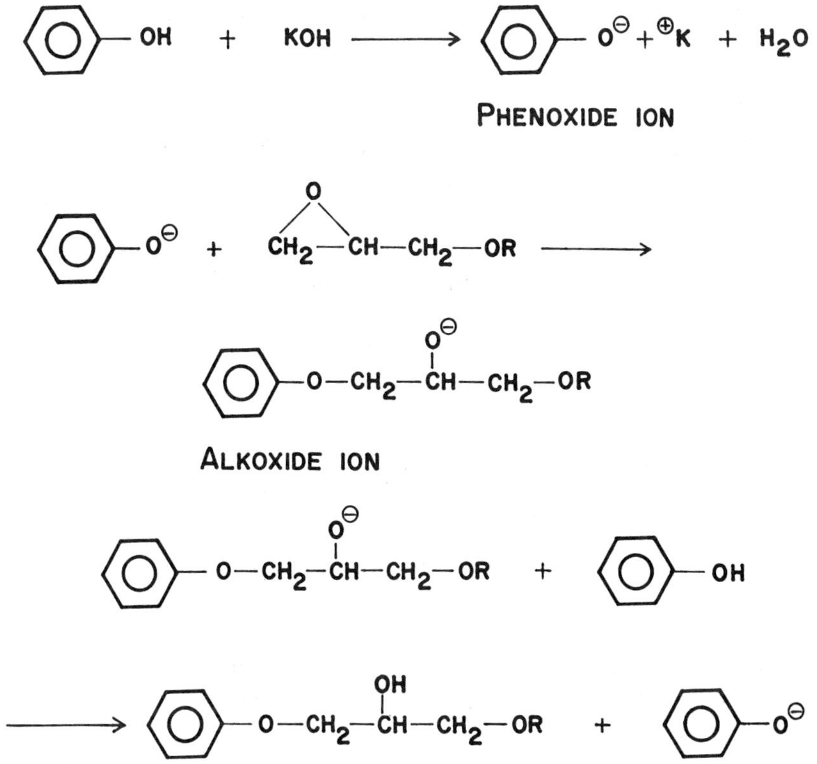

$$\text{Ph-OH} + \text{KOH} \longrightarrow \text{Ph-O}^\ominus + {}^\oplus\text{K} + \text{H}_2\text{O}$$

PHENOXIDE ION

$$\text{Ph-O}^\ominus + \underset{\text{CH}_2\text{-CH-CH}_2\text{-OR}}{\overset{O}{\triangle}} \longrightarrow$$

$$\text{Ph-O-CH}_2\text{-}\underset{|}{\overset{\text{O}^\ominus}{\text{CH}}}\text{-CH}_2\text{-OR}$$

ALKOXIDE ION

$$\text{Ph-O-CH}_2\text{-}\underset{|}{\overset{\text{O}^\ominus}{\text{CH}}}\text{-CH}_2\text{-OR} + \text{Ph-OH}$$

$$\longrightarrow \text{Ph-O-CH}_2\text{-}\underset{|}{\overset{\text{OH}}{\text{CH}}}\text{-CH}_2\text{-OR} + \text{Ph-O}^\ominus$$

The nitrogen-containing Lewis base benzyldimethylamine was found to be a better catalyst than potassium hydroxide. Furthermore, evidence shows that the quaternary compound of benzyldimethylamine was even a better catalyst (14). The phenolic cured epoxy reaction is a first order kinetics. The following Lewis bases have been used as catalysts for the epoxy-phenolic reaction (15):
  (a) Inorganic bases (16)
  (b) Nitrogen bases (17)
  (c) Ammonium salt of strong acids (18)
  (d) Heterocyclics, such as imidazole or triazole (19)
  (e) Alkanolamine (20)

Recently, phosphine compounds have been used as catalysts in the epoxy-phenolic or epoxy-anhydride reactions. There is indication that the mechanism does not involve the decomposition of the phosphonium compound to the free phosphine species. The initiation mechanism probably involves the formation of hydrogen-bonded phosphonium-epoxy or phosphonium-anhydride complexes that rearrange with heat application to form activated species resulting in the polymerization of the epoxy-anhydride components (21).

Mika (22) proposed the complex intermediates in triaryl phosphine catalysis as:

$$\text{Ph-OH} + R_3P: + \underset{O}{CH_2-CH-CH_2-OR}$$

$$\longrightarrow \text{Ph-O}\cdots H:PR_3 + \underset{O}{CH_2-CH-CH_2-OR}$$

$$\longrightarrow \left[\begin{array}{c} \text{Ph-O}\cdots H \cdots :PR_3 \\ \vdots \\ O \\ CH_2-CH-CH_2-OR \end{array} \right]$$

Mika (22) also proposed that salts of tertiary amine or phosphonium complexes are not a dissociation (reaction 1), but a push-pull concerted effect.

$$R_3PY \rightleftharpoons R_3P: + Y$$

OR $$R_3N:Y \rightleftharpoons R_3N: + Y \qquad (1)$$

$$R_3P: \cdots \underset{Y}{\overset{\delta+}{CH_2}-CH-}_{O^{\delta-}}$$

There is considerable evidence for the push-pull concerted effect in the reaction mechanism by using lead salts and DMP-30 salts as catalysts in the phenolic cured epoxy reaction. The complex formation is also supported by Alvey (13) and Son (23).

The acid catalyzed reaction of epoxy-phenolic has not been studied in detail. Oxidized stannous acylate (24) and other stannous salts (25) have been used. From the experimental results available, stannous salts have been found to be particularly useful for phenolic cured cycloaliphatic epoxy systems (25).

Catalysts

From the up-to-date literature and patent review of catalysts used in anhydride and phenolic cured epoxy molding compounds, it is evident that imidazoles and their derivatives predominate (Table I). Metal complex, trialkyl or triaryl phosphines and their complexes, Lewis acids such as zinc or stannous octoate are used to a much lesser extent (Table II). There are a few examples of tertiary amines and urea derivatives used.

Table I. Lewis Base and Lewis Base Salt Catalysts

Catalysts Used	Refs.
Lewis Base:	
Imidazole	
2-heptadecylimidazole	26
Imidazole 16	27
4-methyl-2-phenylimidazole	28
2-ethyl-4-methylimidazole	29,30
1-(2-hydroxy-3-phenoxypropyl)imidazole	31
2-methylimidazole	32
2-phenylimidazole	33
Organic Phosphine	
Triphenylphosphine	34
Urea Derivatives	
Latent sources of dimethylamine	35
Lewis Base Salt:	
Quaternary phosphonium salts	21
Tetrahydrocarbyl phosphonium phenoxide salts	36
Butyl tri-phenylphosphonium salt of 2,2',6,6'-tetrabromobisphenol A	36
Phenylphosphonium tetraphenylborate	37,38
Carbon disulfide-tricyclohexylphosphine adduct	33,39

Table II. Organometallic and Lewis Acid Catalysts

Catalysts Used	Refs.
Organometallic:	
Tetrakis(acetylacetonato)zirconium	40
Bis(acetylacetonato)dipyridine cobalt	41
Lewis Acid: Zinc or stannous octoate	42

The mechanism for organometallics and Lewis acids in phenolic or anhydride cured epoxy molding compounds are still not fully understood. Lewis bases such as imidazoles can be reacted with organic acids to form salts in order to improve latency. Imidazoles are, so far, the most widely accepted as a compatible catalyst family for encapsulating microelectronics.

Generally speaking, catalytic activity should correlate with electron density at the azole nitrogen of the imidazole structure. In addition, the basicity of a given imidazole may determine its corrosivity towards aluminum and, hence, its microelectronic compatibility.

Triaryl and triallyl phosphine complex catalysts are gradually growing in usage and have shown potential for future development in terms of latency and excellent microelectronic compatibility. Further studies are required before the reaction mechanism can be understood more fully.

Literature Cited

1. Reich, B. Solid State Technology Sept. 1978, 82.
2. Reich, B.; Hakim, E. B. Microelectronics and Reliability 1976, 15, 29.
3. Woodard, J. B. Preprints. Society of Plastics Engineers 4th Annual PACTEC. Jan. 31-Feb. 2, 1979, p. 89.
4. Olberg, R. C. J. Electrochem. Soc. 1971, 118(1), 129.
5. Tanaka, Y.; Kakiuchi,H. J. Macromol. Chem. 1966, 1, 307.
6. Tanaka, Y.; Kakiuchi, H. J. Polym. Sci. 1964, A2, 3405.
7. Fischer, R. F. J. Polym. Sci. 1960, 44, 155.
8. Tanaka, Y.; Kakiuchi, H. J. Appl. Polym. Sci. 1963, 7, 1063.
9. Sorokin, M. F. Lakokrasoch. Mater. Ikh. Primen 1967, 5, 67.
10. Feltzin, J. J. Macromol. Sci. Chem. 1969, A3, 261.
11. Mika, T., personal communication.
12. Crandall, E. W.; Mih, W. C. Preprints. American Chemical Society Div. Org. Coatings Plast. Chem. Sept. 1982, 47, 592.
13. Alvey, F. B. J. Appl. Polym. Sci. 1969, 13, 1473.
14. Shechter, L.; Wynstra, J. Ind. and Eng. Chem. 1956, 48(1),86.
15. May, C. A.; Tanaka, Y. "Epoxy Resins: Chemistry and Technology"; Dekker: New York, 1973; Chap. 4.
16. Dynamit Nobel. German Pat. 1,184,496.
17. Aba, Ltd. Netherlands Pat. Appl. 246,535,
18. Berlanger, W.; Cooke, H. G. (to Devoe and Raynolds Co.) U.S. Patent 2,928,803.
19. Union Carbide Corp. Netherlands Pat. Appl. 6,512,268.
20. Ephraim, S. N. (to Reichhold Chemical). U. S. Pat. 3,264,369.
21. Smith, J. D. B. Preprints. American Chemical Society Div. Org. Coatings and Plast. Chem. Sept. 1978, 39, 42.
22. Mika, T., personal communication.
23. Son, P. N.; Weber, C. D. J. Appl. Polym. Sci. 1973, 17,1305.
24. Proops, W. R. (to Union Carbide). U. S. Pat. 3,284,383.

25. Proops, W. R.; Fowler, G. W. (to Union Carbide). U. S. Pat. 3,117,099.
26. Morton-Norwich Products, Inc. Japanese Pat. 82 49 647, 1982; Chem. Abstr. 1982, 97, 39874r.
27. Trautmann, H.; Schillgalies, J.; Martin, M.; Krausche, C.; Eckhardt, I.; Reichardt, L. Ger.(East) DD 154 824, 1982; Chem. Abstr. 1982, 97, 164100m.
28. Nitto Electric Industrial Co., Ltd. Japanese Patent 82 59 365, 1982; Chem. Abstr. 1982, 97, 93634a.
29. Segawa, T.; Suzuki, H.; Kitamura, M.; Numata, S.; Nishi, K. Ger. Offen. DE 3 137 480, 1982; Chem. Abstr. 1982, 97, 7395u.
30. Mitsui Toatsu Chemicals, Inc. Japanese Patent 81 127 625, 1981; Chem. Abstr. 1982, 96, 53239y.
31. Lyalyushko, K. A.; Sorokin, M. F.; Sigunova, O. V. Tr.-Mosk. Khim.-Tekhnol. Inst. im. D. E. Mendeleeva 1980, 110, 76; Chem. Abstr. 1982, 96, 182808g.
32. Ricciardi, F.; Joullie, M. M.; Romanchick, W. A.; Griscavage, A. A. J. Polym. Sci., Polym. Letter Ed. 1982, 20(2), 127; Chem. Abstr. 1982, 96, 105153n.
33. Toshiba Corp. Japanese Patent 82 24 553, 1982; Chem. Abstr. 1982, 97, 57259y.
34. Ikeya, H.; Suzuki, H.; Oguni, T.; Matsumoto, K.; Hatanaka, A.; Wada, M. Eur. Pat. Appl. EP 41 662, 1981; Chem. Abstr. 1982, 96, 114563y.
35. LaLiberte, B. R.; Sacher, R. E.; Bornstein, J. Report 1981, AMMRC-TR-81-30; Chem. Abstr. 1982, 96, 123843s.
36. Doorakian, G. A.; Bertram, J. L. U. S. Patent 4 302 574, 1981; Chem. Abstr. 1982, 96, 86422f.
37. Hitachi Chemical Co., Ltd. Japanese Patent 81 84 717, 1981; Chem. Abstr. 1981, 95, 170427y.
38. Suzuki, H.; Sato, M.; Muroi, T.; Watanabe, Y. Preprint. SPE 19th ANTEC, 1973, p. 6.
39. Toshiba Corp. Japanese Patent 82 23 626, 1982; Chem. Abstr. 1982, 96, 218787r.
40. Toshiba Corp. Japanese Patent 82 51 720, 1982; Chem. Abstr. 1982, 97, 56707f.
41. Shukla, A. K.; Karkozov, V. G.; Nikolaev, A. F.; Vinogradov, M. V. Plast. Massy 1982, 5, 20; Chem. Abstr. 1982, 97, 24629r.
42. Takahama, T.; Geil, P. H. Makromol. Chem., Rapid Commun. 1982, 3(6), 389; Chem. Abstr. 1982, 97, 56602t.

RECEIVED September 2, 1983

23

Thermogravimetric Analysis of Silicone Elastomers as Integrated Circuit Device Encapsulants

CHING-PING WONG

Engineering Research Center, Western Electric Company, Princeton, NJ 08540

>The increasing demand for thermally stable polymers as electronic encapsulants is consistently creating a need for more information on such materials. Thermogravimetric analysis (TGA) is a valuable tool for the thermal analysis of the silicone polymers. The sample's weight loss versus temperature from a TGA measurement reveals crucial information regarding the thermal stability, decomposition volatiles and kinetic properties of the material which are a critical concern in the material's use as an encapsulant.
>   With the new data analysis software, the microprocessor and microelectronic hardware, the thermal analyzer provides rapid and reliable thermal information of the materials.
>   In this paper, the thermal stability of the silicone elastomers, base silicone resins, fillers, and their interactions with each other within the silicone matrix are described. Thermal decomposition volatiles, obtained indirectly through solvent extractables, reaction kinetics of the materials as integrated circuit (IC) devices encapsulants will be discussed.

In the Bell System for bipolar, metal oxide semiconductor (MOS) and hybrid integrated circuitry (HIC), RTV silicone elastomer has proven to be one of the most effective encapsulants for mechanical, moisture and alpha particle protection of the IC devices[1-3]. This RTV material is also one of a few commercial polymer materials that meet most of the Bell System encapsulant specifications. However, the RTV material occasionally experiences some material variations and creates production problems in IC devices shop coating[4].

0097-6156/84/0242-0285$06.00/0
© 1984 American Chemical Society

Wicking and runover of the RTV material on the HIC devices are the major problems of the material. Variation in the rheological properties of the RTV material is the one believed to be the main cause of the wicking and runover of the RTV encapsulant in shop coating performance. Since the rheology of the RTV material is related to its filler content, filler deactivation and silicone (resin)-filler incorporation technique, the analyses of these properties are important to the RTV material. The thermal properties of the cured RTV encapsulant material have also been demonstrated to have a significant effect on thermal cycling and molding performance of the material. It is important to have a good understanding of the properties of this RTV silicone material and their relationship between decomposition temperature, and formulation components. Thermogravimetric analysis (TGA) which measures the sample's weight loss versus temperature is a rapid and easy to use analytical measurement to analyze the RTV elastomers, filler level, degree of incorporation and thermal stability of the final compounded RTV materials. These are the valuable information in the RTV formulation.

Experimental

Sample Preparation. RTV silicone samples were obtained from a thin coating which had been cured on a Teflon-coated aluminum plate at room temperature cured for 16 hours, followed by oven cure at 120°C for 4 hours. For hydroxy-terminated silicone fluids (silanol) and filler (SiO_2), samples were used as received without any further purification.

Thermogravimetric Analysis. Instrumentation and Experimental Parameter: (1) Instrumentation: The thermal analysis technique of thermogravimetric analysis (TGA) is one in which the change in sample mass is recorded as a function of temperature[5]. There are three commonly employed modes of thermogravimetry: 1. Static or isothermal thermogravimetry, in which the loss of a sample mass is recorded as a function of time at a constant temperature; 2. quasistatic thermogravimetry, in which the sample is heated to a constant mass at each of a series of increasing temperatures; 3. dynamic thermogravimetry, in which the sample is heated in an environment whose temperature is heated at a predetermined manner, usually at a linear rate. In our studies, most of the TGA data were obtained by the dynamic thermogravimetry analysis method. The resulting mass-change versus temperature curve provides valuable information regarding the thermal stability and composition of the original sample, the thermal stability and the composition of any intermediate compounds that may be formed, the degree of incorporation between resin and fillers, and the level and composition of the residues.

In a typical TGA curve, four temperatures are characteristic of any single-stage nonisothermal reaction (see Figure 1). Ti represents the temperature at which the first distinguishable decomposition is observed; Tb is the temperature at which a significant bulk reaction begins; Tm represents the maximum decomposition temperature of the bulk sample, and Tf, the final temperature where the mass loss ends. Although the Ti may be the lowest temperature at which a mass-change may be observed in a given temperature, it is neither a transition temperature nor a true decomposition temperature below which the reaction rate suddenly becomes zero. At a constant heating rate, Tf must be greater than Ti, and the difference, (Tf - Ti), is called the reaction interval. For an endothermic thermogravimetric analysis decomposition, Ti and Tf increase with increasing heating rate. Ω(omega) is the angle between the tangents of Tm and Ti which provides an indication of the maximum rate of decomposition of the sample. Besides the TGA curves, the derivatives curve of the TGA spectra is the measurement of mass (m) with respect to decomposition time (dm/dt) or temperature (T). The first derivative thermogravimetry (DTG) curve has the following advantages: (a) It can be obtained in conjunction with TGA measurements, (b) DTG curves clearly indicate the beginning, the end and the maximum temperature of the mass loss during the whole process; (c) It is very sensitive in detecting minor changes, and provides exact quantitative analyses; (d) It shows overlapping changes better than TG. In most of our thermal spectrum, both TG and DTG were listed. These provide valuable complementary information - not only the physical properties of the polymer, but also its kinetic parameters which will be discussed in a later section of this paper.

A DuPont Model 1090 Thermal Analyzer system equipped with the computerized software and hardware was used in this study. (2) Experimental Parameters: The heating rate and the heating atmosphere have effects on the results of the TGA curves. The decomposition temperature varies with different heating rate (see Figure 2). However, the final mass-loss with different heating rates should remain unchanged. The composition of the heating atmosphere effects the mass change, because of the difference in degradation mechanisms and final degradation residues (see Figure 3). However, to minimize the heating condition, nitrogen is used to purge all the samples and quantify the results in this study. For direct TGA comparison of materials, equal sample weights (usually a few mg) are highly preferred, as well as same heating condition and heating rate. For our dynamic thermal measurements, the samples were heated from room temperature (25°C) to 750°C at a uniform heating rate (usually 10-20°C/min) under constant nitrogen atmosphere. (Unless specified, all samples were run at 20°C/min with 15 psi nitrogen pressure.) The weight loss versus temperature data was recorded on a floppy disc during the

288 POLYMERS IN ELECTRONICS

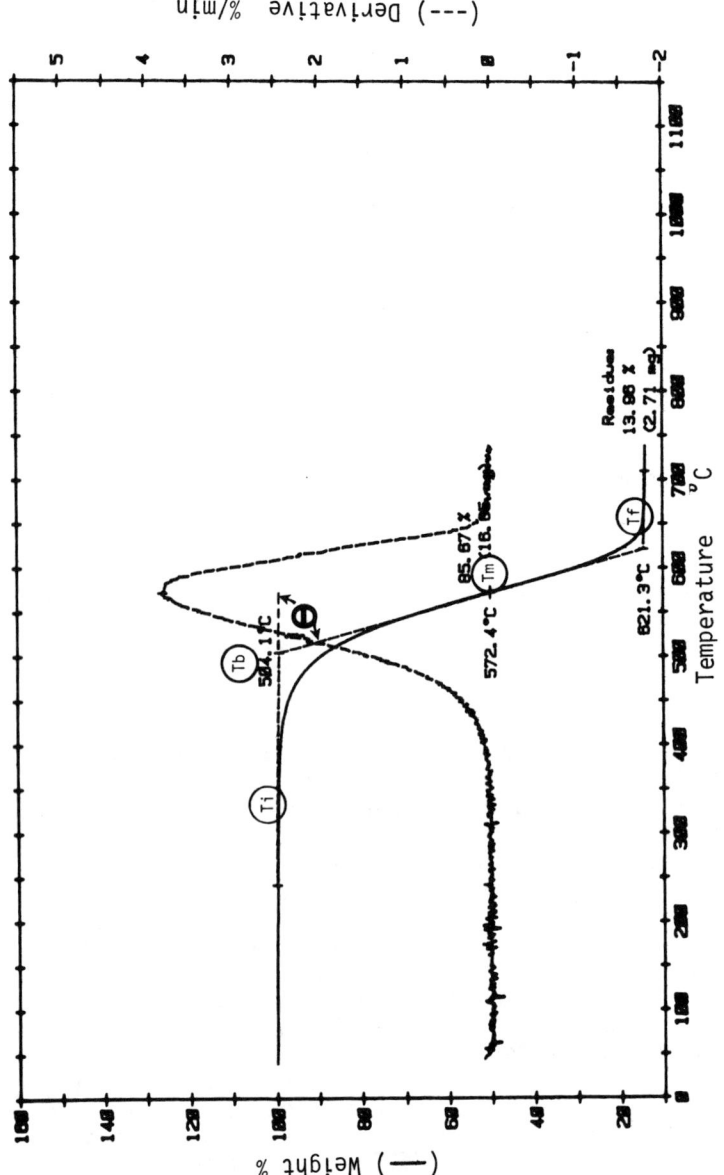

Figure 1 Temperature Region of a Single-Stage Nonisothermal TGA

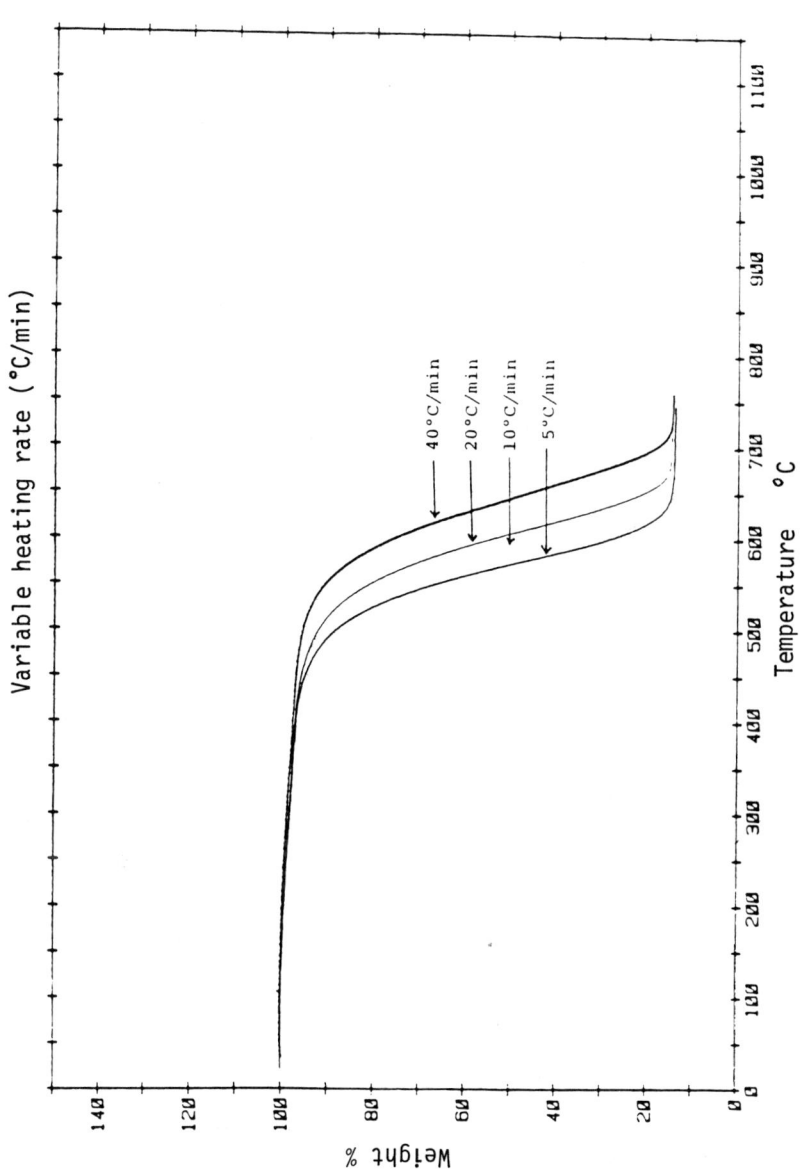

Figure 2. TGA of a Silicone with Various Heating Rates

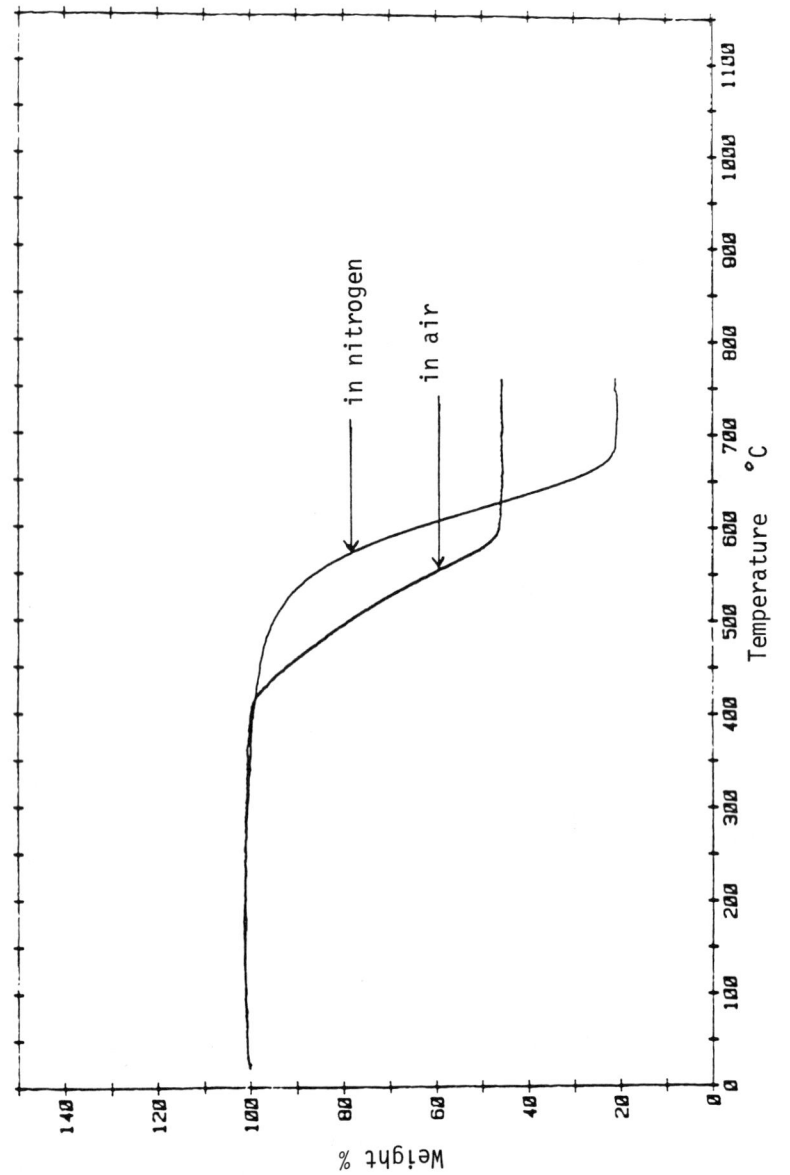

Figure 3 TGA of Silicone with Various Heating Conditions

whole process. This data could be recalled and replotted after the data acquisition. The first derivative of the TGA curve and percent residue which remains after the thermal decomposition of the sample can be printed out by a new DuPont software program. Valuable information which is described in the following section could be obtained from this software program. (C) Fourier Transform Infrared (FT-IR) and Gas Chromatography/Mass Spectrometer (GC/MS) Analyses of Thermal Decomposition (via Solvent Extractables) Products: A Nicolet Model No. 7199 FT-IR was used to analyze the RTV silicone thermal decomposition products (Via solvent extractables). NaCl plates were used to hold the silicone sample. Standard routine procedure was used to obtain the IR spectrum.

GC/MS analyses were performed on a Hewlett Packard Model No. 5993 GC/MS spectrometer. Solvent extractables were further diluted with Freon TA before GC/MS analysis. Six feet long, 0.3% SP2250 OV17 type material was used as the GC column, with 20-30 cc/min helium flow rate as carrier gas. Mass/charge units were scanned from 30-800 atomic mass unit (amu). Three dimensional GC/MS spectrum were printed out at the end of each run.

Results and Discussion

TGA of Silicone OH Fluids, Fillers, and Generic RTV Silicones.
TGA has been shown to be a rapid and reliable technique to determine the thermal stability of the RTV silicones, silicone OH fluids, fillers, and filler level in the RTV silicone material. The percent residue from TGA measurements reveals the filler(s) level of the material. The filler level determination method by density measurement of the RTV material is relatively tedious and insensitive. There is only a minor difference in density between 5 and 10 phr filler level, and further filler loading (> 10 phr) shows very little change in density. By using our generic RTV formulated materials, in which we know the precise filler levels, a standard curve relating filler(s) level used and percent residue obtained by TGA analysis can be constructed. We are then able to detect filler level of any commercially-cured RTV silicone by comparing its TGA percent residues with our standard curve. Please note: when silicone OH fluid was examined by TGA, we were quite surprised to learn that a zero percent residue was observed from the thermal degradation. Since the TGA was operated under an inert N_2 atmosphere, the thermal degradation of the polydimethylsiloxane to cyclic siloxanes were swept out by the N_2 carrier gas. Hence, a zero percent residue from our TGA of the silicone OH fluids were obtained. GC/MS studies of the decomposed volatiles which were collected by a liquid nitrogen trap have confirmed the presence of these cyclic siloxanes. (A proposed thermal degradation mechanism is listed in Figure 4). Since oxidation

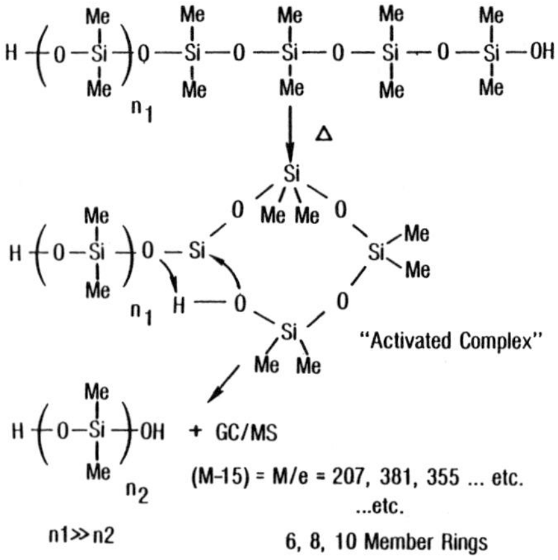

Figure 4 Silicone Thermal Degradation Mechanism

products, such as: $Si(OH)_x$, SiO_2 were not obtained from the degradation of silicone fluid in the TGA analysis, it was not surprising to find that when all the OH fluid materials were degraded, a zero percent residue was observed in the TGA curve (see Figure 5). By using this TGA method, with our generic RTV silicone formulations as filler level standards, we are able to detect the commercial filler level. The information obtained from these thermal stability studies was useful in evaluating and optimizing the RTV materials, in its filler level, and silicone base polymer. These components will be described in detail in the following sections:

OH-Terminated Polydimethylsiloxane (Silicone) Base-Polymers. We have investigated the TGA of all the commercially available silicone OH fluids (with viscosity of 100, 1,000, 3,500, 8,000, 12,500, 18,000 c.s.). We have found that the 12,500 c.s. OH fluid has the best thermal stability. However, by incorporating the right amount of filler, we have improved the generic RTV silicone material which used the 8,000 c.s. OH fluid, resulting in a thermal stability close to the best commercial material. From Figure 5, we can clearly see that there is no direct correlation between the viscosity of a material and its thermal stability. The broad molecular weight distribution of the 8,000 c.s. OH fluid which consists of low molecular weights, unreactive cyclic components may directly contribute to the thermal separation of the material at a lower temperature (see Figure 6 and Table I).

Table I. Molecular Distributions of OH-terminated Silicone Fluid

OH Fluid Samples	M_n	M_w	M_z	Dispersity
8,000 c.s. (as received from Union Carbide-#061381)	21.2	68.5	140	3.23
8,000 c.s. (purified)	46.1	99.8	158.2	2.16
12,500 c.s. (as received Petrarch Systems #SB00382002)	37.4	107.1	162.4	2.86

Filler(s) Level. RTV silicone, without additives of filler, has a very low tensile strength and elongation. Due to its intrinsic physical property, this unfilled silicone material is very brittle and fragile; it breaks easily upon peeling. However, when a fumed silica is added to the RTV silicone, it provides excellent tensile strength and elongation compared to the unfilled material with the same formulations. The chain-linked spherical SiO_2 particles which interact with themselves (filler-filler interaction) and/or the silicone-resin

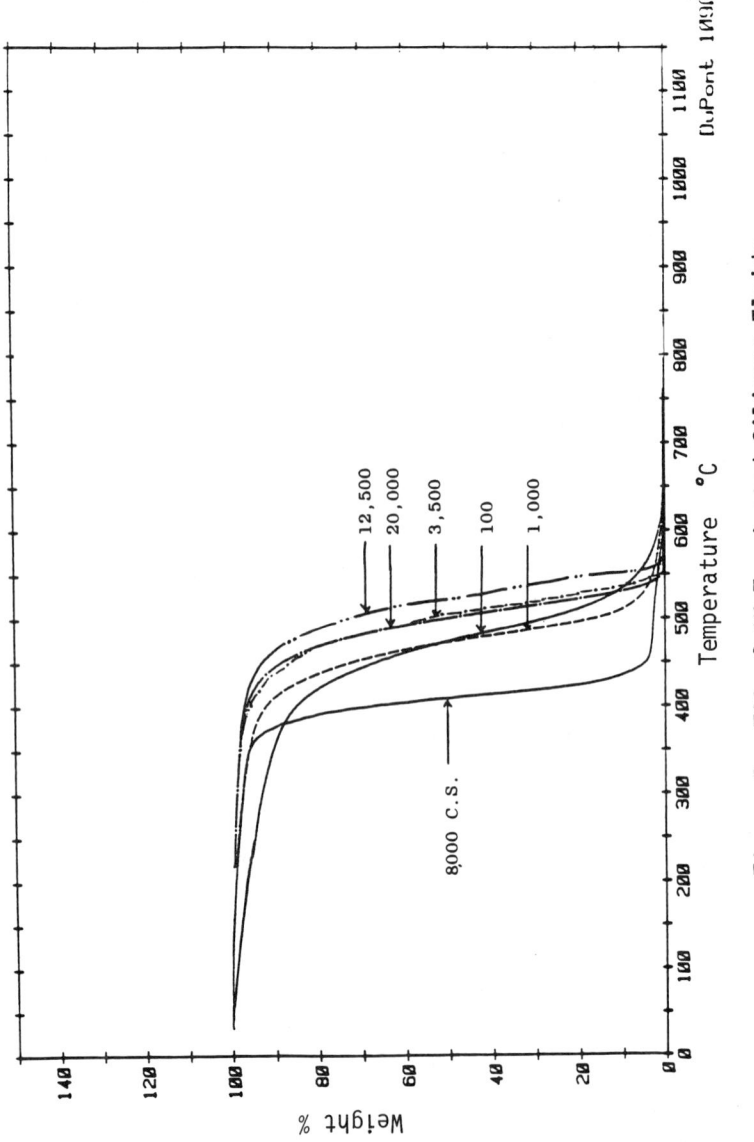

Figure 5. TGA of OH-Terminated Silicone Fluids

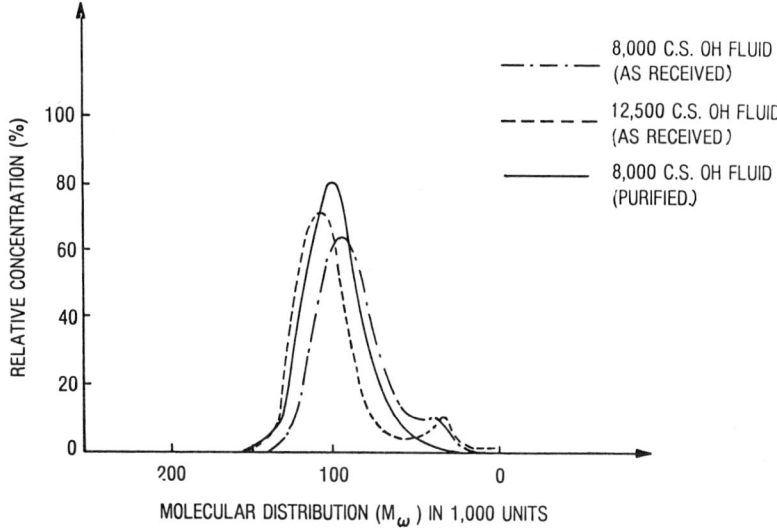

Figure 6 Gel Permeation Chromatography of Silicone Fluids

(polymer-filler interaction) may be the key to this physical property improvement.

Filler level has some correlation between the thermal stability of the material. From Figure 7, we can see that the filled 8,000 c.s. silcone OH fluid loaded with 10 parts per hundred resin (phr) fumed silica is adequate in improving thermal stability of the resin. Either activated (Cab-o-sil HS-5 of Cabot Corp.) or deactivated fumed silica (Tullanox 500D of Tulco Co.) lends similar thermal improvement of the OH fluid, with the activated Cab-o-sil HS-5 possessing slightly better thermal stability than the deactivated Tullanox 500D. Trimethyl-silyl groups which present only on the deactivated fumed silica (Tullanox 500D) surface may be the cause of the lower thermal stability of the silica in silicone. Any excess amount of the 10 phr fumed silica filler loading to the OH fluid does not have any substantial thermal improvement by the TGA. For the protection of the light sensitive IC devices, pigments are usually incorporated into the RTV encapsulant (such as low level of the carbon black and titanium dioxide). The main parameter that affects the RTV rheology may be filler incorporation and filler activity. Through the RTV silicone study, we have learned that the rheology of the RTV silicone is closely related to its coating performance.

Generic RTV Silicones. Our major purpose in the study of the RTV silicone by TGA is to optimize the thermal stability of the formulation. The TGA curve reveals not only the thermal stability of RTV silicone material but also the degree of dispersion of polymer-filler and/or filler-filler incorporation. With the aid of derivative thermogravimetry (DTG) curve we are able to detect any decomposition of the polymer. A well-dispersed filler-filled silicone material will result in one distinct derivative TGA curve (see Figure 8). However, poor dispersion of the polymer and filler(s) or poor polymer resin often reveals more than one derivative from the DTG curve. The phase separation between polymer-filler and/or filler-filler may be the cause of the multiple DTG cure. Even very close decompositions peaks which are hard to distinguish from the TGA curve will also be easily distinguished by the derivative curve.

Fourier Transform-Infrared (FT-IR) and Gas Chromatography/Mass Spectrometry (GC/MS). A TGA curve reveals the intrinsic thermal properties of the material. The weight loss with temperature indicates the removal of low molecular weight volatiles during the heating process at a lower temperature region ($<400°C$) of the generic and commercial RTV silicones. In order to characterize the low volatiles of the RTV material, FT-IR and GC/MS were used to identify these components. IR provides the fingerprint regions absorption for organic

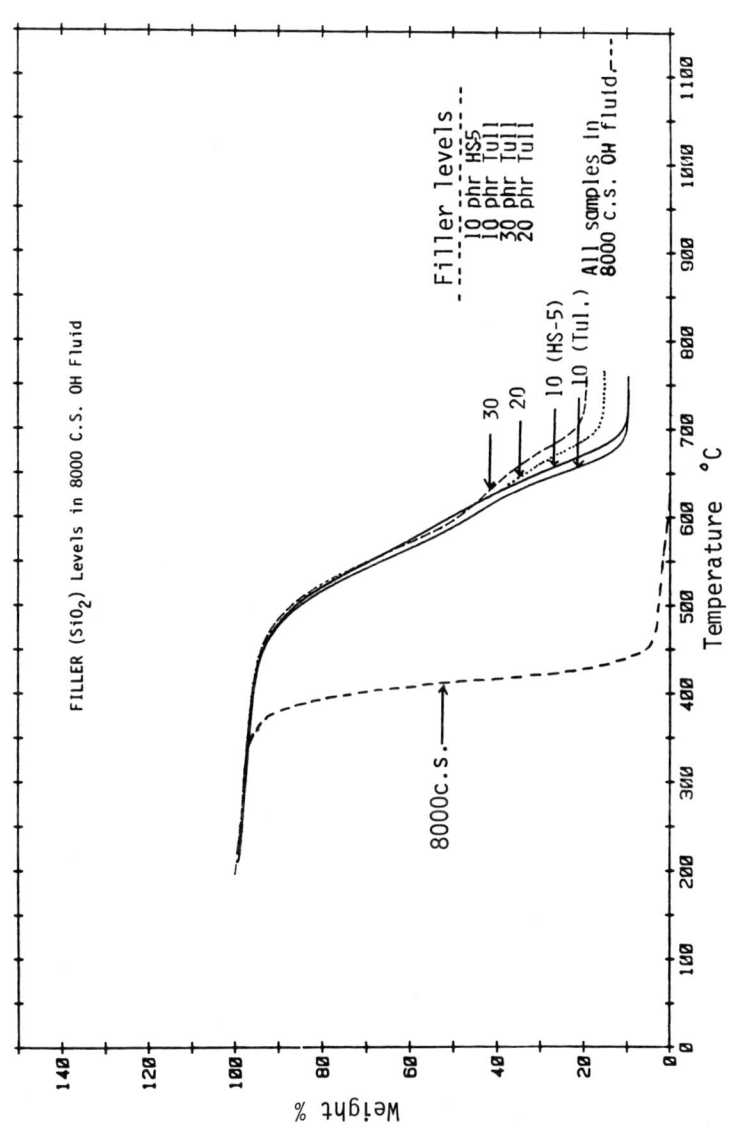

Figure 7. TGA Filler Levels in silicone Fluid

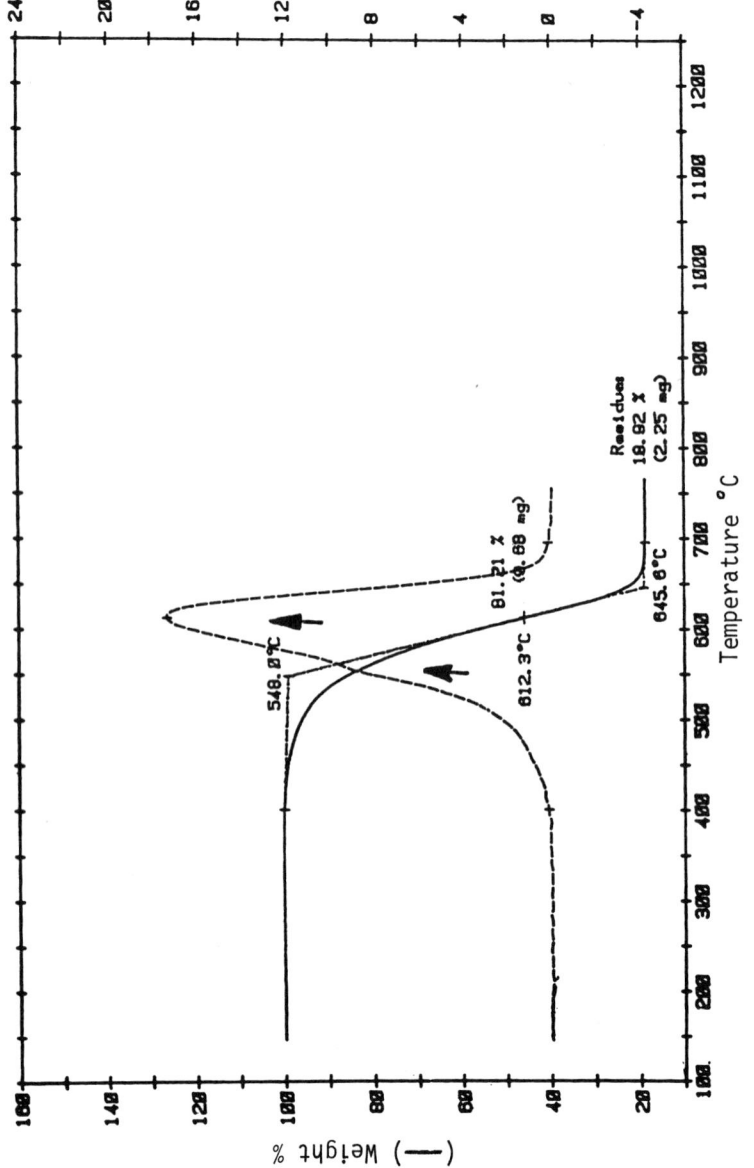

Figure 8   First Derivative Thermogravimetry Spectra of a Silicone

functional group identification. The collected extractables show a strong Si-O-Si stretching vibration between 1100 and 1000 cm^{-1}, this distinct absorption is a characteristic of silicone compounds. Figure 9 shows the results of FT-IR on some of the extracts. Detail IR analyses of the extracts are listed on Table II.

Table II. FT-IR Analysis of Low Volatiles from Cured RTV Silicone Solvent Extractables

Absorption (cm^{-1})	Functional Group
3000-2850	C-H stretching of methyl groups
1412	Asym. deformation of Si-CH$_3$ groups
1258	Sym. deformation of Si-CH$_3$ groups
1100 and 1000	Si-O-Si backbone stretching
800	Si-C stretching

Infrared spectroscopy has identified the volatiles as a mixture of silicones; however, it does not distinguish the type of silicones.

GC/MS is one of the most useful analytical instruments to separate and identify mixtures of organic compounds. Gas chromatography separates each of the components from the mixture by their differences in retention time on a column. Mass spectrometry identifies each component by its fragmentation pattern. The amount of volatiles obtained from TGA is usually sufficient for direct GC/MS analysis. However, the isolation of these low boiling point volatiles of the material is very difficult. We have employed a Soxhlet extraction of a cured RTV material with an organic solvent to obtain their low volatile residues from the RTV. These low volatiles residue were collected from the Soxhlet extractables using Freon TA and evaporated to viscous oils. These viscous oils were redissolved in Freon TA and injected into the Hewlett Packard model 5993 GC/MS. Figure 10 has shown the components of the GC/MS spectrum of the silicone extractables. Table III has listed the GC/MS tentative assigned structural fragmentations.

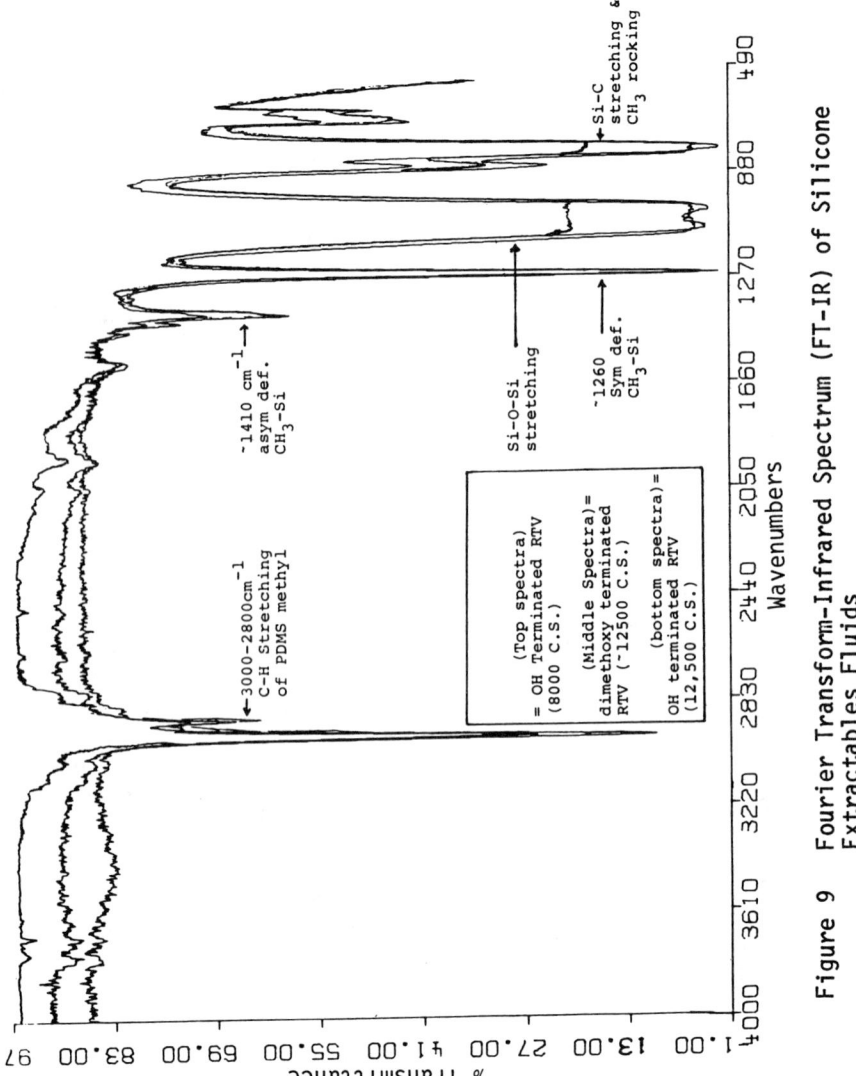

Figure 9 Fourier Transform-Infrared Spectrum (FT-IR) of Silicone Extractables Fluids

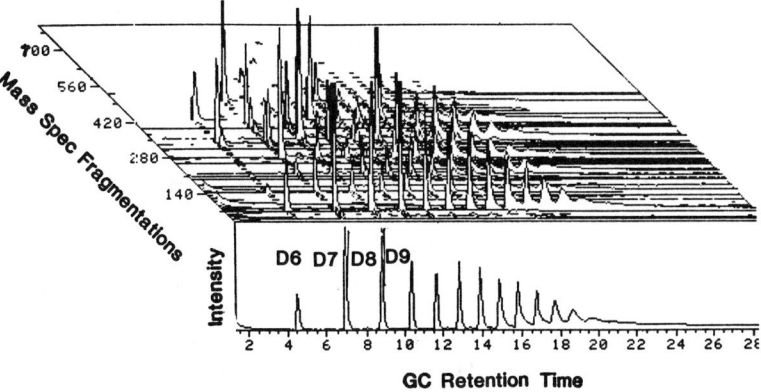

Figure 10 Gas Chromatography/Mass Spectrometer (GC/MS) of Silicone Extractables

Table III. GC/MS Analysis of Low Volatiles from Cured RTV Silicone Extractables

M/e (AMU)	Tentative Assignment of Mass Spec Fragmentations — Fragmentation Structures
73	$(CH_3)_3Si^+$
133	$(CH_3)SiOCH_3Si^+O$
147	$OSi(CH_3)_2O\ CH_3SiCH_2^+$
221	$(OCH_3SiCH_3)_n$-O-$CH_3SiCH_2^+$, n = 2
281	D_4^*-15
327	-
355	D_5-15
357	-
415	-
429	D_6-15
503	D_7-15
577	D_8-15

*D_4 stands for four dimethylsiloxane units of the cyclic compound.

Over ten components were observed from the extractables. Each extractable component was further identified by its mass spectrum. By comparing the TGA spectrum of the before and after solvent extraction of cured RTV samples, we are able to observe the difference due to the loss of unreactive volatiles. With the removal of unreacted low volatiles from the RTV sample, a more thermally stable material is obtained.

Reaction Kinetics. Empirical reaction kinetic information can also be obtained from thermal analyses. Non-isothermal methods rather than conventional isothermal (constant temperature) studies have certain advantages. The reaction kinetics are based on one single sample and calculated over the entire range, therefore, it requires less data. There are numerous methods of calculating the kinetics from thermal analysis[5]. However, the specific rate constant can be calculated from the DTG curve by using the following expression[6]:

$$k = (dx/dt)/(a-x)^n$$

where $(a-x)$ is the mass of remaining reactant (not decomposed portion)
  dx/dt is the height of the curve at time t.
  n = reaction rate order, for RTV silicone n = 1
  k = reaction rate constant.
  E_{act} = Activation Energy
  The E_{act} can be obtained from the slope of an Arrhenius plot of $\ln(k)$ vs. $1/T$ (K^o). For the RTV silicone samples the

E_{acts} are 29.6 and 30.9 Kcal/mole. This is in good agreement with literature value[7] (E_{act} of polydimethylsiloxane PDMS of 30 Kcal/mole).

Conclusions

We have demonstrated that TGA is an easy, rapid and reliable method to detect filler level in commercial RTV materials. It is also an excellent method for defining and choosing thermally stable ingredients (silicone fluid base polymer, and fillers, etc.) for the RTV silicone encapsulant. For silicone OH fluid, the optimal material should consist of minimal level of cyclic compounds. Fillers do improve the mechanical and physical properties of the silicone encapsulant, however, higher filler levels do not necessarily imply a better rheological silicone properties. The silicone OH fluid polymer and filler incorporation is a critical step in manufacturing a good, reproducible silicone encapsulant. Reaction kinetics (such as: reaction rates, enthalpies, activation energies ... etc.) of the polymers can be obtained from the TGA study. Decomposition products, which are the low volatile cyclic silicones are also obtained from this TGA study in conjunction with the GC/MS and FT-IR measurements. These techniques provide valuable information in formulating a thermally stable RTV silicone. A better control of the base silicone polymer with specifications of a minimal level of low molecular weight volatile monomers and a narrow molecular distribution polymer material are essential in formulating a good device encapsulant. Besides, we have shown additives that would improve the encapsulant electrical and coating performances. These modified encapsulants have improved the reliability of the encapsulated devices, and have been reported elsewhere[8]. Nevertheless, the TGA method with the recent software program will provide us with a better RTV material control, compounding procedures, and less material variations.

Literature Cited

1. Wong, C. P., American Chemical Society Symposium Series, 1982, 184, 171-183.
2. Mancke, R. G., IEEE Transaction on Components, Hybrids and Manufacturing Technology, 1981, CHMT-4, 492-8.
3. White, M. L., Proc. IEEE, 1969, 57, 1610.
4. Wong, C. P., Rose, D. M., Proceedings of 33rd Electronic Components Conference, 1983, 602-09.
5. Wendlandt, W. W., "Thermal Methods of Analysis", J. Wiley, New York, NY (1974) and references therein.
6. Papazian, H. A., Pizzolato, P. J., Orrell, R. R., Thermochem. Acta, 1972, 4, 97.

7. Thomas, T. H., Kendrick, T. C., J. Polym. Sci., 1969, A2, 7, 537 (1969) and ibid, 1970, A2, 8, 1823 (1970).
8. Wong, C. P., U.S. Patents 4 271 425, (June 1981); 4 278 784 (July 1981); 4 318 939 (March 1982); and 4 330 637 (May 1982).

RECEIVED September 2, 1983

Removable Polyurethane Encapsulants

K. B. WISCHMANN

Sandia National Laboratories, 7472, Albuquerque, NM 87185

Castable thermoplastic polyurethane formulations with superior upper temperature capability have been developed for use as removable electronics packaging materials. Several formulations have been developed employing a two-step prepolymer processing technique. For example, a 58/42 (pbw) ratio of polyether to polyester diols reacted with bitolylene diisocyanate forms a prepolymer; this step is followed by reaction with a chain extender, 1, 4 butanediol to form the final polymer. This formulation resulted in a semi-crystalline material with a tensile strength of 8.2 MPa and a 275% elongation. Changes in the polyether/polyester ratio yielded materials of differing mechanical properties which have been attributed to differences in the microphase structure between hard and soft segments in these linear polyurethanes. These formulations were essentially free of creep over an 80 hour period at $70^{\circ}C$, the temperature of interest. These casting resins can be processed with a filler such as alumina or glass microballoons to reduce the coefficient of thermal expansion or change density. The formulations are soluble in polar organic solvents such as dimethylformamide. Functional electronic components have been successfully potted and depotted using these solvent removable formulations.

The purpose of this investigation was to improve the upper use temperature of solvent-removable encapsulants as compared with previously reported systems (1). These kinds of materials have been developed to provide a component re-work capability, that is, they allow one to repair a sophisticated electronic package without discarding the entire, often costly assembly. Previous work has demonstrated the feasibility of the removable encapsulant concept (1,2). In order for these encapsulants to be easily removed with solvent, they must either be linear or very lightly crosslinked. Unfortunately, polymers of this nature are generally susceptible to creep, especially at elevated temperatures. Past efforts produced linear TDI-polyether polyurethane encapsulants that were creep resistant at ambient temperature but prone to creep at $>40°C$ (about Tg). To improve these materials use temperature range ($-40°C$ to $100°C$), new polymers were developed which possess differing chemical structure and morphology. Polyurethanes were selected for investigation because of the wide range of physical properties that can be achieved.

The thermoplastic polyurethanes studied here are segmented copolymers composed of alternating flexible polyether or polyester units (soft segments) which are connected to short rigid urethane groups (hard segments) (3,4). By varying the ratio of hard to soft segments, the chemical structure, and the molecular weight, dramatic changes in mechanical properties can be realized. These variables are directly related to the polymer's morphology resulting from phase separation of the hard and soft segments (5). The hard phase regions in these materials act as virtual crosslinks which impart elastomeric-like properties to the system. Unlike thermosetting polymers, however, these materials can be dissolved in solvent or molded upon heating.

Efforts were made to enhance the domain structure by increasing the amount of hydrogen bonding between hard and soft segments and altering the diisocyanate structure. It was felt that these changes would restrict the mobility of the polymer resulting in better creep resistance, increased strength and higher upper use temperature. In this paper optimized polymer formulations, processing techniques and properties are discussed.

Experimental

The materials employed in these formulations consisted of bitolylene diisocyanate (3,3'-dimethyl-4,4'-biphenylene diisocyanate, 136T Upjohn Co.) referred to as TODI, a polyether diol (tetramethyleneoxide polyol, Polymeg 2000, Quaker Oats Co.), a polyester diol (caprolactone polyol, PCP-200, Union Carbide Corp.) and a chain extender, 1,4-butanediol (Aldrich Chemical Co.). The formulations were prepared by a two-step procedure. In the first step the appropriate ratio of polyether to polyester was reacted with the bitolylene diisocyanate (NCO/OH = 2) to form a prepolymer. In the second step stoichiometric amounts of 1,4-butanediol were reacted with the prepolymer. The formulations were designated 100ET20, 64ET33 and 58ET34, where the first number refers to the weight ratio of polyether (ET) in the soft segment and the remainder is assumed to be the polyester, e.g., 58 parts by weight polyether, 42 parts by weight polyester, while the second number designates the weight percent of the diisocyanate based on the overall weight of the elastomer. This nomenclature has been employed elsewhere (6,7). The formulation termed 100ET20 contained only the polyether and was used primarily for comparison purposes. See Tables 1 and 2 for polymer designations and formulations.

Polymers of this nature can be polymerized either in solution or in bulk; in the latter case they are normally reacted at high temperatures, e.g., 100-150°C. Since our goal was a casting resin, the formulations were reacted in bulk and at lower temperatures to protect heat sensitive electronic components; furthermore, low reaction temperatures minimize side reactions that can lead to crosslinking and polymer insolubility. In this process the polyols and diisocyanates were mixed and allowed to react for about 25 minutes at 71°C to form the prepolymer formation while longer times resulted in material too viscous to cast or deaerate. After the indicated time, 1,4-butanediol was added followed by deaeration and subsequent encapsulation of a preheated (71°C) electronic device. A second deaeration of the encapsulated part is usually necessary. Pot life for such a system is about 15 minutes. Final reaction or "cure" was 24 hours at 71°C.

Mechanical testing was accomplished with an Instron Model 1130 using an ASTM tensile specimen D1708, after the specimens were aged for two weeks. Glass transitions, T_g, were obtained from a Perkin-Elmer Thermomechanical Analyzer TMS-1 in the penetrometer mode. Volume resistivity measurements were made according to ASTM D-150 method. Dynamic mechanical properties were obtained from -140°C to +140°C at 11 Hz on a Rheovibron DDV-II direct reading dynamic viscoelastometer. Rheovibron compliance corrections were made following a method suggested by Massa (8).

Results and Discussion

The synthesis of these linear polymers follows a typical two step urethane addition sequence shown as follows:

Step 1

        HO-\/\/\/-OH + 2 OCNRNCO ⟶ OCNRNHCOO-\/\/\-OOCHNRNCO

        Macroglycol   Diisocyanate   Prepolymer

Step 2

$$\text{OCNR'NCO} + \text{HOR''OH} \longrightarrow \left(-\overset{O}{\underset{}{C}}\text{NHR'NHC}\overset{O}{\underset{}{O}}\text{R''O}-\right)$$

Prepolymer   1,4 Butanediol   Polyurethane

These reactions are essentially chain extending reactions which yield the thermoplastic product. Difunctional materials were used throughout in an effort to maintain polymer linearity. Since a casting resin was desired, all reactions were of necessity carried out neat. All materials were used in the "as received" from, consequently, competing reaction such as the reaction of moisture with the diisocyanate may not have been avoided. An attempt was made to control allophanate formation (crosslinking reaction) by maintaining a mix and cure temperature no higher than 71°C (3). Completeness of reaction in the final product was determined by monitoring the disappearance of isocyanate via infrared analysis.

Based on previous efforts to formulate these kinds of materials (1), attention was directed towards increasing rigidity in the hard and soft segment domain structure. This was accomplished by employing bitolylene diisocyanate as part of the hard segment and augmenting the polyether diol soft segment with a polyester diol. The polyester diol adds carbonyl groups to the chain backbone which should promote stronger hydrogen bonding than the polyether diol. The net effect from these modifications was increased mechanical strength, negligible creep at one of the higher use temperatures (71°C) when compared to a polymer system that did not contain any polyester diol. These modifications also introduced crystallinity and a resultant melting transition (Tm) which translates into improved thermomechanical properties. Crystallinity was established by x-ray measurements (9).

Mechanical properties of three formulations designated 100ET20, 64ET33 and 58ET34 are shown in Table 3; the first system does not contain any polyester diol. As polyester diol content was increased tensile strength and modulus increased accordingly. The Tg also increased as well as the crystalline melting point, Tm. Dynamic mechanical properties, reflecting the same trends are shown in Figure 1. It is interesting to note that a wide use temperature range was achieved; that is, little change in modulus between 0° and 125°C.

Since these polymers are thermoplastic, creep becomes an important property consideration in a casting resin. Creep compliance measurements were made on 64ET33 at ambient, 40°C and 70°C employing a load approximately equal to 10% of the material's tensile strength. As shown in Figure 2, some initial

24. WISCHMANN — Removable Polyurethane Encapsulants

Table 1. Component designation

Component	Molecular Weight	Code
Poly(tetramethylene oxide)	~2000	ET
Poly(ε-caprolactone)	530	ES
Bitolylene diisocyanate	264	TODI
1,4 - butanediol	90	BD

Table 2. Elastomer formulations

Sample Designation	Molar Composition TODI/BD/ET/ES	Weight % Hard Segment
100ET20	1/0.5/0.5/0	24
64ET33	1/0.5/0.16/0.34	38
58ET34	1/0.5/0.14/0.36	40

Table 3. Thermoplastic polyurethane physical properties

	100ET20	64ET33		58ET34
	7.8 gm Polymeg 2000 2.1 gm TODI 0.33 gm 1,4-Butanediol	7.0 gm Polymeg 2000 4.0 gm PCP-200 5.8 gm TODI 1.0 gm 1,4-Butanediol		7.0 gm Polymeg 2000 5.0 gm PCP-200 6.8 gm TODI 1.1 gm 1,4-Butanediol
	Unfilled	Unfilled	Al_2O_3 Filled	Unfilled
*Tensile Strength$_{(f)}$, MPa	6.55 (950 psi)	7.58 (1100 psi)	10.3 (1500 psi)	13.1 (1900 psi)
Percent Elongation	600	275	20	500
Modulus, MPa	35 (5000 psi)	124 (18,000 psi)	1275 (185,000 psi)	145 (21,000 psi)
Shore D	24	45		51
Coeff. Expansion (-52 to 72 °C)	$220 \times 10^{-6}/°C$	$160 \times 10^{-6}/°C$	$100 \times 70^{-6}/°C$	$150 \times 10^{-6}/°C$
T_g	-59°C	-53°C		-33°C
T_m (softening)	120°C	147°C		168°C
Volume Resistivity, ohm-cm		3.27×10^{12} (500 v) 3.01×10^{12} (1000 v)		1.3×10^{13} (500 v) 1.04×10^{13} (1000 v)

*Tensile testing conducted at ambient temperature and a crosshead speed of 0.2" min^{-1}.

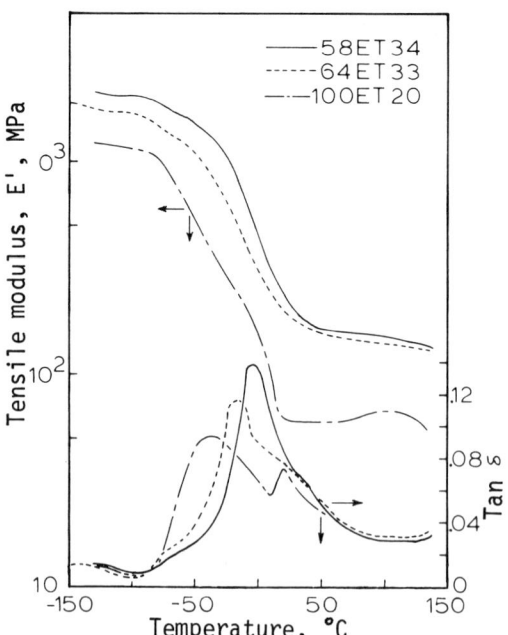

Figure 1. Dynamic mechanical properties of linear polyurethane encapsulants at 11 Hz. Formulation nomenclature: the first number refers to the weight ratio of polyether (ET) in the soft segment and the remainder is assumed to be the polyester; the second number designates the overall weight percent of the diisocyanate.

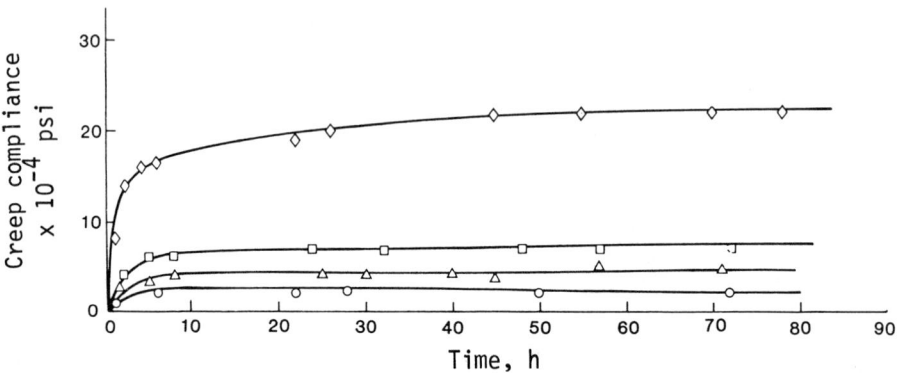

Figure 2. Creep compliance vs. time for linear polyurethane encapsulants at constant load. See Figure 1 for nomenclature.
Key (64ET33): ◇, 70 °C; □, 40 °C; and △, ambient. Key (58ET34): ○, 70 °C.

creep occurs in the first 5-6 hours and then essentially ceases over the next 70 hours observation time. Since 64ET33 showed evidence of creep at 70°C, more polyester diol was added to the formulation in an effort to reduce creep. As a result 58ET34 exhibited negligible creep (< 1% at same load) at all the test temperatures although only the 70°C test results are shown. These improved properties may be related to 1) increased hydrogen bonding due to the increased polyester diol content, 2) increased hard segment content, and 3) evidence of crystallinity in the hard segment domain. All effects result in a higher modulus material. The structure-property relationships of these formulations in terms of their crystalline morphology are discussed in another paper (9).

Addition of a filler such as Al_2O_3 is an alternate method to reduce creep. As shown in Table 3 this filler also increases mechanical strength as well as lowering the coefficient of thermal expansion. The latter aspect can be an important consideration when trying to match material coefficients of expansion in an encapsulating situation. It should also be noted that volume resistivity measurements of unfilled material (see Table 3) indicate these formulations provide satisfactory electrical insulation for encapsulating purposes.

These formulations are soluble in strong polar solvents such as dimethyl formamide and tetrahydrofuran at room temperature thereby permitting the necessary rework capability. The depotting process can be accelerated by vigorous agitation. If large amounts of encapsulant must be removed, periodic changes of solvent will facilitate the operation. Obviously, the depotting time will vary depending on the mass and exposed surface area of the encapsulant. Solvent damage to other components is a possibility and each situation must be carefully considered. Components employing crosslinked polymers as dielectrics, i.e., resistors, capacitors, PC boards, etc., were found to survive two or three depotting sequences. However, swelling occurs in rubbers, and some thermoplastics may be attacked. After depotting, the device is "baked-out" at an elevated temperature to remove residual solvent. Experience has shown that reclaiming and reconstituting an electronic device via this technique is less costly than replacing it with a new one.

Summary and Conclusion

Two solvent-removable polyurethane encapsulants have been formulated with varying properties. The 58ET34 has greater mechanical strength than the 64ET33. The 58ET34 exhibits negligible creep from ambient to 70°C, whereas the 64ET33 appears to exhibit some creep especially at the upper test temperature.

Both formulations have a wide use temperature range
(0° to 125 °C), that is, mechanical properties change very
little. The improved physical properties are the direct result
of formulation modifications which show enhanced domain structure
and polymer crystallinity. Processing characteristics are
similar to an epoxy casting resin. Both formulations can be
filled which results in increased strength and lower coefficients
of expansion, and both formulations are soluble in convenient
organic solvents, thereby allowing a component rework capability.

Acknowledgments

The author wishes to thank R. R. Lagasse for his valuable
assistance in this program. Work supported by U. S. Department
of Energy

Literature Cited

1. K. B. Wischmann, G. L. Cessac, and J. G. Curro, J. of Elastomers and Plastics, 9, 299, 1977.
2. K. B. Wischmann and R. A. Assink, SAMPE Journal, 13, 15, March /April 1977.
3. P. Wright and A. P. C. Cumming, "Solid Polyurethane Elastomers," Gordon and Breach Pub., New York, 1969.
4. J. M. Buist and H. Gridgeon, "Advances in Polyurethane Technology," Wiley and Sons, New York, 1968.
5. S. L. Samuels and G. L. Wilkes, J. Polymer Sci., Symposium No. 43, 149, 1973.
6. N. S. Schneider, C. R. Desper, J. L. Illinger, A. O. King and D. Barr, J. Macromol. Sci. Phys., B11(4), P. 527, 1975.
7. D. S. Huh and S. L. Cooper, Poly. Eng. and Sci., 11, 369, 1971.
8. D. J. Massa, J. Appl. Phy., 44, (6), 2595, 1973.
9. R. R. Lagasse and K. B. Wischmann, Spring 1977 Natl ACS Meeting, Div. of Organic Coatings & Plastics, Coatings and Plastics Preprints, 37, (1), 501, 1977.

RECEIVED September 21, 1983

Improvements to Microcircuit Reliability by the Use of Inhibited Encapsulants

F. W. AINGER, J. BRETTLE, I. DIX, and M. T. GOOSEY[1]

Allen Clark Research Centre, Caswell, Towcester, Northants, England

> A novel method of improving the reliability of plastic packaged semiconductor devices has been developed. One of the major causes of failure in such devices is corrosion due to moisture permeating through the encapsulant, transporting small amounts of ionic impurity and condensing at the chip surface to form an aggressive electrolyte. This can be combatted by compounding corrosion inhibiting chemicals in the encapsulating resin. These inhibitors are transported by diffusing moisture and greatly reduce the aggressiveness of the condensed electrolyte. The basis of choosing suitable inhibitors is described together with the results of accelerated tests on microcircuits encapsulated using this technology. Both liquid epoxy and transfer moulding epoxy novolac systems have been investigated. Ancillary testing has also been performed to investigate possible deleterious side effects of using inhibitors e.g. electro-migration, parasitic MOS formation, effects on resin adhesion and Tg. Such effects were found to be absent.

Over 95% of all the microcircuits made are packaged in plastic, usually a transfer moulded epoxy resin. Changes in packaging technology will occur away from the familiar PDIP (plastic dual-in-line package) to smaller SOT or chip carrier formats but plastics will continue to be the dominant packaging material for cost reasons. At the same time there is a need to improve the reliability of plastic encapsulated devices (PEDs) as they find further use in professional and certain military applications.

[1]Current address: Dynachem Corporation, 1275 Lake Avenue, Woodstock, IL 60098

Conventionally, ceramic or metal packaging devices have been considered to be more reliable than PEDs because, in theory at least, in these types of packaging the microcircuit could be sealed in an hermetic cavity containing dry inert gas and environmental degradation of the microcircuit surface avoided. The situation is different in plastic packaging where a partly cured epoxy powder is fused under pressure and cured in intimate contact with the microcircuit to form a solid monolithic block around the chip and its interconnection leads. Reliability problems arise when atmospheric moisture diffuses through the epoxy resin either through its bulk or along the resin/lead frame interface. This moisture "picks up" ionic impurities remaining in the resin from its manufacture, particularly chloride ions, and transports them to the chip surface. The chip surface carries electrical conductors, usually aluminium alloy tracks perhaps 1μm thick and a few μm wide which interconnect the various active regions of the chip, these conductors will normally be covered with a protective "passivating" layer as part of the chip fabrication process but such layers often contain defects or cracks and in any event the bonding pads required to make contact to the outside world cannot be passivated. During operation of the circuit, closely spaced tracks are often at different voltages and once contact is achieved with the diffusing moisture, electrolytic corrosion can occur leading to rapid destruction of the very thin aluminium layer and circuit failure.

The comparison between the reliability of hermetic and plastic packaging is perhaps not as clear cut as suggested above, the hermetic package cavities are not always as clean and dry as required and often of dubious hermeticity, also plastic packaging materials of low water permeability and high purity have been developed. In addition, plastic packages offer improved performance against other forms of reliability hazard, e.g. mechanical shock, because of their monolithic construction.

Because of the large amount of work which has been performed on reducing the moisture permeability and impurity content of epoxy transfer moulding resins it is difficult to see that they can be further improved from the point of view of minimising microcircuit corrosion without recourse to a novel approach; this is the subject of this paper.

Corrosion of Aluminium and its Inhibition

A well established technique for reducing the rate of corrosion of metals in aqueous solutions is the use of corrosion inhibitors; these are chemicals which interfere with either the anodic or cathodic corrosion reaction and thus slow the overall corrosion rate. These inhibitors may function by various mechanisms but three types are particularly relevant in the case of aluminium:

(i) Oxidation inhibitors. These promote the oxidation of the metal surface and help to heal defects and cracks in the oxide. Ions from these inhibitors may become incorporated in the oxide enhancing its stability. Care must be taken when using oxidation inhibitors in that at the wrong concentration they may actually promote corrosion.

(ii) Adsorption inhibitors. These are often organic materials with polar groups which are strongly adsorbed on any bare aluminium surface in preference to water and thus inhibit the early stages of aqueous corrosion.

(iii) Buffer inhibitors. Aluminium is an amphoteric metal which dissolves (corrodes) readily at both low and high pH. Any small amount of corrosion which occurs at more neutral pH values tends to lower the pH at cathodic sites due to cathodic reduction reactions and increase the pH at anodic sites due to anodic oxidation reactions. Local changes in environment can then lead to other corrosion reactions occurring under these new pH conditions. Buffering chemicals which ameliorate development of this pH differential can therefore inhibit corrosion.

It appeared to us that if we could add aluminium corrosion inhibitors to the encapsulating resin it was possible that diffusing moisture would transport the inhibitor to the chip surface so that the aluminium conductors would experience an inhibited electrolyte rather than an agressive electrolyte and corrosion would thereby be reduced. It should be noted that this mechanism does not operate in the absence of diffusing moisture but that it is not needed in those circumstances because corrosion would not normally occur. For this approach to be successful the following conditions must be met:

1. The inhibitor should be capable of being transported through a cross-linked epoxy resin matrix.
2. The inhibitor should not contain ions which may be deleterious to microcircuit functioning.
3. The enhanced ionic conductivity of the resin caused by these additives should not cause any deleterious side effects on circuit functioning. (It should be noted that for many years the emphasis has been on lowering the ionic content of encapsulating resins).
4. The inhibitor should be capable of functioning in the complex and changing environmental and electrical situation at a chip surface.

To establish if these conditions could be met, the literature of aluminium corrosion inhibitors was searched, suitable inhibitors were chosen for evaluation, and a programme of measurements and testing initiated.

There is a large range of potential inhibitors available from the literature: both inorganic and organic, solid and liquid. Inorganic inhibitors are often cations and it is important to choose the complementary anion so that:

(i) The compound has some solubility in water. (This fulfils, in part, condition (1).)

(ii) The anion is not one of the many metallic ions which are deleterious to silicon microcircuit performance through their ability to diffuse into silicon and silicon dioxide (Condition (2)).

Small cations such as sodium and lithium must therefore be rejected, and in general group 1 elements should not be used although the heavier members may be suitable. Transition elements, particularly group 8, are also unsuitable. The most suitable cations are probably Ca, Sr, Ba, Mo, Cs, W, Al, Pb and Sn.

Different restrictions apply to organic inhibitors; in particular they should not react with the epoxy resin and should be reasonably stable at transfer moulding and post curing temperatures; again, a sufficient solubility in water to maintain a concentration at the chip surface is required.

Based on these requirements the following inhibitors were chosen for experimental investigation:
1. Sodium Silicate
2. Barium Permanganate
3. Calcium Citrate
4. Ammonium Tungstate
5. Furfuraldehyde
6. Salicylaldehyde
7. Dicyandiamide
8. p-toluidine
9. p-cresol
10. p-anisidine.

It should be noted that not all of the inhibitors fulfilled the earlier stated requirements totally but this was relatively unimportant at this stage where the intention was to demonstrate feasibility of concept rather than an optimised solution to the problem.

The experimental programme followed to evaluate potential inhibitors is described below.

Experimental Measurements

These were of four main types:
1. Reliability testing of microcircuits encapsulated via a model epoxy resin/inhibitor system to demonstrate feasibility.
2. Ancillary testing to determine any deleterious side effects of inhibitor additions on circuit function.
3. Ancillary testing of inhibitor/resin properties e.g. solubility of inhibitor in resin and diffusion rates of inhibitor in resin.
4. Larger sacle testing of inhibitors in transfer moulding compounds prior to practical use of the idea.

Feasibility Testing of a Model System. A simple test circuit silicon chip comprising three pairs of parallel aluminium conductors deposited on 0.5µm of SiO_2 on silicon was used. The conductor pairs were 1µm thick, 4µm wide and interdigitated to save space,

Figure 1. This test device was mounted in an adapted 12 pin TO5 header package, Figure 2, so that resin encapsulant could be added over the chip to a predetermined thickness. The resin system was a high purity two part liquid epoxy from Ciba-Geigy i.e. 10 parts of diglycidyl ether of bisphenol A (MY 790) cured with 1 part of triethylene tetramine (MY951) for 3 hrs. at 60°C. The bond pads allowed each pair of tracks to have a polarising voltage applied between them and the resistance of each track (initially about 30Ω) could be monitored. A large number of such devices could be placed on life test and monitored for failure; it was found that once a track began to corrode its resistance rose very rapidly from $\sim$ 30Ω to hundreds of kΩ. To accelerate failure, testing was performed in an 85°C, 85% relative humidity environment with 1V polarising voltage, this is a common technique in microcircuit accelerated testing. Decapsulating the devices by refluxing in hot dimethylformamide after test and examining their surfaces by scanning electron microscopy confirmed corrosion as the cause of failure. Using this technique a large number of inhibitors could be screened for efficacy in terms of numbers of tracks remaining uncorroded after a particular test time (a device resistance > 200Ω was taken as a reasonable measure of failure).

The results of these tests showed that potential inhibitors could be successful, ineffective, or positively deleterious to device reliability (Figure 3, the reference devices contain no inhibitor and failure rates should be judged in comparison to these). Three inhibitors were found to be particularly successful as a result of these screening tests i.e. ammonium tungstate, paracresol and calcium citrate (1). Screening tests were repeated with these materials at different concentrations to determine if there was an optimum concentration for maximum effectiveness. This was generally found to be the case e.g. an optimum concentration of $\sim$800 ppm was determined with ammonium tungstate (Figure 4). These encouraging results led to further tests to check that no deleterious side effects occurred due to the presence of inhibitors.

<u>Ancillary Testing to Determine Side Effects</u>. Two types of side effect were of interest, electromigration and parasitic MOS formation. Electromigration is a diffusion effect operating at high temperatures and current densities in conductors due to a mass transport of material in the direction of current flow, leading to rupture of the conductor in thinned areas. There have been suggestions that electromigration may be accelerated by corrosion or high ionic conductivity at the chip surface; therefore it was necessary to examine the effect of inhibitors on this phenomenon. Special test chips previously developed for electromigration testing of conductor materials were used with the TO5 test vehicle described earlier and exposed to high temperature high current density conditions. Under the most severe conditions electromigration could be induced but no significant differences in failure rate between devices encapsulated in inhibited and uninhibited

Figure 1. Microcircuit used for corrosion monitoring. Contains three interdigitated pairs of conductor tracks.

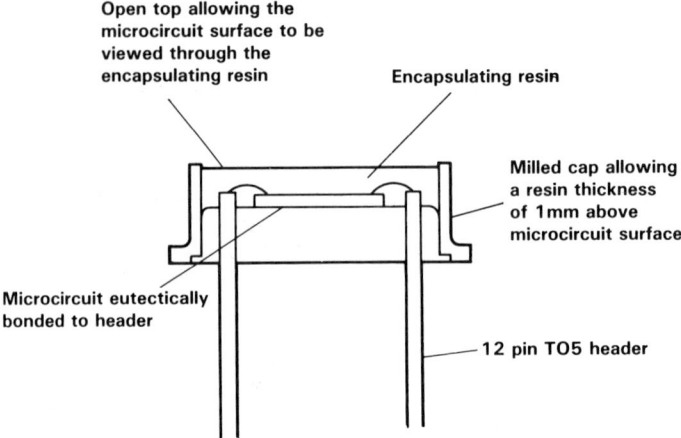

Figure 2. Section through microcircuit mounted in TO5 header test vehicle.

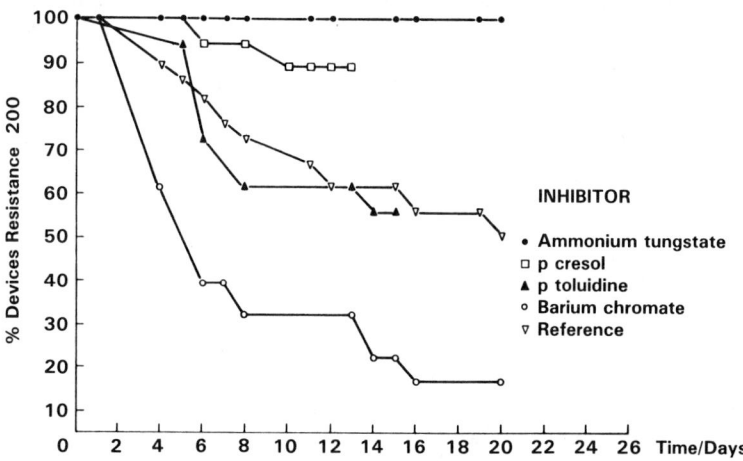

Figure 3. Variation of device reliability with corrosion inhibitor. Reference device contains no inhibitor.

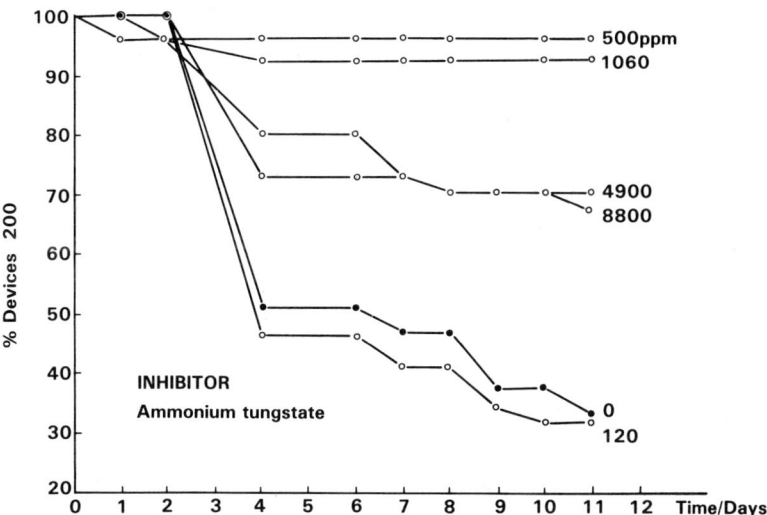

Figure 4. Variation of device reliability with inhibitor concentration.

resins could be detected. Parasitic MOS formation is an effect
which occurs when a conductive channel forms in an encapsulating
or passivating layer over a MOS structure and induces an opposite-
ly charged inverted channel in the device below causing malfunct-
ion. A special test chip comprising a chain of MOS transistors
was used in a similar fashion to the screening reliability tests
discussed above and parametric monitoring of these chips demonstrated that the increased ionic content of the resin due to inhibitor addition did not cause increased parasitic MOS formation.

Ancillary Testing of Inhibitor Resin Properties. It was found
that inhibitor could be introduced fairly readily in the two part
liquid system either by direct dissolution in the resin or harden-
er or, for some of the less soluble inhibitors, as a fine dispersion in the resin. Introduction in powder transfer moulding resins was achieved by ball milling the inhibitor with the powder.
Subsequent analysis of samples of powder showed that the inhibitor
had been uniformly dispersed.

To establish that inhibitor addition had no deleterious effect on resin properties the glass transition temperature, moisture
permeability, and adhesion properties of the two part liquid resin
with and without inhibitor was determined and in no cases were
significant changes in properties observed. Technically important
factors were also determined for transfer moulding resins i.e.
mould staining and spiral flow; again no serious detrimental
effects were observed.

As discussed earlier, it is important that corrosion inhibitors are transportable through the resin matrix and therefore this
transport was investigated using radiochemical, spectroscopic and
conductimetric methods on dry and water immersed resin, different
detection techniques being applicable to different inhibitor species. In all cases it was found that diffusion was negligible in
dry samples but that diffusing water could "pick up" inhibitor
species and transport them sufficiently well to compensate for any
small amount of depletion of inhibitor at the chip surface.

Larger Scale Testing of Inhibitors in Transfer Moulding Resins.
Two of the most successful inhibitors in the earlier screening
tests i.e. ammonium tungstate and calcium citrate were tested on a
larger scale in transfer moulding resins. Large (25kg) batches of
a widely used microcircuit encapsulating transfer moulding epoxy
resin had these inhibitors added and the resistor pattern test
circuit described earlier was encapsulated in the resin at a number of different encapsulation facilities. Circuits were encapsulated in uninhibited resin at the same time as a control and many
hundreds of dual-in-line plastic packaged devices were prepared
for long term reliability testing using $85^{\circ}C/85\%$ RH and autoclave
exposure tests. Table 1 gives some of the results obtained on
long term autoclave test. It appears that reliability of devices
depends on the care taken at the encapsulation facility but that

the use of inhibitors can greatly improve the reliability of devices even when encapsulated in less than ideal conditions.

Table 1: Autoclave Test Results (120°C Over Water)

Hours	Calcium A	Citrate B	% Devices Remaining Unfailed			
			Ammonium A	Tungstate B	Reference A	Reference B
500	100%	100%	100%	100%	100%	100%
600	"	"	95%	"	90%	"
700	"	"	"	97%	"	"
900	"	"	"	"	"	"
1000	"	"	"	"	85%	"
1300	"	"	"	"	"	"
1400	"	"	"	"	"	97%
1500	"	"	"	"	"	"
1600	"	"	"	"	"	93%
1700	"	"	"	"	"	"
1800	"	"	90%	"	80%	"
2100	"	"	"	"	"	"
2200	"	"	"	"	"	85%
2400	"	"	"	"	"	82%
2600	"	"	"	"	"	80%
2800	"	"	"	"	"	78%
3000	"	"	85%	"	"	77%
3200	"	"	"	"	"	58%
3400	"	"	"	"	"	48%
3600	"	"	95%*	"	"	45%
3800	"	"	"	"	"	"
4000	"	"	"	"	"	43%

A – encapsulated by the authors with careful control of conditions in a laboratory encapsulation facility.
B – encapsulated at a FarEastassembly house.
* – spurious value due to equipment fault.

Further Work

The work briefly described above was specifically concerned with transfer moulding resins for monolithic packaging but we have since shown that the technique of inhibitor addition is also applicable to other encapsulation methods and materials e.g. powder coating. Work is continuing in this area.

Acknowledgments

This work has been carried out with the support of Procurement Executive, Ministry of Defence, sponsored by DCVD.

Literature Cited

1. U.K. Patent Application GB 2087 159A. Plastic Encapsulated Electronic Devices.

RECEIVED November 3, 1983

POLYMERS
FOR PRINTED WIRING

26

Photopolymer Dielectrics: The Characterization of Curing Behavior for Modified Acrylate Systems

R. D. SMALL and J. A. ORS

Engineering Research Center, Western Electric Company, Princeton, NJ 08540

B. S. H. ROYCE

Princeton University, Princeton, NJ 08540

The control and characterization of photopolymerization reactions is extremely important for printed circuit applications of photocured dielectrics. Many properties of the cured polymer system are strongly dependent upon degree of cure.

We have been investigating the effects of multifunctional monomers, rubber modification and light intensity upon both the rate and final degree of cure in acrylated epoxy systems.

We have applied a number of analytical techniques to characterize the curing behavior of our systems including;
- Fourier Transform Infrared (FTIR) and FTIR Photoacoustic Spectroscopy (FTIR-PAS) - Initial rates of polymerization, photoinitiator effects and oxygen effects were evaluated using these techniques.
- Thermal Gravimetric Analysis/Solvent Extractables - We have used both methods to characterize the change in curing behavior induced by varying light intensity, type and concentration of crosslink agent.
- Solvent Absorption - The kinetics of solvent absorption were evaluated for a number of systems with varying degrees of cure.

The dual purpose of this study was to characterize the curing behavior of model acrylate photopolymer mixtures and to evaluate analytical techniques to determine which best describe the curing process. The characterization of curing behavior and its effects on the final physical, chemical and electrical properties can be a difficult task.

We have utilized a number of techniques to examine and/or monitor the photopolymerization of model formulations. The techniques presented here include the following:

- Fourier Transform Infrared Spectroscopy (FTIR) and Fourier Transform Infrared - Photoacoustic Spectroscopy (FTIR-PAS)
- Solvent Absorption (SA)
- Thermogravimetric Analysis (TGA)
- Solvent Extraction (SE)

Our emphasis was upon evaluating the utility of the above techniques in examining our model systems.

Within the acrylate formulations we were especially interested in the effects of acrylonitrile-butadiene rubber modification of photopolymers with varying multifunctional monomers, photoinitiator concentration and light intensity.

EXPERIMENTAL

Materials

The materials used in this study were obtained from commercial sources (Table I) and were used without further purification. The Epocryl-370 can be described as the diacrylated derivative of diglycidyl bisphenol A (Epon-828). The VTBN 90-109 is a vinyl terminated acrylonitrile butadiene rubber (27% CN). The photoinitiator (PI), DMPA, is available from Giba-Geigy under the tradename Irgacure-651. This PI has been reported by Osborn[1] to be a more reactive initiator than other members of the benzoin ether series[2]. The uv spectrum of DMPA shows a major absorption peak (λ max) at 335 nm. The band is relatively broad allowing for the use of 313 $\pm$ 10 nm interference filter to isolate the desired irradiation wavelength for the FTIR kinetic measurements.

TABLE I - Commercial Materials

NAME	SOURCE
Ethoxyethyl acrylate (EEA)	Polysciences
1,6-Hexanediol diacrylate (HDODA)	Polysciences
Trimethylolpropane triacrylate (TMPTA)	Polysciences
Epocryl 370	Shell
VTBN 90-190	B.F.Goodrich
2,2-Dimethoxy-2-phenylacetophenone (DMPA)	Ciba-Geigy

Mixtures

All the mixtures were formulated based on weight percent resin and or the equivalent parts by weight (pbw) to total 100%. The compositions are summarized in Table II.

TABLE II - Mixture Compositions (Weight Percent)

Mixture	Epocryl 370	EEA	HDODA	TMPTA	VTBN	DMPA
I	69	29	-	-	-	2
II	69	19	10	-	-	2
III	69	19	-	10	-	2
IV	69	9	20	-	-	2
V	69	-	29	-	-	2
VI	69	9	-	20	-	2
IR	69	19	-	-	10	2
IIR	69	9	10	-	10	2
IIIR	69	9	-	10	20	2
II-0.5	69	20.5	10	-	-	0.5
II-1	69	20	10	-	-	1
II-4	69	17	10	-	-	4
II-8	69	13	10	-	-	8
IIR-4	69	17	10	-	-	4
II-8	69	13	10	-	-	8
IIR-4	69	7	10	-	10	4
IIR-8	69	3	10	-	10	8

Irradiation

The samples used in the solvent uptake, TGA and solvent extraction analysis were cured using a CoLight UV curing system Model UVC24. All experiments, both 'high' and 'low' intensity irradiations, were carried out using a single lamp. Low intensity was achieved by running the lamp at low power and filtering the light with an ACRYLITE acrylic sheet with a cut off at 330 nm. FTIR and PAS-FTIR samples were irradiated using an Oriel arc source model 8500 equipped with a 100 watt Hg arc lamp. The irradiation wavelength was isolated using a 313 nm interference filter. The 'high' flux differed from the 'low' flux by a factor of 15:1.

FTIR and FTIR PAS

FTIR allows for a quantitative analysis of the reaction rates along with the monitoring of subtle changes in the film curing polymerization via substraction techniques[3]. Figure 1 shows the reaction profiles for mixtures I, II and III. Three main observations can be made from this plot. First, the graph shows an induction. This induction or inhibition period may serve as a measure of the susceptability of these mixtures to oxygen inhibition. The similarity may be inherent to the method, since the polymerization reaction is being measured between two NaCl plates hence the amount of dissolved O_2 can be assumed to be similar in each case and the amount of diffused O_2 at the surface negligible[4]. Second, from the dynamic part of the curve the rate of polymerization and the system's efficiency can be derived. This efficiency can be obtained if it is assumed that at the steady state equilibrium the rate of propagation is virtually the same as the rate of acrylate (A) consumption. The rate of polymerization (R_p) expressed in equation (1):

$$R_p = -d[A]/dt = K' k_p [A] \tag{1}$$

where

$$K' = (k_i/2k_t)^{1/2} (I)^{1/2} [PI]^{1/2} \tag{2}$$

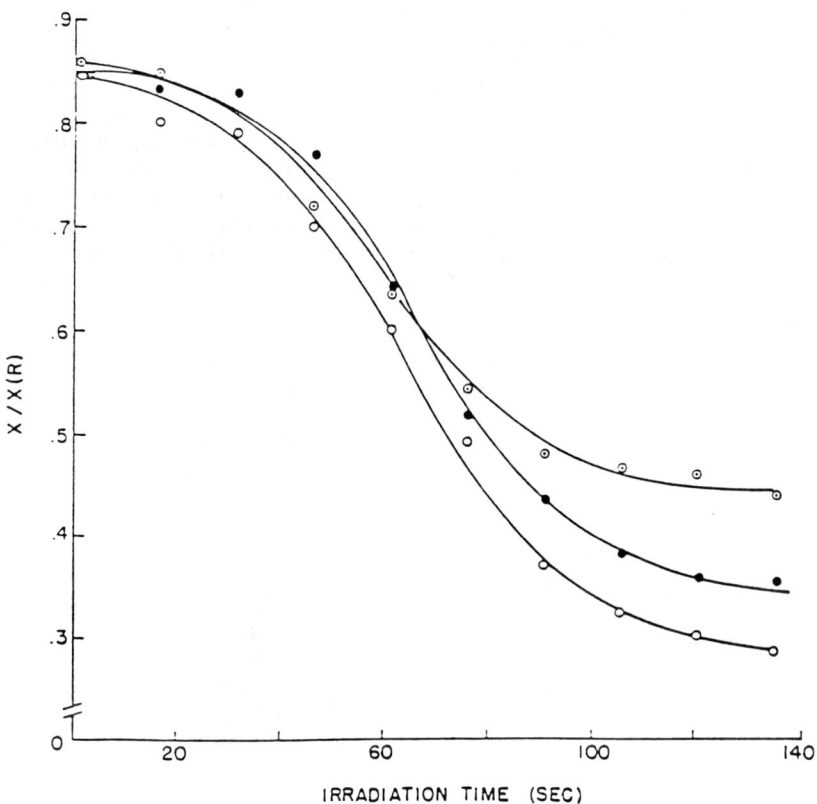

Figure 1. FTIR profile of the curing process for mixtures I (mono, ○), II (di, ●), and III (tri, ⊙).

k_p is the propagation rate constant and [A] is the acrylate concentration. Thus, K' is proportional to the square root of both the initiation (k_i) and termination (k_t) rate constants, the light intensity (I) and the photoinitiator concentration, [PI](<u>5</u>). Table III shows the relative efficiencies of these mixtures. The values are reported relative to the efficiency of mixture I whose value was defined as unity. The trends shown by the initial rates (eg. I > II > III, II > IV > V, III > VI, etc.) can be attributed in part to the increase in viscosity that accompanies the higher degree of functionality. Third, Figure 1 shows the tapering or levelling effect of the curing rate. This effect is shown to occur more rapidly in mixture III than II followed by I, and may be due to film gelation caused by a more rapid growth of high molecular weight branched polymer which should occur in the same order III, II, I, owing to the greater functionality of the starting materials. This levelling can also be used as an estimate of the extent of cure in these systems.

TABLE III - Relative Efficiencies of Acrylate-Based and Rubber Modified

Mixture	Acrylate-Based Mixtures	
	Relative Efficiency[a]	Viscosity[b] (Poise)
I	1.0	7.9
II	0.89	9.5
III	0.65	22.8
IV	0.60	26.4
V	0.44	34.5
VI	0.44	97.9
IR	0.50	57.7
IIR	0.34	115.0
IIIR	0.12	169.0

a - Values represent the ratio of polymerization rate to light intensity measured in mW/cm^2. Mixture I was set equal to unity.
b - Viscosity measurements show that these mixtures behave as Newtonian fluids.

At the onset of reaction, when the O_2 concentration in the film is high, the rate of propagation is much less than the rate of inhibition: $k_p \ll k_{O_2}$ [4]. This is particularly true at the surface of the film where the $[O_2]$ is higher and leads to bulk versus surface cure. The extent of O_2 inhibition of the cure of these films ($\sim$ 4 mils thick), in the presence of air, can be measured using FTIR-PAS [6]. The contrast between the bulk versus surface polymerization rates of Mixture II is shown in Figure 2.

Two options in overcoming this inhibition, when curing in air, are: (1) to increase the light flux and/or (2) to increase the PI concentration. The latter is evident by the decreased induction period with increasing PI, Figure 3. The contrast of the reaction rates, at the surface of the film, both in N_2 versus air and with increased PI is shown in Figure 4. Note that the effect of ambient air (O_2) on the cure rate is reduced with the higher PI concentration from 25% to 60% (at 0.5 to 2% PI) of the rate under an inert atmosphere.

It has been shown previously that the rate of polymerization of acrylate monomers, particularly HDODA, obeys a linear relationship to the square root of the photoinitiator concentration (isopropylbenzoin ether). A similar behavior was observed in this study with HDODA and DMPA. In the case of Mixtures II and IIR however, a deviation from linearity was observed at [DMPA] of 2% by weight. The exact cause for the breakdown of this relationship is not certain although the levelling could be attributed to the initial rate becoming independent of PI concentration [8]. Further studies of monomers and simpler mixtures are being carried out.

Solvent Absorption

The rate of solvent absorption by a cured polymer film has been used previously as a measure of crosslink density [9]. Values for the solvent uptake of these mixtures are shown in Table IV.

Under 'high' intensity cure the expected decrease in solvent uptake with increased degree of functionality (increased crosslink density) is observed. The rubber modified materials follow the same trend (IR > IIR > IIIR) with no significant change in absorption.

Under 'low' intensity cure the trends are not apparent, except for II > IV > V. This may arise from the contrasting effects of rate to degree of cure due to the viscosity of the components. The rubber modified mixtures were more susceptible to DMF uptake in this instance.

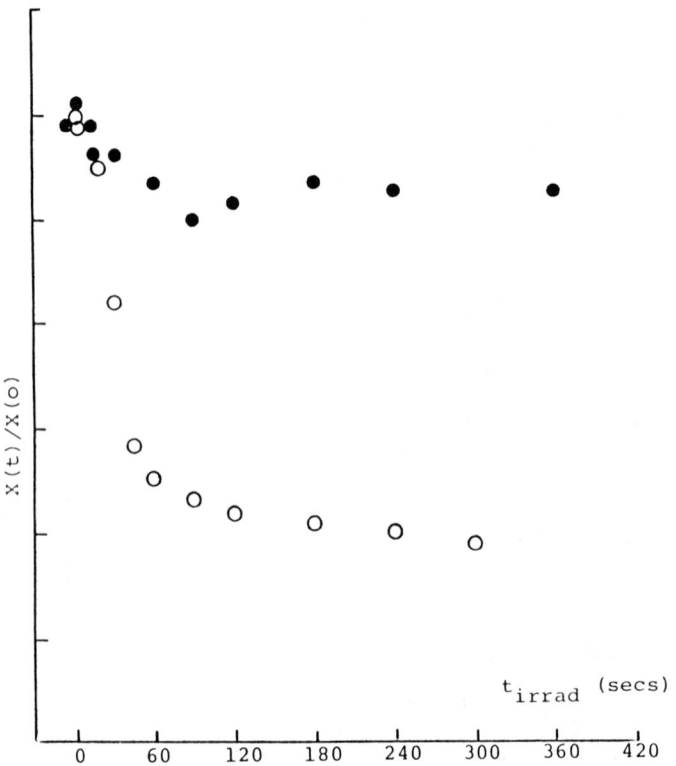

Figure 2. Comparison of surface vs. bulk polymerization rates. Key: ●, FTIR — PAS/air; and ○, FTIR.

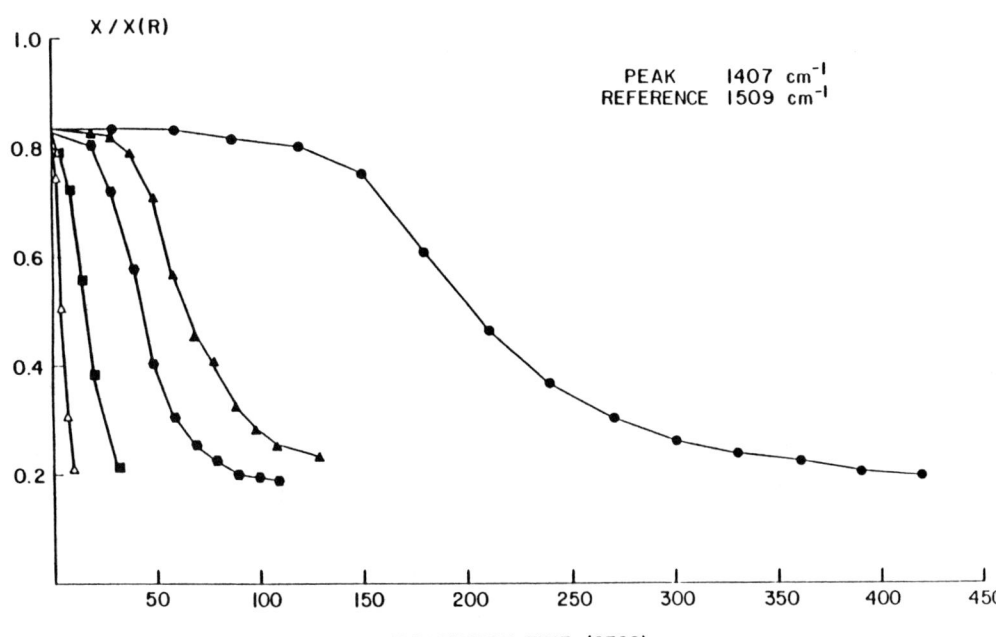

Figure 3. FTIR-derived induction period vs. PI concentration for mixture II. Key (DMPA concentration): ●, 0.5; ▲, 1.0; ⬢, 2.0; ■, 4.0; and △, 8.0%. Lamp flux = 1.2 mw/cm^2.

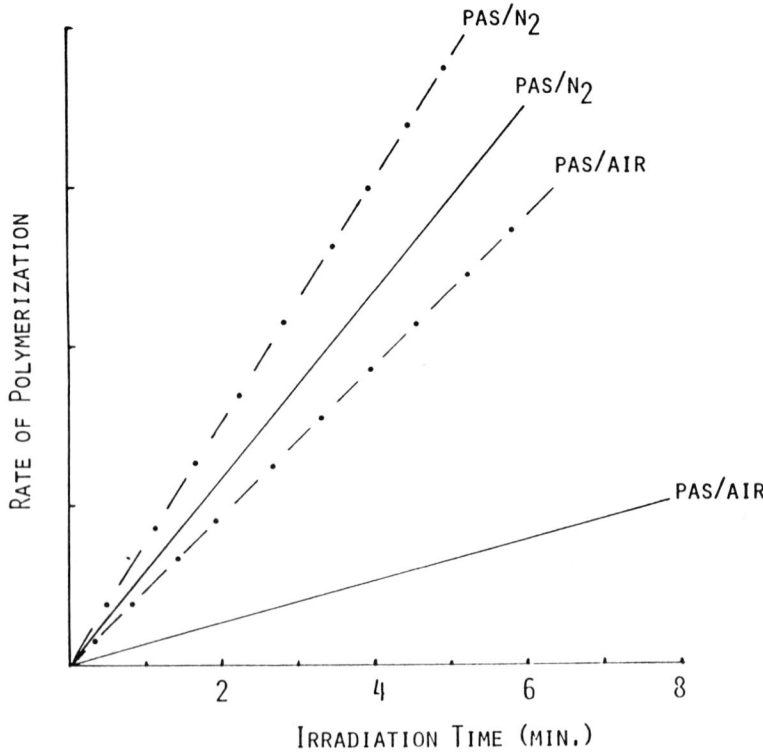

Figure 4. PAS polymerization rates. N_2 vs. air at two PI concentration rates. Key: —, 0.5; and —•—, 2% DMPA.

TABLE IV - Solvent Uptake (DMF 100%/4min/23°C)

DMF ABSORPITON (mg/cm^2)

Mixture	High Cure [a]	Low Cure[b]
I	0.51	1.22
II	0.28	0.96
III	0.18	0.95
IV	0.20	0.62
V	0.12	0.16
VI	0.06	0.86
IR	0.57	1.86
IIR	0.37	1.41
IIIR	0.12	1.49

a - Samples cured for five seconds with 'high' intensity light (75 mW/cm^2).
b - Samples cured for twenty-five seconds with 'low' intensity light (4.9 mW/cm^2).

Variations in the PI concentration for mixture II show a decrease of 10 to 30% with increasing PI, Figure 5, leveling off beyond 4% PI for both high or low intensity cure. The rubber modified mixture IIR is also subject to change with PI concentration, and at 'high' cure the same trend is observed beyond 4% PI. The 'low' cure however shows no levelling effect at 4% PI. In fact, at 8% the uptake is still decreasing.

Thermogravimetric Analysis (TGA)

Thermogravimetric analysis was used to follow the curing of polymer films. This technique is predicated on the volatility of the unreacted components. In the present study the volatile components have been determined to be EEA, HDODA, TMPTA, Irgacure 651, and their volatile reaction products. The Epocryl 370 resin and the VTBN rubber were found not to contribute to the volatiles. Decomposition of the mixtures generally took place about 310-350°C in all cases, including the rubber modified samples.

The curing profiles of mixtures I, II, and III where the monomer functionality has been increased from mono to tri functional show very little differentiation at a wide range of light intensity (Figure 6). Varying HDODA concentration (i.e. mixtures II, IV, and V) and/or adding the rubber modifier showed no measurable effect in the curing profiles as monitored by TGA (Figure 7).

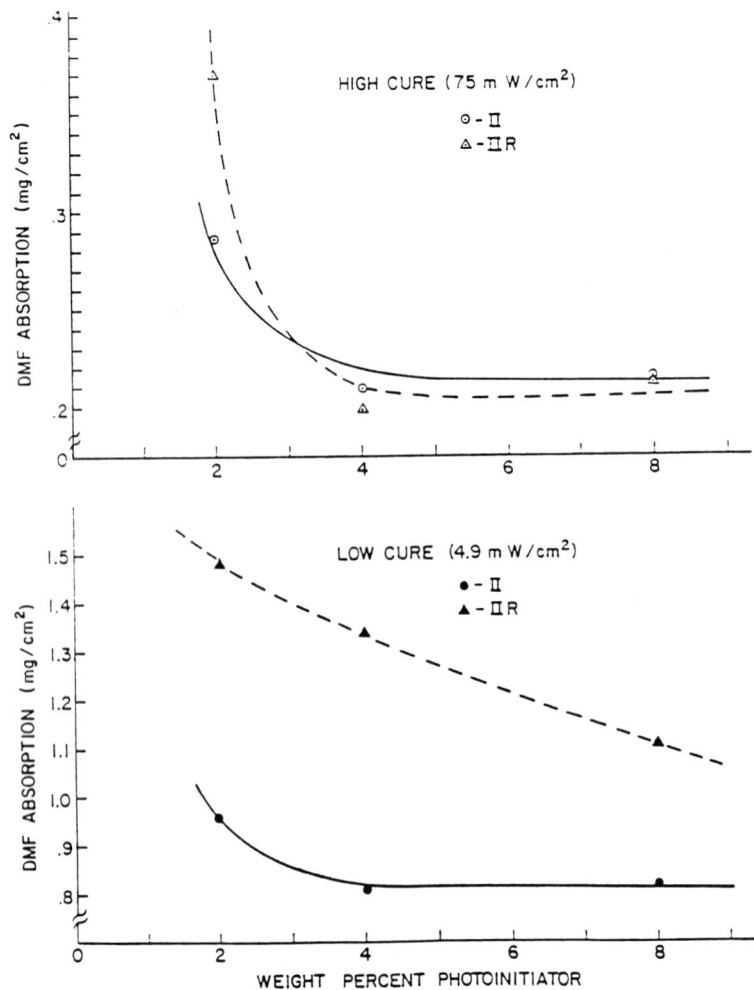

Figure 5. DMF uptake vs. PI concentration of acrylate-based and rubber-modified acrylate-based mixtures II and IIR.

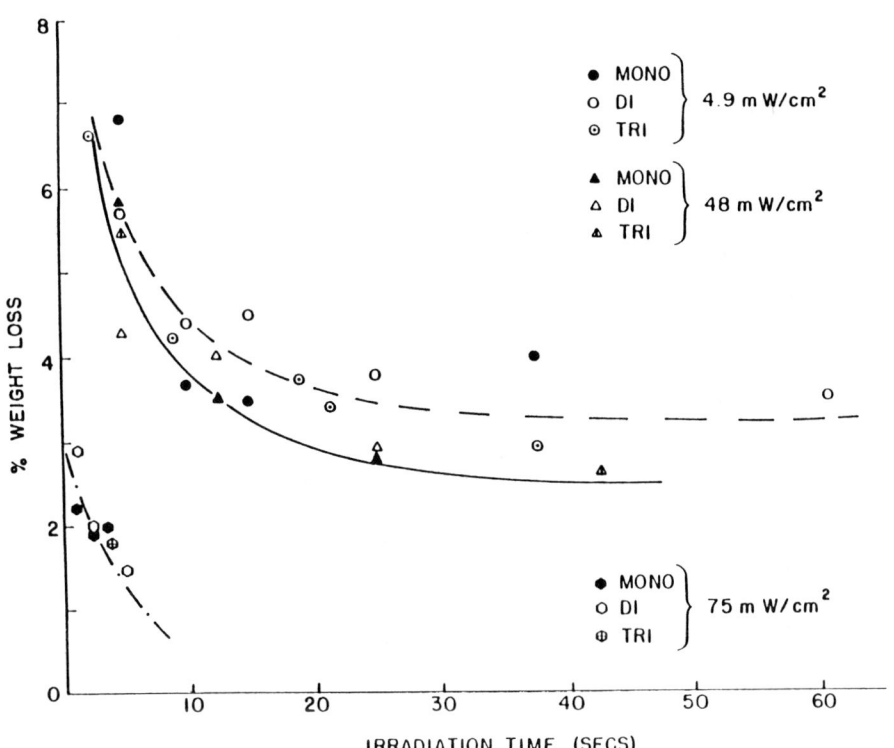

Figure 6. Curing profile of mixtures I, II, and III based on TGA data.

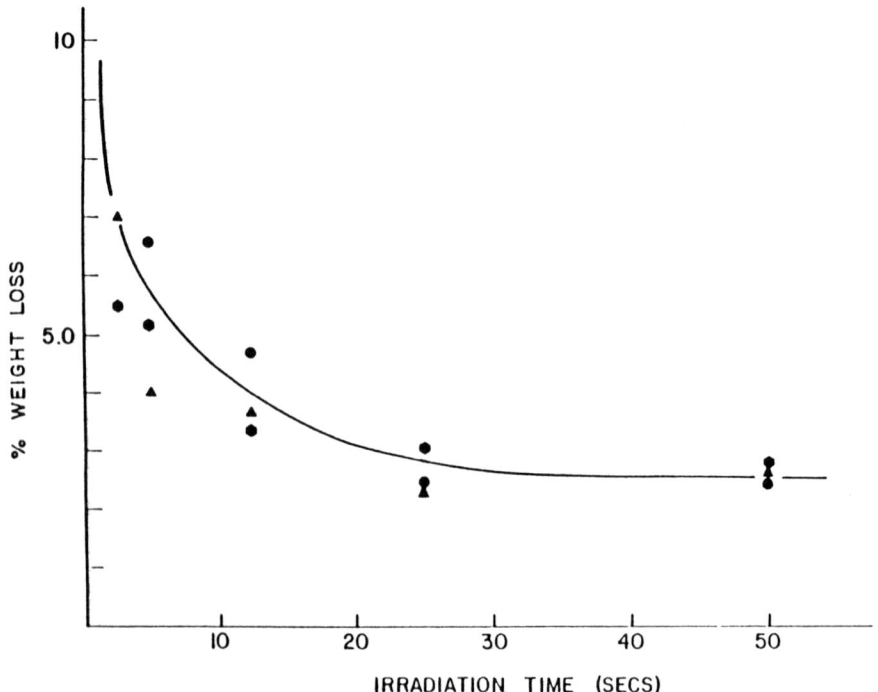

Figure 7. Curing profile of rubber-modified acrylate mixtures IR (●), IIR (▲), and IIIR (⬢) based on TGA data. Lamp flux = 4.9 mw Lamp flux = 4.9 mw/cm^2.

TGA does show substantial sensitivity to photoinitiator concentration in mixtures II and IIR. Figure 8 shows no significant differences between the two mixtures at any given photoinitiator concentration but photoinitiator variations within the same mixture gave significant differences.

Solvent Extraction

Solvent extraction is a technique that can be used to evaluate the curing behavior of photopolymers and as a measure of sol fraction. The method is dependent on the solubility of the unreacted or partially reacted components in the chosen solvent (methylene chloride). The method yields the percentage of unreacted monomers and partially reacted low molecular weight components in the cured film. This is derived from the weight differential of the film before and after extraction.

The data in Figure 9 show no significant differences in solvent extractables as a function of cure time at 'high' intensity. At 'low' intensity an increase in extractables with increased functionality and rubber modification is noted. The extractable levels were always higher at 'low' intensity cure even when the total light dose was equivalent. This suggests that we are beyond the limits of 'reciprocity' as far as the light intensities (high vs low) in these systems are concerned. Variations in the PI concentration show an optimum range of 2 to 4% (pbw). Generally, the rubber modified system yield the same amount of extractables as the straight acrylate system. This implies that the degree of cure under this light exposure, according to this technique, is approximately the same for both systems. At 'high' cure both II and IIR the higher content of PI leads to an increase in the amount of extractables. This could be attributed to a larger degree of low molecular weight oligomers resulting from the enhancement of termination reactions during polymerization and/or PI screening effects, along with excess unreacted PI and reacted by-products.

The extracted solutions of mixture II and IIR were examined by gel permeation chromatography. GPC analysis showed no significant differences between the two mixtures at 'low' intensity cure. This indicates that the components of the mixture appear to react in relatively the same ratio. However, at 'high' light intensity in mixture II and IIR the EEA monomer appears to be the major component of THF soluble fraction. This implies that under these conditions the EEA is the slowest reactant of the mixture. This could be in part attributed to its (EEA) monofunctional character.

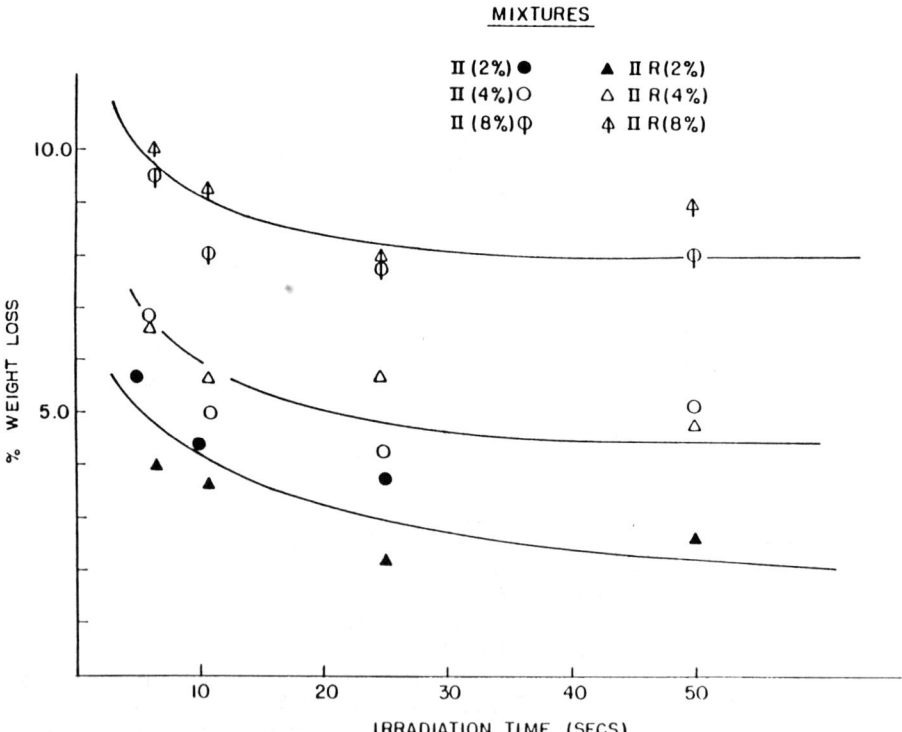

Figure 8. TGA curing profiles vs. PI concentration in acrylate and rubber-modified acrylate mixtures. Lamp flux = 4.9 mw/cm^2

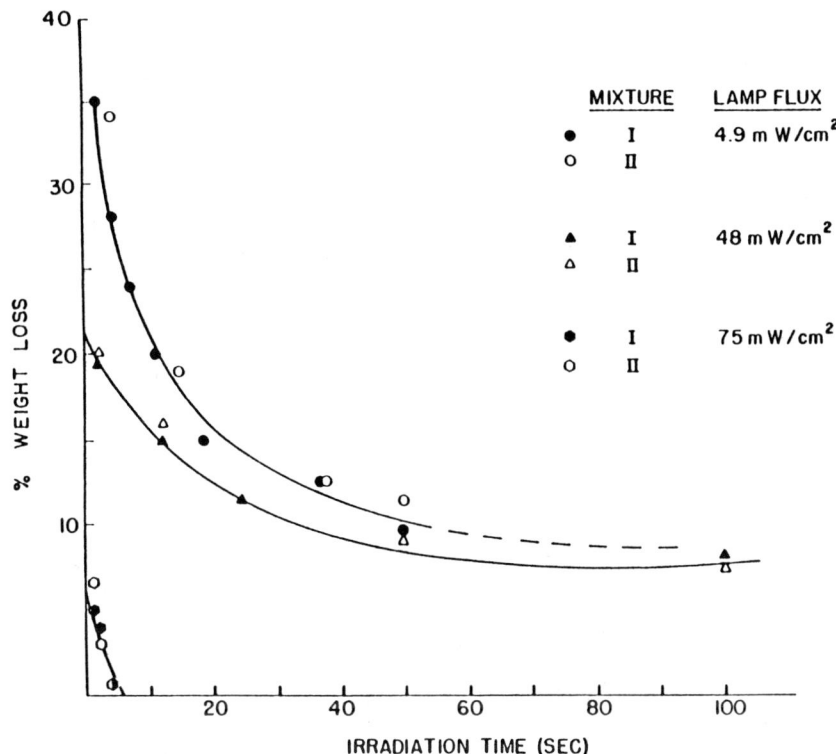

Figure 9. Curing profile for mixtures I and II from solvent extraction data.

DISCUSSION AND CONCLUSIONS

The curing behavior of these mixtures appears to be a function of several parameters. Three of these are the presence of oxygen, the viscosity and the total functionality in the mixture.

In all these systems the curing behavior at 'high' light intensity (versus 'low' intensity) appears to be less susceptible to mixture variations. This can in part be due to the reduction of oxygen inhibition at 'high' flux. Hence, by eliminating or reducing oxygen inhibition the effects of other factors i.e. viscosity and screening effects, can be minimized and a high degree of cure can be obtained.

In contrast, at 'low' intensity viscosity appears to play a major role in the rate of cure (Table 3). Increases in functionality are accompanied by the increase in viscosity of our mixtures. The higher degree of functionality, during the induction period, may lead to the formation of high molecular weight branched chains resulting in an increased viscosity. By the time $k_{O_2} \ll k_p$ the viscosity is higher than at the onset, this results in a decrease of both the cure rate and the degree of cure. This is supported by the FTIR, solvent extraction and TGA data at 'low' intensity. In contrast, the solvent uptake data show greater DMF absorption with decreased functionality. This indicates that the materials with increased functionality yield higher crosslink density films even at a lower degree of cure. Samples irradiated with 'low' intensity can be cured to completion by re-irradiation with high flux.

The induction period at low intensity was shown to be shortened by increasing the PI concentration. However, determination of the PI concentration threshold should be carried out to maximize the cure rate while avoiding excess low molecular weight products or unreacted initiator that could lead to film degradation.

In epoxy resins rubber modification imparts changes in the cured polymer([10]). These changes, reflected by the solvent uptake data, can be attributed to a decrease in the crosslinked density and/or the presence of domains which allow a high degree of solvent uptake.

In the case of these rubber modified acrylate mixtures two main characteristics of the cured films appear to govern the solvent uptake interaction: (1) The percentage of VTBN used in this study is compatible with the acrylate mixtures leading to a homogeneous film upon curing. Staining techniques have shown the VTBN rubber phase to be dispersed throughout the film and no domains of any significant size (greater than 0.1 micrometers) are present, (2) The lack of affinity of DMF to the rubber phase within acrylate matrix. This implies that any increase in DMF uptake should depend on the degree of cure and crosslink density of the film.

In general, the presence of rubber (VTBN) in these acrylate mixtures affects the cure rate due to increases in viscosity, however, no change in curing mechanism is expected.

Figure 10 summarizes the range of the four methods as applied to the systems studied. At the beginning stages of photo-cure the FTIR and FTIR-PAS give the best quantitative results. Values for the kinetic rate, oxygen inhibition and PI dependence are readily obtained via these techniques. Solvent uptake, solvent extractables and TGA are not applicable in obtaining initial polymerization data, however these techniques can be used as a guide for extent of cure. The latter two techniques along with the FTIR do not give an accurate description of the crosslink density. The solvent uptake method is the only one of the above mentioned five that appears to give a measure of the final degree of crosslinking and differentiates highly cured systems.

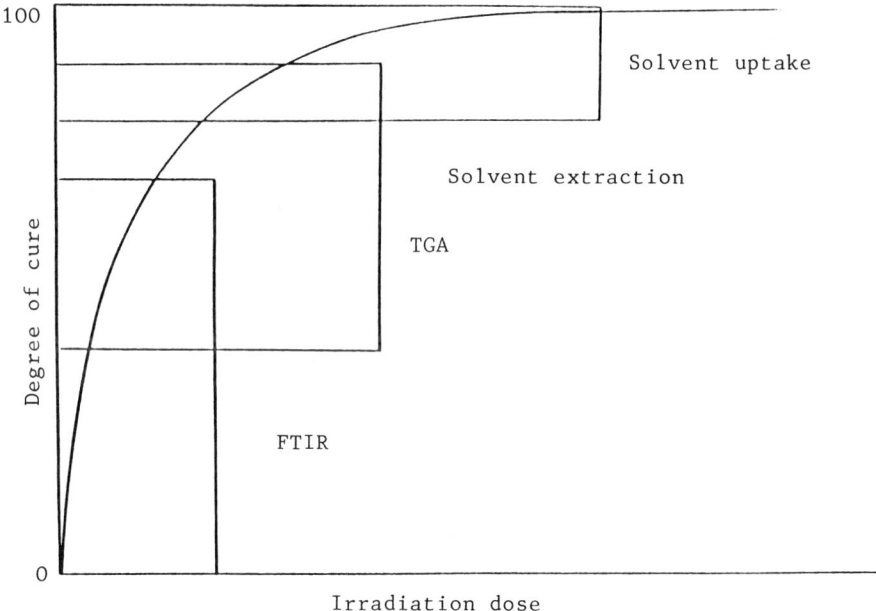

Figure 10. Schematic for applicable ranges of individual analytical techniques.

Acknowledgments

The authors would like to thank Dr. J. F. Geibel for his input on solvent absorption studies, and R. S. Bentson and C. R. Nijander for their assistance. The authors are also indebted to Dr. Y. C. Teng for his contribution in the FTIR-PAS work.

Literature Cited

1. C. L. Osborn, J. of Rad. Cur., p 2, July, 1976.

2. DMPA has been suggested to generate two chain initiating radicals through a Norrish Type I fragmentation. Further disproportionation of the resulting dimethoxy radical gives methyl benzoate and methyl radical.

3. J. L. Koenig, Applied Spectroscopy, $\underline{29}$, 293 (1975).

4. F. R. Wight, J. Polymer Sci., Polym. Lett. Edition, 121 (1977).

5. J. E. Wilson, "Radiation Chemistry of Monomers, Polymers and Plastics", Marcel Dekker, New York, Chapter 5, 1974.

6. a) Y. C. Teng and J. A. Ors, 8th Annual FACSS, Philadelphia, September, 1981, b)B.S.H. Royce, Y. C. Teng and J. A. Ors, IEE Proceedings, Chicago (1981).

7. G. A. Lee and G. A. Doorakian, J. of Rad. Cur., p 2, January, 1977, and references therein.

8. M. E. Joshi, J. of Appl. Polym. Sci., $\underline{26}$, 3945 (1981).

9. W. A. Romanchick and J. F. Geibel, Preprints of the Organic Coatings and Plastics Chemistry Division, ACS, $\underline{43}$, 559 (1980).

10. J. E. Sohn, Preprints of the Organic Coatings and Plastics Chemistry Division, ACS, $\underline{43}$, (1980).

RECEIVED December 19, 1983

Morphology of Rubber-Modified Photopolymers

J. A. ORS

Engineering Research Center, Western Electric Company, Princeton, NJ 08540

J. B. ENNS

BTL-Whippany, NJ 07981

> The morphology of ruber modified epoxy photopolymers was found to depend on the cure conditions as well as the nature and concentration of rubber. The commercially available acrylonitrile-butadiene copolymer rubber modifiers with varying percentages of acrylonitrile content were used. They were polymerized using a photocationic initiator involving a UV exposure followed by a thermal cure. Transmission electron micrographs of osmium tetroxide stained specimens, coupled with dynamic mechanical measurements indicated that phase separation and particle size distribution depended not only on rubber concentration and compatibility, but also on the cure conditions. The resulting morphology is dependent on the rate of polymerization relative to the rate of phase separation preceding gelation.

The adhesive properties of epoxy resins coupled with their dielectric behavior have made them attractive to the electronic industry. The evaluation of thermally cured rubber modified epoxy thermosets has been the subject of recent studies ([1],[2]), which dealt with the dependence of morphology on the curing parameters, e.g., catalyst, cure schedule, time of gelation, etc. This work utilizes one of the new series of photocationic initiators (PCI) developed by Crivello, et al ([3]) which are presently commercially available. These 'onium salts' initiate the reaction by absorbing the actinic radiation, generating radicals and producing a protonic acid. The radicals can lead to polymerization of olefinic moieties ([4]) while the acid initiates the polymerization of the epoxy groups ([3]).

In the thermally initiated cure of rubber modified epoxy, the rubber may be present within in the epoxy matrix as distinct domains. The morphology of the cured resin has been shown to be dependent on: (1) the cure temperature and accelerator concentration, since the extent of particle (domain) size growth appears to be limited by gelation; and (2) the nature (percent acrylonitrile) of the rubber used, since mixture compatibility increases with the acrylonitrile content of the rubbers (1,2).

In a photoinitiated system, in which the reaction proceedes rapidly during exposure (resulting in relatively short gel times), it is thought that the epoxy matrix is set up during the initial irradiation thus influences the final morphology. The final particle size and distribution would then be dictated by the compatibility of the components in the uncured state since the rapid formation of the gel would be expected to preclude the redistribution of the phases.

This study examines the relationship between cure schedule and morphology for a rubber modified, photocationically initiated, diglycidyl ether of bisphenol A resin; using differential scanning calorimetry (DSC) and torsion pendulum (TP) to monitor the glass transition temperature (T_g), transmission electron microscopy (TEM) to look at phase separation and domain sizes, Fourier transform infrared (FT-IR) and sol fraction to compare the extent of cure, and thermogravimetric analysis (TGA) to analyze the specimens for volatilization and/or degradation. In an attempt to explain some of the observations the cure schedule, concentration of rubber, type of rubber, and the concentration of initiator were varied independently.

EXPERIMENTAL

All materials used in this study are listed in Table I and, unless otherwise noted, are commercially available and were used as received. The rubber modifiers used are as epoxy-terminated butadiene-acrylonitrile (ETBN) rubbers prepared as described elsewhere (5). These ETBN's contain approximately 30% by weight of the corresponding carboxy-terminated butadiene-acrylonitrile (CTBN) rubber.

The films (~4 mils in thickness) were coated on an aluminum panel and irradiated by UV light (CoLight Model UVC-24) using a single medium pressure Hg vapor lamp (6). The degree of cure was estimated by a combination

TABLE I

DESCRIPTION OF EPOXY-RUBBER MIXTURES

Mixture	Component (weight %)				
	Epon 828	ETBN-13[a]	ETBN-8[b]	ETBN-15[c]	PCI[d]
I	81	-	15	-	4
II	66	-	30	-	4
III	51	-	45	-	4
IV	81	15	-	-	4
V	66	30	-	-	4
VI	81	-	-	15	4
VII	66	-	-	30	4
VIII	75	-	15	-	10
IX	96	-	-	-	4

(a) epoxy terminated - CTBNX13 (27% CN)
(b) epoxy terminated - CTBNX8 (17% CN)
(c) epoxy terminated - CTBNX15 (10% CN)
(d) PCI - proprietary photocationic initiator; similar materials may be obtained from GE Corporation and/or 3M Corporation.

of solvent extraction, FT-IR, thermogravimetric analysis (TGA) (6) along with differential scanning calorimetry (DSC) (Dupont 1090) and Torsion Pendulum (TP) (7) analysis reported in this study. For the TP analysis the resin was applied to a 1.4 mil copper substrate as a 4 mil film, irradiated for 5.5 sec. and baked for 2 hours at 150°C. The sheet was cut into narrow strips (2" x 1/8") and placed into the helium-purged torsion pendulum sample chamber.

EXTENT OF CURE

To correlate curing behavior and morphology, a measure of the degree of cure is needed. The sol fraction obtained via the solvent extraction method can serve as a measure of the gel time along with an estimate of the extent of cure when coupled with some of the previously mentioned techniques. The cure history for mixture II is shown in Table II. According to the solvent extraction data, a cure schedule of 5.5 or 11.0 seconds followed by a 2 hour bake at 150°C yields a film with less than 5% extractables for this mixture. This data is in agreement with both dynamic and isothermal TGA which consistently show a 2-5% weight loss at a temperature range of 250-350° or at 200°C after 48 hours. Figure 1 shows the decomposition pattern for mixture II before and after cure (5.5 sec UV irradiation followed by a 2 hour bake at 150°C).

The solvent extraction experiments coupled with DSC and FTIR data show that the degree of cure of these mixtures, under identical irradiation and bake conditions, is dependent on the concentration and nature (% acrylonitrile) of the rubber modifier. The sol fractions for PCI cured epoxy films with three different rubber modifiers (5), ETBN-13 (27% CN), ETBN-8 (17% CN) and ETBN-15 (10% CN) at a range of concentrations are shown in Figure 2. The data show that a decrease in extent of cure occurs with increased rubber concentration and that this decrease (ETBN-13 > ETBN-8 > ETBN-15) may be correlated to the percent acrylonitrile in the rubber modifier. This is supported by the FT-IR spectra of two of these mixtures (IV and VI) as shown in Figure 3 and the quantitative measure of the extent of cure as a function of irradiation time for mixtures V (30% TBN-13) and VIII (30% ETBN-15) as compared to mixture IX (no rubber) silicon in Figure 4 (8).

TABLE II

DETERMINATION OF DEGREE OF CURE FOR MIXTURE II

Exposure time[a] (sec)	Post Irradiation Bake[b] (hrs)	Sol Fraction[c] (%)
5.5	0	28.7
11.5	0	19.6
5.5	1	11.5
11.0	1	8.2
5.5	2	5.0
11.0	2	5.0

(a) Residence time under CoLight
(b) Bake Temperature: 150°C
(c) Average of two samples on Al panels

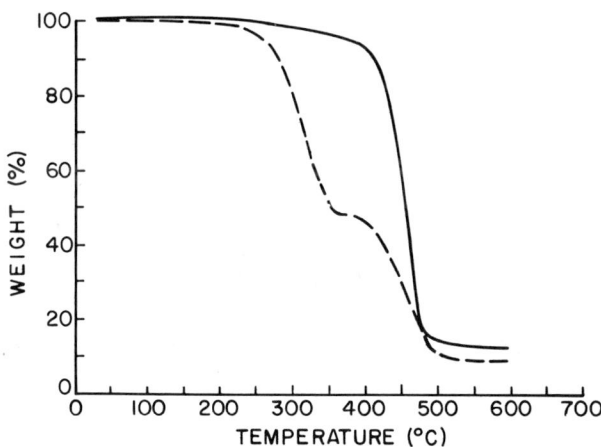

Figure 1. TGA of MIXTURE II from 25°C to 600°C (20°C/min). [----] uncured, [———] cured at 5.5 sec radiation followed by 2 hrs @ 150°C bake.

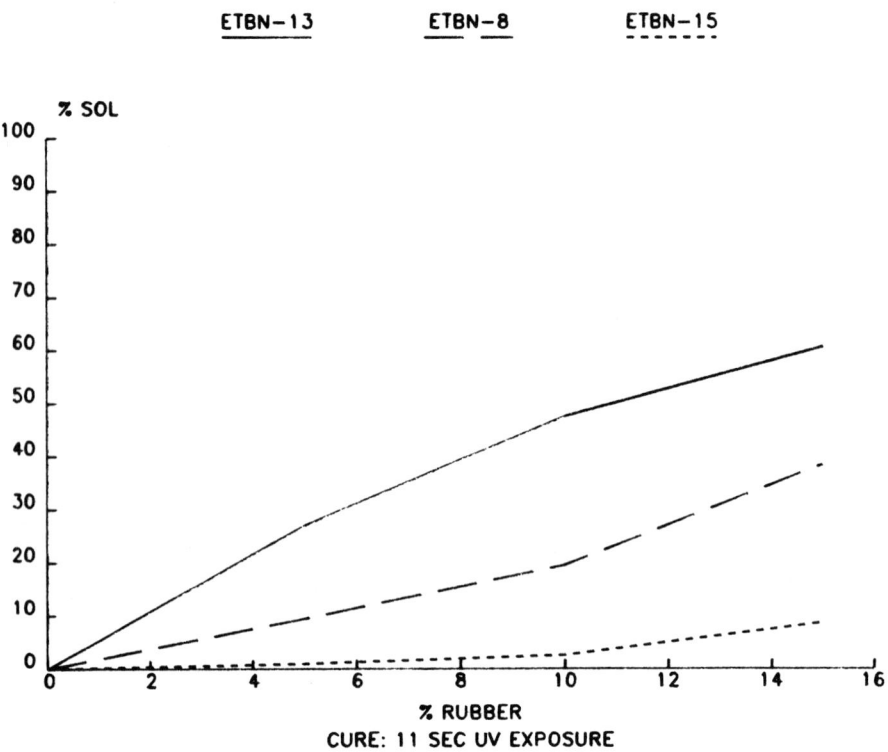

Figure 2. Effect of Rubber Concentration on Sol Fraction.

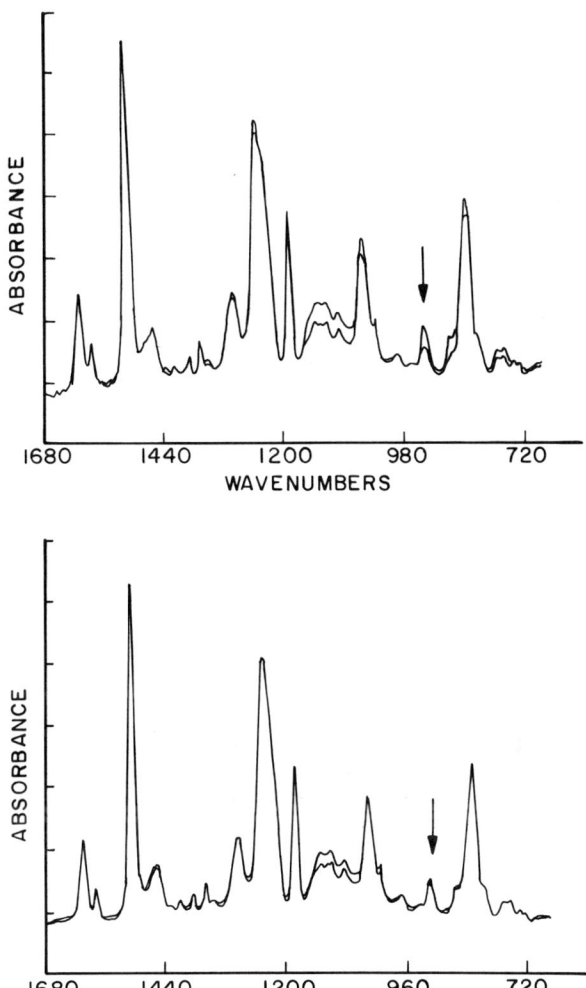

Figure 3. FT-IR Spectra of Mixtures VI (top) and IV (bottom) cured for identical irradiation times.

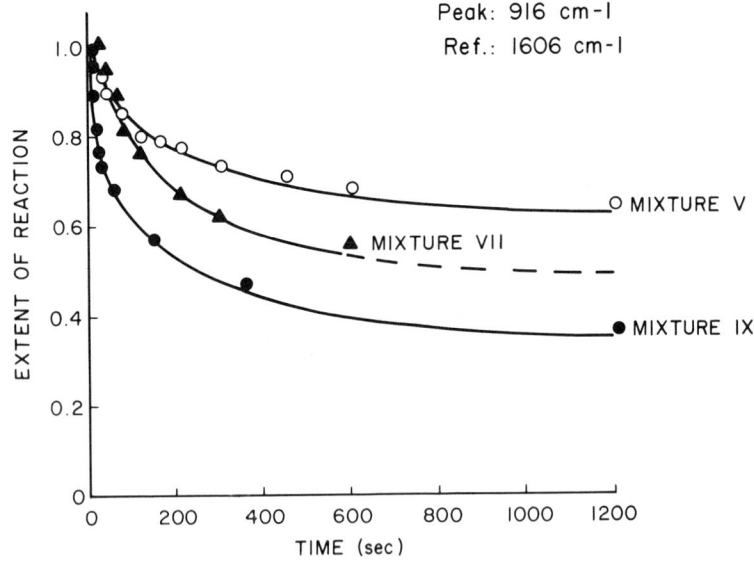

Figure 4. Extent of cure for mixtures V, VII and IX as a function of irradiation time at room temperature. Samples were irradiated using a 150 W Hg/Xe arc lamp. The intensity of the filtered radiation was measured at 4.5 mW/cm^2 @ 365nm.

The effect of the rubber modifier on the cure rate can be attributed in part to interactions between the rubber and the PCI. These interactions may result in the rubber acting as a screen or quencher to the PCI, inhibiting the generation of reactive species that lead to cure. The decrease in degree of cure can in turn be associated with the acrylonitrile content of the modifier due to an increase in interactive sites and/or enhanced compatibility with the resin (increased solubility parameter) (2). These modifiers can also affect the way thick films cure on a substrate and may give rise to a cure gradient through the film. The films showed various degrees of tackiness at the bottom surface adjacent to the substrate, especially with mixtures III, IV and V. Some of these mixtures, particularly II, III, and VII, showed lack of compatibility on a macroscopic scale at room temperature. However, the degree of tackiness at the resin substrate interface did not correlate with the turbid appearance.

THE GLASS TRANSITION TEMPERATURE

The DSC and TP data for mixture II both show dual glass transitions between 40° and 150°C, which merge into one upon heating (cycling up to 200°C), as shown in Figures 5 and 6 respectively. The dual transitions both increase in temperature, and the lower transition decreases in intensity and eventually disappears as cure proceeds. In mixture I (Figure 7), in which the rubber content has been reduced, the intensity of the lower transition is considerably less, and the final Tg is higher than that of mixture II. But mixture III (Figure 8), in which the rubber concentration has been increased to 15% CTBN, shows only a low, broad Tg which remains broad and lower than that of mixtures I and II.

Although these mixtures underwent identical cure schedules, their extent of cure is not necessarily the same. As discussed earlier, the rubber inhibits the photocationically initiated reaction; therefore the extent of cure for mixture I would be expected to be advanced further than that of mixture II under the same cure conditions, because it contains less rubber. The TP spectrum of mixture I (Figure 7, first scan) is similar to the spectrum of mixture II (Figure 6, later scans), suggesting that the differences in the initial spectrum are due to different degrees of cure. In contrast, mixture III apparently already consist of a single phase after the initial cure schedule.

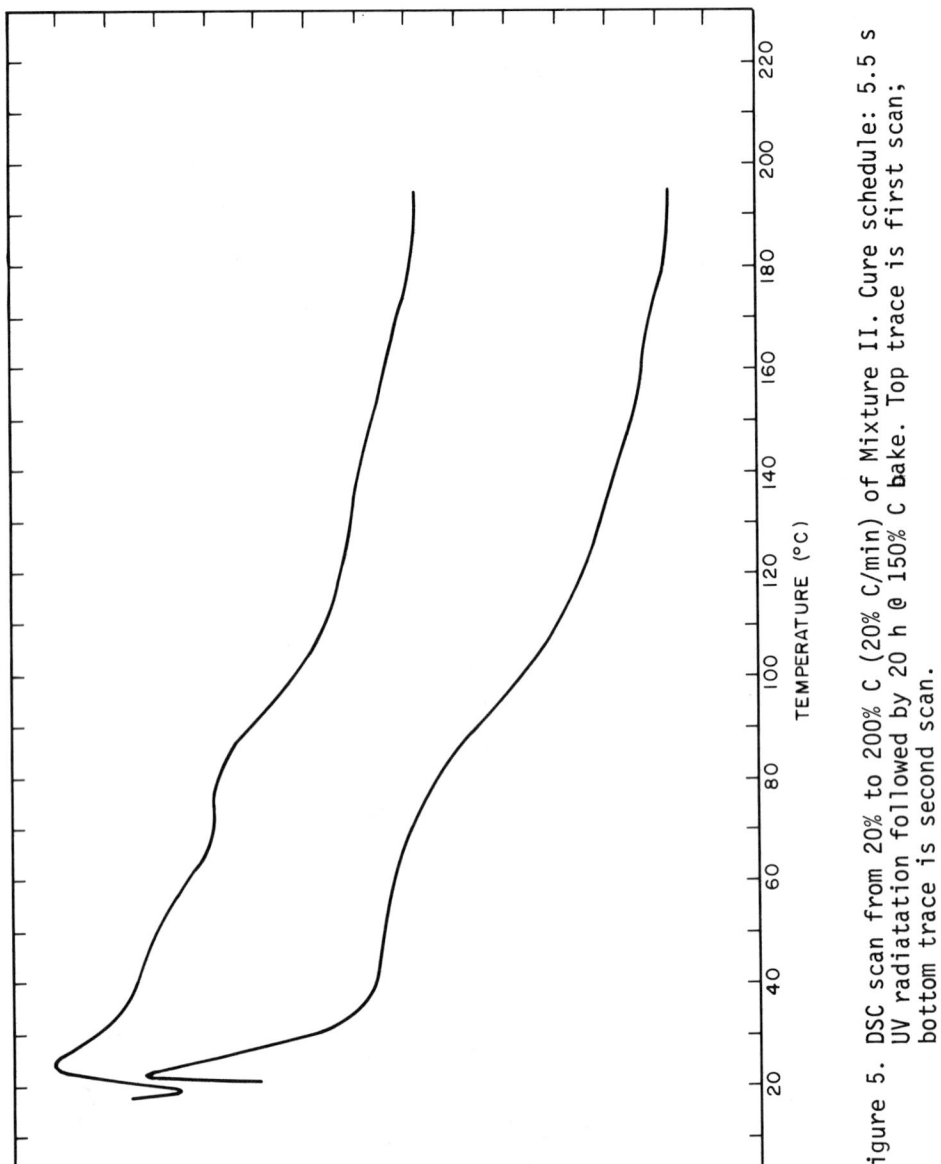

Figure 5. DSC scan from 20% to 200% C (20% C/min) of Mixture II. Cure schedule: 5.5 s UV radiatation followed by 20 h @ 150% C bake. Top trace is first scan; bottom trace is second scan.

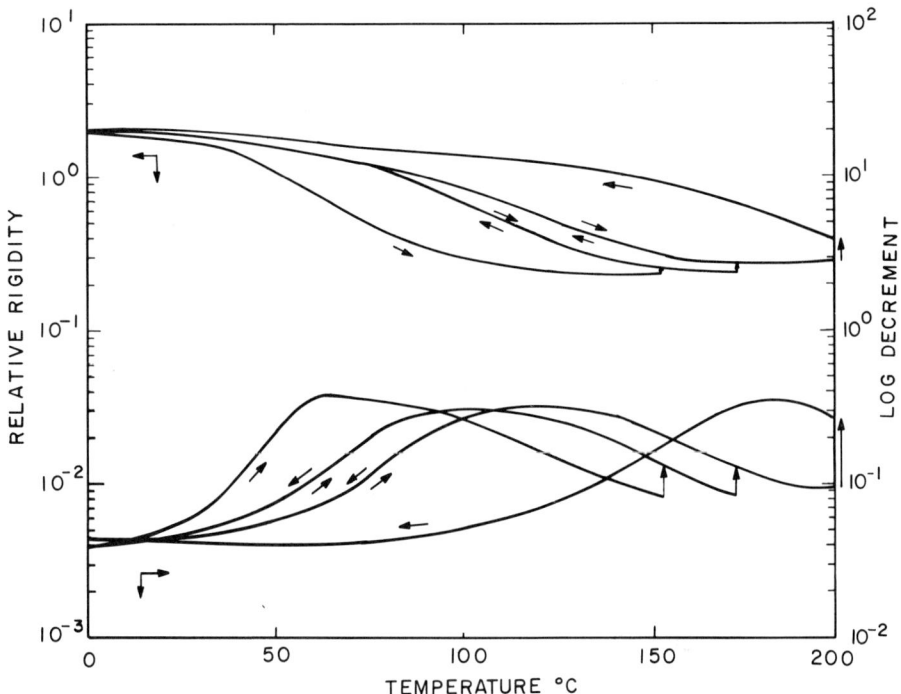

Figure 6. Thermomechanical Spectra (TS) of mixture II. Thermal history:

1. Room temperature to 150%C, isothermal at 150%C for 8 hours.

2. 150°C to -180°C to 170°C, isothermal at 170°C for 8 hours.

3. 170°C to -180° to 200°C, isothermal at 200°C for 8 hours.

4. 200° to -180°C

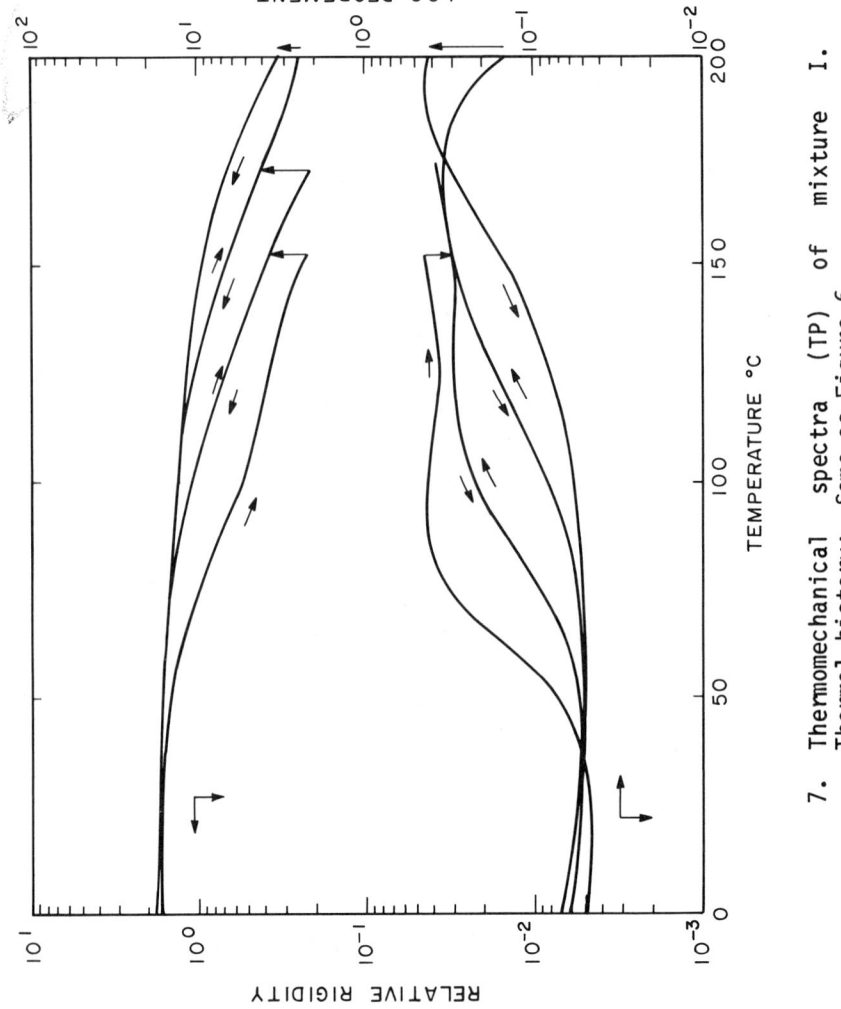

7. Thermomechanical spectra (TP) of mixture I. Thermal history: Same as Figure 6.

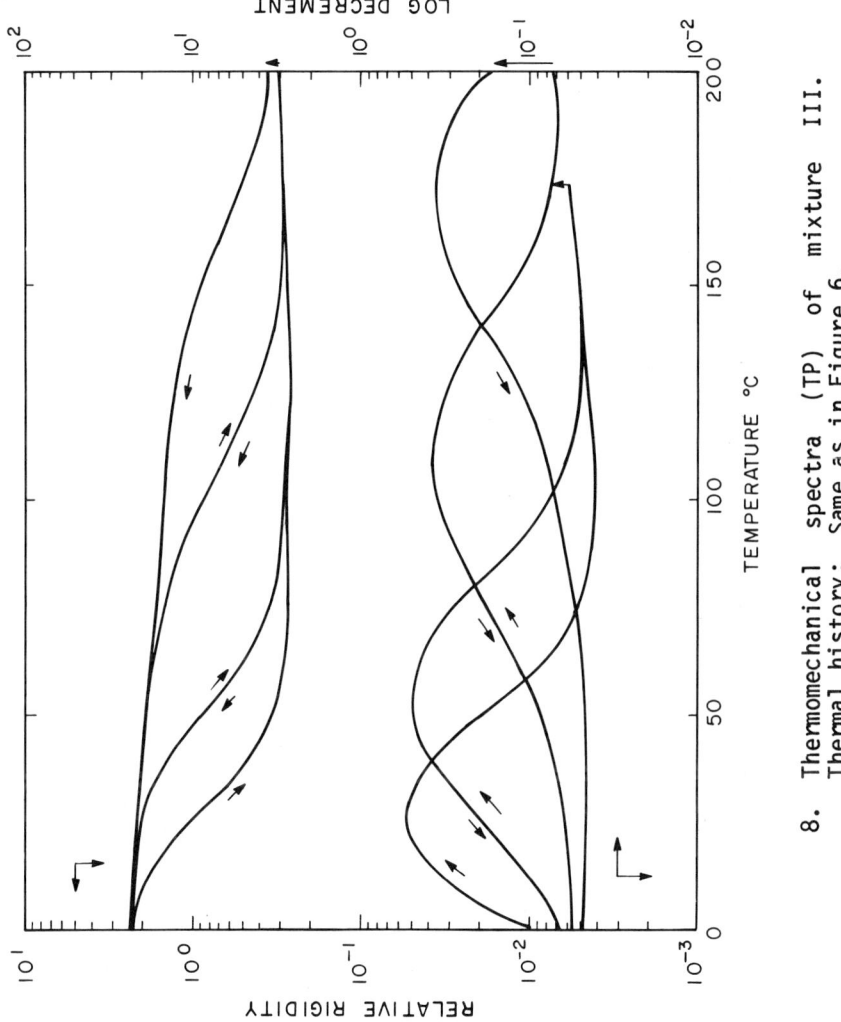

8. Thermomechanical spectra (TP) of mixture III. Thermal history: Same as in Figure 6.

Increasing the initiator concentration in mixture I to 10% (mixture VIII) again results in two distinct transitions (Figure 9), which, on heating, merge into one. The final Tg is very close to that of mixture I.

MORPHOLOGY

It has previously been shown that the morphology of a rubber modified resin is determined at gelation (1). In these photoinitiated systems over 80% of the film has gelled within the first 11 seconds. This would imply that an epoxy matrix has been formed during the irradiation which controls the morphology of the film. The irradiation is followed by a thermal cure during which the unreacted species within the matrix react.

However, since the DSC and TP results were not consistent with this model, i.e., the dual transitions coalesce, implying a more homogeneous mixture after prolonged exposure to higher temperature, the particle size distributions were measured for various cure schedules, initiator concentrations and types of rubber. The Transmission Electron Micrographs (TEM) of osmium tetroxide stained specimens shown in Figure 10 clearly illustrate the effect rubber compatibility has on morphology. Although the same cure schedule (11 sec UV irradiation, followed by 1 hour bake at 150°C) was applied to each of the mixtures, the degree of phase separation and the particle size distribution depend on the type of rubber in the mixture. The mixture with the most incompatible rubber (mixture VIII CTBN x 15) has the largest particles as well as the most phase separation, whereas the mixture with the most compatible rubber (mixture V: CTBN x 13) has the least phase separation. Figure 11 shows the rubber particle distribution for a film of mixture II for various cure schedules, as determined from TEM's similar to those in Figure 10. Rubber particle size while not significantly affected by an increase in irradiation time, shows a change in distribution to larger particle size with post irradiation bake temperature. This trend toward larger particle size is comparable with results in some thermally cured systems, where morphological changes continued to occur even after gelation (2). The results of the dynamic mechanical experiments indicate that these morphological changes occur very slowly, even when baked above the ultimate glass transition temperature.

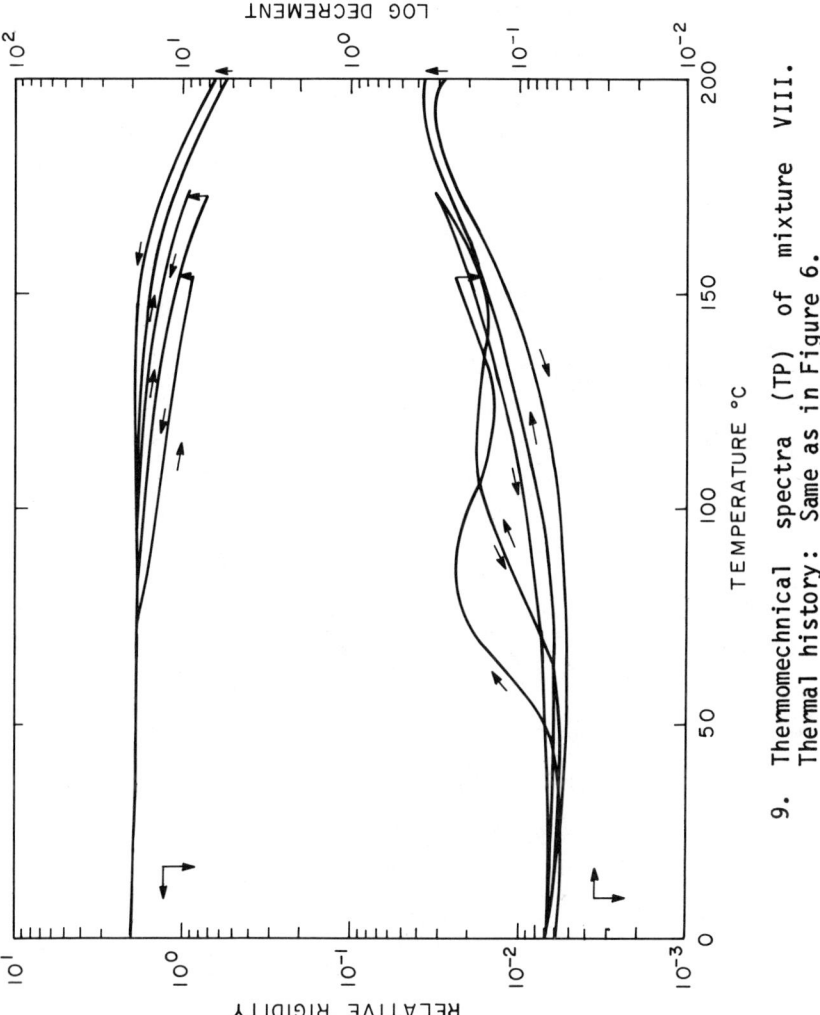

9. Thermomechanical spectra (TP) of mixture VIII. Thermal history: Same as in Figure 6.

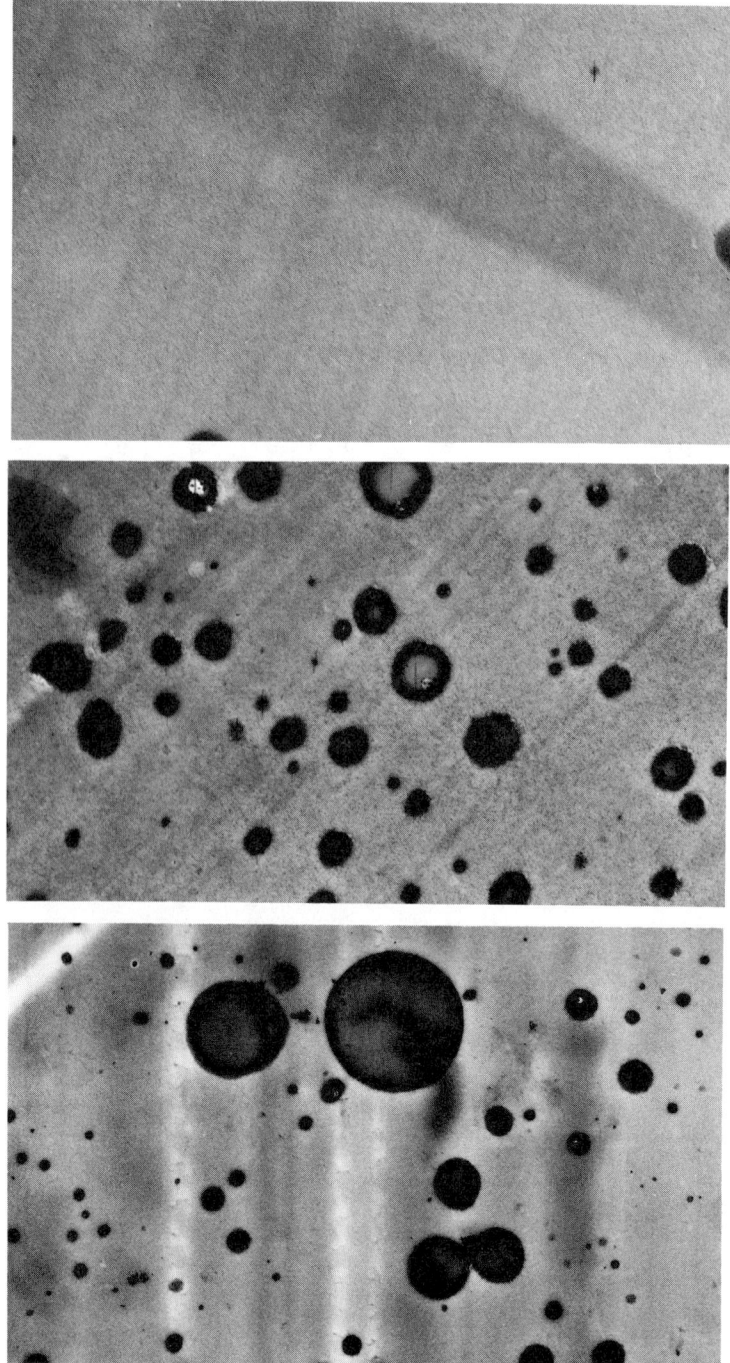

Figure 10. TEM micrographs of PCI-initiated rubber-modified epoxies, showing the effect of three types of rubber: left, Mixture VII; middle, Mixture II; and right, Mixture V.

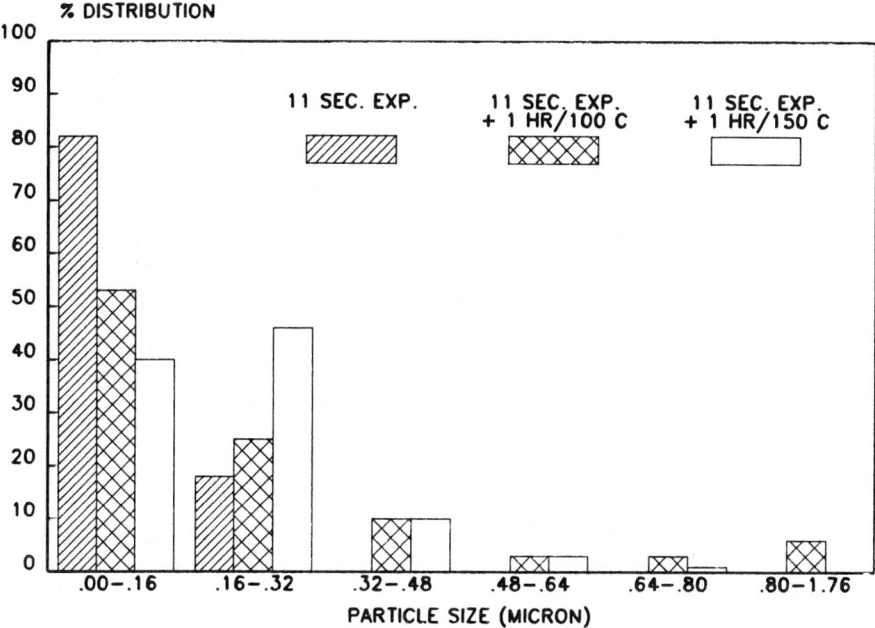

Figure 11. Particle size distribution for three cure schedules (Mixture II).

Experiments are in progress to determine the extent of the morphological changes with long post irradiation bake times.

An increase in the PCI concentration also yields an increase in the rubber particle size (Figure 12). Since doubling the irradiation time for mixture II (Figure 11) did not have a large effect on particle size, an increase in PCI concentration (at constant irradiation dose) may yield a higher number of reactive sites and higher gel fraction without resulting in a higher weight average molecular weight (6).

The effect of rubber concentration on particle size is shown in Figure 13 for mixtures I and II. The increase in particle size with increased rubber content is expected, since it causes a decrease in the cure rate and facilitates aggregation during cure due to the increased heterogeneity of the components (mixture II is more opaque than I). This macroscopic phase separation is even more pronounced in the ETBN-15 mixtures, in which the rubber is less soluble (8).

In addition to the morphological changes observed by electron microscopy and dynamic mechanical experiments, a corresponding increase in transmission of visible light is observed. Figure 14 shows the transmission spectra of mixture I at the early stages in the cure schedule. On prolonged baking a shift to higher wavelengths occurs (yellowing).

SUMMARY

The cure and resulting morphology of photocationically initiated rubber modified epoxy films have been investigated, as well as the effects of varying rubber concentration and post irradiation bake temperature and time on the morphology. Particle size did not change much with increasing irradiation time, but increased appreciably with increasing bake temperature and time. This implies that in these epoxy systems the matrix created during irradiation under the described conditions does not restrict changes in morphology, even though the resin has gelled. The increase in rubber particle size with increased rubber concentration can be attributed to the decrease in cure rate and the tendency to segregate.

Acknowledgement

The authors would like to thank M. H. Papalski for the solvent extraction work.

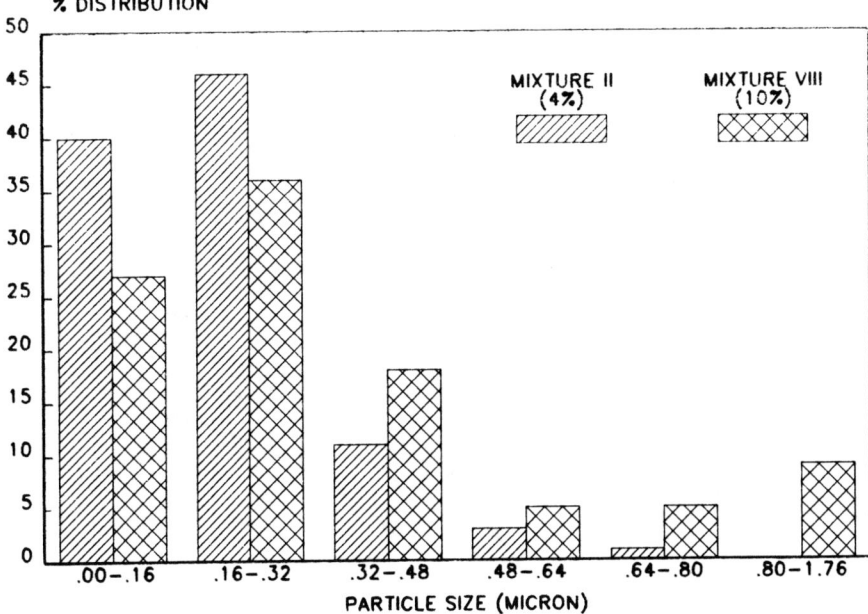

Figure 12. Particle size distribution for two levels of PCI concentration.

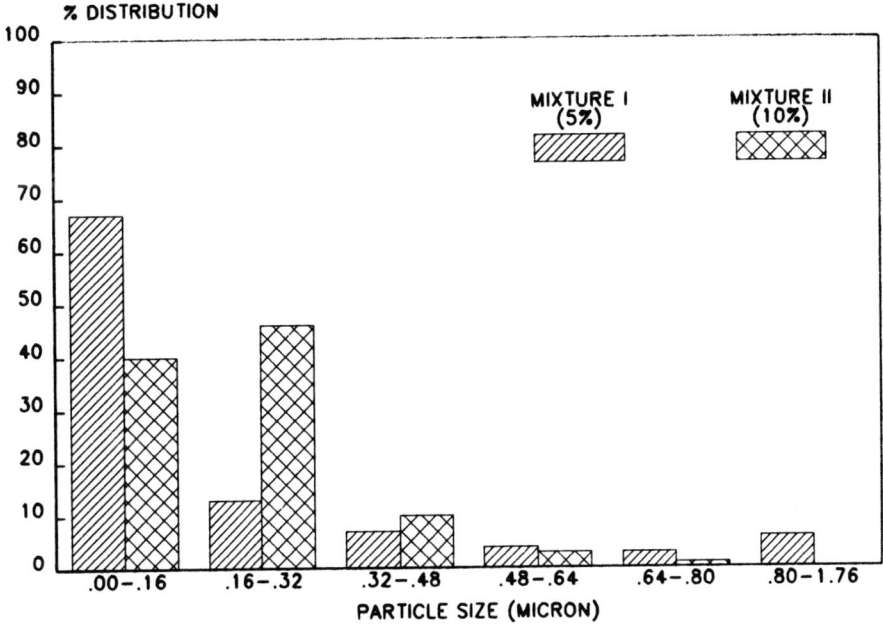

Figure 13. Particle size distribution for two levels of rubber concentration.

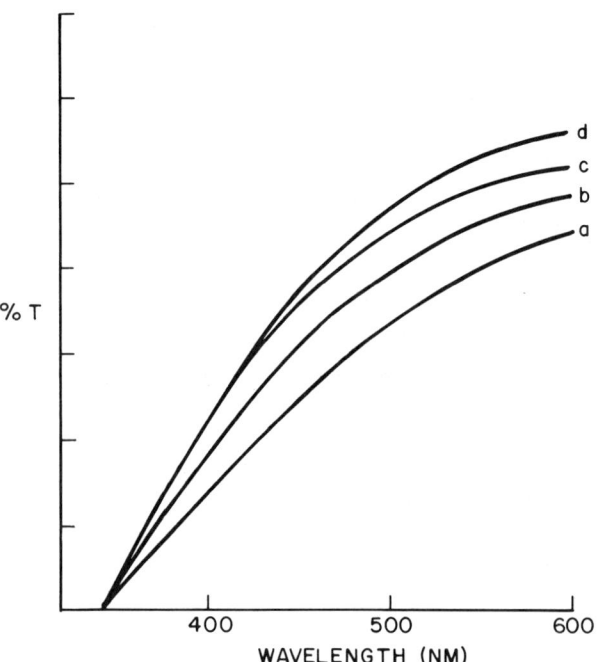

Figure 14. Transmission spectra of a free film of Mixture I at various stages of cure

    a) after 5.5 secs radiation
    b) irradiation plus 20 min @ 150°C
    c) irradiation plus 40 min @ 150°C
    d) irradiation plus 90 min @ 150°C

LITERATURE CITED

1. L. T. Manzione, J. K. Gillham and C. A. McPherson, J. Appl. Polym. Sci., 26, 886 (1981);
2. J. E. Sohn, Am. Chem. Soc. Prepr., Div. Org. Coat. Plast. Chem, 44, 38 (1981); and references therein.
3. J. V. Crivello, Lecture for "Fundamentals on Adhesion", SUNY at New Paltz, N.Y., Oct. 7-9 (1981); J. V. Crivello, U.S. Patent 4,138,255, Feb. 6 (1979) and U.S. Patent 4,139,385, Feb. 13 (1979).
4. W. C. Perkins, J. Rad, Curing, 16, Jan. 1981.
5. W. A. Romanchick and J. F. Geibel, Am. Chem. Soc. Prepr., Div. Org. Coat. Plast. Chem., 46, 410 (1982); J. F. Geibel, W. A. Romanchick and J. E. Sohn, ibid, 46, 416 (1982). These rubber modifiers can also be obtained commercially from Wilmington Chemicals, Wilmington, Delaware.
6. R. D. Small and J. A. Ors, Am. Chem. Soc. Prepr., Div. Org. Coat. Plast. Chem., 48, (1982).
7. J. B. Enns and J. K. Gillham, ACS Symposium Series, No. 197, Chp. 20 (1982).
8. J. A. Ors, unpublished results.

RECEIVED November 7, 1983

UV Solder Masks as Insulators for Printed Circuit Boards

NEIL S. FOX

Dynachem Corporation, Santa Ana, CA 92680

>The effect of temperature and humidity on the insulation resistance of photopolymerized and thermally cured solder masks was determined. Dry film and screen ink solder masks were applied to a test pattern. The insulation resistance of the coated test pattern decreased with exposure to elevated temperature and humidity. Elevated temperature alone, however, had only a small negative affect on the insulation resistance.

Solder masks (solder resists) are protective polymer coatings used on printed circuit boards (PCB's) to prevent solder bridging between conductors during soldering (1-3). The selective application of the solder mask limits the amount of conductor area in contact with the molten solder, minimizing solder usage and slowing the rate of conductor metal contamination in the solder pot.

The ability of a solder mask to protect conductors from physical and chemical deterioration and to insulate adjacent circuitry is a major consideration of PCB manufacturers. The Institute for Interconnecting and Packaging Electronic Circuits (IPC) has defined the requirements for the qualification and performance of solder masks in the Standard Specification IPC-SM-840 (4). The specification defines classes (1, 2, and 3) to reflect progressive increases in sophistication, functional performance, and testing methods. These classes try to provide PCB manufacturers with assurances of reliability. For example, Class 1 requirements provide the reliability needed by commercial boards used in radios, televisions, and small appliances. Class 2 is for computers and Class 3 for military and life-dependent products.

Ultraviolet light (UV) curable screen ink solder masks and photopolymerizable dry film solder masks have been available commercially since 1973 (5). These materials offer the PCB manufacturer many processing and production advantages relative to the conventional solvent-evaporative thermally cured solder masks. These products based on a new emerging technology have helped

0097-6156/84/0242-0367$06.00/0
© 1984 American Chemical Society

foster the explosive growth of the printed circuit industry. Along with this growth is greater concern for printed circuit board reliability. Currently, the IPC-SM-840 is undergoing a revision with major changes in the insulation resistance testing requirements for the three classes (6). Some PCB manufacturers carefully test to their own internal insulation resistance specifications and conduct research to isolate the factors affecting the insulation resistance of their finished PCB's (7). The insulation resistance of a coated test pattern often is measured at elevated temperature and humidity to approximate hostile environments the commercial circuit assembly might be exposed to during its expected life span and to artificially accelerate the aging process. The IPC-SM-840 moisture resistance test conditions are shown in Table I.

Table I. IPC-SM-840 Moisture Resistance Test Conditions

Class	Temperature	Humidity	Time
1	35°C	95% RH	4 days
2	50°C	95% RH	4 days
3	85°C	95% RH	7 days

Many studies have been concerned with electromigration on PCB's (8-11). Metallic (dendritic) growth of metal from one conductor to another can cause microamp current leakage or electrical shorts. Since high insulation resistance lowers the rate of electromigration, the insulation resistance of a solder mask and the factors which influence it are of great interest. This study was performed to isolate the affects of temperature and humidity on the insulation resistance of photopolymerizable solder masks relative to conventional solder masks.

Chemistry

The typical UV curable solder mask formulation consists of the following.

Photoinitiation System. In most cases the curing is initiated by free radicals generated by UV light and photoinitiators. The free radicals can be formed from photocleavage or hydrogen abstraction mechanisms. Benzoin ethers, benzil ketals, and acetophenone derivatives are typical photoinitiators.

Photoinitiators + hv ⟶ Free Radicals

Free Radicals + Crosslinkers ⟶ Three Dimensional Polymer Matrix

Monomers and Oligomers. These materials are polymerized by free radicals and impart much of the ultimate properties to the solder mask. Examples are acrylated epoxy resins, acrylated urethanes, and multi-functional monomers. A typical acrylate monomer is trimethylolpropane triacrylate (TMPTA).

$$CH_3CH_2C\ (CH_2OOCCH=CH_2)_3$$

Functional Fillers. These relatively inert materials are dispersed in the formulation to develop the desired rheology and body needed for the application of the solder mask.

Colorants. Pigments or dyes are used to ease the inspection of the PCB after solder mask application. The colorant content is limited by competition with the photoinitiators for the UV light. Too high a level of colorant can result in a loss of cure speed.

Additives. These specialty materials can be flow modifiers, adhesion promoters, stabilizers, plasticizers, etc.

Experimental

The IPC Multi-Purpose Test Pattern, IPC-B-25 is shown in Figure 1. Test boards were fabricated in-house containing the IPC-B-25 pattern on both sides. The center comb pattern with 12.5 mil copper conductors and 12.5 mil spaces was used to measure the insulation resistance. A total of eight values were measured for each board and the average reported. A General Radio Model 1864 Megohmmeter was used to measure the insulation resistance using 100 volts DC and a 60 second electrification time.

A saturated salt solution humidity chamber with a gold tab mounting fixture was used to produce 95% relative humidity as specified in IPC-SM-840. A controlled temperature Blue M oven was modified to allow external monitoring of insulation resistance.

The test boards were inspected visually, cleaned, and baked dry before application of solder mask. The various solder masks were applied and cured according to methods specified by the solder mask supplier. The solder masks tested are described in Table II. A typical insulation resistance determination is performed by placing five test boards in the test fixture and following the measurement procedure. The test fixture is sealed in the humidity chamber and placed in the oven. The specified conditions (Table I) are controlled by the oven temperature.

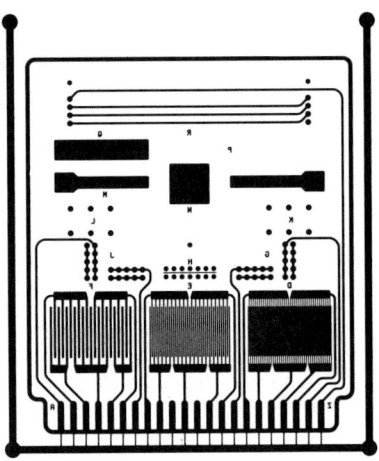

Figure 1. IPC-B-25 test pattern. (Reproduced from Ref. 4. Copyright 1977, IPC.)

Table II. Description of Solder Masks

Code	Description
SM-1	One-part UV cured acrylic for tin/lead circuitry.
SM-2	One-part UV cured acrylic for copper circuitry and hot air leveling.
SM-3	One-part UV cured acrylic for copper circuitry.
SM-4	Two-part epoxy, all purpose.
SM-5	One-part alkyd/amine, for copper circuitry.
SM-6	Dry film solder mask, all purpose aqueous developable.
SM-7	Dry film solder mask, all purpose aqueous developable.
SM-8	Dry film solder mask, all purpose aqueous developable.

Results and Discussion

Generally, it is believed that many plastics classed as insulators become partial conductors when "hot and wet" (12). The UV and thermally cured solder masks in Table III show a dramatic loss of insulation resistance after seven days at 85°C and 95% relative humidity (RH). All the test boards recovered most of their initial insulation resistance after 24 hours at ambient conditions. Table IV shows that an additional exposure to 85°C at ambient humidity produces an increase in insulation resistance.

Table III. Insulation Resistance (Ohms), IPC Class III Conditions, 85°C and 95% RH

Materials	Initial	One Day	Seven Days
SM-2	1.4×10^{13}	3.0×10^{7}	8.9×10^{6}
SM-3	1.3×10^{13}	6.6×10^{7}	1.5×10^{7}
SM-4	6.8×10^{12}	2.3×10^{8}	4.8×10^{7}
SM-5	1.5×10^{13}	8.1×10^{8}	7.8×10^{8}
SM-6	6.8×10^{12}	2.4×10^{6}	1.7×10^{6}
SM-7	7.4×10^{12}	9.0×10^{7}	4.2×10^{7}
SM-8	8.0×10^{12}	2.6×10^{7}	4.5×10^{6}

Table IV. Four Day Dry Oven Test, 85°C

Materials	Initial	Four Days
SM-4	4.4×10^{10}	6.15×10^{11}
SM-5	2.9×10^{12}	4.9×10^{12}
SM-6	1.5×10^{10}	9.6×10^{10}
SM-7	4.7×10^{11}	1.6×10^{12}
SM-8	3.9×10^{11}	2.3×10^{11}

Table V illustrates the effect of cure and soldering on solder masks tested according to Class 2 conditions, 50°C, and 95% RH. Subjecting the test boards to hot solder (510°F) increases the insulation resistance.

Table V. Insulation Resistance (Ohms), IPC-SM-840 Class 2, 50°C and 95% RH

Material	Initial	Day One	Day Four
Blank	1.5×10^{13}	1.3×10^{12}	5.9×10^{11}
Blank	1.3×10^{13}	1.2×10^{12}	5.7×10^{11}
SM-1 (10 FPM)	1.1×10^{13}	1.1×10^{9}	6.9×10^{8}
SM-1 (20 FPM)	6.4×10^{12}	1.0×10^{9}	1.4×10^{9}
SM-1 (20 FPM, Soldered)	1.7×10^{13}	1.8×10^{10}	1.9×10^{10}
SM-2 (10 FPM)	1.0×10^{13}	2.5×10^{9}	9.9×10^{8}
SM-2 (20 FPM)	1.0×10^{13}	3.3×10^{9}	3.9×10^{9}
SM-2 (20 FPM, Soldered)	$> 10^{13}$	1.1×10^{10}	1.3×10^{10}
SM-3 (20 FPM)	5.4×10^{12}	2.2×10^{10}	1.4×10^{10}

Conclusions

The loss of insulation resistance of a solder mask determined at elevated temperature and humidity is dependent mostly on the moisture content of the air. Elevated temperature alone does not cause a dramatic reduction of insulation resistance. The soldering process increases the insulation resistance of the solder masks examined. It is postulated that hot solder seals the surface of the solder mask, rendering a better moisture barrier. This study suggests higher insulation resistance should be obtained from solder masks with higher hydrophobic properties.

Acknowledgments

The author thanks Curtis Lustig and Bill Winkler for running the electrical tests.

Literature Cited

1. Cole, H. "Printed Circuits Handbook", 2nd ed, C. F. Coombs, Jr., Ed., McGraw-Hill Book Company, New York, N.Y., 1979, Chapter 14.
2. Fefferman, G. B. "Using Ultraviolet Radiation Curable Resins for Printed Circuit Coatings", 2nd International Conference on Radiation Curing, Cincinnati, Ohio, May 1975.
3. Tautscher, C. J. "Protective Circuit Coatings", Novatech Research Corporation, Redmond, Washington, 1981.
4. IPC Standard Specification IPC-SM-840, "Qualification and Performance of Permanent Polymer Coating (Solder Mask) for Printed Boards", Evanston, Illinois, 1977.
5. Axon, F., Cleek, R., Custer, W., Lipson, M., and Mestadagh, D. Circuit World 1978, 4, 24.
6. "IPC Standard Specification IPC-SM-840A", PROPOSAL, September 1982.
7. Tautscher, C. J. and Barmuta, M. J. Printed Circuit Fabrication 1982, 5 (10), p 34.
8. Smith, G. A. The IPC Technical Review, May 1980, 227, pp 11-15.
9. Jennings, C. W. in "How to Avoid Metallic Growth Problems on Electronic Hardware", IPC-TR-476, Institute of Printed Circuits, Evanston, Illinois, 1977 p 25.
10. Jennings, C. W. Proc. Printed Circuit World Convention II, 1981, 1, p 47.
11. Mitchell, J. P., Welsher, T. L. Proc. Printed Circuit World Convention II, 1981, 1, p 80.
12. Mathes, K. N. "Encyclopedia of Polymer Science and Technology", Interscience, New York, New York, 1966, 5, pp 531-533.

RECEIVED October 12, 1983

UV-Curable Conformal Coatings

C. R. MORGAN, D. R. KYLE, and R. W. BUSH

W. R. Grace & Co., Columbia, MD 21044

>   This paper discusses a new type of conformal coating that is solventless, one-component and cured by a combination of UV and heat.

Completed printed circuits often require total encapsulation by conformal coatings to insulate the electronic components and to provide protection from moisture, dust, solvents, and other contaminants present in the environment. The current most common conformal coatings are one- or two-component systems whose main constituents are epoxies, acrylics, urethanes or silicones. While these conformal coatings can provide excellent protection for the finished circuit board assembly, they have one or more undesirable features such as high solvent content, need for mixing, or long cure times. For example, the handling times range from 10 min to 5 hr, and the time to reach optimum properties ranges from 25 hr to 1 wk, depending on the system and curing conditions.

To overcome these disadvantages, we have developed a new type of conformal coating that is solventless, one-component, and cured by exposure to UV followed by heating. The UV exposure rapidly sets the coating, thereby allowing for immediate handling and preventing any runoff during the heating step. The heating step completes the cure of the parts of the coating that are not exposed to the UV (e.g., under the components). This paper discusses the chemistry, curing, and properties of these dual UV/thermally curable conformal coatings.

Chemistry

These dual UV/thermally curable conformal coatings are based on the ability to rapidly transform a liquid composition into a cross-linked solid by the free-radical polymerization of a polyfunctional acrylate in the presence of a low concentration of a polyfunctional thiol (1). Under these conditions, the polythiol functions partly by adding across the acrylic double bond and partly as a chain transfer agent in the acrylate homopolymerization (2, 3). The dual UV/thermal curability is obtained by adding both a photoinitiator and a thermal initiator (4, 5, 6).

The most efficient photoinitiators for these thiol/acrylate systems are ones that cleave to free radicals on exposure to UV (7). An example of this type of photoinitiator is alpha, alpha-dimethoxy-alpha-phenylacetophenone which is reported to photocleave as in Equation 1 to form a highly reactive methyl radical (8).

$$\underset{\underset{OCH_3}{|}}{\overset{\overset{O}{\|}}{PhC}}-\overset{\overset{OCH_3}{|}}{\underset{\underset{OCH_3}{|}}{CPh}} \xrightarrow{h\nu} \overset{\overset{O}{\|}}{PhC\cdot} + \overset{\overset{OCH_3}{|}}{\underset{\underset{OCH_3}{|}}{PhC\cdot}} \longrightarrow \overset{\overset{O}{\|}}{PhCOCH_3} + CH_3\cdot \qquad (1)$$

With this type of photoinitiator and with the proper UV system, curing of these thiol/acrylate conformal coatings can be accomplished in 10 sec or less.

The thermal initiator is benzopinacol (4), which is stable at room temperature (shelf-life of the uncured liquid conformal coating is more than 6 months at room temperature), but which undergoes a homolytic cleavage upon heating to form diphenylhydroxymethyl radicals (Equation 2), which are believed to initiate polymerization by hydrogen atom transfer to the double bond of the acrylate (Equation 3) (9).

$$\underset{\underset{}{}}{\overset{\overset{OH}{|}}{Ph_2C}}-\overset{\overset{OH}{|}}{CPh_2} \xrightarrow{\Delta} 2\ \overset{\overset{OH}{|}}{Ph_2C\cdot} \qquad (2)$$

$$\overset{\overset{OH}{|}}{Ph_2C\cdot} + CH_2=CHR \xrightarrow{\Delta} Ph_2C=O + CH_3\dot{C}HR \qquad (3)$$

DSC analyses (Perkin-Elmer DSC-2C) of thiol/acrylate/benzopinacol systems show that polymerization can be initiated at temperatures as low as 85°C. However, the reaction is too slow at 85°C for the conformal coating application and temperatures of 100°C or higher are required to get practical heat cure times of 5 to 30 min.

The uncured conformal coatings are clear, transparent liquids with viscosities of 700 to 1400 cp at 25°C. A typical composition which has a viscosity of 820 cp at 25°C consists of 43 wt % of a high viscosity urethane acrylate oligomer, 49 wt % of a low viscosity monoacrylate diluent, 5 wt % of a polyfunctional thiol, 2 wt % of benzopinacol and 1 wt % of alpha, alpha-dimethoxy-alpha-phenylacetophenone. A trace amount of a fluorescent dye may be added, if desired, to facilitate inspection of the finished coating.

Coating and Curing

Printed circuit boards, cleaned by ultrasonic vapor degreasing with a Freon solvent, were coated by dipping them into the conformal coating composition described above. The dip coating cycle was as follows: immersion rate - 1 in/min; standing time in composition - 2 min; withdrawal rate - 1 in/min; drip time - 0.5 min. (Formulations of this type can also be spray-coated, but only dip-coated boards were used for the work discussed in this paper.) After dipping, the coating was immobilized by rotating the board for 10 sec at about 70 cm from a Berkey Addalux 300 W/in medium pressure mercury lamp. After this short exposure to set the surface, the coating was further UV cured by exposing the board under the Addalux for 1 min on each side under an atmosphere of nitrogen to eliminate surface tack caused by oxygen inhibition. Following the UV cure, the coating was heat-cured by placing the board in a 120°C forced air oven for 30 min.

For certain tests the composition was drawn down on 0.23-mm thick aluminum sheets or glass plates and then cured for 1 min under an atmosphere of nitrogen under the Addalux UV lamp followed by heating at 120°C for 30 min. The coatings on aluminum were used for flexibility testing and those on glass were removed for tests requiring free-standing films.

The curing conditions described above were used throughout this work to maintain standard conditions for preparing various test samples. In actual practice, heat curing times of 30 min at 100°C, 20 min at 110°C, or 10 min at 120°C will probably be adequate, depending on the coating thickness, the size of the circuit board, and the kinds and configurations of the attached components. The UV curing time can be shortened considerably and the nitrogen atmosphere can be eliminated if lamps of higher intensity and/or different spectral distribution are used. For example, in a separate series of experiments, it was shown that this conformal coating can be cured tack-free in 2 passes at 18 fpm (estimated exposure time of 4 sec) under a Fusion Systems Type D 300 W/in doped medium pressure mercury lamp located 8 cm above the sample.

Testing and Properties

Mechanical Tensile Properties, measured according to ASTM D638 on 3-mil thick cured films of the conformal coating, are: modulus of 1750 psi, tensile of 2000 psi, and elongation-at-failure of 285%.

Flexibility was determined on 2-mil thick films of the cured conformal coating on 0.23-mm thick aluminum sheets by bending over a 1/8 in mandrel according to Fed. Std. 141, Method 6221. The film showed no cracking or loss of adhesion at the bend.

Water Vapor Permeability was measured on cured films of the

conformal coating essentially according to ASTM E96 except relative humidities of 0-75% were used instead of 0-50%. The water vapor permeability for this coating at 23°C is 1.72 (g)(mm)/(day)(m^2)(cm Hg).

Volume Resistivity was determined on cured films of the conformal coating in accordance with ASTM D257 and found to be 3.5 x 10^{14} ohm-cm at 23°C and 50% relative humidity.

Solvent Resistance. Printed circuit boards were dip-coated and cured as described above and then immersed in various solvents. The conformal coating was readily removed by toluene, acetone, and methylene chloride but not removed by water, isopropanol or ethylene glycol.

Thermal Stress Resistance. Epoxy-glass printed circuit boards, containing various electronic components were dip-coated and cured as described above and then cycled from -65°C to +125°C in forced air chambers according to the following schedule: 1 hr at -65°C, 2 min at room temperature, 1 hr at +125°C, 2 min at room temperature, 1 hr at -65°C, etc. Such boards survived 20 to >50 cycles without cracking and/or delamination of the coating.

Hydrolytic Stability. Epoxy-glass circuit boards with the conductor pattern shown in Figure 1 were encapsulated in a 2-mil thick cured conformal coating and tested for hydrolytic stability by maintaining them at 100°C and 95% relative humidity and observing changes in appearance and insulation resistance with time. After 1,000 hr in this test, this coating showed only a slight discoloration and essentially no change in insulation resistance.

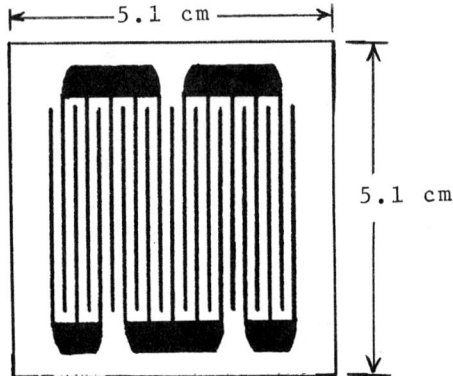

Figure 1. Circuit pattern for hydrolytic stability testing. Circuit: Tin/lead over copper. Conductor width: 0.7 mm. Space between conductors: 1.2 mm.

Summary

We have described a new type of conformal coating which is shelf-stable, solventless, one-component, and rapidly cured by a combination of UV and heat. Once cured, this coating has very good electrical properties, thermal stress resistance, and hydrolytic stability. With the proper choice of solvents, all or a portion of the cured coating can be removed from a circuit board to repair the electronic parts.

Acknowledgments

The authors thank Leslie Schulz and William Ehmann for their technical assistance, Ronald Andrejak for the DSC curves, and John Arnreich for the electrical and mechanical tensile properties.

Literature Cited

1. Kehr, C. L. U.S. Patent 4 008 341, 1977.
2. Walling, C. "Free Radicals in Solution"; John Wiley & Sons, Inc: New York, N.Y., 1957; p. 313 ff.
3. Morgan, C. R.; Ketley, A. D. J. Rad. Curing 1980, 7(2), 10.
4. Morgan, C. R. U.S. Patent 4 020 233, 1977.
5. Morgan, C. R. U.S. Patent 4 288 527, 1981.
6. Morgan, C. R. U.S. Patent 4 352 723, 1982.
7. For a review of photoinitiators for photopolymerization, see Pappas, S. P.; McGinness, V. D. in "UV Curing: Science and Technology", Pappas, S. P., Ed.; Technology Marketing Corporation, Stamford, Conn., 1978, p. 1.
8. Sander, M. R.; Osborn, C. L. Tetrahedron Lett. 1974, 415.
9. Braun, D.; Becker, K. H. Ind. Eng. Chem. Prod. Res. Develop. 1971, 10(4), 386.

RECEIVED September 2, 1983

30

Thermal Expansion Coefficients of Leadless Chip Carrier Compatible Printed Wiring Boards

Z. N. SANJANA, R. S. RAGHAVA, and J. R. MARCHETTI

Research & Development Center, Westinghouse Electric Corporation, Pittsburgh, PA 15235

Packaging densities and reduction of circuit lengths for very high speed integrated circuits have led to the development of leadless hermetic chip carriers. The chip carrier, made of a ceramic material, is soldered directly to the printed wiring board and the flexibility associated with leads is no longer present to accommodate differential thermal expansions. Thus it becomes necessary to match the thermal expansion characteristics of the wiring board in the X-Y plane to that of the chip carrier which is made of a ceramic material having a thermal expansion coefficient of $6 \times 10^{-6}/°C$. Several materials combinations were examined and it was determined that fabrics made from fibers with expansions significantly lower than that of conventional E-glass have to be used to make boards that have the desired thermal expansion characteristics. Aramid reinforced laminates appear to be particularly useful in this application and their thermal expansion characteristics were studied using several experimental techniques and using a computer model to study the effect of resin content, resin modulus and resin T_g on these materials. Data on a quartz fabric reinforced epoxy laminate is also presented.

Present day integrated circuits use dual-in-line packages having leads which are soldered to the printed wiring board (PWB). The presence of such leads provides a degree of flexibility between the packages and the PWB, therefore, a considerable amount of thermal mismatch can be tolerated during thermal cycling of the assembled board.

Requirements of increased density of packaging and very high speed processing of information (generally known as VHSIC) has

0097-6156/84/0242-0379$06.00/0
© 1984 American Chemical Society

led to the development of a package generally known as leadless chip carrier, LCC, or hermetic chip carrier, HCC. The LCC has pads that are directly soldered onto the PWB thus reducing circuit length and increasing component density. Unfortunately, present day PWBs which are flat sheet laminated compositions made from E-glass fabric impregnated with epoxy or polyimide resin are not thermally compatible with the chip carrier material which is a ceramic having linear thermal coefficient of expansion (TCE) of 6 x 10^{-6}/°C. The TCE of glass reinforced PWBs in the X-Y plane, i.e., in the plane of the laminate are of the order of 12-18 x 10^{-6}/°C. During thermal cycling there occurs a strain on the solder joint. Solder being inelastic undergoes plastic flow and after a number of cycles it cracks due to thermal fatigue([1,2](#)). Thus, it is necessary to develop LCC compatible PWBs. The compatibility is primarily a case of matching the TCE of the LCC and the PWB. Obviously advantageous would be a compatible board which could be processed similar to conventional PWBs, i.e., an organic matrix, fabric reinforced sheet material with copper bonded to one of both sides.

Kevlar, or aramid as it is generically known, is a highly oriented polymer - poly(p-phenylene terephthalamide). Kevlar fiber has the interesting property of having a linear TCE in its axial direction of -2 x 10^{-6}/°C. The E-glass used in normal PWBs has a positive linear TCE of +5 x 10^{-6}/°C. It has been found that Kevlar reinforced epoxy or polyimide laminates yield room temperature TCEs in the X-Y plane of around 4 to 8 x 10^{-6}/°C(3) and solder joint cracking is greatly reduced or eliminated(2). Quartz is another reinforcement, available in fabric form, which can be used to control the planar TCEs of a laminate PWB. Quartz has a TCE of +0.54 x 10^{-6}/°C in the axial direction.

The thermal expansion characteristics of a laminate used in PWBs for carrying LCCs have to be obtained over a range of temperatures that are expected to be observed by the assembly. Such a range, at least for military applications, is -60°C to +150°C. We have found that in this range the thermal expansion of Kevlar laminates in the X and Y direction are not necessarily linear. In fact, depending on the T_g of the resin used, the expansion curve as a function of temperature can actually show a change of slope from positive to negative. We have also found that the method used to obtain TCEs may affect the values obtained.

Experimental

The matrix material for the laminates for this investigation was a typical flame retardant epoxy resin system used in PWB technology. The epoxy resin was a brominated diglycidyl ether of bisphenol-A cured with dicyandiamide and catalyzed by benzyl dimethylamine. The weight ratios of epoxy, curative and catalyst used were 100/3/0.25.

The fabrics used were Kevlar and quartz. Data on two
dissimilar styles of Kevlar fabric are presented. Both were
obtained from Clark-Schwebel Corp. with an epoxy compatible surface finish and are as follows:
   CS348 - 8H Satin Weave, 50 x 50 count, .008" thick
   CS352 - Plain Weave, 17 x 17 count, .004" thick.
Quartz fabric was obtained with an epoxy compatible finish from
J. P. Stevens and is as follows:
   Astroquartz Sytle 503 - Plain Weave, 50 x 40 count,
   .005" thick.

Prepreg was prepared by hand-dipping the fabric into the
resin mix dissolved in an appropriate solvent (methyl cellosolve/
acetone). Drying and B-staging was done in a hot-air circulating
oven at 150°C for about 6 minutes. The prepreg was cut and
stacked to provide a nominal 1/8" thickness laminate. The stack
was laminated at 500 psi for 1 hour at 180°C. Laminates were
cooled under pressure and no post-bake or stress relief was
applied to the laminates. While data was collected on laminates
of differing resin contents, most of the data presented here is on
laminates containing 50% by weight resin content.

The TCEs were measured on samples machined from the 1/8"
thick laminates by using two techniques. In a horizontal quartz
tube dilatometer, samples 2" long and 1/4" wide were measured in
the 2" direction. The measurements were made in accordance with
procedures defined in ASTM E-228. Since the dilatometer is horizontal, measurement of shrinkage, if any, requires that the probe
be kept in contact with the specimen. This is done by means of a
light spring which applies a load of about 3 ozs on the sample
at room temperature. Measurement of expansion can also be made
without the spring load.

A Perkin Elmer TMS-2 thermal analysis set-up was used to
measure TCEs on 1/4" x 1/4" samples machined from the laminate.
In this technique, commonly known as TMA, a vertical probe (flat,
0.140" diameter) is placed on the sample and thermal expansion
measured as the sample is heated or cooled. The vertical probe
is balanced in oil suspension and a certain weight is added to the
probe to arrive at a net zero weight on the sample. Additional
weight may be placed on the probe to improve the signal/noise
ratio, but this affects the results as does spring loading in
the horizontal dilatometer. Strain gauge techniques were also
examined by us and were found to be less useful.(4)

A minimum of two cycles were run to measure the thermal
expansion and second cycle values were used.

<u>Mathematical Modeling</u>. The results obtained by various dilatometric techniques were compared to theoretical predictions of thermal
expansion using a simple mathematical model. Thermal properties
of Kevlar fabric/epoxy lamina were simulated by considering the
fabric to be made of two consecutive 0° and 90° lamina of

unidirectional Kevlar/epoxy composite. Influence of fiber crossover present in a fabric on thermal properties was neglected.

Linear TCE for a unidirectional composite in the fiber direction (α_{11}) and transverse to it (α_{22}) have been calculated as suggested by Schapery(5), who used complementary and potential energy principles in their derivations. The following equations give values of α_{11} and α_{22} for a unidirectional lamina:

$$\alpha_{11} = \frac{E_f \alpha_f v_f + E_m \alpha_m v_m}{E_f v_f + E_m v_m} \quad (1)$$

$$\alpha_{22} = (1+\gamma_m)\gamma_m v_m + (1+\gamma_f)\gamma_f v_f - \alpha_{11}\gamma_c \quad (2)$$

where:
E_f and E_m are Young's moduli of fiber and matrix, respectively.
α_f and α_m are linear TCEs of fiber and matrix, respectively. and
γ_f, γ_m and γ_c are Poisson's ratios of fiber, matrix and the composite, respectively.

Laminate properties of Kevlar fabric/epoxy were calculated by using laminate theory.(6) A lamina stacking sequence of 0° and 90° was adopted for the calculations.

Input to the model requires a knowledge of Young's modulus of the resin as a function of temperature. This was obtained using a DuPont DMA 980 dynamic mechanical analyzer and is presented in Figure 1. TCE of the resin as a function of temperature was obtained on the Perkin Elmer TMS-2 and is presented in Table I along with TCEs for the fibers as obtained from the supplier's literature.

Table I. Thermal Coefficient of Expansion of Kevlar, Quartz and Epoxy as a Function of Temperature

| Temperature Range, °C | Kevlar | | Quartz | Epoxy |
	Axial ($°C^{-1}$ x 10^6)	Radial ($°C^{-1}$ x 10^6)	Axial & Radial ($°C^{-1}$ x 10^6)	($°C^{-1}$ x 10^6)
25-100	-2.0	60	0.54	61.0
100-125	-4.0	60	0.54	151.0
125-170	-4.0	60	0.54	174.0

Results and Discussion

The first series of experiments were performed on a horizontal dilatometer with a light spring load. Figure 2 shows the thermal expansion obtained during heating to 170°C and cooling to room temperature. Repeated number of cycles were run and it was found that the second cycle and subsequent provided reasonably

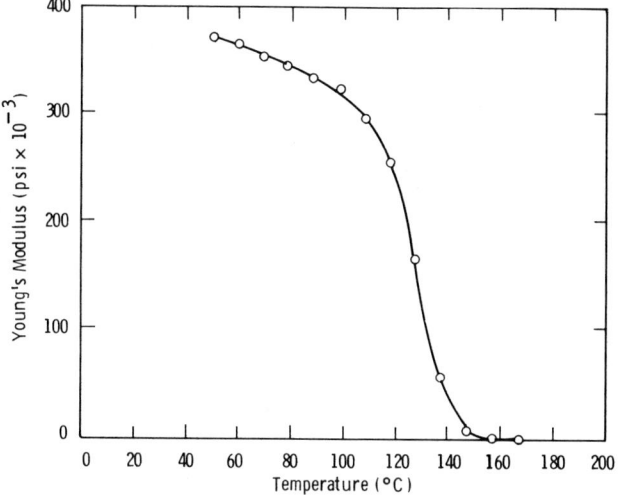

Figure 1. Young's modulus as a function of temperature for the resin system used in Kevlar-reinforced laminates.

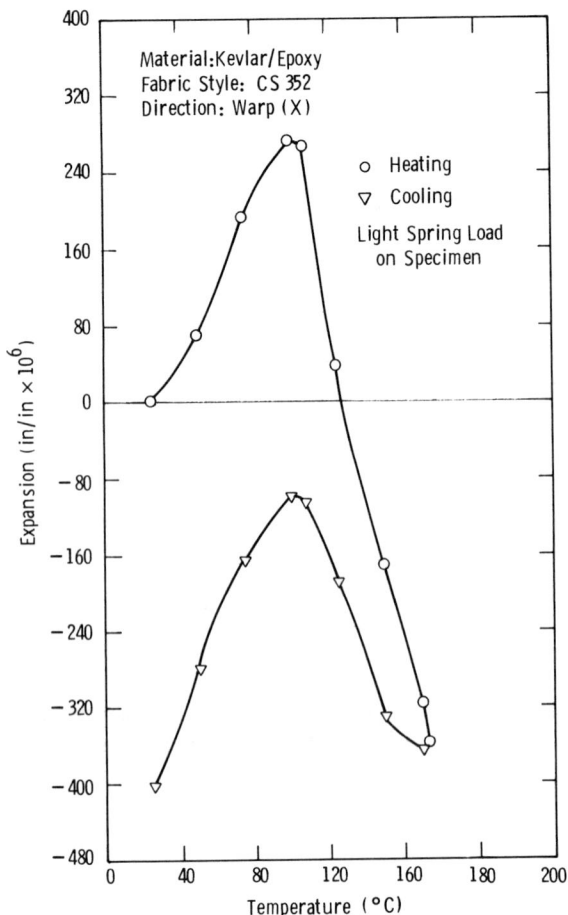

Figure 2. Thermal expansion of Kevlar-epoxy laminate as a function of temperature measured in a horizontal dilatometer.

close results. Thereafter, the second cycle values were used and are reported. As the sample is heated there is an expansion at the rate of $3.53 \times 10^{-6}/°C$ (TCE). After 100°C which is relatively close to the Tg of the resin (110°C), the expansion ceases and the sample begins to contract to 170°C at a rate of $8.8 \times 10^{-6}/°C$. On cooling, the sample expands at a TCE of $3.7 \times 10^{-6}/°C$ to a temperature of 100°C. On further cooling to room temperature, the sample contracts at a TCE of $4 \times 10^{-6}/°C$. There is a residual compression in the sample of 400×10^{-6} in./in. We have found that the magnitude of the expansion coefficients are dictated by the spring load on the sample. The change in slope from positive to negative on heat-up is real, as is the change on cooldown. Figure 3 presents expansion curves for the same laminate with and without spring loading. The spring loaded sample was heated to just above the T_g (115°C) and immediately cooled down; the two curves overlapped. The sample without any spring loading was heated to 170°C, and produced a larger TCE of $5.86 \times 10^{-6}/°C$. Since there was no spring load, no contraction was observed. The large contraction observed during heat-up in Figure 2 between 100° and 70°C is a combination of true contraction and bending or plastic deformation of the sample.

The same material was tested on the TMA apparatus and the expansion curve is shown in Figure 4 along with the predicted expansion curve obtained from the model for a 50 weight percent resin content laminate. Up to 100°C the agreement between the model and the experimental values is excellent. The TCE is $5.8 \times 10^{-6}/°C$ which is also the value obtained on the horizontal dilatometer with no spring load (Figure 3). At temperatures above 100°C, differences are considerable between theoretically predicted expansion values and those obtained from the TMA. The TMA curve changes to contraction at 115°C whereas the model would predict the change from expansion to contraction at about 135°C.

Figure 5 shows experimentally obtained and theoretically predicted values of thermal expansion for the same laminate in the fill or Y direction. Since a balanced fabric (i.e., same number of fiber yarns per unit length in either direction) was used, the expansions should be identical in both warp (X) and fill (Y) directions. Naturally, the predicted curve is the same as in Figure 4. The experimentally obtained expansion curve is also almost identical.

Figure 6 shows the thermal expansion to 170°C for a laminate made with the same epoxy resin but with CS348 fabric. The theoretically predicted curve is the same as in Figures 4 and 5 because the model assumes no weaving. The experimental curve obtained on the TMA with no load on the sample shows close agreement to the predicted expansion up to 115°C. Above 115°C a difference appears between the two, similar to those found in Figures 4 and 5. Figure 6 also shows the result of adding 1.0 gram weight onto the TMA probe. The TCE up to 100°C changes from

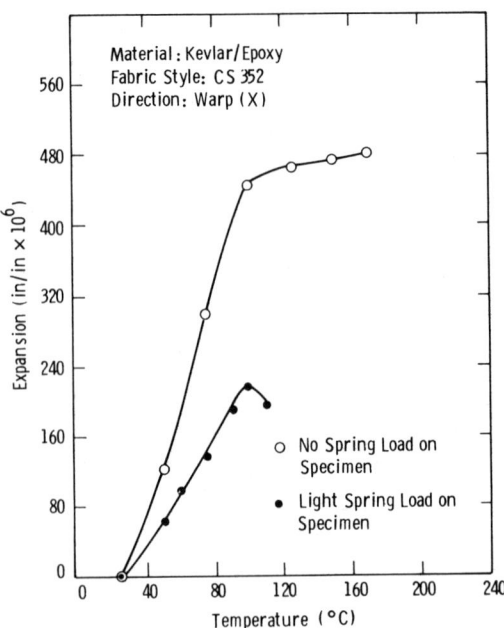

Figure 3. Thermal expansion of Kevlar-epoxy laminate as a function of temperature measured in a horizontal dilatometer.

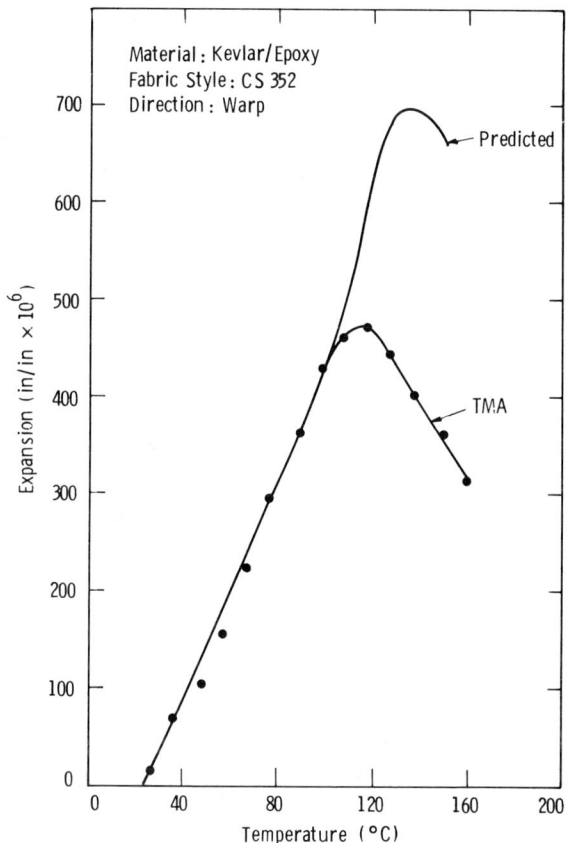

Figure 4. Predicted and experimental values for thermal expansion of epoxy laminate as a function of temperature.

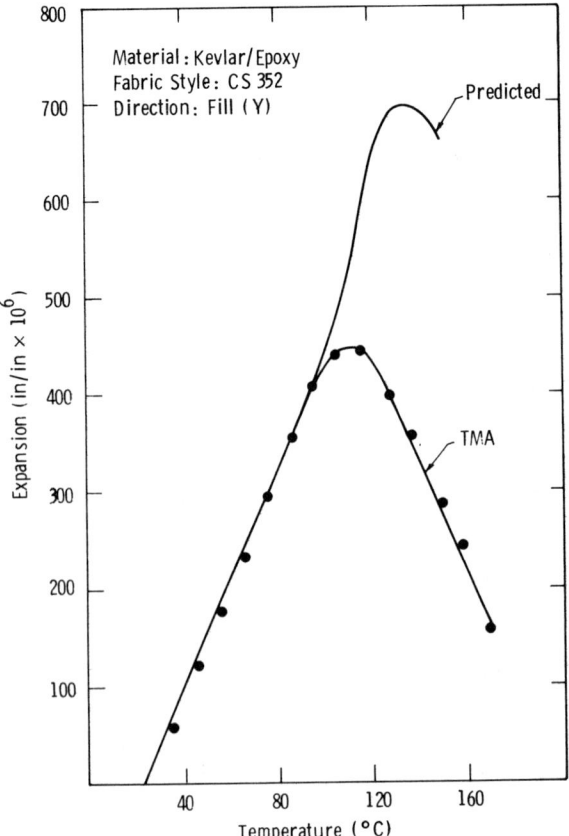

Figure 5. Predicted and experimental values of thermal expansion of Kevlar-epoxy laminate as a function of temperature.

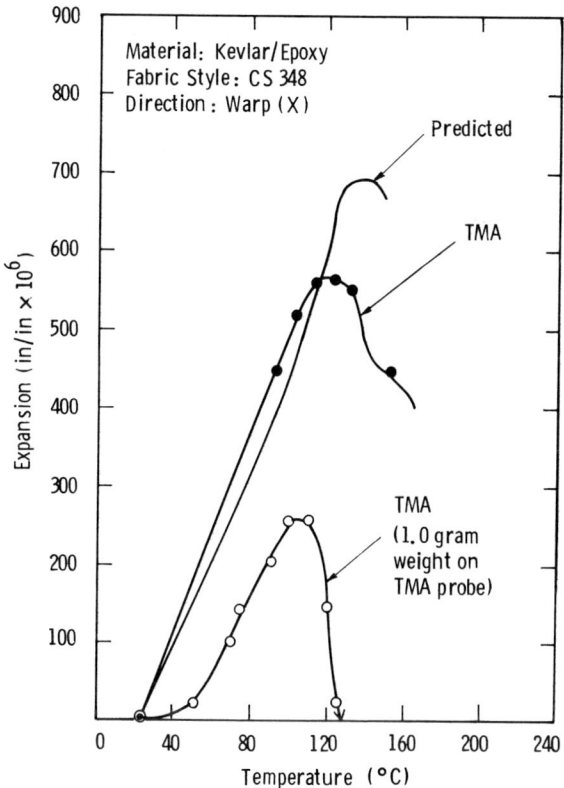

Figure 6. Predicted and experimental values of thermal expansion of Kevlar-epoxy laminate as a function of temperature.

$6.6 \times 10^{-6}/°C$ to $3.3 \times 10^{-6}/°C$. The latter value is similar to that obtained on the horizontal dilatometer with a light spring load (Figure 3).

In Figures 4-6, the predicted expansion curve is not plotted beyond 150°C, because above this temperature the modulus of the resin as input to the model is essentially zero (Figure 1). The computer program gives an algorithmic error and does not calculate values of TCE for the input. The only significant difference between model prediction and experimental expansions obtained with no load on the sample is the temperature at which the expansion changes from a positive rate to a negative rate. On the experimentally obtained curves, this occurs right at the T_g of the resin. The theoretical model would indicate this takes place at 20°C higher temperature. This may be due to two factors: (a) Even with essentially "no load" the probe may still be interacting with the sample, particularly as the resin softens near the T_g. (b) As the composite expands, the matrix wants to expand at a rate larger than that of the fibers. Therefore the fibers are held in tension by the resin. Near the T_g the resin softens and can no longer hold the fibers in tension, therefore the residual stress within the fibers relaxes and causes the composite to begin shrinking. The simple model used here does not account for stress relaxations near the T_g.

The model and the data show that the thermal expansion of a fiber reinforced laminate is controlled by the following factors: T_g of resin (if it is within range in which thermal expansion is measured), volume fraction of fiber and matrix, moduli of fiber and matrix as a function of temperature, and expansion coefficients of fiber and matrix as a function of temperature. For example, the model predicts that for Kevlar reinforced laminates made with the epoxy formulation discussed here, a 60% by weight resin content laminate would have a TCE of $7.5 \times 10^{-6}/°C$ and for a 35% by weight resin content the TCE would be $4.3 \times 10^{-6}/°C$. Data obtained from the model and experimentally obtained data using the horizontal dilatometer are presented in Figure 7 for laminates with 60% by weight resin content and Figure 8 for laminates with 35% by weight resin content. Experimental and predicted data is once again in close agreement.

Due to the large radial coefficient of expansion of Kevlar (about $+60 \times 10^{-6}/°C$) it may be necessary to blend in some fiberglass along with Kevlar to reduce the Z-directional TCE of the laminate. It would then be expected that the X or Y direction expansion would be larger than if Kevlar fabric alone were used. Figure 9 shows data for a case in which Kevlar and glass reinforcement are combined within the laminate. This was done by alternating plies of Kevlar fabric and fiberglass fabric, keeping the overall resin content at 50% by weight. For the model it was assumed that Kevlar and glass fibers were in a 1:1 volume ratio. Figure 9 shows the predicted value of the TCE in the linear range to be $11.7 \times 10^{-6}/°C$. The experimentally measured value is

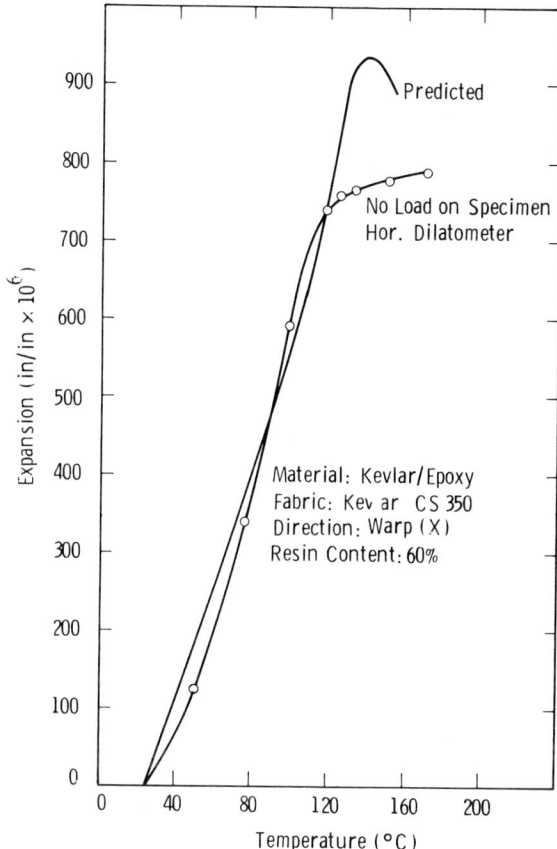

Figure 7. Predicted and experimental values of thermal expansion of Kevlar-epoxy laminate as a function of temperature.

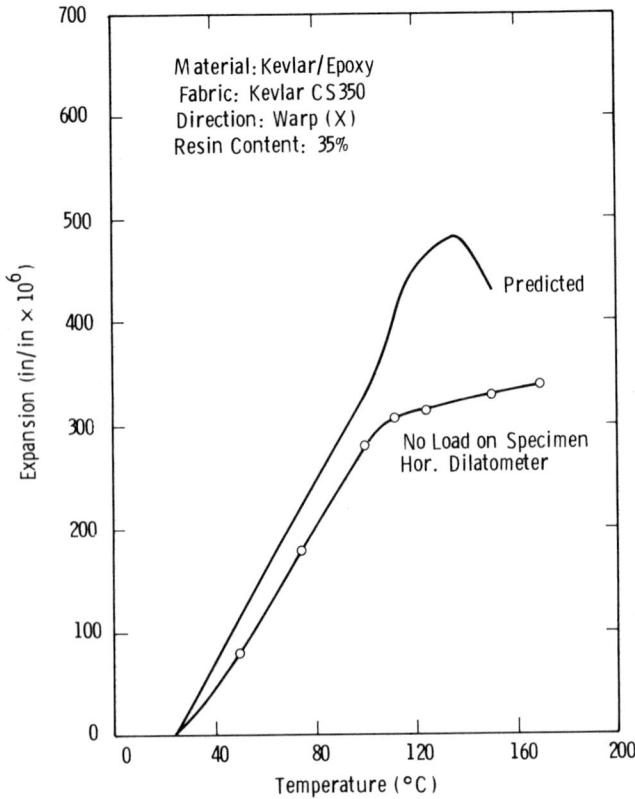

Figure 8. Predicted and experimental values of thermal expansion of Kevlar-epoxy laminate as a function of temperature.

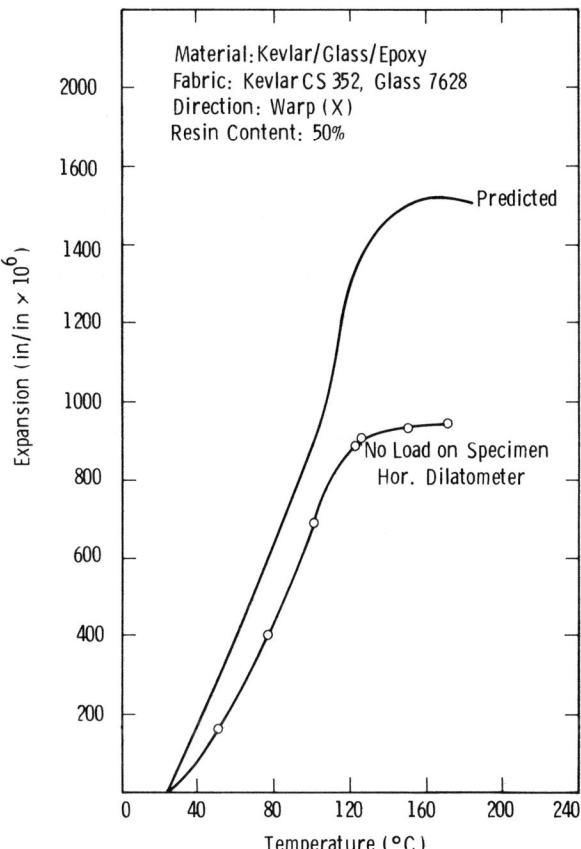

Figure 9. Predicted and experimental values of thermal expansion of Kevlar-glass-epoxy laminate as a function of temperature.

somewhat less: $9.3 \times 10^{-6}/°C$. As expected, the addition of fiberglass to the laminate results in higher TCEs than if Kevlar alone were to be used which would be $5.8 \times 10^{-6}/°C$, according to Figures 3 and 4.

Some work was done with quartz fabric reinforced laminates because as indicated in the Introduction, quartz also has a low axial TCE. The same resin system used with the Kevlar laminates was used with quartz fabric to make a laminate of 50% by weight resin content and the data shown in Figure 10 was obtained. The model predicted a TCE in the X or Y direction of $17.3 \times 10^{-6}/°C$ in the linear range. Expansion data obtained from a horizontal dilatometer indicate a TCE of $13.33 \times 10^{-6}/°C$ in the X direction in the linear range. These values are higher than those reported by us for other quartz reinforced resin combinations(4) but the resin content here is higher (50% by weight, whereas in Reference (4) the resin content was 35%) and the resin system is different. The values reported here are within the range of values obtained by Gates and Reimann(7). Based on the data reported by Gates and Reimann(7)and us, elsewhere(4) and here it appears that the TCE of quartz laminates are much more affected by differences in resin system TCEs than are Kevlar laminates. This is due to the difference in the moduli of the two fibers. Kevlar has a larger Young's modulus in the axial direction than glass or quartz fabric (by a factor of two) and therefore according to Equation 1 the effect of resin differences will be less influential on the α_{11} of Kevlar laminates than quartz or glass laminates.

While further work is continuing with quartz fabric, the data reported here and elsewhere(4,7) indicates that all other factors being equal, a quartz fabric reinforced laminate will produce TCEs that are higher than if Kevlar fabric were used. Further work is also indicated in the general area of hybrid fabrics such as combinations of Kevlar/fiberglass and Kevlar/quartz.

Conclusions

1) The technique used to measure TCEs in the plane of the laminate can strongly affect the values obtained.
2) If dilatometric methods are used, accurate results can be obtained only if no-load is applied to the sample. However, this makes the measurement of contractions more difficult.
3) A mathematical model using laminate theory can reasonably predict values of TCEs in the temperature range in which the effects are linear and elastic.
4) Non-linearity at temperatures approaching the T_g of the matrix, caused by viscoelastic effects, cannot be predicted by the model.

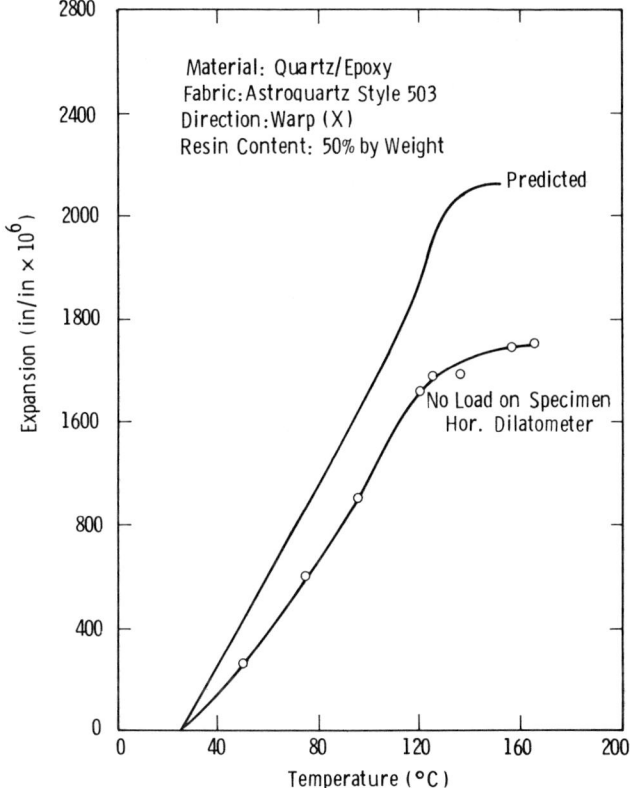

Figure 10. Predicted and experimental values of thermal expansion of quartz-epoxy laminate as a function of temperature.

Acknowledgment

Data obtained on the thermal mechanical analyzer by P. J. Sorokach and on the dilatometer by R. W. Dunning is gratefully acknowledged.

Literature Cited

1. Love, G. F. IPC Technical Rev. Dec. 1981, Pg. 12.
2. Woodruff, W. Electronics January 27, 1982, 55, 2, 119.
3. Green, S. E. IEEE Trans. on Components, Hybrids and Manufacturing Technology 1979, CHMT-2, 1, 140.
4. Sanjana, Z. N.; Valentich, J.; Marchetti, J. R. Proc. of 28th SAMPE Symposium, Anaheim, CA, 1983, SAMPE Ser. V. 28.
5. Schapery, R. A. J. Comp. Mater. 1968, 2, 380.
6. Ashton, J. E.; Halpin, J. C.; Petit, P. H. "Primer on Composite Materials", Technomic Publ., Stamford, CT, 1969; Ch. 3.
7. Gates, Jr., L. E.; Reimann, W. G. Proc. 2nd Annual Intl. Electronics Packaging Conf. 1982, Pg. 605.

RECEIVED September 2, 1983

POLYMERS
FOR SPECIAL FUNCTIONS

Piezoelectric Poly(vinylidene fluoride) in Small-Bore, Thick-Walled Tubular Form
Continuous Production and Properties

P. PANTELIS

British Telecom Research Laboratories, Martlesham Heath, Ipswich, Suffolk, IP5 7RE, England

> The preparation of piezoelectric poly(vinylidene fluoride) (PVDF) from pellets usually involves several batch processes. These are time consuming and inevitably lead to wastage, lack of uniformity in large quantities, and extra costs. In particular, the manufacture of thick film PVDF (> 100 μm) has required expensive preparative equipment and very high poling voltages. This paper describes a low cost continuous process in which pellets are directly converted into small-bore, thick-walled piezoelectric tube. A small commercial extruder was used to produce PVDF tube (crystal form II), and in-line heating, stretching and corona poling between two haul-offs produced a poled tube (crystal form I) of true draw ratio $\sim$ 4:1. Production rates of metres per minute were obtained using conveniently low poling fields. For example, tubing of o.d. 1.4 mm and wall thickness 234 μm showed piezoelectric coefficients d_{31} 19 pC/N and d_h 14 pC/N, and a pyroelectric coefficient λ 29 μC/m^{2}K, with good ageing properties up to 120°C. The tube can be produced with or without a tightly fitting wire core. Applications are suggested as an underwater sonar transducer and as a pyroelectric detector for absorption loss measurements on low-loss optical fibre.

This paper presents details of a new, low-cost, continuous process in which pellets of poly(vinylidene fluoride) (PVDF) are directly converted into piezoelectric tube. Kawai's paper in 1969 (1) described a batch process for making PVDF piezoelectric by stretching films at an elevated temperature and subsequently applying a high electric field to the electroded and heated films (thermal poling). Since then many variations of this technique have been described. Southgate in 1976 (2) demonstrated that poling could be achieved at room temperature by applying a corona discharge at a high potential to one surface of a PVDF sample, with the other surface at earth potential. Whilst this process is more convenient for continuous poling on a commercial scale (3) the rate at which PVDF can be poled is slow (∿20 cm/min) compared to the rate of production of oriented PVDF film by conventional melt extrusion and stretching techniques. The route to commercial piezofilm samples has been based therefore on batch processing, producing oriented films first and then poling them. Recently, independent reports by groups of workers in Holland (4), France (5) and Japan (6) have shown that the processes of stretching and poling can be combined. Essentially what they discovered was that for modest electrical fields, whether applied by corona discharge (6) or by thermal poling (4, 5), the action of poling at the neck (where deformation converts the initial crystal form II PVDF film to uniaxial form I) (4, 6) or poling whilst rolling (5) produced an unusually high piezoelectric activity in the PVDF. Processing rates were however slow (maximum draw rate 2 cm/min), and as described were batch processes using previously produced form II film and a form of tensile or rolling equipment. Above a draw temperature of $80°C$ a decrease in piezoelectric response was noted.

It was from this background that a continuous way of poling at the neck was conceived, using conventional melt processing plant (7). Equipment for producing oriented polymer tubing to act as a loose sleeve in optical fibre packaging had been used for a number of years at the author's laboratories (8). After confirming that PVDF pellets could be converted in this equipment to form I tubing, a corona discharge was experimentally applied to the outside of the oriented tube at or near the neck whilst the inside of the tube was at earth potential. This was achieved by means of an earth wire which was fed through the cross-head die of the extruder. Surprisingly low electrical fields were required for production rates of metres per minute of piezoelectric tubing. It is this continuous process which would allow a low-cost high-volume route to the uniform, thick walled (> 100 μm) PVDF tube that is described in this paper.

Experimental

Material. The PVDF resin grade Solef 1010 from Solvay et Cie was used in this study. This grade had been found to exhibit good melt extrusion properties for blown film production ($\bar{M}_w$ 351,000, MFI 5.6g/10 min at $190°C$, head to head content 3.5% (9).

Tube Production. Figure 1 shows in schematic form the technique for tube production and poling at the neck. Isotropic form II PVDF tube was continuously produced from an 18 mm screw extruder and oriented along the tube axis in a heated region between two caterpillar haul-offs. An earthed wire was fed via the cross-head

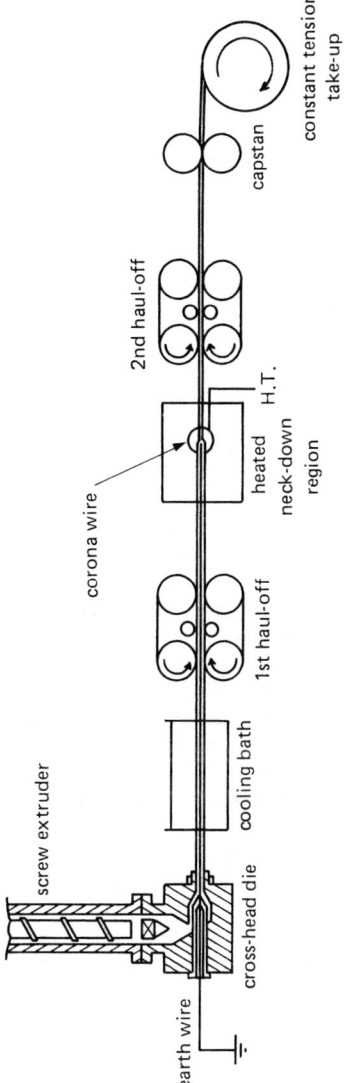

Figure 1 Schematic diagram of the production line for piezoelectric PVDF tube

die to the neck-down region. This region was established using either hot air blowers or a combination of blowers and a tubular pre-heat furnace. The corona discharge was applied by one or more loops of 0.1 mm diameter tungsten wire using a Wallis S303/1 power supply.

Coefficient Test Methods

Sample Preparation. Inner and outer electrodes were applied using conductive silver paint.

Piezoelectric Coefficients. (g_{31}, d_{31}) The piezoelectric coefficients relating to stress applied along the tube axis were measured by vibrating the tube at 20 Hz under ambient conditions using a Bruel and Kjaer mini-shaker type 4810 and monitoring the force applied to the tube at the same time as the voltage or charge developed. The force was measured by a Kistler 9203 quartz force transducer mounted in-line with the tube and feeding a Kistler 5001 charge amplifier. The tube voltage was measured using an Ancom 15A31 high impedance op-amp, and the charge density with a Kistler 5001 charge amplifier; the outputs from these and from the force transducer were fed into a Bryans Southern Instruments 'Transcribe 10' dual transient recorder.

Pyroelectric Coefficient (λ). The sample temperature was changed in a sinuosoidal fashion at 20 mHz by blowing nitrogen gas over the tube. The amplitude of the oscillation was 1 degree about a mean of $20°C$. The charge generated was collected by a Keithley electrometer operating in 'fast feedback' mode with a 10 nF capacitor in the feedback. The ratio of electrometer output to the monitoring thermocouple output was measured using a Solartron 1172 frequency response analyser.

Hydrostatic Coefficient (d_h). The sample was suspended from electrode wires in a pressure vessel containing silicone oil. The charge was measured as in the pyroelectric measurement above while the specimen was subjected to 'square step' pressure cycles of amplitude 0-500 psi and duration 10 sec, applied at 1 minute intervals. It was found necessary to apply a correction to d_h to allow for the pyroelectric response of the sample. This response arose due to the adiabatic heating of the silicone oil on pressurisation.

Relative Permittivity (ε'). The relative permittivity of the sample at 1592 Hz was measured using a Wayne Kerr B221A universal bridge.

Infra-red Spectrum. Attenuated total reflection (ATR) measurements (10) were made using a Perkin Elmer 683 spectrophotometer, to examine the form II and form I bands at 530 and 510 cm^{-1} respectively.

X-ray Diffraction. The relative proportions of form I and form II structures were calculated from the scattering intensities of the (001) form I plane and the (002) form II plane, I_I and I_{II}

respectively. The fraction of form I, X_I is given by $X_I = KI_I/(KI_I + I_{II})$ where K is a constant which takes into account the inherent differences in the scattering power of the two sets of planes. A value of K = 1.46 was used (11).

Density Measurements Sample densities were measured in a Daventest density column using a mixture of ethylene dibromide and carbon tetrachloride at 23°C. The percentage amorphous, form I and form II contents were calculated from their densities at 23°C of 1.674 (12), 1.973 (13) and 1.925 (13) g cm^{-3} respectively, taken together with the X-ray data.

Results and Discussion

Optimum conditions for conversion to form I (as monitored by ATR spectra) were established by preliminary drawing runs using an in-line tubular furnace. It was ascertained that machine draw ratios of between 5:1 and 6:1 at draw rates up to 3m/min were required to achieve substantial conversion. A linear relationship between tube density and furnace temperature was observed for a fixed machine draw ratio of 6:1. Over the temperature range 115°C to 170°C the density varied between 1.7886 to 1.7988 g cm^{-3} which corresponded (assuming complete absence of form II) to a percentage form I content of 38.3% to 41.8%.

Table 1 gives the percentage form I, form II and amorphous content of tube drawn during another set of runs at a furnace temperature of 165°C and machine draw ratios of 4:1, 5:1 and 6:1. The results were calculated from a combination of X-ray measurements which gave the form I to form II ratio and density measurements.

TABLE 1. Crystal form content vs. machine draw ratio

	Machine draw ratio		
	4:1	5:1	6:1
% Form I	5	22	35
% Form II	42	21	6
% Amorphous	53	57	59

The smaller crystalline content of the 6:1 machine draw ratio tube compared to the 4:1 tube is possibly a consequence of the faster kinetics allowing less time for crystal regrowth.

An attempt was first made at poling after the neck. The 6:1 machine draw ratio tube (OD 1.4 mm, wall thickness 290 μm) was poled at a corona voltage of +30KV, 15 seconds after it had emerged from the furnace where necking had occurred, but only a low g_{31} coefficient of 4.9×10^{-2} Vm/N was obtained.

Poling at the neck was far more successful. The conditions used and coefficients obtained are given in Tables 2 and 3 respectively.

TABLE 2. Preparation conditions for piezoelectric tube

Tube No	OD (μm)	Wall thickness (μm)	Machine draw ratio	True draw ratio	Draw speed (m/min)	Corona Voltage (kV)
1	1383	297	5.3:1	3.5:1	2.6	+30
2	1409	234	6:1	4:1	3.0	+25

TABLE 3. Electrical properties of piezoelectric tube

Tube No	ε'	g_{31} (10^{-2} Vm/N)	d_{31} (pC/N)	d_h (pC/N)	λ μC m^{-2} K^{-1}
1	14	10	-	11	25
2	14	14	19	14	29

X-ray results on the tubes showed that in both cases the crystal content was virtually all form I, indicating that possibly form II to form I conversion occurred on poling the lower draw ratio tube 1, where incomplete conversion to form I had been obtained on drawing alone. The densities of tubes 1 and 2 were 1.7887 and 1.7744 g cm^{-3} respectively, corresponding to form I contents of 38% and 35%. Slip had occurred between the caterpillar haul-off rubber tracks and the tubing, and consequently the true draw ratios measured by tube dimensions were less than the machine draw ratios obtained from the haul-off speeds. It is encouraging to note that the coefficients in

Table 3 are at least as high as for high field poled batch-processed material (14) or those obtained by high-draw ratio techniques (15). In addition, there was no decrease in the g_{31} coefficient for poled tube aged isothermally at 120°C for 50 minutes. This was probably due to the fact that the poled tube, as processed, experienced an extra thermal treatment beyond the neck because of the heat generated on drawing. The estimated neck temperature was 115°C.

The earthed wire need not remain stationary but can be allowed to feed through the extruder so that the PVDF stretches down tightly over it to form a wire-cored coaxial structure. Tinned copper wire was used in the example given in Table 4.

TABLE 4. Properties of wire-cored piezoelectric tube

Wire diameter (mm)	Wall thickness (mm)	True draw ratio	Draw speed (m/min)	Corona Voltage (kV)	d_h (pC/N)
0.71	0.54	4.6:1	2.0	+28	11

A possible reason for the good effect of poling at the neck can be inferred from the morphology work reported on oriented linear polyethylene, where the break-up of the original lamellae is completed at ~4:1 effective draw ratio and recrystallization takes place during drawing (16, 17). If this were true in the case of PVDF, then poling at the neck would apply dipole orientation to the molecular chains whilst they were in the process of recrystallization. This ordering could well be more efficient and require a lower field strength than poling a region that was already crystalline.

Applications

A novel research use for this PVDF tube, making use of its pyroelectric response to measure the absorption loss in single-mode optical fibre, has already been reported (18).

The fibre was fed into a 0.5 m length of water-filled PVDF tube, and light from a Nd:Yag laser operating at 1.06 μm was launched into the fibre. The water transmitted the heat generated by the absorption loss processes to the tube wall, so producing a proportional electrical charge. Temperature rises as low as 2 μ°C/s were detectable for an input power of 1W, which corresponded to a loss of less than 0.1 dB/km.

Because PVDF has an acoustic impedance quite close to that

of water, a more obvious area of use would be as an underwater transducer for sonar applications (19, 20). But it is as well to reiterate Bobber's observation (20) that flexible piezopolymers cannot simply be substituted for older materials in conventional designs. Design imagination and innovation will be needed to use PVDF in an ocean environment and in sonar applications. The tubular transducer described in this paper would certainly seem to merit this type of new approach. Some possible coaxial tube configurations are given in Figure 2 which shows how the tube can be potted into cylinder and plate forms. Shielding can readily be effected by earthing the outer electrode or potting compound while allowing the central electrode to be monitored.

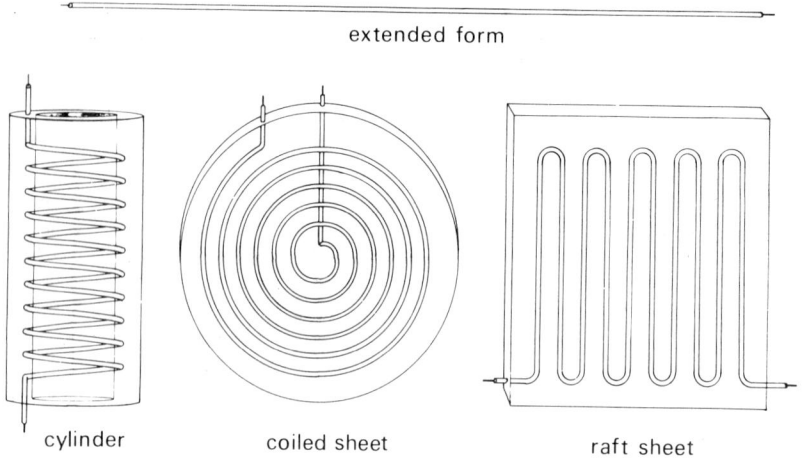

Figure 2 Some PVDF coaxial tube transducer forms

Acknowledgment

The author thanks Dr G R Davies (Leeds University) for measurement of d_h and λ, Dr M Cochran (Metal Box plc) for the X-ray results, and his colleagues at BTRL for their help in realising the initial concept. Acknowledgement is made to the Director of Research of British Telecom for permission to publish this paper.

Literature Cited

1. Kawai, H. Jpn. J. Appl. Phys. 1969, 8, 975.
2. Southgate, P. D. Appl. Phys. Lett. 1976, 28, 250.
3. Pantelis, P.; Morgan A. J. British Patent GB 2 020 483B, 1982.
4. Klaase, P. T. A.; van Turnhout, J. 11th. Europhys. Macromol. Phys. Conf. on Properties of Oriented Polymers, University of Leeds, April 1981.
5. Broussoux, D.; Facoetti, H.; Micheron, F. European Patent Application 0 048 642, 1982.
6. Furukawa, T.; Goho, T.; Date, M.; Takamatsu, T.; Fukada, E. Kobunshi Ronbunshu 1979, 36, 685.
7. Pantelis, P. British Patent Application 8207921, 1982.
8. Jackson, L. A.; Reeve, M. H.; Dunn, A. G. Opt. and Quant. Electron. 1977, 9, 493.
9. Cleaver, A. J.; Pantelis, P. Proc. Plastics in Telecommunications III Plastics and Rubber Institute, London 1982, p 32.1.
10. Doughty, K.; Pantelis, P. J. Mat.Sci. 1980, 15, 974.
11. Cochran, M.; Metal Box plc, Research and Development Division, Wantage, Private Communication.
12. Nakagawa, K.; Ishida, Y. Kolloid-Z. u. Z. Polymere 1973, 251, 103.
13. Hasegawa, R.; Takahashi, Y.; Chatoni, Y.; Tadokoro, H. Polymer J. 1972, 3, 600.
14. Bur, A. J. Polymer 1981, 22, 1288.
15. Nix, E. L.; Holt, L.; McGrath, J. C.; Ward, I. M. Ferroelectrics 1981, 32, 103.
16. Peterlin, A. J. Polymer Sci. 1967, C18, 123.
17. Capaccio, G.; Ward, I. M.; Wilding, M. A. Faraday Disc. Chem. Soc. 1979, 68, 328
18. Kashyap. R.; Pantelis, P; Proc. Symposium on Optical Fibre Measurements (NBS-SP-641) Boulder Co., 1982, p 67.
19. Bloomfield, P.E.; Ferren, R.A.; Radice, P.F.; Stefanou, H., Sprout, O. S. Naval Res. Rev. (USA) 1978, 31, 1.
20. Bobber, R.J., in "Underwater Acoustics and Signal Processing"; Bjorno, L., Ed.; Mathematical and Physical Sciences, NATO Advanced Study Institutes Series, 66, Series C., D Reidel Publ. Co.; Dordrecht, Holland, 1981; p 243.

RECEIVED September 2, 1983

Polymeric Reactions in Magnetic Recording Media

LESLEY J. GOLDSTEIN

Department of Chemical Technology, Graham Magnetics, Inc., North Richland Hills, TX 76118

>Magnetic recording tape is a complex interaction of select chemicals designed for peak recording capabilities in a durable, reliable package. Commonly, magnetic γFe_2O_3 particles are mixed in a binder system composed of solvent, polymer, isocyanate, catalyst and other components. The media is coated onto a base film, oriented, dried, and calendared to enhance recording quality. The oxide adheres to the supporting base film via a polyurethane-isocyanate reaction to yield a strong, flexible material similar to a crosslink thermoset. Actually, a "pseudo" crosslink mesh is formed from interpenetrating units. The main polymer is an ester based polyurethane with hydroxyl termination. Water also reacts with isocyanate and its abundance in a production environment may cause improper cure of the tape. A study was done on the reaction rates of isocyanate and polyurethane including a change in catalysts and alternating functionalities. Looking at these variations, the chemical composition of the recording media can be modeled to a top quality, low contaminant, complete cure system.

"Polymer in Electronics" is designed to discuss innovative ways polymeric materials can be used in the growing electronics field. This paper discusses specific polymer reactions in magnetic recording media. Magnetic properties originate from inorganic transition metal-oxides (γFe_2O_3, CrO_2, cobalt-modified γFe_2O_3) or metal powders. These particles offer the characteristic of acting as tiny bar magnets and containing magnetic moments. When a magnetic field is impressed upon a metal particle, the electron spin will alter the surface charge for the bar magnets to align with the field. Using a strong induced field, the spins orient themselves to a degree ideal for magnetic recording. As shown in Figure 1.

H = 0   H = Weak   H = Strong   H = Saturation   H = 0

Figure 1. Particle alignment in various magnetic fields.

Removing the magnetic field, the particles will either return to the original, nonoriented pattern or continue to remain oriented. In the later case, a remanent magnetization remains. There exists a unique relation between magnetic field strength and degree of particle orientation. For the magnetic particles to remain in a desired pattern, an organic binder system is used.

The magnetic particles are dispersed in a polymer system and coated onto a supporting basefilm. The polymers used can be nitrile rubber, polyurethane, nitrocellulose or polyester. In this investigation, the reactions discussed involve a polyurethane system.

Figure 2 represents a typical polyurethane compound used in magnetic media. This polymer is a product from an isocyanate-glycol ether reaction, and has terminal carboxyl groups.

$$H[O-\overset{O}{C}(CH_2)_m\overset{O}{C}]_nO(CH_2)_mO-\overset{O}{C}-N\sim\sim N-\overset{O}{C}-O(CH_2)_mO[\overset{O}{C}(CH_2)_m-\overset{O}{C}-O]_nH$$

Carboxyl   Urethane   Urethane   Carboxyl

Figure 2. Typical polyurethane structure.

To contain the magnetic particles, the base polymer must be crosslinked. In this study, the polyurethane reacts with a prepolymerized isocyanate. There is a question as to the exact site where the isocyanate reacts. Isocyanate reacts with an "active hydrogen" on a compound as in the following example:

$$R-N=C=O + H-A \longrightarrow R-\underset{H}{N}-\overset{O}{C}-A$$

Aromatic isocyanate tends to react faster than aliphatic isocyanate. The reactivity is also dependent on factors such as steric hinderance, temperature, catalysts and competing reactions. Isocyanate reactions found in a urethane system are shown on Table 1.

Table 1   REACTIONS IN A URETHANE SYSTEM

FUNCTIONAL GROUP	PRODUCT	PREFERRED TEMPERATURE	CATALYST	COMMENTS
$R-N=C=O + R'N\underset{H}{-}OR''$ (urethane)	$\underset{\underset{NHR'}{\overset{C=O}{\mid}}}{R-N-\overset{O}{\overset{\|}{C}}-OR''}$	120-140°C	Tertiary amines. Strong bases. Metal compounds.	Allophanate reaction. Reacts slow.
$R-N=C=O + H\text{-}O(CH_2)_m$ (primary alcohol) $\rightarrow -\overset{O}{\overset{\|}{C}}-\underset{H}{N}-R'$	$R-\underset{H}{N}-\overset{O}{\overset{\|}{C}}-O(CH_2)_m\text{-}O\overset{O}{\overset{\|}{C}}N-R'$	25-50°C	Mild and strong bases. Metal compounds.	Fast reaction.
$R-N=C=O + H_2O \rightarrow [RN-\overset{O}{\overset{\|}{C}}-OH] \rightarrow$	$RNH_2 + CO_2$ (amine)	25-50°C	Determined by desired needs.	Rapid reaction. Expels CO_2.
$R-N=C=O + H-\underset{H}{N}-R'$ (primary amine)	$R-\underset{H}{N}-\overset{O}{\overset{\|}{C}}-\underset{H}{N}R'$ (urea)	0-25°C	Not needed.	Most reactive.
$R-N=C=O + R'\underset{H}{N}-\overset{O}{\overset{\|}{C}}-\underset{H}{N}R''$ (urea)	$R\underset{\underset{NHR'}{\overset{C=O}{\mid}}}{N-\overset{O}{\overset{\|}{C}}-NHR''}$	100°C	Strong bases. Metal compounds.	Biuret reaction. Moderate rate.
$R-\underset{H}{N}-\overset{O}{\overset{\|}{C}}-\underset{H}{N}-R' + R''OH$	$R\underset{H}{N}-\overset{O}{\overset{\|}{C}}-OR'' + R'NH_2$	160-190°C	Determined by desired needs.	Urea dissociates to isocyanate to react with alcohol.

Looking at Figure 2, an isocyanate will react with (1) available H_2O, (2) primary alcohol or carboxyl groups, and (3) urethane groups. Under the right conditions, certain catalysts can influence the rate of the reacting isocyanate. For example, transition metal complexes are used to initiate isocyanate-hydroxyl or isocyanate-urethane reactions. Examples are metal acetylacetonates, napthanates and octoates. The isocyanate will react with terminal carboxyl groups and any available water, creating unwanted side reactions. Without the urethane-isocyanate linkage, a "pseudo-crosslink" occurs as in Figure 3. It is hypothesized that reacting isocyanate with the urethane groups, thus utilizing more reaction sites, would promote a faster cure and a more durable polymer structure.

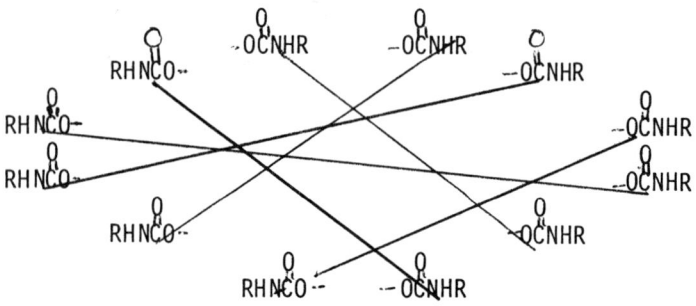

Figure 3. Interpenetrating structure.

Experimental

Saunders and Frisch (2) cite certain catalysts used to induce an isocyanate-urethane (allophanate) reaction. They are zinc octoate, cobalt napthanate and cobalt octoate and are claimed to yield 95% allophanate. An experiment was designed observing the catalytic effect of these metal complexes under varied concentrations and over time. Ferric acetylacetonate, a catalyst known to influence an isocyanate-carboxyl reaction, was included in the study. The catalysts were added individually and in combinations of two into a polyurethane-polyisocyanate system. Concentrations varied from 1.50% to 8.00% by weight. The ingredients were mixed together, cast onto supporting basefilm, and subjected to 70°C for 10, 30 and 60 minute periods.

The test method used was a solvent wipe test where specified pressures and solvent concentrations are applied to each coated sample. Data generated was number of seconds to pinhole distortion which relates to amount of cure. Figure 4 elucidates the results.

Evaluating a single catalyst, the fastest cure rate is when ferric acetylacetonate is used in the system. In high concentrations, cobalt napthanate alone and combined with cobalt octoate, offer a fast reaction rate. Yet, when ferric acetyl acetonate is combined with cobalt napthanate or cobalt octoate, the reaction rate increases. For this system, the increased amount of catalyst yields a decrease in cure time. There is no direct correlation between length of time specimens were was subjected to heat and cure rate.

When using catalysts which influence the allophanate reaction, one would expect a decrease in reaction rate. Yet, as seen in Figure IV, the ferric acetylacetonate offers the fastest reaction rate. It is concluded that the first crosslinking reaction is between isocyanate and the terminal carboxyl groups. The cobalt or zinc complexes do not have the catalytic effect of $Fe(AA)_3$. Surprisingly, when combined with either of the cobalt complexes, the ferric acetylacetonate's catalytic effect is reduced. This suggests that cobalt napthanate and cobalt or zinc octoate may indeed influence the allophanate reaction leaving a linkage so weak it impairs the physical character. Another explanation is that the metals cobalt and zinc and their ligands interfere with the ferric acetylacetonate catalytic effect and very little crosslinking occurs.

To draw a conclusion, one needs to observe the allophanate formation, but unfortunately there is not a test method available to identify this reaction.

To evaluate whether the ligands do effect catalytic cure rate, an experiment was designed keeping the base metal constant while varying the ligands. Ferric acetylacetonate and ferric napthanate were used under the same test conditions as the previous experiment. Results are given in Figure 5. $Fe(AA)_3$ has a greater catalytic effect, yet when combined with ferric napthanate, its reactivity decreases. Thus, the ligands do influence how the catalysts drive a reaction. In this system acetylacetonate yields the fastest cure rate. The napthanate may induce the allophanate reaction, but again there is no way to prove this.

An experiment was done keeping the ligand constant and varying the base metal. This was to evaluate whether the metal has an influencing factor on the overall cure rate. The catalysts used were cupric, manganic and ferric acetylacetonates. The concentration of catalyst varied from 1.50% to 8.00% by weight and were added alone and in combinations of two. The experiment consisted of viscosity measurements over time using a Brookfield Viscometer.

Figure 6 illustrates the change in viscosity of 1.5% by weight catalyst concentrations. $Cu(AA)_2$ had virtually the same effect

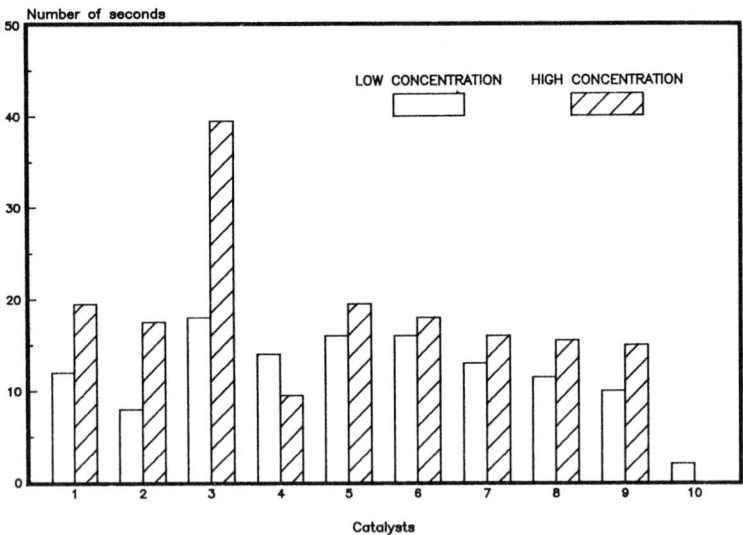

Figure 4. RXN with metal catalysts. Key: 1, Co napthanate; 2, Co octoate; 3, Fe(AA)$_3$; 4, Zn octoate; 5, Co napthanate + Co octoate; 6, Co napthanate + Fe(AA)$_3$; 7, Co napthanate + Zn octoate; 8, Co octoate + Fe(AA)$_3$; 9, Co octoate + Zn octoate; and 10, no catalyst.

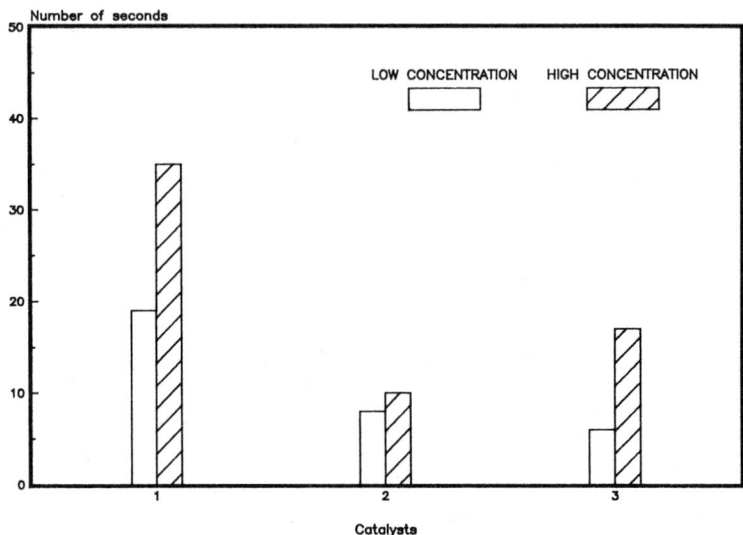

Figure 5. RXN with iron catalysts. Key: 1, Fe(AA)$_3$; 2, Fe napthanate; and 3, Fe(AA)$_3$ + Fe napthanate.

as no catalyst. At 1.5% concentration, $Mn(AA)_3$ caused a rapid increase in viscosity, so fast that data points could not be collected.

The effect of varied ferric and manganic acetyl acetonate concentrations with respect to viscosity is shown on Figures 7 and 8. Of note is the limiting factor $Fe(AA)_3$ has on the system, where 3.0% reacts faster the 4.5% concentration. This indicates a unique catalyst concentration is needed for this system to have an optimum effect. Manganic acetylacetonate has even a greater catalyst influence in this system. At 0.75%, $Mn(AA)_3$ reacts faster than elevated concentrations of either of the acetylacetonates.

Discussion

In this polyurethane-prepolymerized isocyanate reaction, the fastest catalyst are the acetylacetonate complexes. Looking at the molecular structures involved, an octoate is a 2-ethylhexoate, having the structure

$$C_2H_5-O-\overset{O}{\underset{}{C}}-C_5H_{11}$$

The napthanate is derived from naphthenic acid and is a mixture of molecular weight acids. No chemical structure is available.

Metal acetylacetonates are illustrated as:

(3)

Studying the acetyl acetonate mechanism in the presence of isocyanate, a carboxyl will exhibit an interchange of oxygen atoms. The acetylacetonate ligand separates from the base metal and is replaced with isocyanate to form a charge distribution between the oxygen and carbon. Having a cationic character, the carbon readily accepts the ionic carboxyl group. At completion of the reaction, the isocyanate-carboxyl group dissociates from the metal complex and the ligand resumes its original structure (Figure 9).

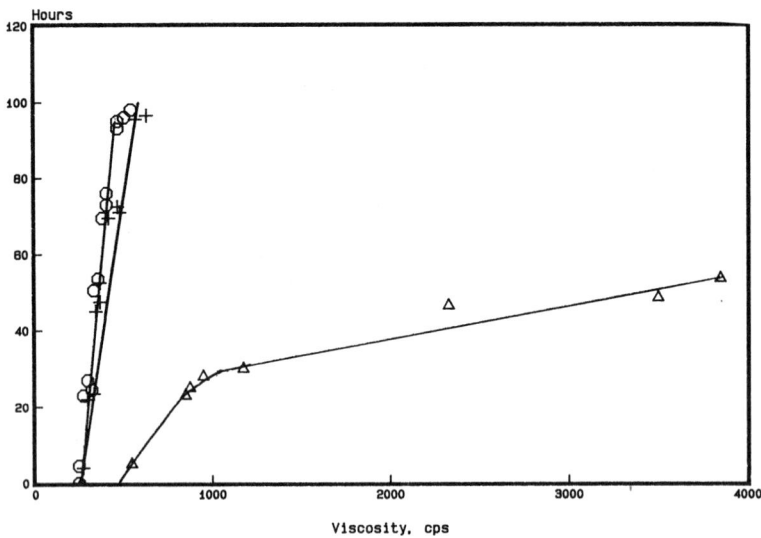

Figure 6. RXN with acetyl acetonate catalysts. Key: ◯, $Co(AA)_3$; +, no catalyst; and △, $Fe(AA)_3$.

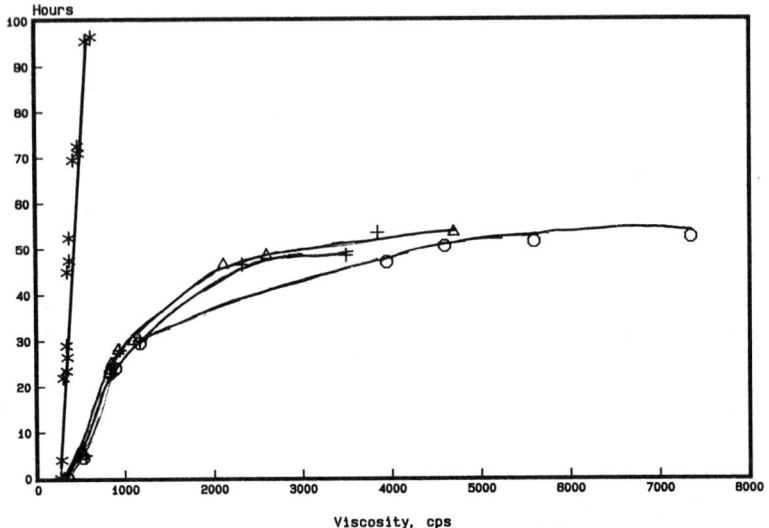

Figure 7. RXN with $Fe(AA)_3$ catalyst. Key(% $Fe(AA)_3$): +, 1.5; △, 3.0; ⬡, 4.5; and ✶, no catalyst.

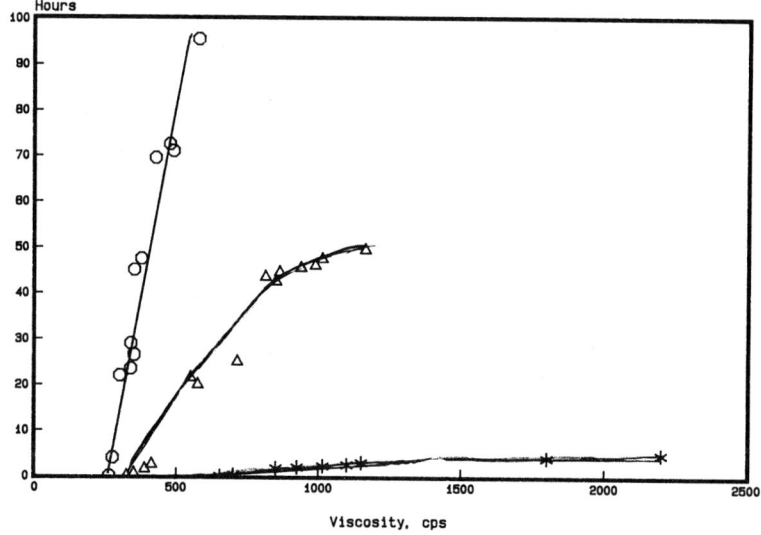

Figure 8. RXN with Mn(AA)$_3$ catalyst. Key(%Mn(AA)$_3$): △, .375; ✶, .75; and ⬡, no catalyst.

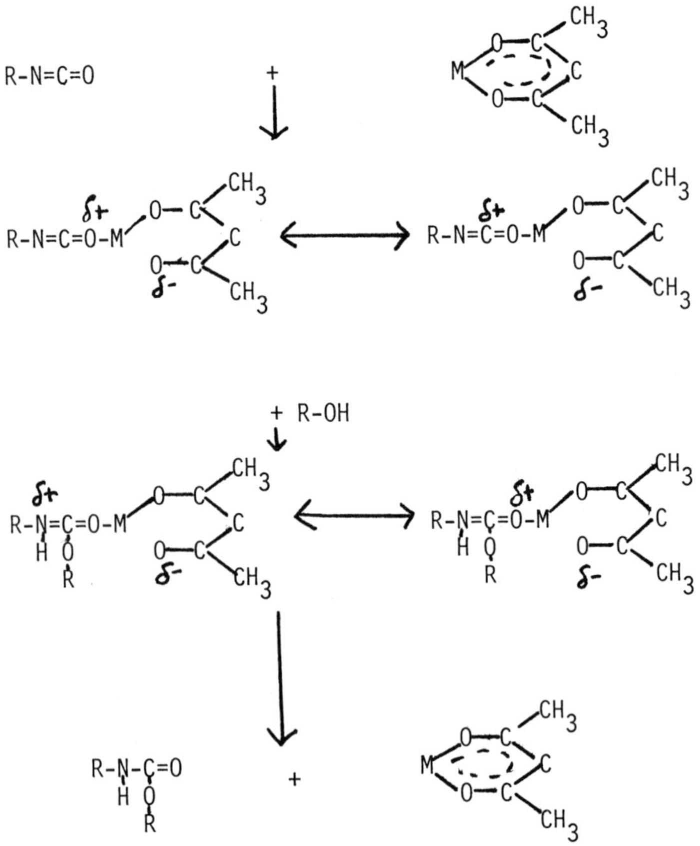

Figure 9. Reaction mechanism

In the system studied, the order of reactivity for the acetylacetonate complexes is:

$$Mn > Fe > Cu$$

In the late 1950's, Weisfeld (4) studied certain metal acetylacetonate in a polyester/disocyanate system. He found the order of reactivity of the metal complexes to be:

$$Cr > Cu > Co > Fe > V > Mn$$

Weisfeld's order of reactivity is directly opposed to the order studied in this work, although the same test method was used in both experiments. Weisfeld used monomeric isocyanate and polyester while this study included carboxyl terminated polyurethane and prepolymerized isocyanate.(5) This distinction between monomeric and polymeric materials could explain the distinct contrast in catalytic activity.

In line with this study, Mellor and Maley proposed a list of divalent ions and their tendency towards coordination. Here, Cu^{+2} tends to chelate stronger than Fe^{+2} which is a stronger chelating agent than Mn^{+2}. This neatly relates to the second ionization potentials Cu^{+2} = 20.2ev, Fe^{+2} = 16.2ev, and Mn^{+2} = 15.7ev. (6) This also correlates to the catalytic activity of the metal ions used with acetylacetonate in this work.

Conclusion

The onset of this study includes promoting the allophanate reaction in a polyurethane-polyisocyanate system. Using catalysts sited to influence this reaction (metal napthanates and octoates), test results show either (1) the allophanate reaction did not occur, or (2) the reaction did occur resulting in a weak linkage. Of interest, the allophanate promoting catalysts disrupted reactivity of catalysts known to influence the isocyanate-carboxyl reaction. Further studies were done to isolate whether the base metal or the ligand dictates the catalyst's effectiveness. Results show the catalyst reactivity is dependent on the metal, ligand and chemical composition of the ingredients.

Literature cited

1. Jorgensen, F., The Complete Handbook of Magnetic Recording; Tab Books, Inc: Blue Ridge Sumitt, PA, 1980; pp. 34-36.

2. Saunders, J. H. and K. C. Frisch. "Polyurethane Chemistry and Technology"; Robert Krieger Publishing Company: Huntington, NY, 1978, p. 197.

3. Allinger, Norman L., *Organic Chemistry*; Worth Publishing; N.Y., NY, 1976, p. 174.

4. Farkas, A. and G. A. Mills. "Advances in Catalysis and Related Subjects." 1962, 13.

5. Weisfield, L.B. "The Estimation of Catalytic Parameters of Metal Acetylacetonates in Isocyanate Polymerization Reaction." Journal of Applied Polymer Science; 1961, vol. 5, pp. 424-427.

6. Gould, Edwin S. *Inorganic Reactions and Structures*; Hole, Rinehart and Winston, N.Y., NY, 1965, pp. 342-3.

RECEIVED December 19, 1983

CONDUCTIVE POLYMERS

Fabrication of Conductive Polyimide-Gate Transistors

DAVID R. DAY

Department of Electrical Engineering and Computer Science, and Center for Materials Science and Engineering, Massachusetts Institute of Technology, Cambridge, MA 02139

> One route to achieve a useful conducting polymer is to combine the conduction properties of a graphite-like material with the processability of common polymers. Such a material can be realized by first processing a common polymer and then pyrolyzing it in its final form. An organic material which exhibits both excellent stability and high conductivity is heat-treated or pyrolyzed polyimide. Pyrolysis is normally carried out at temperatures of 650°C or above and conductivity typically increases from 10^{-18} to 10^2 $(ohm-cm)^{-1}$. Conductive polyimide may be useful in integrated circuits as a gate material because of its conductive properties and its ability to withstand high temperatures. Conductive polyimide gates for field effect transistors (FETs) were fabricated and have been found to be similar to conventional gate transistors. The processing steps and transistor characteristics are discussed.

Polyimide irreversibly forms a conductive material when heated above 650°C in an inert ambient (1-5). During the heating process, the conductivity increases from that of a good insulator, 10^{-18} $(ohm-cm)^{-1}$, to 100 $(ohm-cm)^{-1}$ (2). This final conductivity is comparable to that of doped polyacetylene. Unlike polyacetylene, however, conductive polyimide (CPI) is stable, i.e., it retains conductivity when exposed to air and moisture, and it can withstand temperatures greater than 1200°C in an inert ambient. Elemental analysis of the CPI indicates that significant amounts of oxygen and nitrogen remain in the carbonaceous matrix following pyrolysis and that it is not simply graphite (6). Several conduction mechanisms have been proposed including charging-energy-limited tunneling and variable-range hopping (2-4). In addition to its conductive properties, CPI is easily processed and lithographically patterned by either wet chemical or plasma etching before heat-treatment.

These properties make CPI useful as a gate for field-effect transistors in integrated circuits. Although CPI is not sufficiently conductive to be considered for high-current conduction lines in a circuit, it is an excellent candidate as a gate material. Since the main function of a gate is to deliver a voltage but pass only low amounts of current, extremely high conductivities such as those characteristic of common metals are not required, although the conductivity of the gate material does affect the speed of the device. The ability to withstand high temperatures is another important factor in the choice of a gate material as will be described below.

Background

Conventional Metal-Oxide-Semiconductor (MOS) transistors typically include polysilicon as a gate material. Polysilicon is often used instead of aluminum or some other metal to take advantage of the so-called self-aligned-gate process. In this process the gate is used as a mask during the source-drain implant so that the edge of the source and drain align to the edge of the gate (Figure 1). This minimizes the overlap capacitance between the gate and either source or drain. During the required subsequent anneal of the source-drain implant (1000 to 1200°C), a passivation oxide can be grown over the polysilicon (Figure 1).

CPI can also be used in a self-alignment gate process. Although CPI is not self-passivating, <u>non-conducting</u> polyimide can be deposited over the CPI following the source-drain anneal to act as an insulation layer (7). Thus CPI can be incorporated in a process sequence to provide the same benefits as polysilicon technology. The CPI technology, however, is more easily carried out and demands fewer process steps. A conventional polysilicon gate process proceeds as follows:

1. vapor deposit polysilicon on wafer
2. deposit photoresist and pattern
3. etch polysilicon
4. remove photoresist
5. implant wafer to dope source, drain and polysilicon
6. high temperature anneal in oxygen to form source, drain, and conductive polysilicon gate
7. deposit and pattern photoresist
8. etch contact cuts in oxide
9. remove photoresist

This is not an entire process but it suffices for comparison to the CPI process. In the CPI process described below, a photosensitive polyimide (currently available from EM Chemical Co.) is proposed to simplify processing. An analogous CPI process

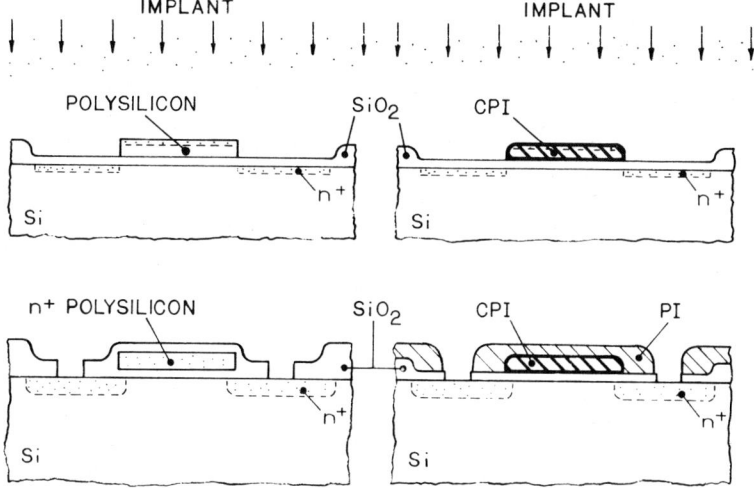

Figure 1. Comparison of conventional polysilicon-gate process to proposed CPI-gate process.

starting at the same point as above for the polysilicon would be as follows:

1. deposit and pattern polyimide (to become gate material)
2. implant to dope source and drain (polyimide is implanted but at this stage remains nonconductive)
3. high temperature anneal in nitrogen to form source, drain, and conductive polyimide gate
4. deposit and pattern polyimide (this will serve as insulation)
5. etch contact cuts

Note that the proposed CPI process segment contains four fewer steps. A schematic of both structures after this partial processing is indicated in Figure 1.

The purpose of this research was to determine the feasibility of incorporating CPI into field-effect transistors. Due to time limitations and availability of various equipment, source and drains were formed by thermal diffusion rather than by ion implantation as proposed in the preceded discussion. As a result of this slight change in process sequence, metal-gate FETs could be fabricated as a control for the CPI devices. In addition, a non-photosensitive PI was used thus requiring extra photoresist deposition and patterning steps.

Experimental

P-channel field-effect transistors were fabricated both with CPI gates and with aluminum gates as controls. After conventional processing up to and including growth of the gate oxide, test wafers were coated with polyamic acid (DuPont PI-2545) while the controls were coated with aluminum. The polyamic acid was imidized with a final bake at $390^\circ C$. A photoresist mask was deposited and patterned on both the test and control wafers. The polyimide was plasma etched in oxygen and the aluminum was etched in a phosphoric-acetic-nitric solution (PAN ETCH). After photoresist removal, the polyimide test wafers were heat treated at $700^\circ C$ for 15 minutes. Both CPI and aluminum gate wafers were then coated with polyamic acid again and imidized at $390^\circ C$. Contact cuts were plasma etched through the polyimide (using a photoresist mask) and buffered HF was used to etch the silicon dioxide underneath. After aluminum was deposited and patterned, both test and control wafers were sintered in forming gas at $400^\circ C$ for 10 minutes. A CPI strip was fabricated beside each FET with four aluminum contacts to allow CPI ´4 point´ conductivity measurements.

A summary of the wafer process is given below:

1. CLEAN WAFERS (organic, oxide removal, inorganic)
2. GROW FIELD OXIDE($1100^\circ C$, O_2, H_2O, O_2)

3. A) DEPOSIT PHOTORESIST, PREBAKE
   B) EXPOSE AND DEVELOP, POSTBAKE
   C) ETCH SILICON DIOXIDE   PHOTOLITHOGRAPHY
   D) STRIP PHOTORESIST
   E) CLEAN WAFER
4. DOPANT DEPOSITION (boron)
5. BORON GLASS REMOVAL (HF)
6. DOPANT DRIVE-IN (1200°C) AND OXIDIZE SOURCE-DRAIN
7. PHOTOLITHOGRAPHY - ETCH GATE REGION
8. GROW GATE OXIDE (1100°C), O_2)   CONTROL WAFERS
9. POLYIMIDE DEPOSITION   9. AL DEPOSITION
10. PHOTOLITHOGRAPHY - POLYIMIDE   10. PHOTOLITH- AL
11. HIGH TEMPERATURE POLYIMIDE BAKE (700°C) 11. CLEAN WAFER
12. POLYIMIDE deposition
13. PHOTOLITHOGRAPHY - POLYIMIDE
14. SOURCE-DRAIN CONTACT ETCH
15. ALUMINUM DEPOSITION
16. PHOTOLITHOGRAPHY
17. ANNEAL (400°C)

RESULTS AND DISCUSSION

The patterned and subsequently pyrolyzed polyimide films on silicon wafers adhered well to the substrate without cracking or peeling (Figure 2). The measured conductivity of the CPI was 10 $(ohm-cm)^{-1}$. This value is about ten times lower than that reported in reference (2) and is likely due to the relatively short test wafer heating period of 30 minutes. It is suspected that if higher temperatures or extended times during heat treatment are used, conductivity should increase by a factor of 10.

A schematic and photograph of the fabricated transistor are shown in Figure 3. Note that the control devices are identical to the test devices except that the CPI is replaced with aluminum. As a test for integrity of the second polyimide insulation layer, the source electrode was made to cross directly over the CPI gate. In both the test and control devices, no shorting of source to gate electrodes was observed, an indication of good insulation between the aluminum and the underlying CPI. Although most contacts from aluminum to CPI and aluminum to silicon were good, a few were totally open. This observation suggests that some residue may have remained in the bottoms of the vias after the plasma contact etch (9).

The electrical characteristics of the CPI- and aluminum-gate transistors are illustrated in Figure 4. The drain current-gate voltage characteristics are similar for both devices. The CPI device, however, has a slightly higher threshold voltage and never completely turns off. This may result from contamination of the gate oxide from impurities in the polyimide or from the pyrolysis chamber in which the heat treatment was performed.

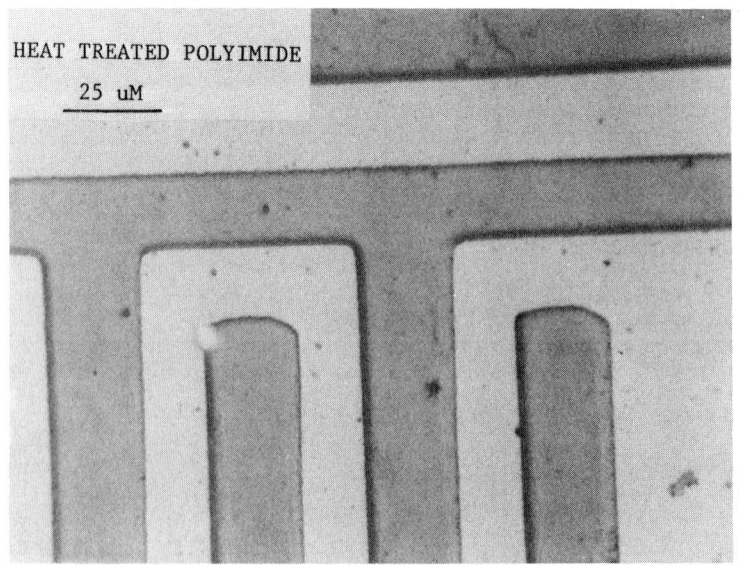

Figure 2. Patterned and pyrolyzed CPI on SiO_2 substrate.

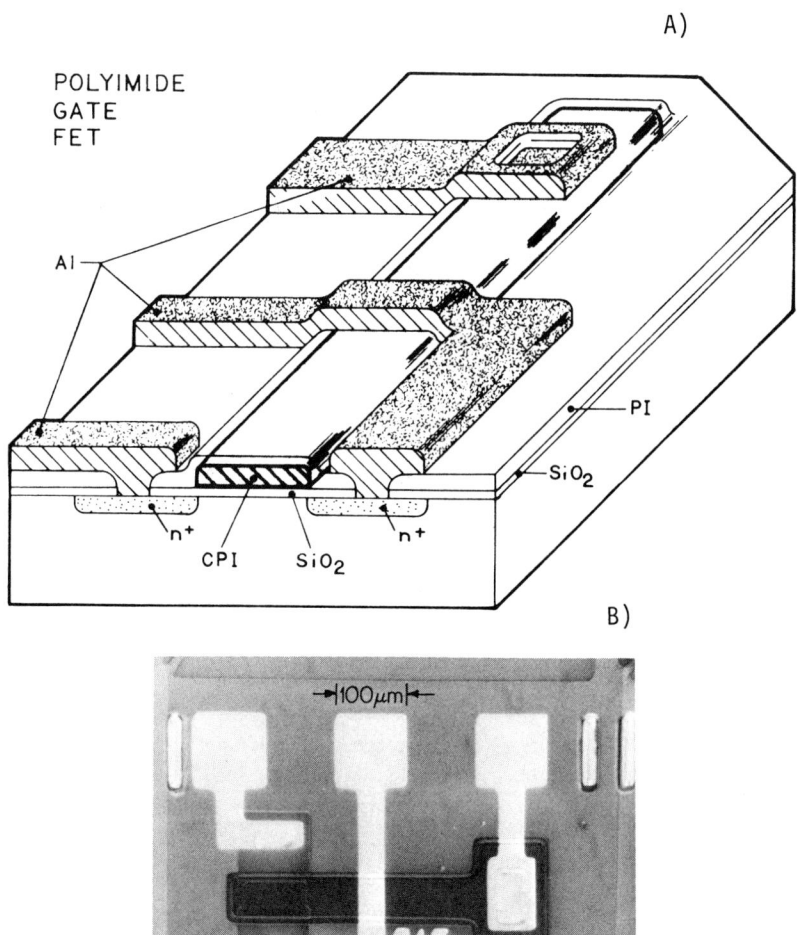

Figure 3. A) Schematic of CPI-gate FET.
B) Optical micrograph of CPI-gate FET (dark material is CPI).

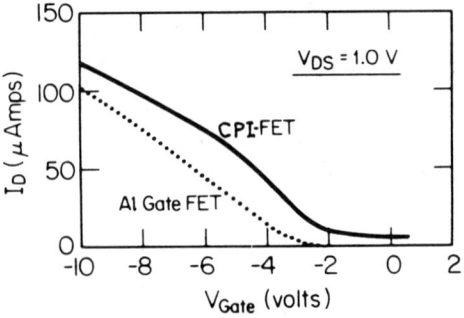

Figure 4. Comparison of drain current for both CPI- and aluminum-gate FETs.

The difference between the CPI- and aluminum-gate devices is minimal and suggests that CPI is a good candidate as a gate material. Although the absolute conductivity of CPI is less than the more conventional polysilicon, CPI processing is much simpler and cheaper than polysilicon technology. To become competitive in device ´speed´ with polysilicon (and the current heavily researched silicides), CPI would have to have equivalent conduction levels. If the conductivity of CPI could be increased by 2 or 3 orders of magnitude through alteration of polyimide chemistry, incorporation of more conductive graphite particles, or possibly by doping, CPI gate technology could become extremely useful.

Acknowledgments

The author thanks E.W. Maby and S.D. Senturia for their helpful comments and discussion and DuPont de Nemours and Company for providing the polyamic acid solutions.

Literature Cited

1. Bruck, S.D. J. of Polymer Sci. (part C) 1967, 17, 169
2. Brom, H.B.; Tomkiewitz, Y.; Aviram, A.; Broers, A. Solid State Comm. 1980, 35, 135
3. Gittleman, J.I.; Sichel, E.K. J. Electronic Mat. 1981, 10, 327
4. Sichel, E.K.; Emma, T. Solid State Comm. 1982, 41, 747
5. Lin, J.W.P.; Epstein, A.J.; Dudek, L.P.; Rommelmann, H.; Organic Coatings and Plastics Chemistry (ACS) 1980, 43, 446
6. Conley, R.T.; Guadiana, R.A. ´´Thermal Stability of Polymers´´; M. Dekker Inc.: New York, 1970, chap. 10
7. See for example Samuelson, G.; Organic Coatings and Plastics Chemistry (ACS) 1980, 43, 446
8. Day, D.R.; Senturia, S.D. J. of Electronic Mat. 1982, 11, 441

RECEIVED September 2, 1983

34

Redox Properties of Conjugated Polymers
A Successful Correlation of Theory and Experiment

R. R. CHANCE and D. S. BOUDREAUX

Allied Corporation, Morristown, NJ 07960

J. L. BREDAS[1] and R. SILBEY

Department of Chemistry, Massachusetts Institute of Technology, Cambridge, MA 02139

> The electrochemical properties of conductive
> polymer systems are important with regard to
> understanding the electrochemical doping process
> and in applications of conductive polymers as
> battery electrodes. We have developed a computa-
> tional method, based on the Valence Effective
> Hamiltonian technique, which is remarkably
> effective in the computation of oxidation and
> reduction potentials of a variety of conjugated
> polymers (polyacetylene, polyphenylene,
> polythiophene, polypyrrole) and their oligomers.
> For example, the VEH results for polyacetylene
> yield an oxidation potential of 0.4 volts versus
> SCE and a reduction potential of -1.1 volts versus
> SCE, both of which are in good agreement with
> experiment. Similar agreement is found with the
> other systems. Of special importance is the
> successful prediction of the surprisingly low
> oxidation potential of polypyrrole (-0.4 volts
> versus SCE).

A number of organic polymers become electrically conducting on
addition of electron donors or acceptors.(1-5) Despite the enor-
mous interest in these conducting polymer systems, many theoretical aspects of the problem remain poorly understood,
especially with regard to the electronic properties of the
"doped" (partially ionized) polymers. Progress is being made,
however, in understanding the undoped polymer precursors. In a
series of recent papers, we have demonstrated the utility of the

---

[1]Current address: Lab. Chemie Theorique Appliquee Fac. Univ. NOTRE-DAME de la Paix, Namur, Belgium.

Valence Effective Hamiltonian (VEH) method in describing the
ground state properties of conjugated polymers—in particular
those which become highly conducting upon doping.(6-9) The VEH
method employs atomic potentials derived from double zeta
quality ab initio computations on small molecules in calculations on large molecules. With this method, x-ray photoelectron spectra, ionization potentials, and optical band gaps
for a number of polymers have been computed in good agreement
with experiment.(6-9) The successful calculation of polymer
ionization potentials is especially important, since the
ionization potential determines whether or not a particular
acceptor dopant will be capable of ionizing the polymer.

In this paper we present VEH calculations on oligomers of
polyacetylene (PA), poly(p-phenylene) (PPP), polythiophene
(PTP), and polypyrrole (PPY) and compare the results to VEH
polymer calculations. For PTP and PPY, considerable uncertainty exists for their geometry-particularly in the carbon-
carbon bonds between monomer units. For this reason we have
employed a semiempirical quantum mechanical technique designed
for geometry prediction, MNDO (Modified Neglect of Differential
Overlap)[10]. The MNDO predicted geometry is used as input to
the VEH program. Our calculations concentrate on optical "band
gaps" and ionization potentials and use optical and electrochemical data for comparison. We emphasize the successful prediction of electrochemical oxidation and reduction potentials,
which is especially important considering recent activity
in the application of conjugated polymers in rechargeable
batteries[11,12].

Theoretical and Computational Technique

In this section we discuss briefly the VEH technique and
our approach to the problem including the use of MNDO as a
geometry input device. The methodology for obtaining molecular
one-electron Hamiltonians from first principles has been
worked out by Nicolas and Durand.(13) The effective Fock
operator of the molecule is assumed to be the sum of the
kinetic energy and the various atomic potentials in the
molecule:

$$F_{eff} = -\frac{\Delta}{2} + \sum_A V_A \qquad (1)$$

where V_A is the effective potential of atom A. For
computational ease, simple nonlocal atomic potentials are
chosen of the form of Gaussian projectors:

$$V_A = \sum_\ell \sum_m \sum_{ij} C_{ij,\ell m} | \chi^A_{i\ell m} \quad \chi^A_{j\ell m} | \qquad (2)$$

where the summations over ℓ and m define the angular dependence of V_A. The numerical coefficients $C_{ij,\ell m}$ are independent of m in the case of spherical symmetry, which we usually consider. The functions are normalized Gaussians:

$$\chi_{i\ell m} = N_i \exp[-\alpha_i r^2] Y_{\ell m}(\theta,\phi) \qquad (3)$$

N_i is the normalization factor and $Y_{\ell m}$ denotes the usual spherical harmonics. Note that only 1s and 2p Gaussian Cartesian functions are used.

The parameterizations of the linear coefficients, C, and the nonlinear exponents, α, first require valence SCF calculations on model molecules by a theoretical pseudopotential method([14](#)) with an STO-3G minimal basis set and a double zeta basis set. The model molecules chosen to parameterize carbon and hydrogen atomic potentials were ethane, transbutadiene, and acetylene([6](#)); for sulfur and carbon linked to sulfur, dimethyl sulfide and thiophene([8](#)); for nitrogen and carbon linked to nitrogen, dimethylamine and pyrrole.([15](#)) For each molecule, the Fock operator is constructed as:

$$F = \sum_\upsilon \varepsilon_\upsilon | \phi_\upsilon \quad \phi_\upsilon | \qquad (4)$$

where the summation is over all occupied states; the valence orbitals ϕ_υ are taken from the minimal basis set calculation and the corresponding monoelectronic energies ε_υ from the double zeta calculation. The choice of this theoretical Fock operator leads to valence effective Hamiltonians providing double zeta accuracy for monoelectronic energies when solved with a minimal set. The parameterization of the atomic potentials is then determined by minimizing the quantity

$$\sum_B (F-F_{eff} | F-F_{eff})_{\text{molecule B}} \qquad (5)$$

where the summation runs over the model molecules used for a given set of atomic potentials. $(F-F_{eff} | F-F_{eff})$ denotes the scalar product of $F-F_{eff}$ with itself in the subspace of the occupied valence orbitals.([13](#)) On the model molecules, standard deviations between the ε_υ energies produced by using the valence effective Hamiltonians and double zeta energies are of the order of 0.015 a.u., and in no case larger than 0.2 a.u.([6,8,15](#)). The position of the highest occupied orbital is especially well-reproduced - a result which lends confidence in obtaining good ionization potential estimates.

No information pertaining to the excited states is included in the atomic potentials. As a result, no special attention should in principle be given to the unoccupied levels. However, for the planar systems considered previously, surprisingly good agreement between experiment and theory has been obtained for the lowest optical energy transitions.(7)

The extension of the VEH method to polymer calculations is straightforward(6,16). The effective operator takes the form:

$$F_{eff} = -\frac{\Delta}{2} + \sum_g \sum_A V_A \tag{6}$$

the summations over g and A running, respectively, over the polymer unit cells and the atoms present in one cell. The band structure E(k) of the polymer, where k is a point in the first Brillouin zone of the polymer, is obtained from eigenvalues of the set of secular equations:

$$\underline{F}(k) \ \underline{C}(k) = \underline{S}(k) \ \underline{C}(k) \ \underline{E}(k) \tag{7}$$

$\underline{F}(k)$ and $\underline{S}(k)$ are the Fock and overlap matrices between Bloch functions and $\underline{C}(k)$ collects the coefficients of the linear combinations of Bloch functions that provide the crystalline orbitals.

The main advantages of the VEH technique are that it is completely theoretical and gives ab initio double zeta quality results with negligible computer time, since only one-electron integrals need to be evaluated and SCF iterative cycles are completely avoided.

It must be pointed out that the VEH atomic potentials have not been parameterized for geometry optimization purposes and should be used with geometric parameters close to equilibrium. For systems whose geometries are experimentally unknown (as is the case of the majority of the large oligomers and polymers studied in this paper), we must make use of other techniques in order to obtain reasonable input geometries. Ab initio techniques, even with small basis sets, rapidly become too expensive when large compounds are considered. As a result, we have chosen to optimize the geometries of all oligomers and polymers studied here with the MNDO (Modified Neglect of Differential Overlap) semiempirical procedure.(10) This method has been thoroughly tested on organic compounds containing carbon, hydrogen, nitrogen, oxygen, and sulfur and reproduces fairly well the experimental geometries.(17)

We find that the VEH results are qualitatively unaffected by small differences in input geometries. This is illustrated for polypyrrole in Table I where we present the VEH ionization potentials and band gaps for four different geometries: the MNDO optimized geometry, the STO-3G optimized geometry(18), and the experimental geometry of pyrrole(19) with inter-ring bond lengths of 1.45A and 1.49A. The ionization potential values fall between 5.68 eV and 5.98 eV and band gaps between 3.0 eV and 4.0 eV.

TABLE I: Evaluation of the VEH ionization potential (IP) and band gap (E_g) values for polypyrrole as a function of the geometry: MNDO optimized geometry (this work), STO-3G optimized geometry(18), and experimental ring geometry (for pyrrole)(19) with inter-ring bond lengths of 1.45A and 1.49A.

geometry:	MNDO	STO-3G	MONOMER EXP.	
R_{C-N}(A)	1.399	1.385	1.38	1.38
$R_{C=C}$(A)	1.410	1.363	1.37	1.37
R_{C-C}(A)	1.426	1.420	1.43	1.43
C-N-C(°)	110.8	109.2	109.0	109.0
C-C-N(°)	106.5	107.4	108.0	108.0
$R_{inter-ring}$(A)	1.453	1.474	1.45	1.49
IP (eV)	5.68	5.96	5.82	5.98
E_g (eV)	3.0	3.9	3.6	4.0

Results and Discussion

Our results are summarized in Table II. Included in the table are VEH computed ionization potentials (IP) and band gaps (E_g), and experimental band gaps and gas phase ionization potentials when available. We have shown previously that the IP values from VEH theory for the polymers, after subtracting ~1.9 eV for a solid state polarization correction, are in good agreement with solid-state IP values. From Table II the IP values for monomers and dimers are also seen to be in reasonable agreement with gas phase values. The agreement of VEH theory and experiment for optical band gaps is also quite good. (For convienence, we use the term "bandgap" even when referring to the optical absorption peak of the oligomers.) As we have noted previously, no attempt has been made to design VEH theory to treat excited states. Thus the relatively good agreement between theory and experiment evident in Table II and in previous work(7) is unexpected.

TABLE II: Gas Phase Ionization Potentials (IP) and Band Gaps (E_g). All energies are given in eV.

System	Chain Length	VEH IP	E_g	Experiment IP	E_g	Ref.
Polyacetylene	1	10.16	7.85	10.5	–	20
(PA)	2	8.78	5.29	9.1	5.7	20,21
	3	8.12	4.09		4.6	21
	∞	6.67	1.45		1.8	22
Poly(p-phenylene)	1	9.30	6.69	9.2	5.9	23,24
(PPP)	2	8.34	4.88	8.3	4.9	23,24
	3	7.97	4.18	8.2	4.4	23,24
	∞	7.45	3.23		3.6	2
Polythiophene	1	9.29	5.51	9.0	5.4	20,24
(PTP)	2	8.14	3.65		4.1	24
	3	7.71	2.93		3.5	24
	∞	7.01	1.71			
Polypyrrole	1	8.13	7.01	8.2	6.0	25,24
(PPY)	2	6.89	5.24		4.4	25,24
	3	6.43	4.35		3.6	25,24
	∞	5.68	2.99		3.0	1

In Figure 1, we have plotted the experimental and theoretical data for the band gap of the polyenes versus $1/n$. A linear relationship is expected for large n. Also included in Figure 1 are data for diphenylpolyenes (DPP), $C_6H_5-(HC=CH)_x-C_6H_5$, which are more stable than polyenes, $H-(HC=CH)_n-H$; consequently, there

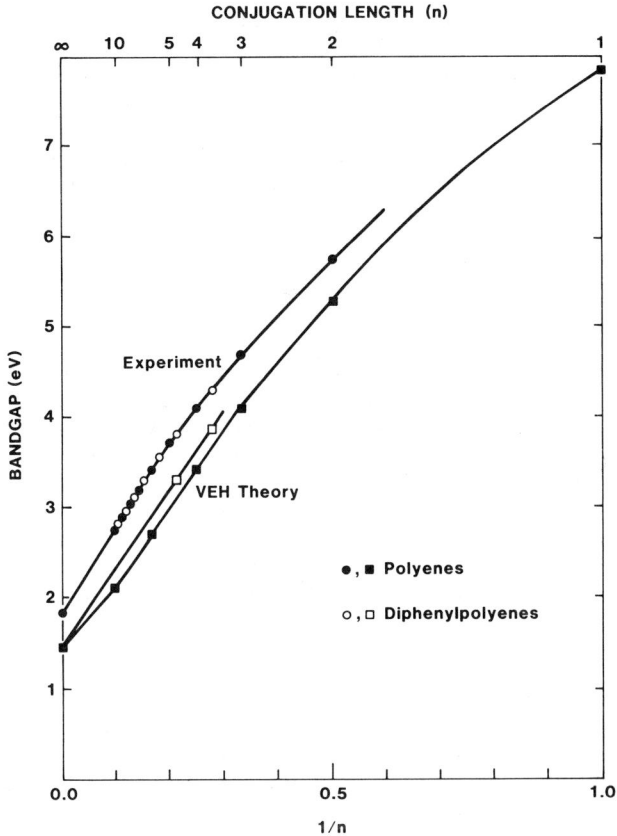

Figure 1. Band gap for polyenes, (21) $H(HC=CH)_nH$ and diphenylpolyenes (DPP), (27,28) $C_6H_5-(HC=CH)_x-C_6H_5$, plotted against reciprocal conjugation length. The highest band gap DPP result is for stilbene (x=1); the lowest DPP result is for x=7. For DPP, n is the effective conjugation length computed as x+2.7. All experimental data refer to absorption peaks in nonpolar organic solvents.

is more extensive literature(26-28) on DPP molecules. When considering DPP as a model for the polyacetylene system, the effect of the phenyl end groups on the electronic properties must be considered. Accordingly, we define an effective conjugation length, $n_{eff}=x+A$, so that A describes the extension of the conjugation length by the phenyl end groups beyond the x-unit polyene sequence. The value of A is established by adjusting it until the experimental band gap of DPP with x double bonds is equal to that of a polyene with x+A double bonds.[29] The results shown in Figure 1 use A=2.7 and provide an excellent correspondence between the polyene and diphenylpolyene data. We have used this approach previously to show that the three optical transition energies observed for radical anions in DPP extrapolate to yield a good description of the near-infrared ("midgap") absorption found with donor-doping of polyacetylene.(29)

VEH calculations have been carried out for two DPP molecules (x=1 and x=2). As can be concluded from Figure 1, the VEH computed band gaps are in reasonable agreement with the polyene data. IP results are summarized in Figure 2. First note that the theoretical IP values for the two DPP molecules are in near perfect agreement with those derived from VEH polyene values with use of the same A=2.7 correction factor noted above. It is also clear from Figure 2 that theoretical IP values for the DPP system, including those inferred from the polyene results, are in excellent agreement with the experimental IP data of Hudson et al.(30) The agreement is as good as that obtained with the CNDO/S2 semiempirical procedure,[31] but without any adjustment of energy scales and, more importantly, without any input of experimental information (except, of course, the geometry). In fact, including our MNDO geometry determination, the excellent agreement with experiment displayed in Figures 1 and 2 and Table II is obtained with input of only the atom connectivity.

A principle reason for considering the DPP molecules in such detail is that a fairly complete set of oxidation(32) and reduction potentials(26) is available. We are interested in the extent to which such data can be predicted theoretically and in applying theory and oligomer extrapolations to understanding the electrochemical properties of polyacetylene. The latter is especially important because of the large interest in battery applications of polyacetylene[11] and other conjugated polymers.(12)

In Figure 3, we have plotted the theoretical values of the ionization potentials for the polyenes and DPP versus 1/n; on the same plot are experimental oxidation potentials (E_{ox}) for DPP. An approximately linear correlation between IP and E_{ox} is expected.(33) Furthermore, we have plotted the theoretical

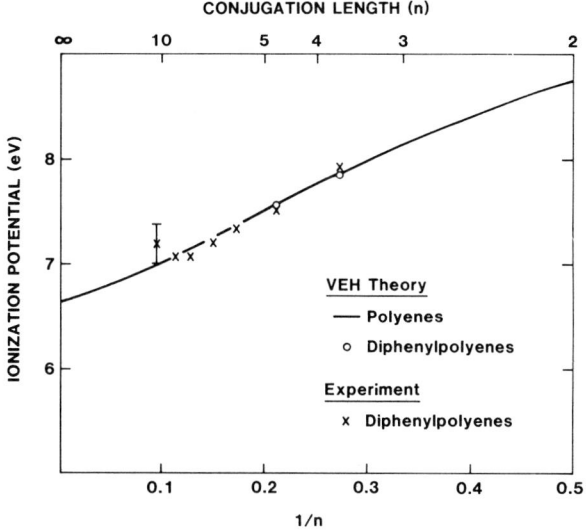

Figure 2. VEH computed ionization potentials for polyenes (solid line) and diphenylpolyenes (open circles) compared to experimental data(X) from reference 30. Diphenylpolyene results, data, theory, and experiment are plotted with n defined as in Figure 1.

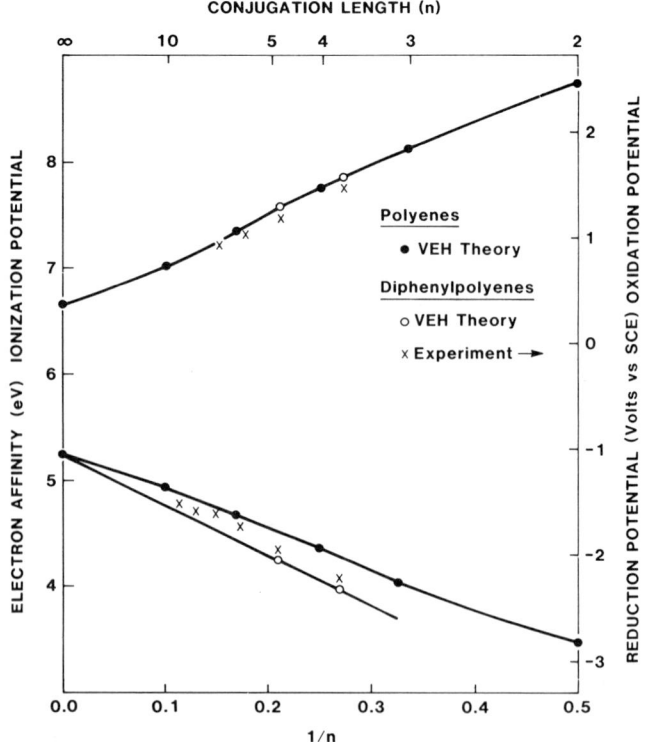

Figure 3. Electron affinity (lower curve) and ionization potential (upper curve) from VEH theory versus 1/n for polyacetylene system; experimental reduction potentials (26) (lower X's) and oxidation potentials (32) (upper X's) versus 1/n with n defined as in Figure 1.

values of the electron affinity ($\chi \equiv IP-E_g$) versus 1/n for these compounds and compared these to the DPP reduction potentials. The oxidation and reduction potentials are given with respect to the saturated calomel electrode, SCE. Best agreement between theory and experiment is obtained by adjusting the zero on the SCE scale to 6.3 volts on the IP scale. This is in excellent agreement with previous work.(33)

The agreement between experiment and theory is remarkably good, especially since we have inherently assumed a chain length independent polarization energy correction to the theoretical values of IP. (Other work showing a nearly linear relation between IP and E_{ox} also suggests a relatively constant polarization energy for a wide variety of organic molecules[33].) Since the agreement in Fig. 3 is so convincing, we may now consider the implications regarding the polymer ($n \rightarrow \infty$). Results for the four polymers considered here are given in Table III. E_{ox} is computed as IP-6.3 and E_{red} as IP-E_g-6.3. (The difference is precisely E_g by construction). Note that these redox values refer to removal or addition of <u>one electron</u> from the polymer. Thus, the theoretical E_{ox} or E_{red} values would correspond to <u>onset</u> values in terms of the observed oxidation or reduction of the polymers. For comparison to theory, experimental values must therefore be taken from redox onsets.(24,34-36) In all cases agreement between theory and experiment is quite satisfactory. Note especially the successsful calculation of the surprisingly low E_{ox} for PPY. The bonding patterns in both

Table III: Polymer oxidation and reduction potentials (volts versus SCE): Onset values from VEH theory and experiment.

	Oxidation Potential		Reduction Potential	
	VEH Theory	Experiment(ref)	VEH Theory	Experiment(ref)
Polyacetylene (PA)	0.4	0.2(34,35)	-1.1	-1.2(34,35)
Poly(p-phenylene) (PPP)	1.2	0.9(34)	-2.1	-2.1(34)
Polythiophene (PTP)	0.7	0.6(24)	-1.0	
Polypyrrole (PPY)	-0.6	-0.4(36)	-3.6	--

PPY and PTP mimic that in cis PA. In fact, the highest valence band of PPY (PTP) contains no contribution from the nitrogen (sulfur) atoms. Although it is surprising that the IP value for PPY is so low (0.9 volts lower than that of cis PA), this result is supported by available experimental data.(36)

E_{red} for PPY in Table II is predicted to be quite negative, -3.6 volts (0.3 volts more negative than lithium) and would be difficult to observe due to electrolyte stability problems. Attempts to dope PPY with sodium naphthalide have been unsuccessful(37)--a result which is quite consistent with our calculations, since it suggests a E_{red} more negative than the -2.9 volt E_{red} of naphthalene.

In conclusion, the results of our calculations indicate that we may now predict with confidence the onset potential of oxidation and reduction for conjugated polymers. This capability is particularly important given the current interest in polymer batteries. However, more needs to be done. For example, the open circuit voltage of acceptor-doped polyacetylene/Li battery cell varies from 3.2 V at low charge to about 3.8 V at high charge (~6% doping). The present theory predicts the voltage at low charge extremely well. In order to predict the variation in voltage with charge, theory must include multiple ionization steps, i.e. knowledge of the complete band structure including all relaxation effects. Though such a theory is currently unavailable, we believe that the results of the present theory and some knowledge of the π electron bandwidth will prove sufficient in understanding the electrochemical properties of these polymeric systems.

Acknowledgments

We thank L. W. Schacklette, J. F. Wolf, and J. E. Frommer for helpful discussions.

Literature Cited

1. K. K. Kanazawa, A. F. Diaz, R. H. Geiss, W. D. Gill, J. F. Kwak, J. A. Logan, J. F. Rabolt, and G. B. Street, J. Chem. Commun. 1979, 854.
2. L. W. Shacklette, R. R. Chance, D. M. Ivory, G. G. Miller and R. H. Baughman, Synth. Metals 1, 307 (1979).
3. H. Shirakawa, E. J. Louis, A. G. MacDiarmid, C. K. Chiang, and A. J. Heeger, J. Chem. Soc. Chem. Commun. 1977, 578.
4. J. F. Rabolt, T. C. Clarke, K. K. Kanazawa, J. R. Reynolds, and G. B. Street, J. Chem. Soc. Chem. Commun. 1980, 347.
5. R. R. Chance, L. W. Shacklette, G. G. Miller, D. M. Ivory, J. M. Sowa, R. L. Elsenbaumer, and R. H. Baughman, J. Chem. Soc. Chem. Commun. 1980, 348.
6. J. L. Bredas, R. R. Chance, R. Silbey, G. Nicolas, and Ph. Durand, J. Chem. Phys. 75, 255 (1981).

7. J. L. Bredas, R. R. Chance, R. H. Baughman, and R. Silbey, J. Chem. Phys. 76, 3673 (1982).
8. J. L. Bredas, R. R. Chance, R. Silbey, G. Nicolas, and Ph. Durand, J. Chem. Phys. 77, 371 (1982).
9. J. L. Bredas, R. L. Elsenbaumer, R. R. Chance, and R. Silbey, J. Chem. Phys. 78, 5656 (1983).
10. M. J. S. Dewar and W. Thiel, J. Am. Chem. Soc. 99, 4899 (1977).
11. D. MacInnes, M. A. Druy, P. J. Nigrey, D. P. Nairns, A. G. MacDiarmid, and A. J. Heeger, J. Chem. Soc. Chem. Commun. 1981, 317.
12. L. W. Shacklette, R. L. Elsenbaumer, R. R. Chance, J. M. Sowa, D. M. Ivory, G. G. Miller, and R. H. Baughman, J. Chem. Soc. Chem. Commun. 1982, 361.
13. G. Nicolas and Ph. Durand, J. Chem. Phys. 70, 2020 (1979); 72, 453 (1980).
14. Ph. Durand and J. C. Barthelat, Chem. Phys. Lett. 27, 191 (1974).
15. J. L. Bredas, B. Themans, and J. M. Andre, J. Chem. Phys. 78, 6137 (1983).
16. J. M. Andre, L. A. Burke, J. Delhalle, G. Nicolas, and Ph. Durang, Int. J. Quantum Chem. Symp. 13, 283 (1979).
17. M. J. S. Dewar and W. Thiel, J. Am. Chem. Soc. 99, 4907 (1977).
18. J. L. Bredas, B. Themans, and J. M. Andre, unpublished results.
19. B. Bak, D. Christensen, L. Hansen, and J. Rastrap-Anderson, J. Chem. Phys. 24, 720 (1956).
20. G. Herzberg, Electronic Spectra of Polyatomic Molecules (Van Nostrand, New York, 1976) pp. 629,654, and 656.
21. B. S. Hudson, B. E. Kohler, and K. Schultan in Excited States Vol 5, ed. by E.C.Lim (Academic, New York, 1982) p.1.
22. T. C. Chung, A. Feldblum, A. J. Heeger, and A. G. MacDiarmid, J. Chem. Phys. 74, 5504 (1981).
23. J. Maier and D. W. Turner, Farad, Disc. 54, 1419 (1972).
24. A. F. Diaz, J. Crowley, J. Baryon, G. P. Gardini, and J. B. Torrance, J. Electroanal. Chem. 121, 355 (1981).
25. W. Ford, C. B. Duke, and W. R. Salaneck, J. Chem. Phys. 77, 5020 (1982).
26. C. J. Hoijtink and P. H. van der Meij, Phys. Chem. Neue Folge 20, 1 (1959).
27. E. D. Schmid and R. D. Topson, J. Am. Chem. Soc. 103, 1628 (1981).
28. Rodd's Chemistry of Carbon Compounds Vol III, ed. S. Coffey (Elsevier, New York, 1974) p. 222.
29. J. L. Bredas, R. R. Chance, and R. Silbey, Mol. Cryst. Liq. Cryst. 77, 319 (1981); Phys. Rev. B 26, 5843 (1982).
30. B. S. Hudson, J. Ridyard, and J. Diamond, J. Amer. Chem. Soc. 98, 1126 (1976).
31. K. L. Kip, N. Lipari, C. Duke, B. Hudson, and J. Diamond, J. Chem. Phys. 64, 4020 (1976).

32. *CRC Handbook Series in Organic Electrochemistry*, Vol. 1, (CRC Press, Cleveland, 1977) pp. 614, 678, 736, 778.
33. L. L. Miller, G. D. Nordblom, and E. A. Mayeda, J. Org. Chem. $\underline{37}$, 916 (1972).
34. L. W. Shacklette, unpublished results.
35. P. J. Nigrey, A. G. MacDiarmid, and A. J. Heeger, Mol. Cryst. Liq. Cryst. $\underline{83}$, 309 (1982).
36. G. B. Street, T. C. Clarke, M. Krounbi, K. Kanazawa, V. Lee, P. Pfluger, J. C. Scott, and G. Weiser, Mol. Cryst. Liq. Cryst. $\underline{83}$, 253 (1982).
37. G. B. Street, personal communication.

RECEIVED September 23, 1983

A Novel Phase of Organic Conductors: Conducting Polymer Solutions

J. E. FROMMER, R. L. ELSENBAUMER, and R. R. CHANCE

Allied Corporation, Morristown, NJ 07960

>Arsenic trifluoride has been shown to be an active participant in the doping of poly(paraphenylene sulfide) to the electrically conducting state with AsF_5. When introduced in the vapor phase, arsenic trifluoride greatly increases both the doping rate and the homogeneity of dopant dispersion throughout the polymer sample. As a liquid, it serves the role of solvent in providing the first example of a conducting polymer solution. These solutions can be cast into films with properties that excel those of polymers doped by conventional gas/solid techniques. Improved mechanical behavior, homogeneous distribution of dopant, and higher conductivities characterize the electrically conductive materials that result from this process. This discovery of a conducting polymer solution has made available a new liquid phase of these novel materials, as well as a means of processing them into useful designs.

A primary goal in designing conducting polymers is to blend the optimal properties of plastics and metals into one material. Plastics offer facile processibility, flexibility, and stability - characteristics attractive for insulators of electrically conducting materials. The emergence of "doped" plastics which can conduct electricity dates back to 1977. Much activity has ensued in designing new electro-active polymeric systems from both a computational, theoretical approach and a synthetic, laboratory approach. The latter has produced several organic polymers of various backbone repeat units, e.g., olefin, phenyl, phenyl sulfide, phenyl oxide, and pyrrole. These polymers, once doped to the conducting state, all have in common that the doping process renders them intractable and unprocessible. This final,

0097-6156/84/0242-0447$06.00/0
© 1984 American Chemical Society

unwieldly state not only hinders design for conducting polymer
applications, but also prevents analyses by most standard procedures. In this article we report the discovery of <u>conducting polymer solutions</u> and the successful development of a method for the facile processing of certain conducting polymers which produces materials that are amenable to scientific analysis and practical application. This process yields electrically conductive materials with improved mechanical behavior, more homogeneous distribution of dopant, and higher conductivity.

Most polymeric organic conductors are formed in a process of treating the neutral polymer with electron donors or acceptors. For example poly(p-phenylene sulfide) (PPS), on exposure to strongly oxidizing dopants, forms a conducting complex($\underline{1}$). PPS is unique among the presently known conducting polymer precursors in that it is commercially available (from large-scale production). This resin will be the focus of our attention in this article.

$$\mathrm{+(\phi)-S+}$$

PPS

Traditionally, solid polymers have required exposure to gaseous dopants over several days to achieve considerable conductivities. Table I compares these conductivity values for four polymers. With PPS as an example, a melt-molded film (0.25 mm thick) of the resin exposed to AsF_5 (400 Torr) over four days at room temperature displays a final conductivity of ca. 10^{-2} S/cm:

$$+(\phi-S)+ \quad + \quad AsF_5 \quad \xrightarrow[\text{4 days}]{\text{room T°}} \quad +(\phi-S^+)+ \; AsF_6^-$$

(0.025cm film)   (400 Torr)   ($\sigma \approx 10^{-2}$ S/cm)

However, microprobe analysis of a cross-section of the film reveals that only the outer surfaces of the film have been reacted($\underline{2}$). Conductivity values recorded from these samples reflect an averaging of the doped and undoped regions and are therefore spuriously low. The physical form of the doped polymers is usually an intractable, crumbly solid with undetectible mechanical attributes. Insolubility limits characterization to a select group of solid state techniques.

Table I.
Conductivities Of Doped Polymers (S/cm)

Polymer	Dopant		
	AsF_5	I_2	Potassium
Polyacetylene	1200 (7)	500 (7)	50 (7)
Polyphenylene	500 (8)	$<10^{-7}$ (a)	50 (a)
Polyphenylene Sulfide	1 (1)	$<10^{-7}$ (a)	chemical degradation
Polypyrrole	100 (b)	600 (10)	—

a. Unpublished results.
b. AsF_6 introduced electrochemically (9).

Gas Phase AsF$_3$-AsF$_5$ Synergism

Arsenic trifluoride has been used as a 'codopant' in the gas-phase doping of PPS with AsF$_5$(2). The effect of the AsF$_3$ is a thousandfold enhancement in the doping rate. The enhancement is achieved by exposure of the PPS substrate to AsF$_3$ vapor during the treatment with AsF$_5$. This procedure is also shown to lead to nearly uniform doping throughout relatively thick PPS films.

The effect of using AsF$_3$ as a co-dopant is dramatically illustrated in the conductivity data of Figure 1. As seen in the lower curve, with AsF$_5$ alone (400 torr) the conductivity increases rapidly to ~10^{-5} S/cm and then slowly increases from that point on. The final conductivity achieved after 4 days of exposure to AsF$_5$ is ~10^{-2} S/cm. When AsF$_3$ is first added at ~180 torr (the vapor pressure of AsF$_3$ at room temperature) to the reactor containing the PPS film, no measureable effect on the film conductivity ($\sigma < 10^{-7}$ S/cm) is noted. On addition of AsF$_5$ to bring the total gas pressure in the cell up to 300 torr, the film's conductivity rises to 10^{-2} S/cm within 10 minutes and then to 0.3 S/cm after 90 minutes (upper curve, Figure 1). This AsF$_3$-induced improvement in both doping rate and final conductivity is attributed to a homogeneous doping of the entire polymer sample in contrast to superficial doping by AsF$_5$ alone. Evidence for this homogeneity is provided by the following scanning electron microscopy-microprobe analysis.

Microprobe analyses have been performed on cross-sections of PPS films doped with AsF$_5$ in both the absence and presence of AsF$_3$ vapor (Figure 2). A film exposed to AsF$_5$ alone (solid line) displays arsenic signals in the outermost 10% of the sample; the remaining interior contains no detectible arsenic. Arsenic trifluoride alone (dashed line) also has minimal interaction with the bulk of the polymer film, being concentrated at the film's surfaces. Whereas neither AsF$_3$ nor AsF$_5$ alone effectively penetrates the PPS films, when introduced together arsenic penetration is nearly complete and uniform. The cooperative interaction of AsF$_3$ and AsF$_5$ on PPS appears to be achieving homogeneous doping throughout the sample. This homogeneity is reflected in the higher conductivities measured on samples doped in the presence of both arsenic species: the measured conductivity of a PPS film exposed only to AsF$_5$ is spuriously low because only a minor portion of the material has been doped, whereas a sample exposed simultaneously to AsF$_3$ and AsF$_5$ exhibits a conductivity more representative of the true conductivity of AsF$_5$-doped PPS. From the microprobe analyses, it can also be concluded that the doping rate enhancement induced by AsF$_3$ results from gaining rapid access to the entire polymer sample.

The unusual affinity of PPS towards a combination of AsF$_3$ and AsF$_5$, in view of its impenetrability by either arsenic species alone, can be explained in terms of sequential doping and solvation reactions. Without added AsF$_3$, once the outer layers of the

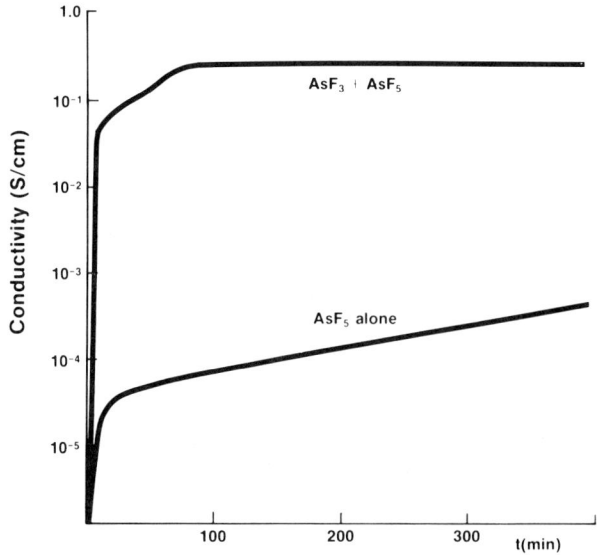

Figure 1 – Conductivity vs. time in the AsF_5 doping of a PPS film (0.025cm): lower curve, AsF_5 alone (400 torr); upper curve, AsF_3 present at 180 torr with AsF_5 added to a total pressure of 300 torr. (Reproduced with permission from Ref. 2. Copyright 1983, J. Polym. Sci. Polym. Lett. Ed.)

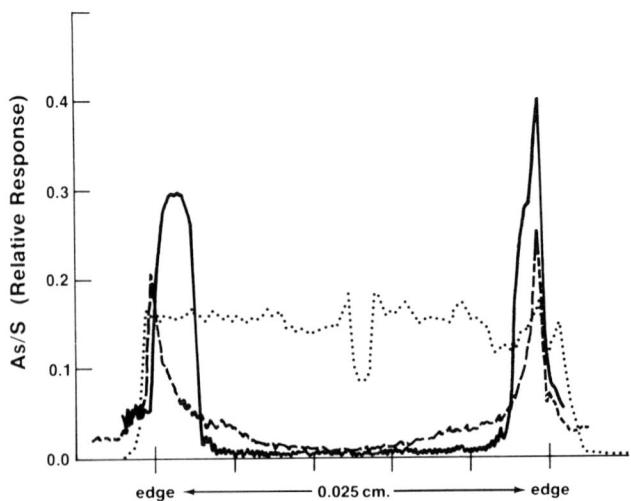

Figure 2 - Distribution of arsenic in a PPS film (0.025cm): solid line, AsF_5 alone (3 days); dashed line, AsF_3 alone (13 days); dotted line, AsF_3 and AsF_5 (3 days). The profiles are taken from scanning electron micrographs and normalized to the average sulfur signal in the midfilm regions. (Reproduced with permission from Ref. 2. Copyright 1983, J. Polym. Sci. Polym. Lett. Ed.)

polymer film are doped by AsF$_5$, the oxidized surface presents a barrier to further penetration of AsF$_5$ into the undoped, neutral interior. In the presence of AsF$_3$, the <u>reacted</u> polymer region becomes solvated by the trifluoride - plasticized for further penetration of AsF$_5$ to unreacted polymer regions. Additionally, the rate of reaction of AsF$_5$ with PPS should be enhanced by the presence of AsF$_3$ as a solvating species for charged intermediates in the PPS oxidation process. That AsF$_3$ would preferentially solvate the oxidized polymer while barely penetrating the undoped polymer (Figure 2) is not surprising when the different polarities of the charged and neutral media are taken into account. Apparently the ionic polymer demonstrates a greater solubility in AsF$_3$ than its neutral precursor. This solvation by AsF$_3$ of AsF$_5$-doped PPS occurs on a time scale observable in Figure 1, and affords a means of homogeneously doping PPS samples.

Solutions Of Doped PPS in AsF$_3$ Liquid

We have extended the use of AsF$_3$ from the gas to the liquid phase. PPS, when doped with AsF$_5$ in the presence of AsF$_3$ liquid, readily dissolves to produce a deep-blue <u>conducting polymer solution</u>. The dissolved conducting polymer remains electro-active in solution indefinitely. From this solution we have been able to cast films of varied and controlled thickness. The films sport a shiny blue-black appearance and possess good mechanical properties of strength and flexibility. Their conductivities range from 5×10^{-3} to 200 S/cm. Standard application of this technique routinely yields a film of 25 S/cm, one to two orders of magnitude larger than that previously obtained under conventional gas/solid doping conditions.

In a typical experiment, AsF$_5$ (~500 Torr) is introduced to a suspension of PPS solids (powder or film) in AsF$_3$ solvent. In the neutral undoped state PPS shows no sign of dissolving or swelling in AsF$_3$. Immediately after exposure to the dopant, however, the oxidized blue polymer readily dissolves to give a deep blue solution. These solutions can be diluted or concentrated to any concentration with no sharp phase transition being observed. On slow reduction of solvent volume under vacuum, the solution thickens until a viscous coating remains on the reactor walls. Further slow removal of solvent results in a free standing film. The film, once cast in this manner, is not soluble in AsF$_3$. This irreversible transition from solution to solid is likely accompanied by covalent bond formation within the polymer that is induced by removal of a stabilizing solvation sphere. Inter- or intrachain crosslinking on decreasing volume is a process by which this could be occurring. Dilute solutions of doped PPS prepared as described above are indefinitely stable in a dry atmosphere. The fact that clean, non-viscous solutions result from this process argues against there being appreciable crosslinking in the polymer on doping, even at room temperature, as long as AsF_3 is present to solvate the polymer as it dopes. The stability of the solution (i.e., no precipitation) with time also suggests that no substantial degree of crosslinking occurs in solution. At high concentration, intentional gelling of the polymer solutions can be induced, however.

A conductivity of 2×10^{-2} S/cm has been measured on a solution of AsF_5-doped PPS in AsF_3 (0.2 PPS monomer equivalents/ℓ AsF_3) with a conductivity bridge operating at an AC frequency of 400 Hz. When measurements were made using a DC source some polarization occured, possibly indicating an ionic component to the conductivity. Films cast from this solution displayed a D.C. conductivity of 200 S/cm. This rise of four orders of magnitude in conductivity reflects a marked transition to predominantly electronic conductivity on passing from the liquid to the solid state. Increased interchain contact on collapsing of the polymer into a space-filled solid is a reasonable explanation for the increase in conductivity; a decrease in solvent separation of polymer chains might cause an increase in intermolecular electron transport. Intermolecular transport should also be aided by concomitant cross-linking which in the PPS case would maintain the aromaticity of the rings and introduce a "conjugated bridge". No crystallinity is detected in the cast films by wide angle x-ray diffraction.

A remarkable feature of the conducting polymer solution is the nature of the radical signal observed by EPR spectroscopy (Table II, Figure 3). The lineshape of the strong paramagnetic resonance is distorted, a Dysonian lineshape(3), indicative of a highly electrically conducting sample. Similar behavior is observed in EPR of solutions of alkali metals in liquid ammonia(4). The observed lineshape is expected to be dependent on the exact shape and geometry of the sample as well as the overall conductivity. A sample of PPS film (25μ thick) which was cast from a $PPS-AsF_5-AsF_3$ solution gives an EPR spectrum very similar to that of the frozen $PPS-AsF_5-AsF_3$ solution; Dysonian lineshapes were not observed in either case. Strong Dysonian lineshape appears to be unique to the room temperature doped PPS solutions, not being observed in the cast films, frozen solution, or in PPS powders doped conventionally with AsF_5 only(5). This is not surprising since the thickness of the cast film (~25μ), for example, is much less than the effective skin depth (~100μ) expected for a sample with a conductivity of ~25 S/cm(3).

The EPR spectrum of an AsF_3 solution of the AsF_5-doped "tetrameric" oligomer Ph-S-Ph-S-Ph-S-Ph, in contrast to that of the PPS solution, exhibits normal lineshape (Table II, Figure 3). The observation of the tetramer's symmetric and the polymer's assymmetric lineshapes recorded under similar conditions of temperature, concentration and doping level, together with the knowledge that the doped "tetramer" shows low conductivity, supports the assignment of the Dysonian lineshape to the transport of the conducting electrons in solution.

Table II. EPR Spectral Parameters for Doped PPS and Doped Oligomer, Ph-S-Ph-S-Ph-S-Ph

Sample	g-value[a]	Γ_H[b]	A/B[c]	Temperature
AsF$_5$-doped PPS				
AsF$_3$ Solution	2.006	12	5.1	RT
" "	2.006	9	1.1	77°K
AsF$_5$-doped Cast Film	2.006	12	1.0	RT
AsF$_5$-doped Ph-S-Ph-S-Ph-S-Ph				
AsF$_3$ Solution	2.006	4	1.0	RT
" "	2.006	12	0.96	77°K

a) Measured from strong pitch; dual cavity. Recorded at microwave powers ≤ 1 mW. Corrected for Dysonian lineshape.
b) Peak-to-peak separation in the first-derivative EPR curve.
c) Asymmetry parameter - the ratio between the maximum and minimum of the lobes in the first-derivative EPR curve.

Infrared spectra of a solution of the doped polymer and of an undoped PPS film are compared in Figure 4. The two spectra are remarkably similar (aside from the solvent peaks at 340-250 and 750-600 cm^{-1}). The frequencies of the C=C stretching modes of the phenyl rings of the undoped polymer (1575, 1470, and 1390 cm^{-1}) shift to lower energy in the doped polymer solution spectrum (1550, 1460, and 1385 cm^{-1}). The phenyl-sulfur stretch at 1090 cm^{-1} in the undoped polymer shifts to a higher energy of 1125 cm^{-1} for the doped, dissolved polymer. These shifts are consistent with the shortening of the C-S bond and lengthening of the C-C ring bonds as expected in the charge delocalized doped polymer. The C-C stretch modes of the phenyl rings (1600-1400 cm^{-1}) appear to have split into doublets in the solution spectrum. The same is true of the ring breathing mode at 1005 cm^{-1}, the new peak appearing at 990 cm^{-1}. The peak-splitting of the phenyl modes could suggest well-defined doped and undoped regions of the polymer chains or a lifting of symmetry restrictions on the phenyl ring modes due to ionization. Some bridging of adjacent phenyl rings to form dibenzothiophene structures(1) is observed after neutralization of the films cast from solution (characterized by a 865 cm^{-1} "isolated H wag"):

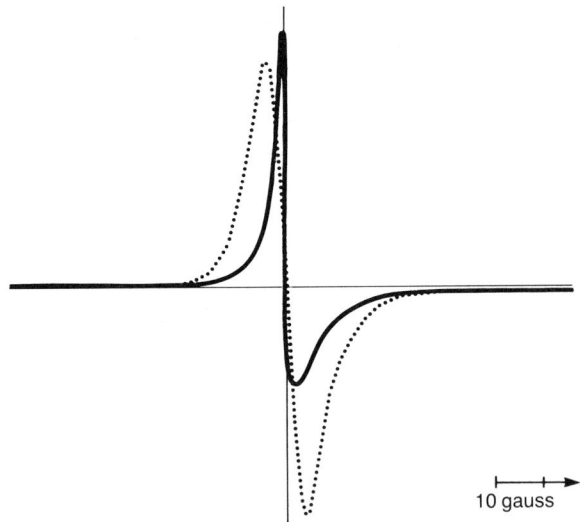

Figure 3 - EPR spectra of AsF_3 solutions: solid line, AsF_5-doped PPS (ca. 0.2M in -PhS- equivalents); dashed line, AsF_5-doped Ph-S-Ph-S-Ph-S-Ph (ca. 0.2M in -PhS- equivalents).

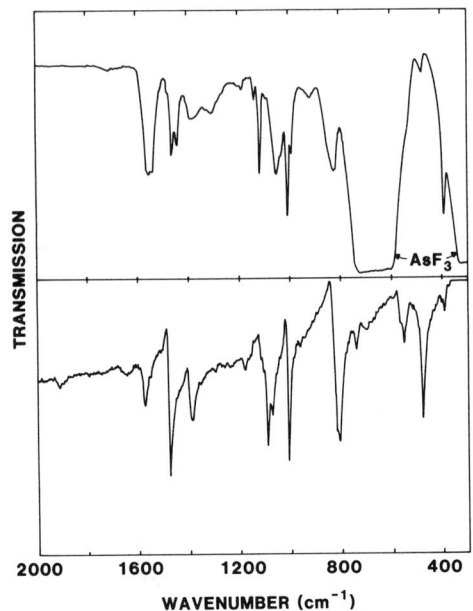

Figure 4 - Infrared spectra of PPS: lower curve, undoped in a solid KBr matrix; upper curve, AsF_5-doped in AsF_3 solution.

Evidence for the occurrence of this coupling in solution is obscured by a broadening of the signal from the phenyl substitution region, 850-800 cm^{-1}.

Electronic spectra have been recorded in the UV-vis-near infrared region on both solutions and films of doped PPS (Figure 5). The main feature of both spectra is a near infrared absorption assigned to the radical cation of PPS located at 926nm in the solution and 1030nm in the film. The "blue" shift of the absorption in going from the solid to the solution phase may be the result of a solvent shift or localization of charge along the polymer chain. Localization could be occuring by AsF_3 solvation of charged carriers. This is consistent with an observed decrease in conductivity on exposing doped melt-molded films to AsF_3(2). Alternatively, more highly disordered chain configurations attainable only in the solution phase could also lead to shorter effective conjugation lengths on the ionized PPS chain, and hence a blue shift. The basic similarities between optical spectra (both electronic and vibrational) obtained on conventionally doped PPS solids and on doped PPS solutions confirm our description of the AsF_3 solution as a "conducting polymer solution".

A color change from deep blue on immediate dissolution of the doped complex to greenish-blue over the following ~30 min at room temperature has been noted(11). This color change might be indicative of chemical modification of the polymer backbone, e.g., bridging between adjacent phenyl rings to give dibenzothiophene moieties. Such a resulting polymer is predicted to have a lower band gap according to recent calculations(6). We do detect some chemical modification of the polymer backbone after doping and compensation: crosslinking as well as intramolecular bridging. The extent and rate of crosslinking is expected to depend on both the doping level and the concentration of polymer in solution. In fact, by IR analysis the cast films of higher conductivities (25→200 S/cm) show more extensive chemical modification, whereas lightly doped PPS (conductivity ≈10^{-2} S/cm) from solution closely resembles undoped starting material.

Elemental analyses on the cast films range from an empirical formula of $C_6H_4S_{1.0}(AsF_3)_{0.5}$ (lightly doped) to $C_6H_{3.8}S_{1.0}$ $(AsF_{5.1})_{1.45}$ (heavily doped). The exact nature of the inorganic anion in these films (and in solution) is still being debated. The IR of a cast film contains bands at 395 and 690 cm^{-1}, characteristic of AsF_6^- (or solvated AsF_6^-). Low F/As ratios for the cast films might be indicating extensive complexation of the inorganic anion with AsF_3. Inclusion of excess AsF_3 in lightly doped samples may be essential to the stability of the doped complex. The solvating effects of AsF_3 would be less important at high doping levels where screening by neighboring charged species would lessen the requirements for AsF_3 as a solvating species. This could explain the low F to As ratios observed in lightly doped samples as well as the range of arsenic to sulfur ratios in AsF_5-doped PPS. The solvating effect of AsF_3 on the $PPS^+AsF_6^-$ complex is important not only in stabilizing the solution phase but also in the initial formation of the conducting complex. The AsF_3 solvation energy probably accounts for a significant fraction of the AsF_3-induced (gas phase) enhancement of the AsF_5 doping rate of PPS films discussed earlier. In that case, the AsF_3 would serve to both lower the kinetic barrier to complex formation and solvate (or partially solubilize) the doped surface of the PPS film for more facile diffusion of AsF_5 into the interior of the film.

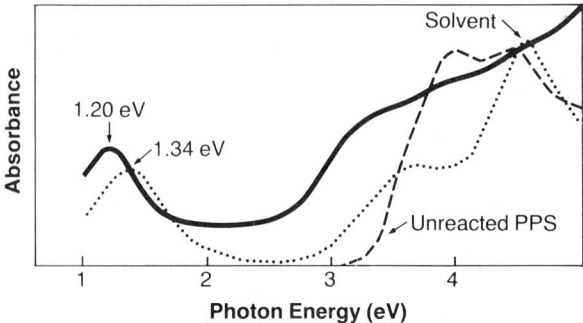

Figure 5 - Optical absorption of AsF_5-doped PPS: solid line, melt-molded film (ca.5 x 10^{-5}cm thick) doped with AsF_5; dotted line, AsF_5-doped PPS in AsF_3 solution.

Conclusion

Arsenic trifluoride has been shown to be an active participant in the doping of PPS with AsF_5. When introduced in the gas phase, it greatly increases both doping rate and homogeneity. As a liquid, it serves the role of solvent in providing the first example of a conducting polymer solution. Research is presently continuing to expand the scope of this solubility phenomenon to other polymer, dopant, and solvent systems. This discovery of a conducting polymer solution has made available a new, liquid phase of these novel materials, as well as a means of processing them into useful designs.

Acknowledgments

The authors thank L. Shacklette, J. Macur, L. Reinhardt, and C. Forbes.

Literature Cited

1. Shacklette, L.W., Elsenbaumer, R.L., Chance, R.R., Eckhardt, H., Frommer, J.E., and Baughman, R.H., J. Chem. Phys. 1982, 75, 1919.
2. Frommer, J.E., Elsenbaumer, R.L, Chance, R.R., and Eckhardt, H., J. Polym. Sci., Polym. Lett. Ed. 1983, 21, 39.
3. Dyson, F.J., Phys. Rev. 1955, 98, 349.
4. Peck, R.J. and Glavnsinger,W.S., J. Magnetic Res. 1981, 45, 48.
5. Kispert, L.D., Files, L.A., Frommer, J.E., Shacklette, L.W., and Chance, R.R. J. Chem. Phys. 1983, 78, 4858.
6. Bredas, J.L., Elsenbaumer, R.L., Chance, R.R. and Silbey, R., J. Chem. Phys. 1983, 78, 5656.
7. MacDiarmid, A.G. and Heeger, A.J., Syn. Met. 1980, 1, 101.
8. Shacklette, L.W., Chance, R.R., Ivory, D.M., Miller, G.G., Baughman, R.H., Syn. Met. 1979, 1, 307.
9. Diaz, A., Chimica Scripta 1981, 17, 145.
10. Street, G.B., Clarke, T.C., Krounbi, M., Kanazawa, K., Lee, V., Pfluger, P., Scott, J.C., and Weiser, G., Mol. Cryst. Liquid Cryst. 1982, 83, 253.
11. We have not unambiguously identified the UV-vis spectra corresponding to this visual color change. Preliminary results on one solution showed additional strong peaks at about 1.0 and 1.9 eV. The latter peak could be responsible for the blue color and would seem to correlate to the 1.9–2.0 eV peak we assigned previously to an AsF_3 - induced transition involving only arsenic species(2).

RECEIVED September 19, 1983

Poly-*p*-phenylene Selenide and Poly-*p*-phenylene Telluride

Characterization and Assessment as Active Elements

L. A. ACAMPORA, D. L. DUGGER, T. EMMA, J. MOHAMMED, M. F. RUBNER, L. SAMUELSON, D. J. SANDMAN, and S. K. TRIPATHY

GTE Laboratories, Inc., Waltham, MA 02254

> We report the first syntheses of the selenium and tellurium analogs of the thermoplastic poly-p-phenylene sulfide (PPS), i.e., PPSe and PPTe, by reaction of p-dihalobenzenes with new alkali chalcogenide reagents under conditions significantly milder than those reported for PPS. The chemical, thermal, structural, and electrical properties of PPSe and PPTe are compared to those of PPS. While PPSe exhibits good thermal stability, PPTe is decomposed thermally under conditions significantly milder than PPS and PPSe. X-ray and electron diffraction studies of PPSe reveal a crystal structure isomorphous to that of PPS, while PPTe has a different structure. Analogous to PPS, exposure of PPSe to AsF_5 results in an insulator-to-conductor transformation and structural cross-linking with a room temperature conductivity in the range 10^{-3} to 10^{-2} $(ohm-cm)^{-1}$ being reached.

In order for a macromolecular based system to exhibit high dark conductivity, it is necessary that the material have an electronic structure with a partially filled energy band which is either intrinsic or readily created by thermal means. Such electronic structures arise in two ways. The first is that it would occur in the material as synthesized, and the best characterized example of this situation is the metallic and superconducting polymeric sulfur nitride, $(SN)_x$, a material produced by a solid state polymerization and fully ordered crystallographically (2). In contrast, the electronic carriers in most current examples of conducting polymers are created by exposing an insulating, usually partially crystalline (3) polymer, indicating a wide band gap or at least localized states, to appropriate electron transfer reagents.

A useful variant of this technique involves electrochemical oxidation of a molecule such as pyrrole to give a conductive polymer film containing the anion of the support electrolyte (4). While successful application of such electron transfer techniques for carrier creation, commonly termed "doping," dates at least to the use of $SbCl_5$ to render poly-p-phenylene conductive (5), it was the successful doping (6) of films of polyacetylene, $(CH)_x$, with both oxidizing and reducing reagents which triggered the emergence of polymeric conductors as a class of electronic materials. With features ranging from the long-standing theoretical interest in the electronic structure of $(CH)_x$ through potential applications in lightweight batteries (7), the experimental and theoretical results from $(CH)_x$ and its doped forms have stimulated considerable interest not only in a deeper understanding of $(CH)_x$ but also in the synthesis and characterization of other polymer systems with different intrachain structures in the insulating form (8-9). Motivations for such work include not only the search for better polymeric metals, but also materials with improved stability and processability compared to $(CH)_x$ and its doped forms.

With respect to processability, the commercial thermoplastic poly(p-phenylene sulfide) (PPS) has received considerable attention as an example of a polymer without a continuous conjugated carbon Π-system which becomes highly conducting on exposure to strong redox reagents (10) and ion implantation (11). Since the selenium analog of the metallic anisotropic molecular ion-radical solid TTF-TCNQ exhibits not only a crystal structure isomorphous to the sulfur material, but also remains metallic to lower temperatures (12), it is of interest to inquire if the advantages of heavier chalcogen substitution noted in molecular conductors (1b,12) would also accrue to chalcogen-containing polymers, particularly in view of the structural disorder inherent in the types of polymers under discussion. Hence our initial task reduces to the synthesis of the selenium and tellurium analogs of PPS, i.e., PPSe and PPTe, and an initial comparison of their structural properties to those of PPS.

Synthesis and Composition

Since PPS is prepared by reaction of p-dichlorobenzene (1) and Na_2S in N-methyl-2-pyrrolidone (NMP) at ca, 240° (13), the reaction of p-dihalobenzenes with Na_2Se and Na_2Te appeared to be at attractive approach. We note that an unsuccessful attempt to synthesize PPSe from p-dibromobenzene (2) and Na_2Se has been reported. (14) As we have recently found (1a-b) that alkali metals react directly with selenium and tellurium in 1:1 and 2:1 atomic ratios in dipolar aprotic solvents, thus avoiding the use of liquid NH_3 commonly used for the preparation of these reagents, and that these reagents react readily with aromatic halides, which are not activated in the usual sense for a nucleophilic aromatic substitution, to provide a direct synthetic route to both new (e.g., tetratellurotetracene (15)) and previously reported aromatic selenium and tellurium molecules, it was of interest to extend this chemistry to polymeric systems.

In initial work, we found that 2 reacted readily with Na_2Se in NMP at 170-175° to give PPSe in ca. 10% yield. We subsequently found that this reaction could be carried out in N,N-dimethylformamide (DMF) at ca. 140° to reproducibly give PPSe in 65-80% yields free of low molecular weight oligomers. Under similar conditions, 1 reacts with Na_2Se to give PPSe in only 5% yield. When prepared as noted above and with K_2Se in DMF, PPSe is a yellow powder with a melting point (capillary tube) of ca. 220° and a compositional range, from elemental analysis, of $C_{6.0}H_{3.9-4.2}Se_{1.06-1.15}Br_{0.015}$. Assuming that the bromine is present as chain ends, its percentage suggests a molecular weight of ca. 10000. The sparing solubility of PPSe in all solvents precluded a molecular weight determination using solution techniques. The presence of Se in excess of stoichiometry is attributed to diselenide linkages and will be discussed below.

The conditions which are used for the synthesis of PPSe are significantly milder than those used for PPS (13). This point is deemed mechanistically significant, and we suggest that an aromatic $(Ar)S_{RN}1$ process (16) may be involved in the present chemistry. The $S_{RN}1$ process is a radical chain mechanism in

$$Ar - X + e \rightarrow (Ar - X)^{\overline{\cdot}} \quad \text{(Initiation)} \quad (1)$$

$$(Ar - X)^{\overline{\cdot}} \rightarrow Ar\cdot + X^{-} \quad (2)$$

$$Ar\cdot + Y^{-} \rightarrow (Ar - Y)^{\overline{\cdot}} \quad \text{(Propagation)} \quad (3)$$

$$(Ar - Y)^{\overline{\cdot}} + Ar - X \rightarrow Ar - Y + (Ar - X)^{\overline{\cdot}} \quad (4)$$

which the initiation step and one of the propagation steps (Equations 1 and 4) involve single electron transfer (SET) (16). Our observation of higher yields of PPSe from 2 than 1 is consistent with an $S_{RN}1$ process. A PPSe propagation step analogous to Equation 4 is given in Equation 5 in which the dianion-radical 3 donates an electron to 2. If an external

Br-(⟨O⟩-Se)$_n$-⟨O⟩-Se$^{\overline{\cdot}}$ + 2 $\xrightarrow{\text{SET}}$ Br-(⟨O⟩-Se)$_n$-⟨O⟩-Se^{-}

$$+ \underline{2}^{\overline{\cdot}} \quad (5)$$

acceptor can successfully compete with 2 in this SET process, then the yield of PPSe should be reduced. The addition of benzophenone (17) (Equation 6) equimolar to 1 in a PPSe synthesis routinely depresses the yield of high molecular weight PPSe from greater than 65% to less than 25%.

$$\underline{3} + \langle O \rangle - \overset{O}{\underset{\|}{C}} - \langle O \rangle \xrightarrow{SET} \left(\langle O \rangle - \overset{O}{\underset{\|}{C}} - \langle O \rangle \right)^{\overline{\cdot}} \quad (6)$$

The use of a Li_2Se reagent prepared from $Li(C_2H_5)_3 BH^{\cdot}$ (18) for PPSe synthesis was also investigated. The Li_2Se reagent was prepared in THF and reacted with 2 in DMF as above. In contrast to the yellow PPSe isolated above, this experiment gave a 31.5% yield of a white solid found by elemental analysis to have the approximate oligomeric composition $Br(C_6H_4-Se)_{10}Br$.

Our best approach to PPTe (19) involves reaction of p-diiodobenzene with Na_2Te in DMF at 110-120° to give a tan solid, m.p. 162-170°(dec) in 70-75% yield, found by analysis to have the composition $C_6H_{4.6}Te_{1.35}I_{0.065}$. The percentage of iodine, as chain ends, suggests a molecular weight of ca. 8000, and the excess Te is partly due to contamination by the element (cf. X-ray diffraction, see below).

Spectra and Structure

We have used vibrational and electronic spectroscopy to learn about the intrachain aspects of PPSe and PPTe and X-ray and electron diffraction to deduce information about the three dimensional features of these materials.

The infrared (IR) spectra of PPSe and PPTe were recorded with the samples dispersed in cesium iodide using Fourier transform techniques. Between 4000 and 600 cm^{-1}, the IR spectra of PPSe (both yellow and white forms) and PPTe are superimposable on that of PPS. Below 600 cm^{-1}, bands at 550 and 475 cm^{-1} in PPS are found at 500 and 475 cm^{-1} in PPSe (both forms) and at 489 and 465 cm^{-1} in PPTe. Other relevant features between 600-200 cm^{-1} are discussed below.

The electronic spectra of PPSe and PPTe have been studied by diffuse reflectance and transmission through KBr pellets with consistent results observed with the two techniques. PPSe (yellow and white) shows a maximum at 300 nm while PPTe has a maximum at 310 nm, with both tailing into the visible. These results are expected since the absorption maximum of diphenyl selenide is at 280 nm (20). The white form of PPSe exhibits less tailing into the visible than the yellow form. This observation is consistent with the presence of diselenide linkages in the yellow form, (see below) since diphenyl diselenide has its lowest absorption maximum at 332 nm (20).

A diffractometer tracing of yellow PPSe, shown in Figure 1, revealed partial crystallinity with a strong peak observed at $2\theta = 20.2°$. Since this peak, as well as other features, bears a strong resemblance to PPS, whose strong peak is at $2\theta = 21.0°$, we allowed the unit cell of PPS to expand to accommodate the larger Se van der Waals radius and tentatively assign an isomorphous orthorhombic unit cell $\underline{a} = 8.7$, $\underline{b} = 5.7$, $\underline{c} = 10.5 A$, $V = 520 A^3$, $Z = 4$, $\rho_{calc} = 1.98$ g cm^{-3}. Electron diffraction from PPSe was performed on samples prepared by casting a film onto gold electron microscope grids from boiling phenyl ether solution. The film was annealed at 200° for 72 hours and was quite crystalline. The morphology is of lamellar polycrystalline nature, and the diffraction pattern, obtained from a one-micron selected area aperture, led to a superimposed diffraction pattern of 2-3 single crystals. The cell symmetry appears to be that obtained from the X-ray studies. The growth morphology appears similar to that observed for PPS prepared under similar conditions (21).

For polymer prepared from either $\underline{1}$ or $\underline{2}$, the observed flotation density of yellow PPSe is $2.05 - 2.15$ g cm^{-3}, exceeding the theoretical density. Since our diffractometer tracings revealed no indication of the presence of elemental Se, we speculated that the Se in excess of stoichiometry revealed by elemental analysis would be due to diselenide groups in the non-crystalline regions of the polymer which would enhance the density. Support for this hypothesis came from two sources in addition to the electronic spectra noted above. Since diselenides exhibit IR absorption near 290 cm^{-1}, the observation of weak absorption at this wavelength in all yellow PPSe samples is consistent with the suggestion. Additionally, white PPSe, which has the same X-ray diffraction pattern as the yellow form, exhibits an observed density of 1.98 g cm^{-3} and its IR spectrum lacks the 290 cm^{-1} absorption of the yellow form.

Expansion of the PPS - PPSe lattice to accommodate the van der Waals radius of Te leads to a unit cell volume of $548 A^3$ and a calculated density of 2.46 gm/cm^3. X-ray diffraction of PPTe reveals contamination with small amounts of elemental Te and reflections at $d = 5.09$, 4.27, 3.57, and 3.04A. This pattern is sufficiently different from those of PPS and PPSe that a different structure appears to be involved. The flotation density of PPTe is 2.46 gm/cm^3.

Thermal Properties

Thermal characterization of PPSe and PPTe was done using thermogravimetric analysis (TGA), TGA followed by gas chromatography/mass spectrometry (TGA/GC/MS), and by differential scanning calorimetry (DSC).

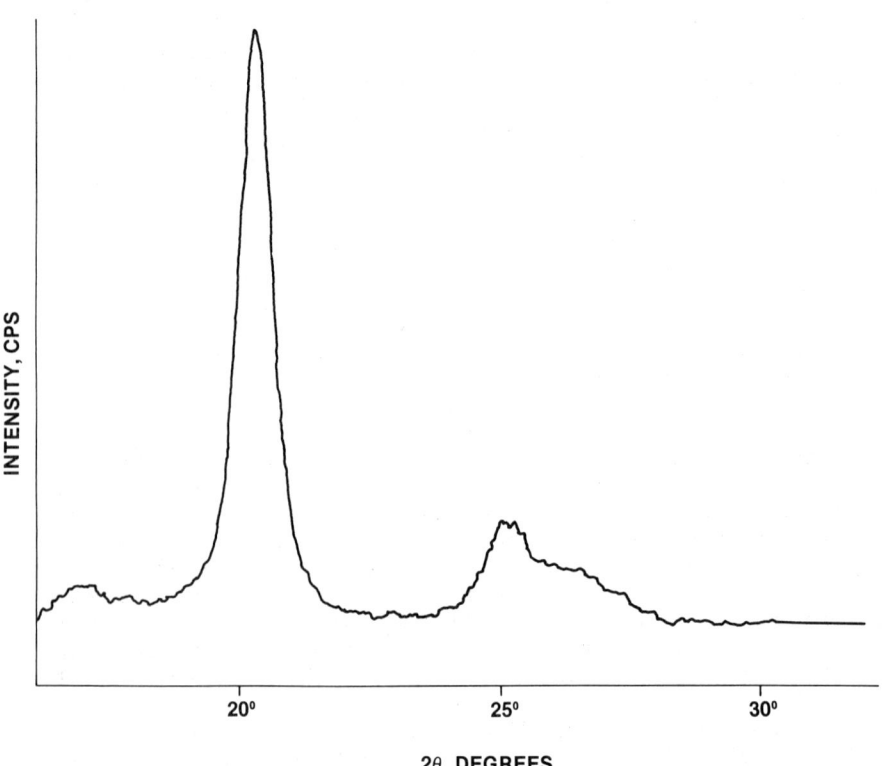

Figure 1. Diffractometer tracing of PPSe (yellow) using CuK$_\alpha$ radiation.

TGA of PPSe and PPTe in a nitrogen atmosphere at a heating rate of 10° per minute indicates that the materials are thermally stable to 400° and 330° respectively. PPSe shows a rapid weight loss (70%) from 400°C to 450°C and then loses weight gradually, for a total weight loss of 88.3% at 1070°C. PPTe undergoes a rapid weight loss (14%) from 330°C to 375°C and then loses weight more slowly from 375°C to 920°C for a total weight loss of 40.6%, slightly more than the C-H weight percentage. In air, both PPSe and PPTe show the same thermal stability as in nitrogen. But, while PPSe undergoes a very rapid weight loss from 440°C to 550°C, losing 98.8% of its weight, PPTe shows four distinct stages: I. 330°C to 370°C (8.4%); II. 370°C to 470°C (4.0%); III. 470°C to 505°C (27.6%); and IV. 756°C to 920°C (60%). Between 505°C and 756°C there is no weight loss. Although PPSe and PPTe are less thermally stable than PPS, the behaviour of PPSe in nitrogen and air and of PPTe in nitrogen is very similar to that of PPS. PPS is thermally stable to 450°C both in nitrogen and in air. In nitrogen it loses 64.5% of its weight from 450°C to 525°C and then loses weight more gradually. In air, PPS loses almost all its weight from 450°C to 750°C (98.3%) and our TGA results of PPS compare very favorably with those reported in the literature (22).

TGA/GC/MS was performed on PPSe and PPTe as well as PPS to determine the nature of the degradation products. The three materials show similarities as well as differences. All show the presence of diphenyl, diphenyl - X (X = S, Se, or Te), dibenzo - X - phene, diphenyldi-X. The dissimilarities are the presence of halogenated benzenes in PPSe and PPTe, and the presence of dibenzfuran in some PPSe samples. PPS on the other hand shows a large amount of thiophenol. It appears that the breakdown mechanism of PPSe and PPTe is, in principle, similar to that of PPS. DSC of PPSe and PPTe was performed to determine the melting point and the heat of transition. The DSC scan of PPSe is shown in Figure 2. PPSe appears to melt in several stages with the major portion melting at 217°C. A second heating of the same sample shows a shift in the melting points to higher temperatures, the major portion melting at 234°C. The melting point endotherm in the second heating is not as broad as during the first heating. PPTe shows just one melting point at 163°C which does not reappear on a second heating. Crystalline PPS shows a relatively sharp melting point at 276°C which decreases slightly to 270°C on a second heating. The appearance of several melting points in PPSe and their shifting to higher temperatures during thermal cycling is reminiscent of the behavior of PPS of relatively low crystallinity (23). The heat of transition obtained by integrating the area enclosed by the endotherm shows that the energy required to melt PPSe and PPTe is less than that required for PPS. ΔH_{PPSe} 46 J/g; ΔH_{PPTe} 19.7 J/g; ΔH_{PPS} 61.1 J/g.

There was no difference between the DSC scans in nitrogen and air for all these materials. This is not unexpected, since

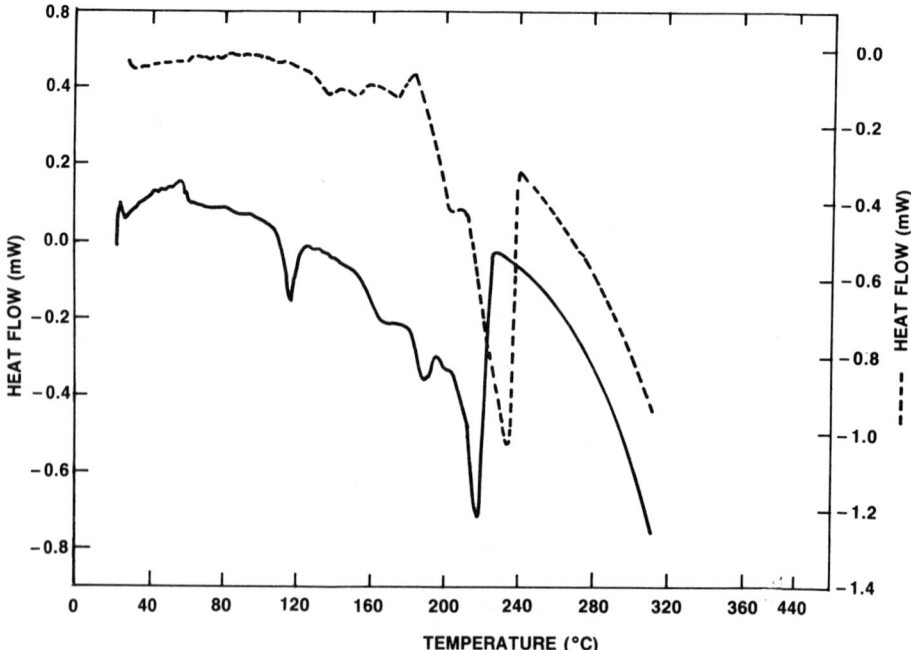

Figure 2. Differential Scanning Calorimetry of PPSe (yellow) at a heating rate of 5°/min. in air (—) First heating; (----) Second heating.

TGA showed the thermal stability in nitrogen and in air to be identical.

Exposure to Dopants and Electrical Properties

PPSe and PPTe, prepared as described above, are insulators whose room temperature dc conductivity, measured on compacted powders, is less than $10^{-11}(ohm-cm)^{-1}$. We have attempted to render these materials conductive by exposure of powders to charge transfer dopants, since we found that, in contrast to PPS (10), exposure of compacted powders to vapors of AsF_5 led to no increase of either weight or dc conductivity.

Both PPSe and PPTe turn black and undergo weight increases of 20 and 150%, respectively, on exposure to iodine, with the conductivity of the PPSe samples being no more than $10^{-11}(ohm-cm)^{-1}$ while the PPTe samples had a conductivity range of 10^{-8} - 10^{-6} $ohm-cm)^{-1}$. Electron transfer does not appear to be involved in the interaction of I_2 with PPSe and PPTe since the solid state spectra show the maxima of the original polymers with long tails into the near infrared. In the case of PPTe, the interaction with iodine largely involves formation of a tetracoordinate diaryltellurium diiodide (24). Reduction of this material with sodium sulfide (24) leads to recovered PPTe, identified by its IR spectrum.

Exposure of yellow PPSe to AsF_5 (100 torr) for 5 hours at 40° leads to a weight increase of 40-45 percent and formation of a black solid with a room temperature conductivity of 10^{-2} - 10^{-3} $(ohm-cm)^{-1}$, a value comparable to that observed for PPS powders(10f) but lower than that for amorphous films (10a-c,e). The role of sample crystallinity in PPS doping has been discussed (10d). Exposure of the black conducting solid to moist air or compensation by NH_3 vapor leads to a yellow orange amorphous solid whose solid state spectrum shows broad maxima at ca. 300 and 400 nm, and whose infrared spectrum shows absorption at 705 and 395 cm^{-1}, suggestive of AsF_6. Analogous to PPS (10), exposure of PPSe and AsF_5 leads to an insulator-conductor transformation which is accompanied by structural crosslinking and apparent dibenzchalcogophene formation. When PPTe is exposed to AsF_5 under conditions analogous to those used for PPSe, the material initially turns black, then tan, and has a conductivity of ca. 10^{-10} $(ohm-cm)^{-1}$. The IR spectrum of this material after NH_3 compensation and salt removal reveals that structural modification has occurred in the PPTe. PPTe is also decomposed by $SbCl_5$ in CH_2Cl_2 at 0°.

Exposure of PPSe to a CH_2Cl_2 solution of $NO^+PF_6^-$, a reagent which renders PPS conductive, leads to dissolution of the polymer to give a red solution. Treatment of this solution with NH_3 precipitates a yellow powder which has undergone considerable chain scission. Analogous to PPS (10), PPSe is decomposed by exposure to sodium naphthalenide in THF.

A recent patent (25) claimed that a conductivity of 10^{-1} (ohm-cm)$^{-1}$ is reached by treatment of PPSe with picrylsulfonic acid. We have found that addition of PPSe to a solution of picrylsulfonic acid (1 molecule per polymer repeat) in THF, kept either at room temperature or at reflux for 16 hours, led to the recovery of unchanged PPSe.

Summary

We have reported herein the first syntheses of PPSe and PPTe and our studies to date of their chemical, thermal, structural, and electrical properties. For the sake of comparison to PPS, it is noteworthy that PPSe has thermal properties comparable to PPS and a crystal structure isomorphous to the sulfur polymer, while PPTe does not have comparable thermal properties and has a different crystal structure. While analogous to PPS, AsF$_5$ brings about an insulator-to-conductor transformation and crosslinking in PPSe, our studies to date of PPSe, PPTe, and other related polymers (1,19) which require change transfer doping to become conductive reveal all of the complexities of existing conducting polymer systems.

Experimental Section

General. Synthetic experiments were carried out under argon atmosphere. PPS was obtained from Polysciences, Inc. Elemental analyses were performed by Galbraith Laboratories, Knoxville, Tenn. Electronic spectroscopy was performed on a Varian Cary 17 recording spectrophotometer and a Nicolet Fourier Transform Infrared instrument was used to record spectra between 4000 and 200 cm^{-1} of samples in CsI. Electrical measurements were performed at room temperature on compacted samples by a four probe technique. TGA was performed on a DuPont 990 Analyzer using a DuPont 951 TGA module. TGA for the GC/MS analysis was performed on a DuPont 950 instrument. Volatile products were collected in a Tenax containing tube which was attached to the GC inlet port, GC/MS was performed on a Hewlett Packard 5982A coupled to a 5934A data system. DSC was performed on a DuPont 1090 analyzer using a DuPont 910 DSC module.

Synthesis of PPSe (yellow) and PPTe p-dibromo- and p-diiodobenzene in DMF (5-8 ml/gm) were reacted with Na$_2$Se and Na$_2$Te, prepared analogously to Na$_2$Te$_2$ (15), respectively, at 110-120° (PPTe) or 120-140° (PPSe) for twenty hours. After cooling to room temperature, the reaction mixtures were poured into brine, and the polymers were isolated by suction filtration. PPSe was washed with Na$_2$S solution, followed by water and methanol. Both crude polymers were extracted with THF and vacuum dried at 100° before further study.

PPSe Analysis. Calcd for $(C_6H_4Se)_x$: C, 46.48; H, 2.60; Se, 50.92. Found: C, 41.36; H, 2.27; Se, 52.15; Br, 1.58. The observed analysis corresponds to $C_{6.0}H_{3.9}Se_{1.15}Br_{0.015}$.

PPTe Analysis. Calcd for $(C_6H_4Te)_x$: C, 35.38, H, 1.98, Te, 62.64. Found: C, 27.96; H, 1.79; Te, 66.88; I, 3.20. The observed analysis corresponds to $C_{6.0}H_{4.6}Te_{1.35}I_{0.065}$.

Synthesis of PPSe (white). To a suspension of Se (0.79 gm, 0.10 gm atom) in THF (10 ml) at 0°C was added Li $(C_2H_5)_3$BH (20 mmole). After warming to room temperature and reflux for one hour, the resultant milky white suspension was added to a solution of 2 (2.359 g, 10 mmole) in DMF (100 ml). This mixture was heated at 140-145° for forty hours when it was worked up as described above for PPSe. A white solid, 0.49 g (31.5% yield), mp 209-214° was isolated. The infrared spectrum and X-ray powder pattern are in agreement with those of PPSe.

Analysis. Calcd for $(C_6H_4Se)_x$: C, 46.48; H, 2.60; Se, 50.92. Found: C, 43.54, H, 2.55; Se, 41.45; Br, 10.17. The observed analysis corresponds to $C_{6.0}H_{4.2}Se_{0.9}Br_{0.21}$ or $Br(C_6H_4Se_{0.9})_{10}Br$.

Literature Cited

1. Earlier accounts of portions of this work have been presented: Sandman, D. J.; Stark, J. C.; Acampora, L. A.; Gagne, P. presented at the 12th Northeast Regional Meeting of the American Chemical Society, Burlington, Vermont, June 27-30, 1982, Paper 158; (b) Sandman, D. J.; Stark, J. C.; Rubner, M.; Acampora, L. A.; Samuelson, L. A. presented at the Sixth International Conference on the Chemistry of the Organic Solid State, Freiburg, German Federal Republic, October 4-8, 1982, Mol. Cryst. Liq. Cryst. 1983, 93, 293; (c) Sandman, D. J.; Rubner, M.; Samuelson, L. J. Chem. Soc. Chem. Commun. 1982, 1133.
2. Labes, M. M.; Love, P.; Nichols, L. F. Chem. Rev. 1979, 79, 1.
3. Sandman, D. J. J. Electronic Materials 1981, 10, 173.
4. Diaz, A. F.; Kanazawa, K. K.; Gardini, G. P. J. Chem. Soc. Chem. Commun. 1979, 635.
5. Kanda, S.; Pohl, H. A., in "Organic Semiconducting Polymers"; Katon, J. E., Ed.; Marcel Dekker: New York, 1968; p. 176.
6. Chiang, C. K.; Heeger, A. J.; MacDiarmid, A. G. Ber. Bunsenges Physical. Chem. 1979, 83, 407.
7. Etemad, S.; Heeger, A. J.; MacDiarmid, A. G. Ann. Rev. Phys. Chem. 1982, 33, 443 and MacDiarmid, A. G. et al., This Symposium.

8. Wegner, G. Angew. Chem. Internat. Edit. 1981, 20, 361.
9. Baughman, R. H.; Bredas, J. L.; Chance, R. R.; Elsenbaumer, R. L.; Shacklette, L. W. Chem. Rev. 1982, 82, 209.
10. (a) Rabolt, J. F.; Clarke, T. C.; Kanazawa, K. K.; Reynolds, J. R.; Street, G. B. J. Chem. Soc., Chem. Commun. 1980, 347; (b) Chance, R. R.; Shacklette, L. W.; Miller, G. G.; Ivory, D. M.; Sowa, J. M.; Elsenbaumer, R. L.; Baughman, R. H. J. Chem. Soc., Chem. Commun. 1980, 348. (c) Shackelette, L. W.; Elsenbaumer, R. L.; Chance, R. R.; Eckhardt, H.; Frommer, J. E.; Baughman, R. H. J. Chem. Phys. 1981, 75, 1919. (d) Rubner, M.; Cukor, P.; Jopson, H.; Deits, W. J. Electron. Mater. 1982, 11, 261. (e) Clarke, T. C.; Kanazawa, K. K.; Lee, V. Y.; Rabolt, J. F.; Reynolds, J. R.; Street, G. B. J. Polym. Sci., Poly. Phys. Edit. 1982, 20, 117. (f) Miller, G. G.; Ivory, D. M.; Shacklette, L. W.; Chance, R. T.; Elsenbaumer, R. L.; Baughman, R. H., European Patent Application 0031444, 1980.
11. Mazurek, H.; Day, D.; Maby, E. W.; Abel, J. F., Senturia, S. D.; Dresselhaus, G.; Dresselhaus, M. S. J. Poly. Sci. Poly. Phys. Edit. 1983, 21, 537.
12. Engler, E. M.; Patel, V. V. J. Am. Chem. Soc. 1974, 96, 7376.
13. Edmonds, Jr., J. T.; Hill, Jr., H. W. U.S. Patent 3,354,129, 1967.
14. Okamoto, Y.; Yano, T.; Homsany, R. Ann. N. Y. Acad. Sci. 1972, 192, 60.
15. Sandman, D. J.; Stark, J. C.; Foxman, B. M. Organometallics 1982, 1, 739.
16. Bunnett, J. F. Acc. Chem. Res. 1978, 11, 413.
17. Scamehorn, R. G.; Bunnett, J. F. J. Org. Chem. 1977, 42, 1449.
18. Gladysz, J. A.; Hornby, J. L.; Garbe, J. E. J. Org. Chem. 1978, 43, 1204.
19. Sandman, D. J.; Stark, J. C.; Acampora, L. A.; Gagne, P. Organometallics 1983, 2, 549.
20. Kuder, J. E., in "Organic Selenium Compounds: Their Chemistry and Biology", Klayman, D. L.; Gunther, W. H. H., Eds.; Wiley-Interscience: New York, 1973; p. 865 ff.
21. Emma, T.; Tripathy, S. unpublished experiments.
22. Lenz, R. W.; Handlovits, C. E.; Smith, H. A. J. Polym. Sci. 1962, 58, 351.
23. Rubner, M. unpublished experiments.
24. Irgolic, K. J., in "The Organic Chemistry of Tellurium", Gordon and Breach Science Publishers: New York, 1974; Chapter 7.
25. Blinne, G.; Naarmann, H.; Penzien, K. U. S. Patent 4,344,869, 1982.

RECEIVED October 20, 1983

Electrochemical Synthesis and Characterization of Poly(2,2′-bithiophene)

M. A. DRUY, R. J. SEYMOUR, and S. K. TRIPATHY

GTE Laboratories, Inc., Waltham, MA 02254

> 2,2′ Bithiophene can be simultaneously electrochemically oxidized and polymerized to yield conducting films of composition $[C_4H_2S(ClO_4)_{0.19}]_x$. The conductivity of the polymer varies from 0.04–1.0 $(\Omega\,cm)^{-1}$ depending on polymerization conditions. The polymer has interesting electrochromic properties. Switching speeds of 500 ms are obtainable for an electrode of area 2.4 cm^2. A variety of growth patterns emerge as one goes from a thin film to a thicker film, and these are dependent on the polymerization conditions.

Currently, much work is devoted to the synthesis of conducting polymers for use in a variety of applications. Polyacetylene, the prototype conducting polymer, has been successfully demonstrated to be useful in constructing p-n heterojunctions, (1) Schottky barrier diodes, (2,3) liquid junction photoelectrochemical solar cells, (4) and more recently as the active electrode in polymeric batteries. (5) Research on poly (p-phenylene) has demonstrated that this polymer can also be utilized in polymeric batteries. (6)

The improved electrochemical synthesis (7) of poly pyrrole has led to its use as coating for the protection of n-type semiconductors against photocorrosion in photoelectrochemical cells. (8,9) Recently, it was announced that pyrrole was not the only five-membered heterocyclic aromatic ring compound to undergo simultaneous oxidation and polymerization. Thiophene, furan, indole, and azulene all undergo electrochemical polymerization and oxidation to yield oxidized polymers of varying conductivities (5 x 10^{-3} to 10^2 Ω^{-1} cm^{-1}). (10-13) The purpose of our research was to synthesize oxidized poly 2,2′ bithiophene via simultaneous oxidation and polymerization. During the course of this research, we discovered that this material was electrochromic, i.e., a reversible color change occurred when it was switched between the oxidized and neutral states.

Experimental

Electrochemical synthesis and switching studies were performed using a signal programmer built in our laboratory, a H-P 3310 B function generator, a PAR Model 173/179 potentiostat/galvanostat, and a Houston Instruments Model 2000 X-Y recorder. Visible absorption spectra were obtained on a Cary 17D spectrophotometer. Electro-optical studies were performed using a ISA, Inc. light source and monochromator and a silicon photodiode with the response recorded on an oscilloscope. Infrared spectra were recorded on a Nicolet Instruments Fourier Transform IR spectrometer. Chemical analysis was performed by Schwarzkopf Microanalytical Laboratories, Woodside, NY. Scanning electron micrographs were obtained on a Jeol microscope. Transmission electron microscopy was performed using a Phillips 400 EMT microscope.

Samples for transmission electron microscopy were prepared in the following manner. Films were grown with different current rates up to various thicknesses on platinum working electrodes. The film thickness was controlled by the period of current flow. The films were transferred onto carbon coated electron microscope grids by stripping with formvar and subsequently removing the formvar with methylene chloride. As-synthesized films were directly used for scanning electron microscopy.

Acetonitrile (MCB Chromatoquality) was distilled in an argon atmosphere over P_2O_5, with only the middle fraction being collected, and subsequently degassed on a vacuum line. $LiClO_4$ (G.F. Smith) was dried for ca. 12 hours at 105° C while under dynamic vacuum. 2,2' Bithiophene (Kodak) was melted under dynamic vacuum and allowed to cool. All materials, after purification, were handled in an inert atmosphere dry box (Vacuum Atmospheres). Electrochemical synthesis and switching studies were either performed in the dry box or outside the dry box in tightly stoppered argon-filled cells. Free-standing films of poly 2,2' bithiophene perchlorate were synthesized by inserting a platinum foil as the working electrode in an acetonitrile solution containing 0.1 M $LiClO_4$ and 0.010 M 2,2' bithiophene. A nickel foil served as the counter electrode and a Ag/Ag^+ (Ag/0.1 M $AgNO_3$) reference electrode was used. The working electrode was separated from the counter electrode by a medium porosity frit. The reference electrode made contact with the solution via a Lugin capillary. Both potentiostatic and galvanostatic conditions were used to synthesize the free-standing films. Thin films of poly 2,2' bithiophene perchlorate were also grown on SnO_2 coated glass electrodes using galvanostatic conditions. Neutral films were produced by reversing the direction of current flow until the potential of the working electrode indicated the polymer was no longer being reduced to the neutral state. During this reduction, the films underwent a color change from green to red.

When potentiostatic conditions were used, the working electrode was held at a potential of +0.7 V vs the Ag/Ag$^+$ reference electrode. A smooth green film formed immediately on the surface of the working electrode. The current density was ca. 0.5 mA/cm^2. After ca. 30 minutes, the texture of the film became quite rough and globular in appearance with a powdery surface. After polymerization was halted, the film was removed from the electrode, washed repeatedly in fresh acetonitrile, and sent for chemical analysis. The dc electrical conductivity (4-probe) of a washed and dried film synthesized by the above procedure was 0.04 (ohm-cm)$^{-1}$.

Infrared spectra of the oxidized film were obtained by grinding some film with dried KBr in the globe box. The pellet was pressed in the globe box and the spectra recorded. An infrared spectrum of the neutral polybithiophene was obtained by scraping off a 4000 angstrom thick film from the surface of a SnO$_2$ coated glass electrode. Subsequently, the film was ground with KBr and a pellet was made as described above.

Both the visible absorption spectra and the optical response were recorded in situ. Films grown on SnO$_2$ coated glass served as the working electrode. The counter electrode was a nickel foil with several 3 mm diameter holes in it so that the transmitted light would pass through to the detector.

Results and Discussion

The chemical analysis of the oxidized polymer is given in Table I. The analysis indicates there are approximately 0.2 ClO$_4$ anions/ thiophene ring. Also, there is excess H, ca. 1H/2 thiophene rings; thus some hydrogenation may have occurred during the simultaneous oxidation and polymerization. Other researchers have found similar occurrences in polypyrrole and polythiophene. (1,10,11)

Table I. Chemical Analysis for Poly Bithiophene Perchlorate

	C(%)	H(%)	S(%)	Cl(%)	O (%) (By Diff.)
Calculated:	47.57	1.98	31.71	6.68	12.06
Found:	48.07	2.50	31.21	6.81	11.41

Films grown under low current density (galvanostatic conditions), i.e., 0.010 mA/cm^2, tend to have superior electrical properties and mechanical integrity to those grown under a higher current density, i.e., 0.1 mA/cm^2. The conductivity of a freestanding film grown under a low current density was 1 (ohm-cm)$^{-1}$. When this film was exposed to laboratory air, the conductivity decreased as illustrated in Figure 1. After ca. 10 days exposure, the conductivity appeared to level off at 0.2 $(ohm-cm)^{-1}$. The reason for this is not clear; however, it is possible that a barrier is formed on the surface of the film. This barrier may prevent further degradation of the polymer.

Figure 2 shows the infrared spectra of both the as-grown oxidized polybithiophene (a) and the neutral polybithiophene (b). The peak which appears at 800 cm^{-1} is due to a C-H out-of-plane vibration (14) and is characteristic of 2,5 disubstitution. The C-H stretching mode at 2950 cm^{-1} is also characteristic of 2,5 disubstitution. Peaks present in the 1400-1500 cm^{-1} range are probably connected with a ring-stretching absorption. (14) The oxidized polybithiophene has a weak absorption at 617 cm^{-1}. This is assignable to the ClO_4 anion, also present, but obscured by the C-H in-plane deformation is a stronger ClO_4 anion absorption at 1080 cm^{-1}. Both spectra contain spurious peaks at 3480 cm^{-1}, most likely due to residual moisture in the KBr.

Figure 3 shows the visible absorption spectra of both the neutral (a) and oxidized (b) film (obtained in situ). The neutral film is red in color as indicated by the absorption at 480 nm. The peak at 480 nm can be attributed to a $\pi-\pi^*$ transition of the aromatic rings. It should be noted that in going from 2,2' bithiophene to 2,2'-5,2" terthiophene, the resulting $\pi-\pi^*$ transition shifts from 301 nm to 350 nm. (14) Therefore, it is not unreasonable to assume that the peak at 480 nm is assignable to the $\pi-\pi^*$ transition of the polymer. The absorption spectrum in Figure 3(b) is obtained by anodically passing 10 mC/cm^2. During this process the film changes in color from red to green. It is interesting to note that the peak at 480 nm, corresponding to the interband transition, decreases considerably in going from the neutral polymer to the oxidized polymer. This suggests the near closing of the band gap. The absorption at 750 nm for the oxidized polymer is in the same region as a peak assigned to charge transfer in complexes of 2,2' bithiophene and TCNE. (15)

A 4000 angstrom film may be grown on a SnO_2 coated glass electrode by passing 35 mC/cm^2. In general, the thickness of the film can be controlled by varying the current density and/or electrolysis time. The as-grown oxidized film may be reduced to the neutral polymer in a solution containing 0.1M $LiClO_4$ in acetonitrile, as described above. A cyclic voltammagram of neutral polybithiophene is shown in Figure 4. The substrate electrode was SnO_2 coated glass, and a 4000 angstrom thick film of poly bithiophene is on its surface. As indicated in the figure, the peak oxidation potential is +0.9V vs Ag/Ag^+, and the peak in reduction is +0.1V vs Ag/Ag^+. By stepping between 0.0V and +1.0V vs Ag/Ag^+ for a 1 s pulse duration and then returning to 0.0V, the film may be made to switch from red to green and back to red.

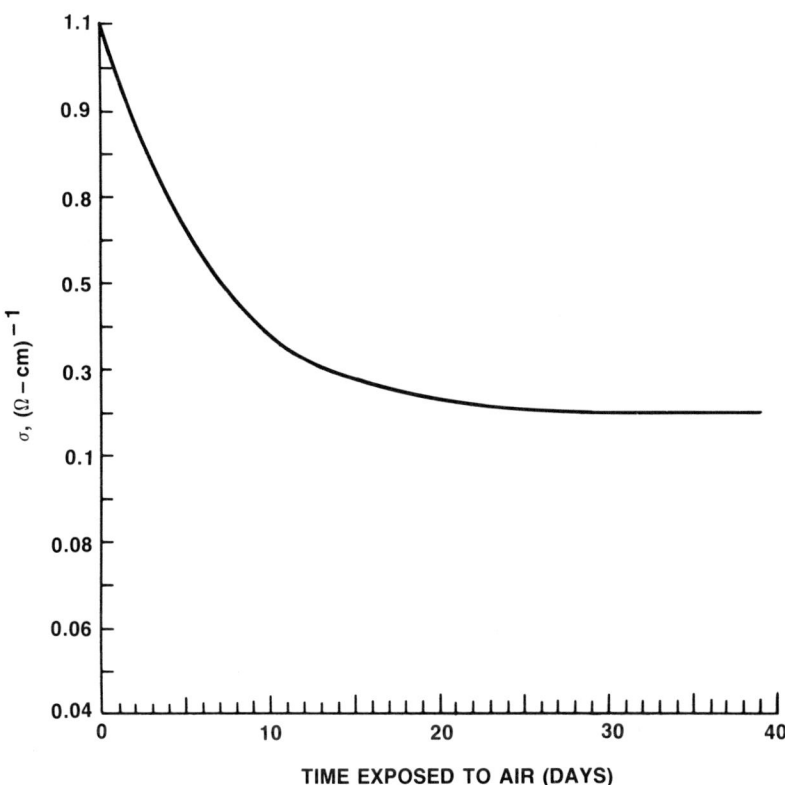

Figure 1. Conductivity of free-standing film of oxidized poly 2,2' bithiophene vs time exposed to air.

478 POLYMERS IN ELECTRONICS

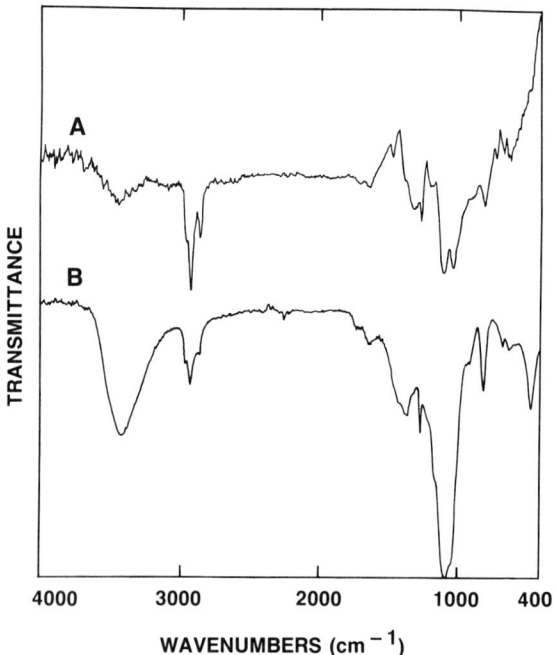

Figure 2. Infrared spectra of (a) as-grown oxidized polybithiophene, (b) and the neutral polybithiophene.

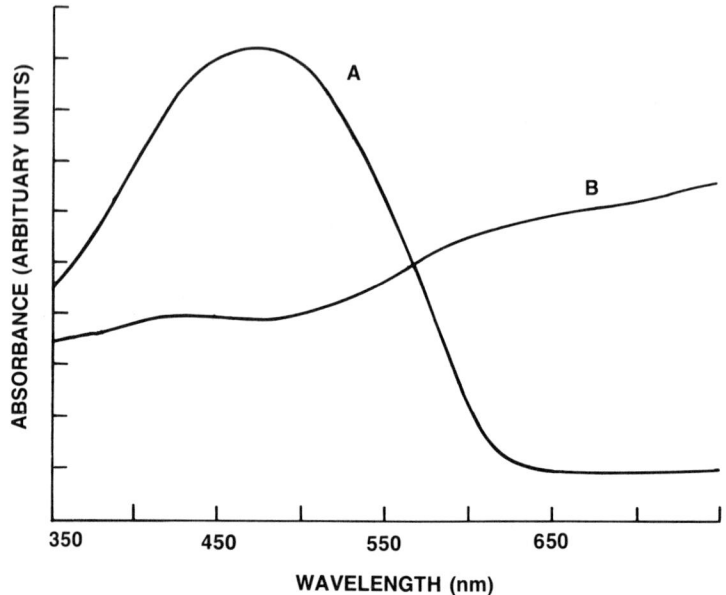

Figure 3. In situ visible absorption spectra of (a) neutral and (b) oxidized poly(2,2' bithiophene).

The optical response of the switching process was measured in situ by measuring the transmission at 640 nm. Figure 5 shows this response. The upper signal corresponds to the applied voltage switching from 0.0 V vs Ag/Ag$^+$ to +1.0 V vs Ag/Ag$^+$. The lower signal represents the transmission as detected by the photodiode. As the applied voltage changes from 0.0 V to +1.0 V, the film changes from red to green and the optical response represents the change in the amount of transmitted light. As illustrated in Figure 5 (for an electrode surface 2.4 cm^2 and 4200 A thickness), switching speeds of 500 ms are obtainable. Approximately 10 mC/cm^2 are required to switch from red to green. The response time of an electrochromic display was shown to be dependent on the surface area of the active electrode. (16) Larger surface areas result in slower response times. Given that the area of the electrode is 2.4 cm^2, the switching speed of 500 ms is not unexpected. Presumably, faster switching speeds could be realized with smaller electrode areas.

The morphology of the poly(2,2' bithiophene) films is quite complex and appears to be a function of several factors such as the nature of the growth substrate, the growth stage, and the rate of growth as controlled by the current rate. When a platinum electrode is used as the growth substrate, a smooth thin layer several hundred angstroms thick initially covers the entire electrode surface.

The transmission electron micrographs (Figure 6a, b) show this layer consisting of continuously overlapping platelet-like structures. Inorganic crystallites (trace amounts from the electrolyte solution) can also nucleate on the bare electrode at this stage of growth. Their replica is seen in the carbon coating (arrow). Because these crystallites did not strip off with the formar, they were not observable by electron diffraction. The back scattered X-ray energies properly revealed the presence of sulfur and chlorine (from the polybithiophene perchlorate) in these platelet structures.

Once a continuous ∿ 1 μm thick film forms on the electrode surface, further growth occurs in a very uneven manner. Mushroom-like growths at isolated places are initiated. As the films grow thicker, the entire area becomes covered by these globular growths. Figures 7 and 8 (scanning electron micrographs) show this type of growth both in the early stages and in the late stages. We speculate that, once a continuous thin layer of the polybithiophene perchlorate is formed, further growth occurs preferentially at the point of high electric fields and where electron transfer can take place with relative ease (the grown film is not an ideal conductor). This is shown in Figure 9 in which a selected area scanning electron micrograph was taken from a film grown under the above conditions.

With slower growth rates (achieved by using lower current densities), thicker uniform films are obtained. Beyond a certain point (several microns) nodular growths again begin to appear and

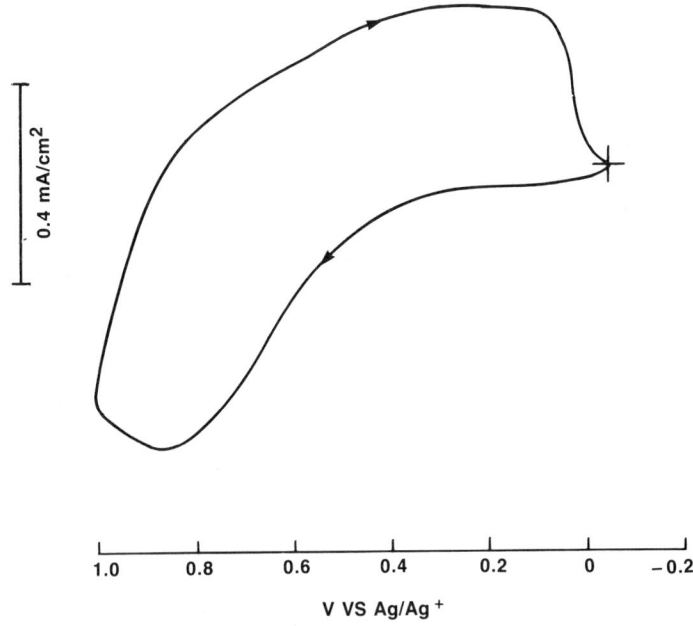

Figure 4. Cyclic voltammagram of neutral polybithiophene.

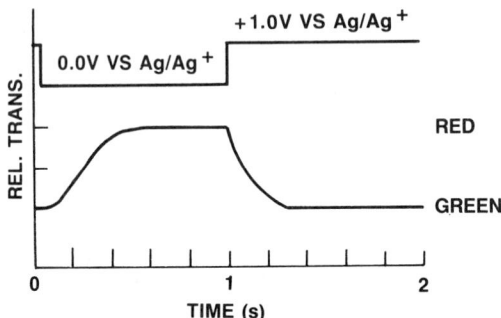

Figure 5. In situ response time of poly(2,2' bithiophene). Relative transmission at 640 nm vs time.

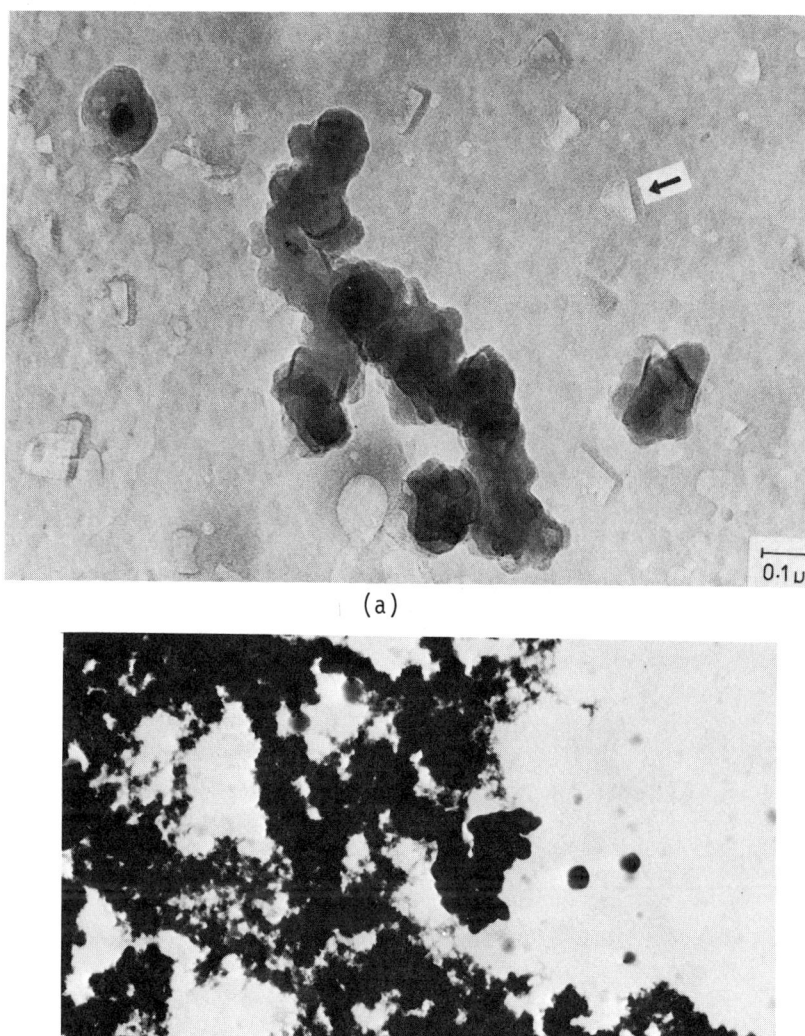

Figure 6. Transmission electron micrograph of oxidized polybithiophene, arrow indicates replica of trace amounts of inorganic crystallites (a. small area and b. large area).

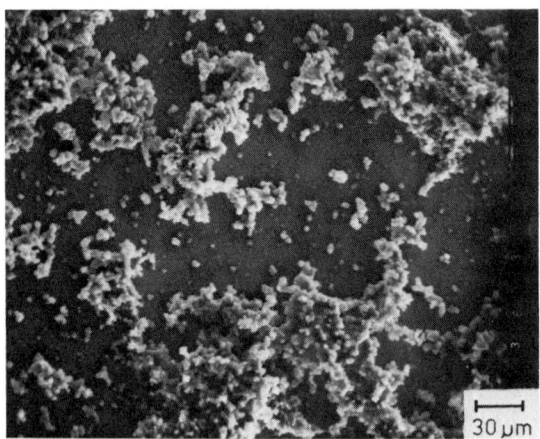

Figure 7. Scanning electron micrograph of oxidized polybithiophene showing globular growth occurring during early stages of polymerization.

Figure 8. Scanning electron micrograph of oxidized polybithiophene showing globular growths occurring during later stages of polymerization.

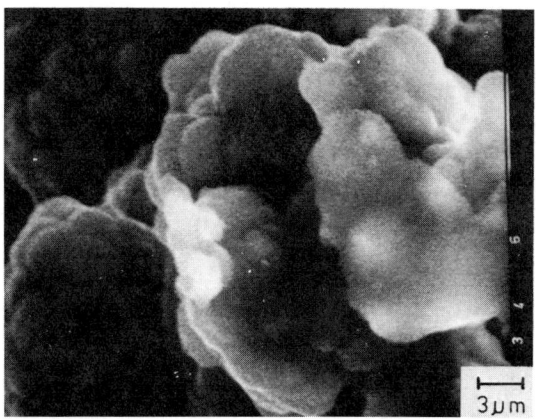

Figure 9. Scanning electron micrograph of a selected area showing large growths obtained under high current density conditions.

further growth occurs at those locations. Thus, by decreasing the current density, it is possible to modify the growth conditions so that thicker, more uniform films (than obtained from high current density conditions) are produced. Even under low current densities, however, nodules begin to grow once a certain thickness is reached.

The fact that the growth morphology is related to the growth stage and the current conditions (stretching the polymerization process over days using low current densities) opens up the interesting possibility for structural manipulation in a manner analogous to "living" polymers. We are presently exploring this and other structurally similar monomers.

Electron diffraction from thin films of polybithiophene perchlorate indicates some crystallinity, and yet the crystallites are so small and the degree of crystallinity so low that an unambiguous structural determination from the diffraction pattern was not possible.

Conclusions

We have shown that 2,2' bithiophene undergoes simultaneous electrochemical oxidation and polymerization to yield a conducting polymer of composition $[C_6H_4S(ClO_4)_{0.19}]_x$. The polymerization occurs through positions α to the S atom. Conductivities of 0.04 - 1.0 $(\Omega \text{ cm})^{-1}$ are obtained depending on the conditions of growth. The oxidized film may be electrochemically reduced to the neutral polymer. When this occurs, the polymer changes in color from green to red. This process is reversible and may serve as the basis for an electrochromic display. Electron microscopy indicates that the growth process is very complicated, and a variety of growth patterns emerge as one goes from a thin film to a thicker film. Electron diffraction reveals some crystallinity; however, the crystallites are very small.

Acknowledgments

The authors wish to acknowledge Tom Emma and Kim Ostreicher for their assistance with the transmission electron microscopy, Michael Salmon for his assistance with the scanning electron microscopy, and Paul Martakos for his assistance with the Fourier transform infrared.

Literature Cited

1. See, for example, Ozaki, M.; Peebles, D.; Weinberger, B. R.; Heeger, A. J.; and MacDiarmid, A. G. J. Appl. Phys. 1980, 51, 4252.
2. Waldrop, J. R.; Cohen, M. J.; Heeger, A. J.; MacDiarmid, A. G. Appl. Phys. Lett. 1981, 38, 53.
3. Grant, P. M.; Tani, T.; Gill, W. D.; Krounbi, M.; Clarke, T. C. J. Appl. Phys. 1981, 52, 869.
4. Chen, S. N.; Heeger, A. J.; Kiss, Z.; MacDiarmid, A. G.; Gau, S. C.; Peebles, D. L. Appl. Phys. Lett. 1980, 36, 96.
5. MacInnes, D., Jr.; Druy, M. A.; Nigrey, P. J.; Nairns, D. P.; MacDiarmid, A. G.; Heeger, A. J. J. C. S. Chem. Comm. 1981, 317.
6. Elsenbaumer, R. L.; Shacklette, L. W.; Sowa, J. L.; Chance, R. R.; Ivory, D. M.; Miller, G. G.; Baughman, R. H. Polym. Preprints 1982, 23, 132.
7. Kanazawa, K. K.; Diaz, A. F.; Gill, W. D.; Grant, P. M.; Street, G. B.; Gardini, G. P.; Kwak, J. F. Synth. Met. 1979, 1, 329.
8. Noufi, R.; Frank, A. J.; Nozik, A. J. J. Amer. Chem. Soc. 1981, 103, 184.
9. Skotheim, T.; Lundstrom, I.; Prejza, J. J. Electrochem. Soc. 1981, 128, 1625.
10. Tourillon, G.; Garnier, F. J. Electroanal. Chem. 1982, 135, 173.
11. Kaneto, K.; Yoshino, K.; Inuishi, Y. Jap. J. Appl. Phys. 1982, 21, L567.
12. Bargon, J.; Mohmand, S.; Waltman, R. Sixth ICOSS, Freiburg, West Germany, 1982.
13. Waltman, R. J.; Bargon, J.; Diaz, A. F. J. Phys. Chem. 1983, 87, 1459.
14. Curtis, R. F.; Phillips, G. T. Tetrahedron 1967, 23, 4419.
15. Abu-Eittah, R.; Al-Sugier, F. Can. J. Chem. 1976, 54, 3705.
16. Viennet, R.; Randin, J. P.; Raistrick, I. D. J. Electrochem. Soc. 1982, 129, 2451.

RECEIVED September 2, 1983

The Influence of Microstructure on the Properties of Polyacetylene/Polybutadiene Blends

S. K. TRIPATHY and M. F. RUBNER

GTE Laboratories, Inc., Waltham, MA 00254

> The mechanical and electrical properties of polyacetylene (PA) were modified by blending it with polybutadiene (PB). Further enhancement of the electrical conductivity of the blends was obtained by stretch elongation of the blends prior to doping. A structural model, based on a complex process of stretch induced ordering in the polyacetylene domains, was proposed to account for these observations. Support for this model was obtained using electron microscopic techniques. Low polyacetylene content blends (<20% PA) were found to consist of discrete polyacetylene domains dispersed in a continuous polybutadiene matrix. In the high polyacetylene content blends (>70% PA), both phases were simultaneously continuous, forming an interpenetrating network structure. Blends with intermediate compositions consist of both continuous and isolated domains of polyacetylene distributed throughout the polybutadiene matrix.

In the past decade, several polymers have been characterized that exhibit high electronic conductivity upon exposure to oxidizing or reducing dopants. (1) In spite of a wide variety of molecular compositions, these polymers have certain structural similarities which account for their unique electrical properties. For example, all known conducting polymers have an extended chain conformation and either exhibit a planar structure or adopt a planar structure subsequent to doping. (2) The polymer backbone also has a degree of conjugation considered necessary for high conduction. These same structural features, however, lead to very poor mechanical properties. The prototype conducting polymer, polyacetylene (PA), for example, can be doped with iodine to a conductivity of ca. 500 ohm^{-1} cm^{-1} (3) and yet the doped polymer is an intractable, brittle material. The modest mechanical properties of undoped polyacetylene, on the other hand, can only

0097-6156/84/0242-0487$06.00/0
© 1984 American Chemical Society

be observed when the polymer is handled with the rigorous exclusion of oxygen. (4) Under these conditions, it was shown that a maximum extension ratio of only 3.3 (L/Lo) could be obtained for cis-PA. Even at these extension ratios, the basic chain organization in the morphological subunits remains largely unaltered.

It is possible, however, to blend these intrinsically brittle polymeric conductors with polymers that enhance their mechanical properties. In the case of polyacetylene, this has been accomplished by polymerizing acetylene gas in the presence of a suitable host polymer. (5-7) Since polyacetylene is actually grown in the matrix of the host polymer, and not simply physically dispersed, the resultant morphology of the polyblend (and, hence, the electrical and mechanical properties of the system) can be manipulated by adjusting the reaction conditions. In addition, by proper selection of the blending component, it is possible to further modify the properties of the polyblend by physical means.

It is therefore of interest to investigate the morphological features of these blends, especially in relation to their electrical and mechanical properties. It is also important to explore further whether the nature of chain organization of polyacetylene can be influenced subsequent to synthesis. Model calculations indicate that the backbone single bond in polyacetylene is quite flexible, (2) and there is evidence that the extent of crosslinking may not be high (at least in pristine cis-polyacetylene). (8) Under these conditions, stress induced chain alignment in the morphological subunits may be possible if high extension ratios can be achieved.

In this paper, we explore the morphological features of polyacetylene/polybutadiene (PA/PB) blends in detail via electron microscopic techniques. We have also extended our original study of stress induced property enhancement in blends with polyacetylene compositions in the 40-60% range (Blend paper 1, ref. 5). The PA/PB blends have been singled out for detailed morphological studies for the following reasons. First, while the physical properties of the two systems are drastically different and as a result complement each other, their chemical structures are not markedly different. Thus, we expect the two phases in this heterogeneous blend to be in intimate physical contact possibly with a narrow interphase. Hence, the polybutadiene phase can act as a template for stress induced ordering of the intertwined polyacetylene phase. Second, the polybutadiene phase is readily soluble in non-polar solvents such as toluene. Thus, the elastomeric phase can be extracted and scanning electron microscopy can yield valuable information regarding the morphology of unextracted phase and hence that of the whole system.

Experimental

The methods used to prepare polyacetylene/elastomer blends have been described in our previous publications. (5,6) The following procedure highlights the preparation of PA/PB blends. Cis-1,4 polybutadiene (98%; average M.W. 200,000 to 300,000) is first dissolved in a toluene solution containing 1 to 5 ml of the stock catalyst solution (1.7 ml Ti(OBu)$_4$ and 2.7 ml of Et$_3$AL in 20 ml of toluene). After thorough mixing, the solvent is removed by dynamic vacuum leaving a film of polybutadiene impregnated with the catalyst on the walls of the reaction vessel. After the vessel is cooled to $-78^{\circ}C$, acetylene gas is introduced until a pressure of 1 atm is reached. The final composition of the blend is a function of the exposure time to acetylene gas and the amount of catalyst added to the elastomer. Once prepared, the resultant blend is washed with cold ($-78^{\circ}C$) heptane and dried at 10^{-3} torr for 8 hours. The procedures used to dope the blends have been described in previous publications. (9) DC conductivities were measured using standard four-point probe techniques. Samples were stretched in an ambient atmosphere using a "homemade" bench-top machine capable of operating at a constant strain rate. Sample dimensions were typically 0.35 cm wide and 0.01 cm thick.

Samples for transmission electron microscopy were prepared in the following manner. A 50/50 blend was potted in an acrylic resin (London Resin Co. Ltd.) and subsequently microtomed in thin films of approximately 600 angstrom thickness and transferred onto electron microscope grids. These sections were then doped with iodine vapor (stained) and were ready for transmission electron microscopy. The blends of other compositions were not examined in this manner for reasons discussed later.

For extraction studies, blends of high and low polyacetylene content were extracted with dry oxygen-free toluene for several days. Extracted samples were removed from the wash at reasonable intervals and dried in vacuum. The surfaces of the samples were coated with approximately a 50 angstrom layer of gold by sputter coating. This provides a conducting surface to avoid charging effects and results in an excellent yield of secondary electrons enabling high contrast from the surface topography. Small grid size samples were sandwiched between TEM grids to be studied under the SEM mode of a Phillips electron microscope (EM 400T).

Results and Discussion

When acetylene gas is polymerized in a solid solution of the Shirakawa catalyst and polybutadiene, a heterogenous blend consisting of a amorphous polybutadiene phase and a crystalline polyacetylene phase is formed. (5) The mechanical and electrical properties of this composite are critically dependent on the composition of the blend components and on their relative arrangement. In our initial Blend paper, (5) for example, we showed that the mechanical properties of PA/PB blends are a function of the blend composition, with low polyacetylene compositions exhibiting rubbery elastic properties (low tensile strength and reversible strain) and high polyacetylene compositions behaving more plastic-like (high tensile strength and plastic strain). As expected, the electrical conductivity of the doped blend is also a function of the polyacetylene composition of the material. (5) Furthermore, stretch induced elongation of the blends leads to a dramatic increase in conductivity subsequent to doping, further confirming that the electrical properties are also very sensitive to the arrangement of the respective phases.

This observation, however, is restricted to blends with polyacetylene compositions in the 40-60% range where each discrete phase seems to influence the structural features of the adjacent domains of the second phase. For example, table 1 lists the electrical conductivity of the PA/PB blends in this compositional range before and after stretch elongation and subsequent doping with iodine. As can be seen, by stretching the blend before doping, conductivity levels an order of magnitude higher than the as prepared blends can be achieved, and in some cases, the conductivity has been increased 40 times that of the unstretched material. The variation in the dopant weight uptake between the stretched and unstretched blends is due to excess iodine which remains trapped in the polybutadiene phase and contributes to the final weight uptake of the doped blend. The presence of any iodine in the polybutadiene phase does not affect the electronic properties of the blend. All blends that exhibit this stretch enhanced conductivity undergo complete plastic deformation on stretching. Improved order in the crystalline phase is also indicated.(5)

X-ray diffraction patterns of blends with a 40-60 percent PA composition show an increase in the intensity of the interchain reflection of crystalline cis-PA ($2\theta=23.9°$) after stretching. (5) Additionally, analyses (DSC,FTIR) of blends prior and subsequent to stretching clearly indicate that the composition of the material remains unchanged. This suggests that stretch elongation has resulted in an improvement in the degree of crystallinity and the extent of order in the normally intractable polyacetylene phase. In order to account for these results, the organization of the polyacetylene and the polybutadiene phases in the blend must be considered. We speculate that in the 40-60% compositional range, both phases are simultaneously continuous or quasi-continuous. Interdiffusion of the polymer chains of the two components of the blend at their interface, most likely in the amorphous regions results in intimate contact (strong adhesion) between the two domains. Upon stretching, the stress is directly transferred to the polyacetylene domains which are subsequently stretch oriented, deforming the polybutadiene matrix in tandem (which results in a complete plastic deformation). Thus, a complex process of stress induced ordering is taking place in the polyacetylene phase.

Further direct evidence for this hypothesis comes from elec-

Table 1. Electrical Conductivity of Polyacetylene/Polybutadiene Blends as a Function of Blend Composition and Mechanical Treatment.

Blend composition (PA/PB)	Maximum wt.% dopant uptake [a]	Conductivity ohm^{-1} cm^{-1}	Extension ratio L/Lo [b]	Maximum wt.% dopant uptake [c]	Conductivity ohm^{-1} cm^{-1}
1. 40/60	35	15	6.0	91	575
2. 50/50	79	12.4	3.8	84	223
3. 40/60	23	4	4.2	19	50
4. 50/50	34	15	7.2	62	110
5. 50/50	--	50	6.5	--	325

[a] Doped with iodine

[b] Lo is the starting length, L is the final length at break

[c] Doped with iodine after stretching

tron microscopic studies. In our proposed model, blends with intermediate polyacetylene compositions (40-60%) exhibit phases that are simultaneously continuous or quasi-continuous. For high polyacetylene content blends, we expect an interpenetrating network of the two phases. Thus, extraction of the polybutadiene phase will leave behind a network structure of the interconnected polyacetylene domains. For blends with low polyacetylene content on the other hand, we propose a discrete polyacetylene phase dispersed in a continuous polybutadiene matrix. Extraction of polybutadiene in this case is expected to reveal a particulate morphology of the embedded polyacetylene domains. However, since these domains can fall off the surface subsequent to the rubber extraction, their distribution as seen in the micrographs is not an accurate measure of the distribution statistics.

Both of these interpretations are clearly supported by scanning electron microscopy studies of extracted PA/PB blends. For example, Figure 1 shows a sequence of micrographs of a low polyacetylene content blend (<20% PA) extracted for different times with toluene. Figure 1(a) shows the surface texture of a typical unextracted PA/PB blend. The surface exhibits features characteristic of polybutadiene cast from a toluene solution. The sample extracted for three days (Figure 1(b)) reveals an uneven surface topography resulting from removal of polybutadiene from predominantly rubbery regions. The darker regions are the areas where the rubber has been preferentially extracted. Under the imaging conditions, the regions of brighter illuminate are "convex" to the surface. Finally, Figure 1(c) shows the surface topography of a blend extracted for eight days. The surface is dominated by a distribution of particulate domains of a few hundred angstroms. The exact density of this distribution and the domain sizes are not to scale in this micrograph for the following reasons. First, as suggested by the surface topography, the particulate structures are not interconnected and, therefore, several of the smaller less-strongly entangled or embedded regions can fall off the surface. Second, the 50 angstrom coating adds approximately 100 angstroms to the physical dimensions of the elevated domains. However, the conclusion that the polyacetylene forms a particulate discrete morphology in a continuous rubbery matrix is convincing.

A similar extraction study of a high polyacetylene content blend (>70% PA) is shown in Figure 2. Again, a featureless surface is observed for the unextracted samples.(the assorted geometric shapes shown in this micrograph are simply loose fragments of polybutadiene that adhere to the surface of this rubber). However, extraction for three days with toluene reveals a matted arrangement of the polyacetylene domains (Figure 2(b)). The interconnected polyacetylene network is clearly evident in the final scanning electron micrograph (Figure 2(c)) of the surface of a blend that has been extracted for two weeks with toluene. Again, given the added layer of the sputtered gold, the exact

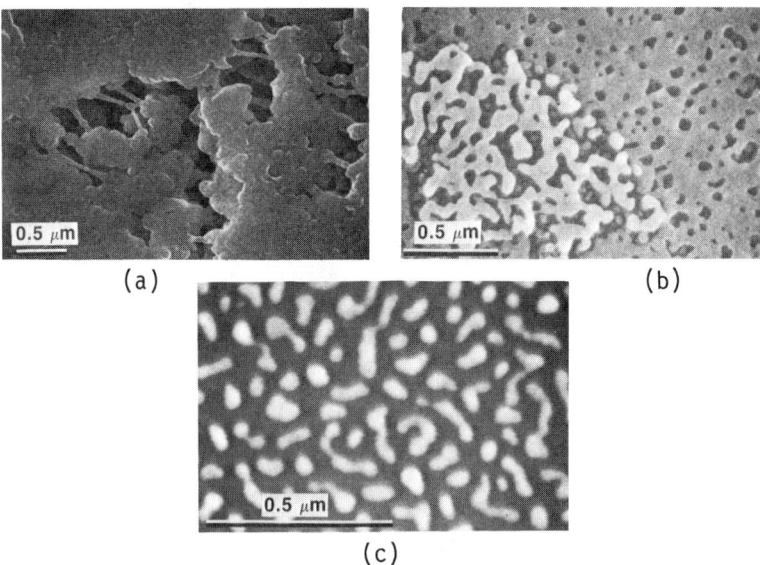

Figure 1. Scanning electron micrographs of a low polyacetylene content blend (20% PA) (a) as prepared, (b) after extraction with toluene for three days, and (c) after extraction for eight days.

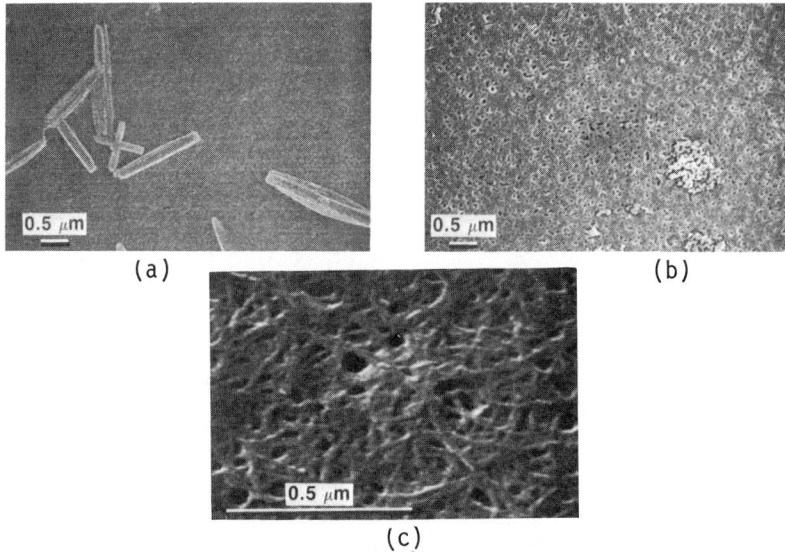

Figure 2. Scanning electron micrographs of a high polyacetylene content blend (70%) PA) (a) as prepared, (b) after extraction with toluene for three days, and (c) after extraction for fourteen days.

dimensions of the morphological subunits are not clear. It is evident, however, that at high polyacetylene compositions both phases are continuous, forming an interpenetrating network.

For all compositions, the polyacetylene domains are crystalline as is revealed by X-ray diffraction. This is further confirmed by electron diffraction from thin microtomed sections of blends with polyacetylene compositions of 40% or higher. Microtoming of samples with less than 40% polyacetylene without sample cooling is difficult due to the rubbery nature of the composite. The polycrystalline nature of the polyacetylene domains is established from the observation that even selected area aperture of a few hundred angstroms produces Debye ring patterns.

To summarize, for low polyacetylene compositions, the blend consists of isolated polyacetylene domains dispersed in a continuous matrix of polybutadiene. At high polyacetylene compositions both components form distinctly separate but continuous phases. From these observations regarding the distribution of polyacetylene in the polybutadiene matrix, we infer that, at the intermediate blend compositions, the crystalline polyacetylene domains must start interconnecting, forming a quasi-continuous network. Evidence for this morphological arrangement, however, cannot be obtained using the extraction technique because, unlike the extreme compositions where removal of the polybutadiene phase qualitatively determines the continuity of the polyacetylene phase, at intermediate compositions much of the nonconnected polyacetylene domains may be removed during the extraction process. The resultant skeletal structure, therefore, may not be representative of the true microstructure.

As a result of this, microtomed blends were studied under the transmission electron microscopy mode to provide a more accurate description of the microstructure in this compositional range. The thin sections were mildly doped (stained) with iodine to provide the requisite phase contrast and moderate conductivity to avoid charging effects. Figure 3 is a representation transmission electron micrograph obtained from such a specimen. Since iodine absorption by the polyacetylene and polybutadiene domains is not entirely selective, due to their somewhat similar chemical structures, a quantitative assessment of the domain shape and size is unrealistic. The amount of iodine incorporated in the polyacetylene domains as a result of doping, however, is much greater than the amount of iodine trapped in the polybutadiene domains. This significant concentration difference provides the phase contrast necessary to assess the organization of the two phases within the blend. In this bright field transmission electron micrograph the darker regions are polyacetylene domains and the lighter regions are polybutadiene domains. Thus, the two phase nature of the blend is clearly evident in this micrograph with polyacetylene forming continuous and isolated domains within the polybutadiene matrix. The two phases are also in the same proportion as dictated by the composition of the blend (determined by weight uptake). The transmission electron micrograph also indicates that the polyacetylene domains have a distribution of sizes, a feature that might have been clouded in an extraction study of this blend composition. Thus we can conclude that substantial evidence for the proposed structural model for PA/PB blends of different composition has been provided through the electron microscopic studies.

Conclusion

Blending of polyacetylene with polybutadiene provides an avenue for property enhancement as well as new approaches to structural studies. As the composition of the polyacetylene component is increased, an interpenetrating network of the polymer in the polybutadiene matrix evolves from a particulate distribution. The mechanical and electrical properties of these blends are very sensitive to the composition and the nature of the microstructure. The microstructure and the resulting electrical properties can be further influenced by stress induced ordering subsequent to doping. This effect is most dramatic for blends of intermediate composition. The properties of the blend both prior and subsequent to stretching are explained in terms of a proposed structural model. Direct evidence for this model has been provided in this paper based upon scanning and transmission electron microscopy.

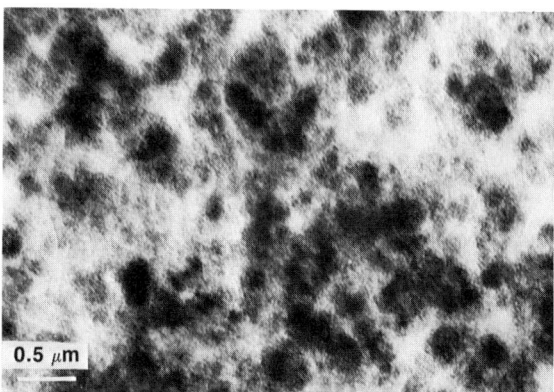

Figure 3. Transmission electron micrograph from a microtomed thin section of a 50/50 PA/PB blend stained with iodine.

Acknowledgments

The authors would like to thank Jacque Georger and Patrica Morris for their experimental contributions and Tom Emma, Kim Ostreicher, Mike Downey, Jack Mullins, and Paul Martarkos for their valuable analytical support.

Literature Cited

1. See, e.g., MacDiarmid, A.G.; Heeger, A.J. Synth. Metals 1980, 1, 101.
2. Tripathy, S.; Kitchen, D.; Druy, M.A. Macromolecules, 1983, 16, 190.
3. Deits, W.; Cukor, P.; Rubner, M.; Jopson, H. J. Electron. Mater. 1981, 10, 683.
4. Druy, M.A.; Tsang, C.H.; Brown, N.; Heeger, A.J.; MacDiarmid, A.G. J. Polym. Sci., Polym. Phys. Ed. 1980, 18, 429.
5. Rubner, M.F.; Tripathy, S.K.; Georger, J. Jr.; Cholewa, P. Macromolecules, 1983, 16, 870.
6. Lee, K.; Jopson, H.; Tripathy, S.; Rubner, M.F. Org. Coat Appl. Polym. Proc. 1983, 48, 598.
7. Galvin, M.E.; Wnek, G.E. Polymer 1982, 23, 795.
8. Bernier, P.; Schue, F.; Sledz, J.; Rolland, M.; Giral, L. Chemica Scripta 1981, 17, 151.
9. Sichel, E.K.; Knowles, M.; Rubner, M.F.; Georger, J. Jr.; J. Phys. Rev. B. 1982, 25, 5574.

RECEIVED September 2, 1983

39

Ethylene-Propylene-Diene Terpolymer/Polyacetylene and Styrene-Diene Triblock Copolymer/ Polyacetylene Blends

Characterization and Stability Studies

KANG I. LEE and HARRIET JOPSON

GTE Laboratories, Inc., Waltham, MA 00254

>Two different blends, EPDM/polyacetylene and Kraton/polyacetylene have been prepared by various blending techniques. The characterization of two blends have been carried out by IR, X-ray and electron microscopic studies. Upon doping of the blends with various electron accepting agents, such as I_2 and $FeCl_3$, conductivities of the blends were found to be in the range of $10 - 100 \Omega^{-1} cm^{-1}$. The conductivity stabilities of the blends have also been studied.

The burgeoning research activity in polyacetylene is apparent upon perusal of journals of various scientific disciplines. Unfortunately, polyacetylene, because of its unsaturated structure, is intractable and unstable to the environment.

In order to circumvent these problems, some workers [1,2] attempted to copolymerize acetylenes with substituted acetylenes. The resulting copolymers, however, were found to have inferior electrical properties compared to homopolyacetylenes. As an alternative approach, Wnek and Galvin [3] prepared a composite of polyacetylene and polyethylene film. In order to introduce polyacetylene into the polyethylene matrix, a high polymerization temperature ($100°C - 110°C$) was employed to break the crystallinity of polyethylene. Such high polymerization temperature might lead to side reactions, such as crosslinking and chain scission. Furthermore, in certain applications, the blend of polyethylene and polyacetylene is still a rigid material because the host polyethylene is partially crystalline. In particular, the stability of polyacetylene is still problematic, although the mechanical properties of polyacetylene were improved somewhat by blending with a processable polymer.

In this paper, we wish to report two different types of elastomer blends, ethylene-propylene-diene terpolymer/polyacetylene and styrene-diene triblock copolymer/polyacetylene, in the hope that the stability of polyacetylene might be improved by such blending techniques. Ethylene-propylene-diene (EPDM) rubber has been chosen since its highly saturated structure affords a material having resistance to attack by heat, light, oxygen, and ozone (4). Furthermore, the presence of low level of unsaturation in the EPDM rubber enables us to crosslink the blends subsequent to their preparation. Such crosslinked structures should lead to localization of polyacetylene chains by preventing molecular movement. It is our hope that immobilization of polyacetylene in the hydrocarbon network of the polymer may contribute to the stabilization of polyacetylene.

The styrene-diene triblock copolymer consists of individual chains of three blocks, an elastomeric diene block in the center and a thermoplastic styrene block on each end. This polymer is called a thermoplastic elastomer. It exhibits some of the physical properties of elastomers at use temperature and is as processable as conventional plastics (5). The styrene/diene triblock copolymer has the unique morphology of glassy polystyrene domains in the rubbery diene matrix. Therefore, such an elastomer does not require conventional vulcanization since the glassy polystyrene domains act as physical crosslinks.

Experimental

All solvents were purified according to the literature methods (6). Sulfur monochloride (Aldrich Chemical Co.) was used as received. Ethylene-propylene-diene terpolymers (Epcar 346 & 585) were obtained from Polysar Incorporated (7). The terpolymers have been purified by dissolving them in heptane and then precipitating in methanol. Three different types of Kraton (a trade name of Shell Chemical Co.) thermoplastic elastomers, Kraton 1107 (styrene-isoprene-styrene triblock copolymer), Kraton 1101 (styrene-butadiene-styrene) and Kraton 4609 (styrene-ethylene-butylene-styrene) were obtained from Shell Chemical Co. (8). The polymers were purifed by dissolving in toluene and precipitating in methanol. IR spectra were obtained using a Perkin-Elmer Model 299B spectrophotometer. Electron micrographs were taken using a Phillips EM 400T instrument. Samples for transmission electron microscopes were either microtomed or casted from toluene.

In a typical experiment, 2g of ethylene-propylene-diene terpolymer was dissolved in freshly distilled toluene in a 3-necked flask under an argon atmosphere. Two ml of Shirakawa catalyst ($Ti(OBu)_4/ALEt_3$) (9) were added to the flask by means of a syringe. Subsequently, all solvents were slowly evaporated under vacuum by rotating the flask to ensure a uniform film of the polymer on the wall of the flask. Next, acetylene gas was introduced into the flask at room temperature. The polymerization of acetylene was evident from the color change of the film (brown → black). The flask was left closed and filled with acetylene overnight at room temperature. Subsequently, the reaction flask was flushed with argon to remove monomer residue. The film was washed with freshly degassed heptane at low temperature. The film was peeled from the side of the flask, and subsequently dried overnight under vacuum. As an alternative procedure, the polymerization of acetylene was carried out by bubbling acetylene gas in the toluene solution of EPDM rubber. In this way, the gel form of the polyacetylene/EPDM rubber blend was produced. After evaporating toluene under vacuum, the gel was pressed and dried under vacuum. Upon doping the film with iodine, the conductivity was measured by standard four-point probe techniques. The crosslinking experiment was carried out by immersing the blend into a 5% S_2Cl_2/toluene solution for 5 to 30 minutes. For radiation curing, a sealed tube containing the blend film was placed in the γ-ray (^{60}Co source) radiation chamber at the dose rate of 1.32×10^5 R/min. for 30 minutes.

The experimental procedure for the preparation of the polyacetylene/styrene-diene triblock polymer blend was essentially the same as that of the EPDM/polyacetylene blend. The polyacetylene/styrene-diene triblock polymer was doped with either I_2 or $FeCl_3$ in nitromethane.

Results and Discussions

EPDM/Polyacetylene Blends. EPDM terpolymer used has ethylidene nobornene as a diene unit. It has a completely saturated hydrocarbon backbone, but with double bonds located on the side chains. EPDM terpolymer is known to have excellent mixing, extruding, and molding characteristics. Thus, the EPDM/Polyacetylene (PA) blend was found to be quite a homogeneous film having excellent flexibility and toughness. The degree of elasticity in the material could be easily controlled by adjusting the ratio of polyacetylene to EPDM polymer in the blends.

IR spectra of the PA/EPDM blends indicated that the blends contain both EPDM and polyacetylene moieties. It was found that the polyacetylene was present in predominantly trans-configuration, as evidenced by a characteristic infrared absorption band at 1015 cm^{-1} (10). Furthermore, there was no evidence that any polyacetylene moieties were grafted onto the unsaturated sites of EPDM rubber. This was corroborated by an extensive extraction experiment. Virtually quantitative amounts of EPDM could be extracted from the blend with toluene. IR spectra of the fully recovered EPDM were identical to those of the virgin EPDM.

Some electron micrographs were taken of PA/EPDM blends (5 wt. % PA) using OsO_4 as a staining agent for the polyacetylene phase. As shown in Figure 1, the polyacetylene phase appears to be discontinuous, while the elastomeric EPDM material forms the continuous matrix. Exposure of the films to I_2 vapor for 24 hours resulted in ultimate conductivities of $10 - 90$ Ω^{-1} cm^{-1}, depending upon polyacetylene contents. It

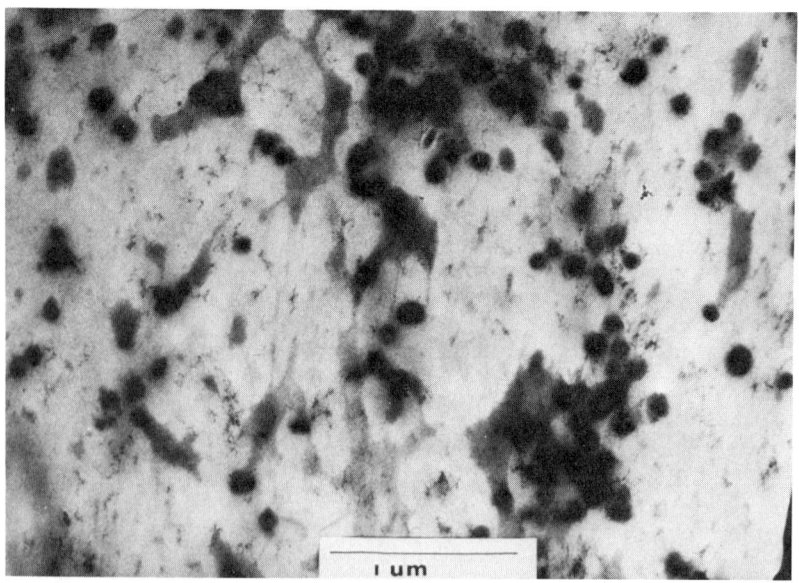

Figure 1. Transmission electron micrograph of the OsO_4-stained EPDM Polyacetylene blend.

was interesting to note that the conductivities of EPDM/PA blends were increased by several orders of magnitude (10 Ω^{-1} cm^{-1} to 7 x 10^2 Ω^{-1} cm^{-1}) upon stretching the sample (600% elongation). Such an increase in conductivity can be attributed to the fact that polyacetylene domains can be elongated and aligned upon stretching the sample. Examination of X-ray data clearly indicated that the degree of crystallinity was increased as the film of the blend was stretched.

The blend of EPDM and polyacetylene was crosslinked using sulfur monochloride in a toluene solution. It should be noted that the crosslinked blend could not be doped with iodine. Samples containing more than 2% of sulfur did not pick up any iodine even after a 72-hour period. The completely saturated EPDM portions of the blend seem to prevent any iodine molecules from permeating into the polyacetylene moieties. In order to circumvent this problem, we have doped the blend with iodine prior to the crosslinking procedure. Subsequently, the doped material having a conductivity of 60 Ω^{-1} cm^{-1} was reacted with sulfur monochloride in a toluene solution for 10 minutes. The color of the solution turned from pale yellow to dark red while the polymer film remained insoluble in the toluene solution. But the film lost its conductivity (less than 10^{-6} Ω^{-1} cm^{-1}) after the S_2Cl_2 treatment. The loss of conductivity can be attributed to the fact that sulfur monochloride essentially removes all iodine from the blend. The complete absence of iodine in the blend after S_2Cl_2 treatment was shown by elemental analysis.

Since the chemical approach was too harsh to prepare a crosslinked conducting blend, we turned our attention to other methods of crosslinking. We found that γ-rays from ^{60}Co can readily crosslink the blend within a short period of time. Even after 30 minutes of γ-radiation (dose rate ≃ 1.32 x 10^5 R/minutes) EPDM/PA blends became completely insoluble in almost all hydrocarbon solvents. However, unlike the chemically crosslinked material, the irradiated blend can be doped with I_2 to produce a a material having conductivity as high as 100 Ω^{-1} cm^{-1}. It was found from infrared data that only EPDM double bonds participated in the crosslinking reaction. It should be noted that conductivities of the irradiated EPDM/PA blend upon doping were consistently high (≃ 100 Ω^{-1} cm^{-1}) regardless of the length of γ-ray radiation. The conductivity of the material did not change even after 8 days of exposure to γ-ray radiation. The radiation does not appear to damage the polyacetylene units of the blend.

We have compared the conductivity stability of the crosslinked EPDM/PA blend with the uncrosslinked blend and homopolyacetylene. As shown in Figure 2, the conductivity of the I_2-doped EPDM/PA blend decays more slowly upon air exposure as compared to that of homopolyacetylene. The increase in stability of the blend reflects the high oxygen impermeability of EPDM rubber due to its highly saturated character. Figure 2 also

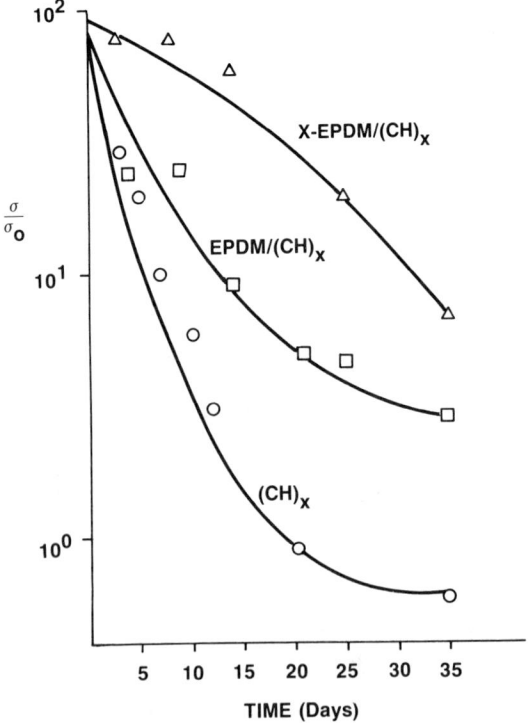

Figure 2. Normalized conductivity vs days of air exposure for EPDM/Polyacetylene blends.

shows the conductivity of the crosslinked EPDM/PA blend decays even more slowly than that of the uncrosslinked blend. Such stabilization may be due to the fact that polyacetylene chains are immobilized by tightly crosslinked EPDM networks.

Kraton/Polyacetylene Blends. As has been mentioned in the previous section, the stability of polyacetylene can be somewhat improved when EPDM rubber is blended with polyacetylene. The crosslinking of EPDM moieties of EPDM/PA blends certainly improved the stability further. However, most conventional elastomers such as EPDM may not be good host polymers for polyacetylene due to the fact that such polymers need post-curing subsequent to blend preparation. The heat and harsh chemicals associated with the curing process would cause scission and/or crosslinking of the highly conjugated double bonds of the polyacetylene. For this reason, we have turned our attention on the thermoplastic elastomer, styrene/diene triblock copolymers. As was the case in the EPDM/PA blend, all of the resulting polyacetylene/Kraton thermoplastic elastomer blends were found to be flexible and metallic-looking films. As shown in Figure 3, the stress-strain curve of Kraton blends containing 4% polyacetylene content behaved almost like pure Kraton rubber. However, as one incorporates more polyacetylenes into the blends, the properties of the materials become more thermoplastic. The infrared spectrum of the blend indicates that the material definitely contains both Kraton and polyacetylene moieties. Polyacetylene was present in the predominantly trans-configuration, as evidenced by exhibiting a characteristic infrared absorption band at 1015 cm^{-1}. Examination of X-ray data clearly indicated that polyacetylene moieties in the blend retain high crystallinity by showing a sharp peak at $2\theta=23°$ to $25°$ in the X-ray diffraction pattern (11).

In order to investigate the detailed morphology of the blend, a considerable amount of time has been devoted to electron microscopic studies. It is well known that, in the ABA-type triblock copolymer, the outer glassy A blocks form spherical domains while the central rubbery blocks are dispersed as the matrix. As shown in Figure 4, the polyacetylene is clearly incorporated into the rubbery matrix rather than the glassy polystyrene domain. Since the polymerization of acetylene in the presence of the triblock polymer was carried out below the glass transition temperature of the polystyrene block, the above result appears to be quite reasonable. As in the case of the polyacetylene/ elastomer blends, the polyacetylene/Kraton blends become less elastic with increasing polyacetylene content. Currently, work is in progress to incorporate polyacetylene in the polystyrene domain of the Kraton polymer. We have also studied the elongation behavior of the blend material upon doping. As shown in Table 1, the undoped blend was found to be extremely elastic and can be stretched up to 1100% of its original length. However, upon doping with iodine, the elongation of the material was con-

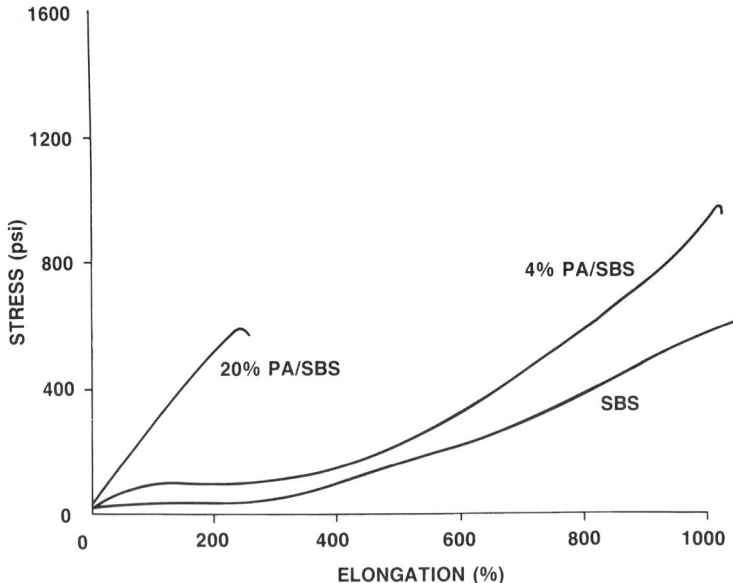

Figure 3. Stress-strain behavior of SBS/Polyacetylene blends.

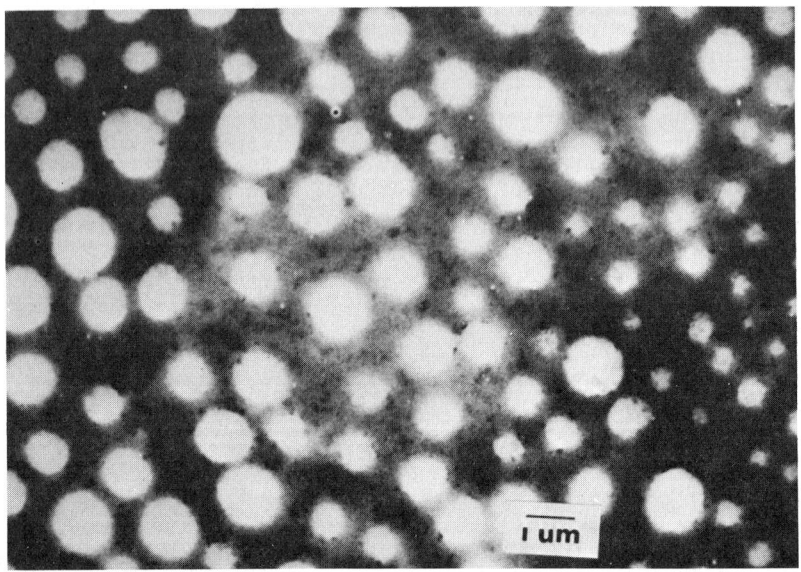

Figure 4. Transmission electron micrograph of the OsO_4-stained SBS/Polyacetylene blend.

siderably reduced. This corroborates the electron microscopic result that polyacetylene moieties indeed incorporate into the rubbery region of the triblock polymer. The conductivity decay of Kraton/polyacetylene is much slower than that of homopolyacetylene (Figure 5).

Table I. Effect of I_2 Doping on the Elongation of SBS Triblock Copolymer/Polyacetylene Blends*

Time Doped (minutes)	Conductivities (Ω^{-1} cm^{-1})	Elongation at Break (%)	Wt. % I_2 Uptake
0	--	1127	--
15	0.95	335	14.8
30	0.54	314	15
60	2.49	300	24.3
120	2.35	255	29.3
180	3.00	167	29.5

*SBS Triblock Copolymer used was Kraton D1101 (Styrene/Butadiene Ratio 30/70).

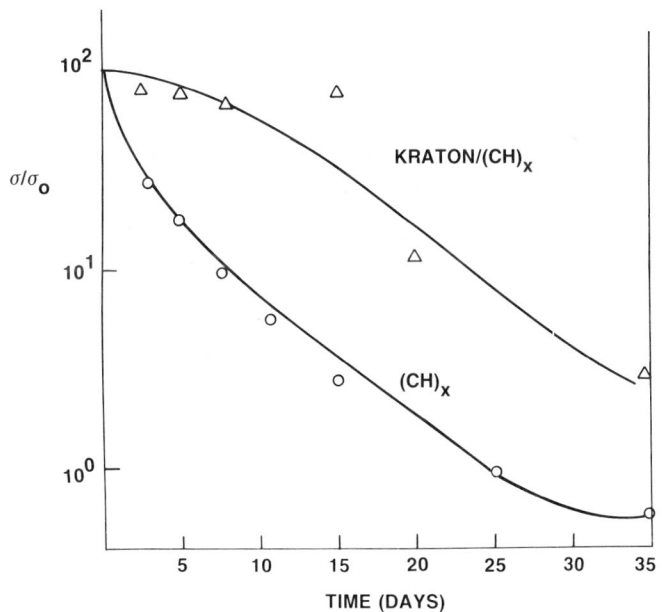

Figure 5. Normalized conductivity vs days of air exposure for Kraton/Polyacetylene blends.

Conclusion

In order to improve the stability of polyacetylene, polyacetylene was blended with ethylene-propylene-diene terpolymer. Subsequently, the resulting EPDM/PA blend was crosslinked with γ-radiation. Upon doping with iodine, the conductivity of the blend was found to be in the range 10-100 Ω^{-1} cm^{-1}. It was found that the conductivity of the crosslinked blend decayed more slowly compared to that of either the uncrosslinked blend or homopolyacetylene. The thermoplastic elastomer/PA blend was also prepared in order to avoid extra crosslinking steps. Electron microscopic results indicated that polyacetylene moieties incorporated into the rubbery matrix rather than into the polystrene glassy domains.

Acknowledgments

We would like to thank S. Tripathy for electron micrographs.

Literature Cited

1. Deits, W.; Cukor, P.; Jopson, H. Synthetic Metals 1982, 4, 199.
2. Chien, J.C.W.; Wnek, G.E.; Karasz, F.E.; Hirsch, J.E. Macromolecules 1981, 14, 479.
3. Galvin, M.E.; Wnek, G.E. Polymer 1982, 23, 795.
4. Borg, E.L., in "Rubber Technology"; Morton, M., Ed.; Van Nostrand Reinhold. New York, 1983; p. 220.
5. Legge, N.R.; Holden, G.; Davison, S.; DeLaMare, H.E. in "Chemistry and Technology of Block Polymers"; Craver, J.K.; Tess, R.W., Ed.,; APPLIED POLYMER SCIENCE, Organic Coatings and Plastics Chemistry Division, American Chemical Society: Washington, D.C. 1975.
6. Riddick, J.A. "Organic Solvents; Physical Properties, Methods of Purification"; Wiley-Interscience: New York 1970.
7. "Epcar(R) EPM and EPDM Rubbers are Versatile," B.F. Goodrich Co., 1981.
8. "Kraton(R)" Product Bulletin of Shell Chemical Co., 1980.
9. Ito, T.; Shirakawa, H.; Ideda, S. J. Polym. Sci., Polym. Chem. 1974, 12, 11.
10. Ito, T.; Shirakawa, H.; Ikeda, S. J. Polym. Sci,, Polym. Chem. Ed. 1975, 13, 1943.
11. Tsuchida, E.; Shih C.; Shinohara I.; Kambara, S. J. Polym. Sci. 1964, A2, 3347.

RECEIVED December 19, 1983

40

Conductive Hybrids Based on Polyacetylene: Copolymers and Blends

M. E. GALVIN, G. F. DANDREAUX, and G. E. WNEK

Department of Materials Science and Engineering, Massachusetts Institute of Technology, Cambridge, MA 02139

> Several approaches used to prepare hybrid polymers in which polyacetylene is an electroactive component are presented. Specifically, these involve the preparation of (1) composites by in-situ polymerization, (2) graft copolymers utilizing carbanions in n-type $(CH)_x$ as polymerization initiators, and (3) A-B diblock copolymers exploiting anionic-to-Ziegler-Natta transformation reactions.

The high level of interest in potential applications of polyacetylene (1), $(CH)_x$, is tempered in many instances by the prospects of intractability and poor physical and mechanical properties. In an attempt to mitigate such undesirable characteristics, we have attempted to prepare copolymers and blends (or composites) in which the electroactive component is $(CH)_x$.

Copolymerization has traditionally been a fruitful approach for the construction of organic materials possessing specific chemical, physical and mechanical properties. Copolymers are classified as random, alternating, graft or block depending upon the structural arrangement of the comonomer units. Random (or nearly so) copolymers of acetylene and methylacetylene have been prepared (2) which exhibit, as expected, electrical conductivities upon doping which are highly dependent upon comonomer concentration. The decrease in conductivity with increasing methylacetylene concentration has been rationalized (2,3) as being due to disruption of planarity of the backbone. A similar argument apparently applies to the acetylene/phenylacetylene system(4). Interest has developed in this laboratory concerning the possibility of attaching a wide variety of polymer chains to the $(CH)_x$ backbone (i.e., grafting). A motivation for such work is the possibility of tailoring the surface properties of $(CH)_x$ for electrode applications. The ability of alkali metal graphitides (5) to initiate polymerization of several monomers suggested the use of doped $(CH)_x$ in this capacity with the prospect of covalently

0097-6156/84/0242-0507$06.00/0
© 1984 American Chemical Society

binding polymer chains to $(CH)_x$ backbones. Block copolymers are also of interest because the planarity of the $(CH)_x$ block is expected to remain essentially intact, allowing the electrical properties to be primarily determined by percolation of $(CH)_x$ domains. Furthermore, soluble and tractable derivatives could be envisioned by judicious choice of the remaining block(s). Since acetylene polymerization will necessarily be initiated by coordination catalysts, we have considered the possibility of using the "alternate feed" method to produce, for example, ethylene-acetylene block copolymers from appropriate transition metal systems. However, such catalytic systems are typically not suitable for the synthesis of well-defined blocks (6) due to variable initiation rates of the catalytic centers and undesirable termination reactions. The exploitation of active polymer-transition metal bonds for the synthesis of block copolymers (7,8) suggested a potentially viable approach. Specifically, we have attempted to use "living" polystyrene to alkylate $Ti(OBu)_4$ followed by acetylene polymerization, viz

$$Bu\text{-}(CH_2CH)_n\text{-}CH_2\overset{+}{C}H\ Li\ +\ Ti(OBu)_4\ \longrightarrow\ "R\text{-}Ti\text{-}"$$

$$R\text{-}Ti\text{-}\ \xrightarrow{mC_2H_2}\ R\text{-}()_m\text{-}Ti\text{-}$$

A potential advantage of such a transformation reaction (8) is that at least the first block (polystyrene) can be synthesized with a well-defined chain length. Different approaches to grafts and blocks based on termination of acetylene polymerization by appropriate carrier polymers have recently been reported (9).

Composites have been prepared (10) through the polymerization of acetylene in low density polyethylene (LDPE) impregnated with the Shirakawa (11) catalyst. This approach may be potentially useful in that the (1) in-situ polymerization could provide a more intimate molecular mixing of the components as compared with mechanically prepared dispersion, (2) a variety of matrices with desirable physical properties may be employed, (3) articles of the selected matrix materials may be pre-fabricated to a desired structure followed by catalyst impregnation, polymerization and doping, and (4) $(CH)_x$ chains may be isolated in a matrix in order to study the influence of the local molecular environment (chemical composition, morphology, etc.) on, for example, soliton mobility.

Experimental

Composites. The $(CH)_x$/LDPE composites were prepared using the $Ti(OBu)_4$/Et_3Al Ziegler-Natta catalyst system as previously described (10). The amount of $(CH)_x$ incorporated was determined by monitoring the acetylene uptake during the polymerization. Electrically conductive derivatives were prepared by immersion of the composites in a saturated I_2/pentane solution for 24-48 hours. Electrical conductivities were measured by standard four-probe techniques.

Grafts. Polyacetylene films were synthesized at -78°C using techniques similar to those developed by Shirakawa and coworkers (11). Reductive doping was carried out in a dry box by immersion of $(CH)_x$ films in 1 M sodium naphthalide/THF solutions for 2 minutes. The films were then washed several times with dry, O_2-free THF and allowed to stand in fresh THF for approximately 1 hour. The conductivities and compositions of the films were in the range 5-50 S/cm and $[CHNa_{0.20-0.25}]_x$, respectively.

Exposure of the n-type films to either liquid (styrene, methyl methacrylate) or gaseous (ethylene oxide, isoprene) monomers resulted in polymerization. Much of our initial work has focused on grafting of poly(ethylene oxide) (PEO) to $(CH)_x$ in an effort to render the $(CH)_x$ surface more hydrophilic and to provide covalent attachment of a material capable of functioning as a solid electrolyte (12). Films of n-type $(CH)_x$ were exposed to dry (CaH_2-treated), gaseous ethylene oxide in the range 55-75°C with initial pressures being ca. 500 torr. Reaction times were typically 5 hours. The films were washed with dry, O_2-free methylene chloride to remove non-covalently bound PEO and then with deaerated H_2O to protonate oxyanions and remove the NaOH byproduct. The presence of bound PEO after extraction was confirmed by IR spectroscopy.

Blocks. The synthesis of block copolymers were attempted in the following manner. "Living" polystyrene was first prepared by conventional anionic techniques using n-BuLi as the initiator in THF at -78°C. In initial experiments, this polystyryl lithium product was treated with $AlCl_3$ to afford a polystyryl Al species which could be capable of alkylating $Ti(OBu)_4$. However, we find it to be more useful to directly aklylate Ti(OBu)$_4$ with the polystyryl Li. The orange-red polystyryl Li solution became deep red upon treatment with Ti(OBu)$_4$ at -78°C (Li/Ti = 2:1). The solution was warmed to room temperature and dry, O$_2$-free acetylene (initial pressures ca. 700 torr) was introduced on a vacuum line. The solution became deep blue in color and in some cases, fine blue-black particulates were observed. The polymerizations were terminated by addition of MeOH. The precipitated polymers were extracted with THF/acetone which removed homopolystyrene. The residues were brominated in THF and then subjected to GPC analysis.

Results and Discussion

Composites. The relationship between the four-probe electrical conductivity, σ, and the wt.% (CH)$_x$ in iodine-doped (CH)$_x$/LDPE composites is shown in Figure 1. An apparent percolation threshold exists between 2-4 wt.% (CH)$_x$. Many additional data points have been collected since our previous communication which show (as expected) a continuous increase in σ with (CH)$_x$ content beyond the "knee" in Figure 1. In Figure 2 we replot the data to include a value (1) for σ of I$_2$-doped, trans (CH)$_x$; extrapolation of our data to this value affords a reasonable line. The apparent threshold at such a low loading is somewhat surprising although it is interesting that thresholds of ca. 4 vol.% carbon black in LDPE have been observed (13). This has been rationalized as being due to a combination of particle size and wettability effects. It should be noted that the (CH)$_x$ entities (as revealed by transmission electron microscopy of thin films) in the composites are irregular in shape with sizes in the range of 600-2,000 Å. We find that (CH)$_x$ "powder" (11) consists of much larger entities (hundreds of microns in size) and that simple dispersions of powdered (CH)$_x$ in LDPE (made by casting films from hot toluene in a dry box) of much higher loading levels (≥40 wt.%) generally fail to yield highly conductive materials upon I$_2$ doping. Thus, a virtue of the in-situ polymerization approach appears to be the ability to form much smaller (CH)$_x$ domains and is presumably responsible, at least in part, for the observed percolation threshold. It is possible that melt extrusion of LDPE/(CH)$_x$ particulate systems could afford materials having lower percolation thresholds (compared with solvent-cast systems) although we have not attempted to perform such experiments. The interesting properties of (CH)$_x$/elastomer composites prepared by the in-situ approach have recently been described (14).

Grafts. The primary reaction of n-type (CH)$_x$ and ethylene oxide is presumably attack of a carbanion of a methylene carbon resulting in ring opening and oxyanion formation, followed by successive monomer additions:

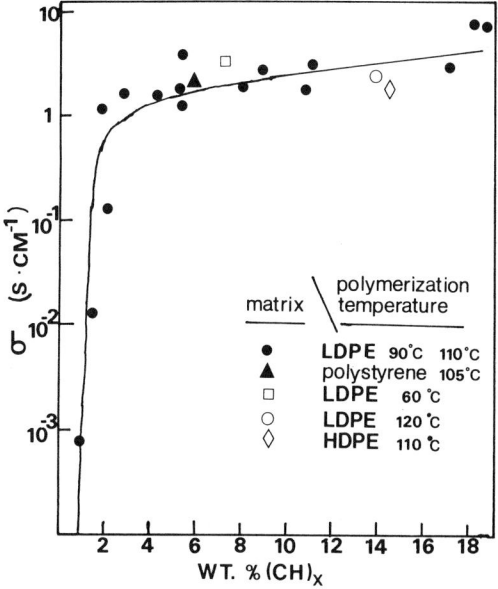

Figure 1. Conductivity/composition data for $(CH)_x$/polymer composites doped with I_2.

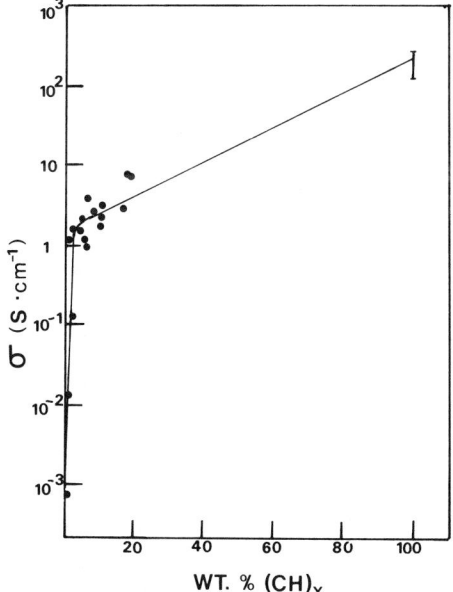

Figure 2. Data for Figure 1 extrapolated to conductivity of I_2-doped, <u>trans</u> $(CH)_x$.

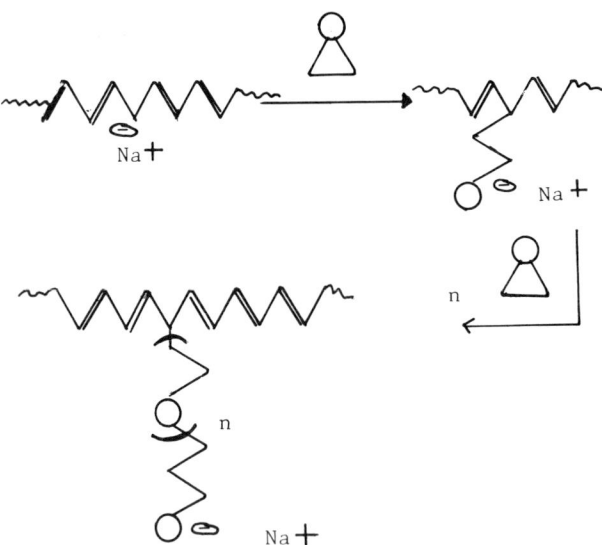

The fact that approximately half of the product can be removed by CH_2Cl_2 extraction may suggest that electron transfer from a $(CH)_x$ radical anion to ethylene oxide occurs leading to polymerization without attachment to the $(CH)_x$ or that some attached PEO suffered mechanically induced scission. The oily character of the extracted PEO suggests a low molecular weight. We have no information regarding the molecular weight of the covalently bound PEO or the number of PEO branches. The $(CH)_x$/PEO materials can be doped with I_2 to afford conductivities of approximately 1 $\Omega^{-1}cm^{-1}$. Problems with electrical contacts due to an insulating PEO layer may be a factor in this lower conductivity compared with I_2-doped $(CH)_x$ but it should be noted that the $(CH)_x$ conjugation is broken at the graft points and this is expected to reduce the intrachain hole mobility.

At present, we do not believe that Na metal in the films is responsible for initiation since n-type $(CH)_x$ films which contained a 2:1 naphthalene/sodium ratio were still highly active as polymerization initiators. A comparison of the number of $(CH)_x$ carbanions which are initiators with the number, if any, of electron transfer sites may yield important information about the distribution of charge carrier reactivities in doped $(CH)_x$.

Blocks. The IR spectra of films cast from the soluble, colored THF fraction are consistent with the presence of both polystyrene and $(CH)_x$ segments, although do not prove the existence of a block copolymer. The required proof is derived from GPC studies (Figure 3). A sample of homopolystyrene (taken from the "living" polystyrene prior to reaction with $Ti(OBu)_4$) was found to have an $\bar{M}_n$ of 27,730. Also, a sample of homopolystyrene (from the THF/acetone extraction, treated with Br_2 in THF) yielded an $\bar{M}_n$ of 27,470 indicating that the extraction does not merely fractionate the polystyrene itself and that Br_2 treatment has a negligible effect on $\bar{M}_n$. However, the GPC trace of the colored residue which had been previously brominated clearly shows a higher molecular weight. An absolute $\bar{M}_n$ cannot be determined since the hydrodynamic volume of poly(1,2-dibromoethylene) is unknown although the apparent $\bar{M}_n$ is 72,360. It is difficult to reconcile this result as being due to anything but the presence of a true diblock copolymer. For the sample discussed above, the yield of block copolymer based on the total weight of polystyryl Li used initially is ca. 5.5%. The observation that the soluble diblocks slowly yield particulates upon standing suggests that crystallization of $(CH)_x$ domains occurs, although at a much slower rate compared with precipitation during the preparation of $(CH)_x$ powders (11). This is presumably due to slower crystallization kinetics by virtue of the presence of the solubilizing polystyrene "tail". Studies of the efficiency of the synthesis and the morphology and electrical properties of the diblock copolymers are in progress.

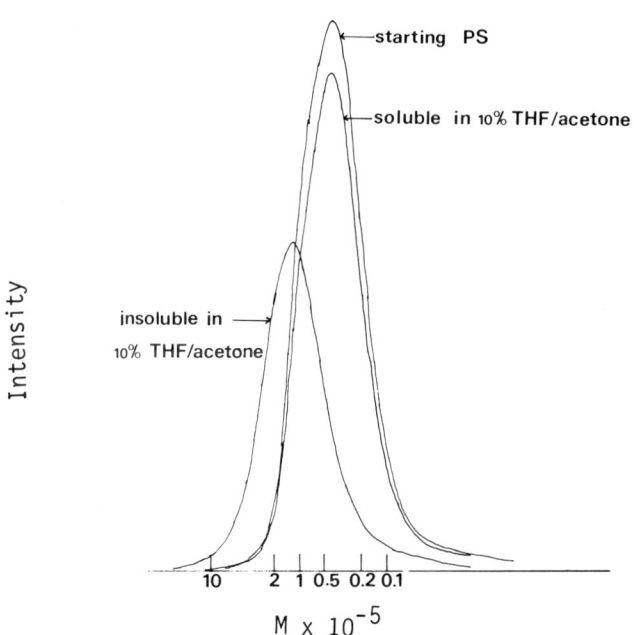

Figure 3. GPC traces for aliquot of "living" polystyrene (starting PS), PS extracted from acetylene/styrene block material (soluble in 10% THF/acetone), and brominated acetylene/styrene block (insoluble in 10% THF/acetone).

Acknowledgments

We are grateful for support from the MIT Center for Materials Science and Engineering (NSF-MRL Core Fund DMR 78-24185), a DuPont Young Faculty Award (to G.E. Wnek) and a fellowship from Polaroid Corporation (to M.E. Galvin).

Literature Cited

1. MacDiarmid, A.G.; Heeger, A.J. Synth. Met. 1979, 1, 101.
2. Chien, J.C.W.; Wnek, G.E.; Karasz, F.E.; Hirch, J.A. Macromolecules 1981, 14, 479.
3. Tripathy, S.K.; Kitchen, D.; Druy, M.A. ACS Polym. Preprints, 1982, 23, 109.
4. Deits, W.; Cubor, P.; Rubner, M.; Jopson, H. Synth. Met. 1982, 4, 199.
5. Gole, J.; Merle, G.; Pascault, J.P. Synth. Met. 1982, 4, 269.
6. Borr, J., Jr. in "Ziegler-Natta Catalysts and Polymerizations"; Academic Press: New York, 1979; chap. 21.
7. Sioue, A.; Fontanille, M. Eur. Polym. J. 1981, 17, 1175.
8. Richards, D.H. Br. Polym. J. 1980, 12, 89.
9. Bates, F.S.; Baker, G.L. Macromolecules 1983, 16(4), 704.
10. Galvin, M.E.; Wnek G.E. Polym. Comm. 1982, 23, 795.
11. Shirakawa, H.; Ikeda, S. Synth. Met. 1980, 1, 172.
12. Armand, M.B.; Chabagno, J.M.; Duclot, M.J. in "Fast Ion Transport in Solids"; Vashista; Mundy; Shenoy, Eds.; Elsevier North Holland, 1979, p. 131.
13. Miyasaka, K.; Watanabe, K.; Jojima, E.; Aida, H.; Sumita, M.; Ishikawa, K. J. Mater. Sci. 1982, 17, 1610.
14. Rubner, M.F.; Tripathy, S.K.; Georger, J.,; Cholewa, P. Macromolecules 1983, 16(6), 870.

RECEIVED December 19, 1983

41

Electrically Conductive Polymer Composites of 7,7,8,8-Tetracyanoquinodimethane (TCNQ) Salt Dispersion
Influence of Charge-Transfer Interaction and Film Morphology

OH-KIL KIM

Naval Research Laboratory, Washington, DC 20375

> Electrically conductive polymer composites were made by dispersing TCNQ salt in polymer matrices. Composite film conductivity and stability are discussed in terms of charge-transfer interaction between TCNQ salt and matrix polymer and the resulting film morphology. The extent of CT interaction and tendency of microcrystallization of TCNQ salt in the matrices were determined by visible spectra. Conductivity and stability are morphology dependent of the film; high conductivity requires a uniform, densely packed dispersion of TCNQ salt microcrystallites. The highest conductivity was always attained at $[TCNQ^0]/[TCNQ^-] \simeq 1$. Extra $TCNQ^0$ doping is needed to make up this stoichiometry and to enhance the the composite stability.

The most desirable properties for electrically conductive polymeric materials are film-forming ability and thermal and electrical properties. These properties are conveniently attained by chemical modification of polymers such as polycation-7,7,8,8-tetracyanoquinodimethane (TCNQ) radical anion salt formation (1-3). However, a major drawback of such a system is the brittle nature of the films and their poor stability (4,5) resulting from the polymeric ionicity. In recent years, polymeric composites (6-8) comprising TCNQ salt dispersions in non-ionic polymer matrices have been found to have better properties. In addition, the range of conductivities desired can be controlled by adjusting the TCNQ salt concentration, and other physical properties can be modified by choosing an appropriate polymer matrix. Thus, the composite systems are expected to have important advantages for use in electronic devices.

This chapter not subject to U.S. copyright.
Published 1984, American Chemical Society

For the study of semi-conducting polymer composites, poly(vinyl acetal) was selected as the matrix because of its easy modification by varying side groups differing in electron-donor strength, in addition to a good film-forming property. In our earlier study (7) we observed that film resistivity and electrical stability were strongly influenced by the matrix polymers which seemed to control the film morphology of TCNQ salt dispersions. We considered the matrix effect on the conductivity to be possibly a result of a charge-transfer (CT) interaction between the conductive components and the insulating polymer matrices. For a better understanding of the interaction, we studied two typical TCNQ salts, namely, N,N,N,-triethylammonium TCNQ ($Et_3NH^+(TCNQ)_2^-$), and N-methylphenazinium TCNQ ($NMP^+TCNQ^{\cdot -}$), in which neutral TCNQ ($TCNQ^0$) and cations were expected to be influential components in the composite conductivity. Of particular interest was $NMP^+TCNQ^{\cdot -}$. A high loading of the salt can be made in the composites to achieve low resistivity and a higher stability.

In the present report, we discuss the relative importance of the matrix effect on the semi-conductivity and the stability of TCNQ salt/polymer composites through the studies of CT interactions and film morphology.

Experimental

Salts of TCNQ were prepared according to procedures described by Melby and coworkers (9,10). Characterization was made by elemental analysis, IR and visible spectra.

Poly(vinyl acetals) were synthesized from poly(vinyl alcohol) (PVA) (Polyscience, 99% saponification, molecular weight 25,000) and aldehydes such as N-ethyl-3-carbazole-carboxaldehyde (ECZA), 9-anthraldehyde (ANT), and 1-naphthaldehyde (NA). The synthetic conditions and analytical results of the poly(vinyl acetals) are given in the previous report (8). Poly(vinyl butyral) (P(BA), 64% acetalization), poly(methyl methacrylate) (PMMA) and polycarbonate (PC) are commercial products. Details of the conducting film preparation is also described in our previous report (8). Films were cast on a teflon substrate by evaporating the solvent DMF, under reduced pressure at about $30°C$. For high electrical resistance samples, films were made up at a surface-type cell by vapor-depositing Au on both sides of the film and the resistance was measured with a Keithley 610C electrometer. A four-probe method was applied to high conductivity samples with a Keithley 164TT digital multimeter.

Spectral studies of $Et_3NH^+(TCNQ)_2^-$ composite films and the casting solutions were carried out with a Shimadzu Double-Beam UV-200S spectrophotometer. A small amount of DMF solution of a

TCNQ salt and a polymer was cast on the quartz cell to give a very thin transparent film after vacuum evaporation of the solvent at room temperature. The concentrations of $TCNQ^o$ and $TCNQ^-$ present in the solid films were determined from the absorbance at 398 nm and 857 nm, respectively. The details of the spectral determination is described in the previous paper (8). The TCNQ salt association in the composites were studied with NMP^+TCNQ^- by visible spectra in a similar manner as described above. Absorbances in the regions of 390 nm and 850 nm were measured and the absorbance ratios are applied to evaluate the salt aggregation behaviors. Relative intensity of the composite films absorption at 390 nm was determined by normalizing the absorptions at 850 nm as a reference point. The absorption intensity of the films was linearly proportional to the film thickness at a fixed salt concentration. Scanning electron micrographs (SEM) of the conductive films were taken using an AMR model 1000 scanning electron microscope.

Results and Discussion

It was found that with composites of complex salt, $Et_3NH^+(TCNQ)_2^-$, the film resistivity decreased with an increase of the salt concentration, then gradually increased with an excess salt concentration after reaching a minimum (Fig. 1). The minimum resistivity and the salt concentration were strongly dependent on the matrix polymers probably as a result of differences in the degree of the interactions between the salt and polymers. With composites of the simple salt, NMP^+TCNQ^-, however, the resistivities of the composite remained high for all polymers with increasing concentration of the salt (Fig. 2). Such a contrasting behavior of the simple salt composites compared to the complex salt composites may result from the poor stability of the simple salt $TCNQ^-$ in DMF and/or somewhat higher solubility of the simple salt in the matrix polymers. It was inferred that the differences in the stability and the solubility of the salts result from the extent of interactions of individual salts with solvents and polymers; TCNQ salts are more stable in weak electron-donor solvents and polymers. Consequently, if the film resistivity of TCNQ salt/polymer composites is controlled by the CT interactions, it should be reflected in the film morphology.

Recently Schulz et al. (11) reported a CT interaction of $TCNQ^o$ with a P(NA) in methylene chloride. We have confirmed this observation and also found CT interaction between $TCNQ^o$ and P(ECZA). There are present two types of CT complexes between a polymer donor (PD) for example, P(NA), and $TCNQ^o$; one is a π-complex (565 nm) and the other one is a radical anion, $TCNQ^-$ (848 nm):

$$PD + TCNQ^o \underset{}{\overset{K_1}{\rightleftharpoons}} [PD \rightarrow TCNQ^o] \underset{}{\overset{K_2}{\rightleftharpoons}} (PD)^+ + TCNQ^-$$
$$\text{Neutral CT Complex}$$

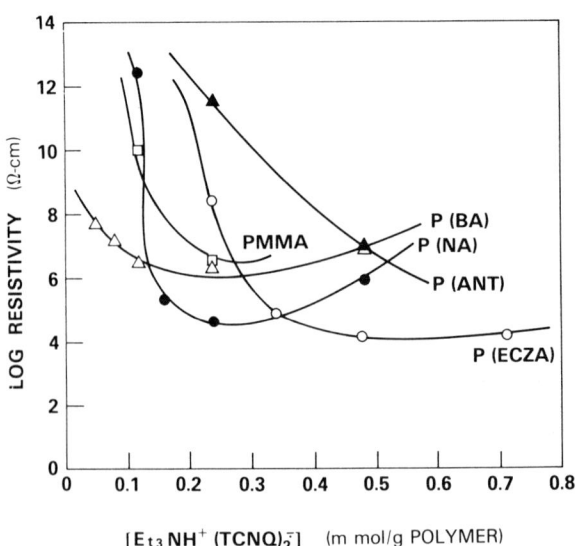

Figure 1. Electric resistivity of composite films as a function of $Et_3NH^+(TCNQ)_2^-$ added to DMF casting solutions of polymers. (Reproduced with permission from Ref. 8. Copyright 1982, J. Polym. Sci. Polym. Chem. Ed.)

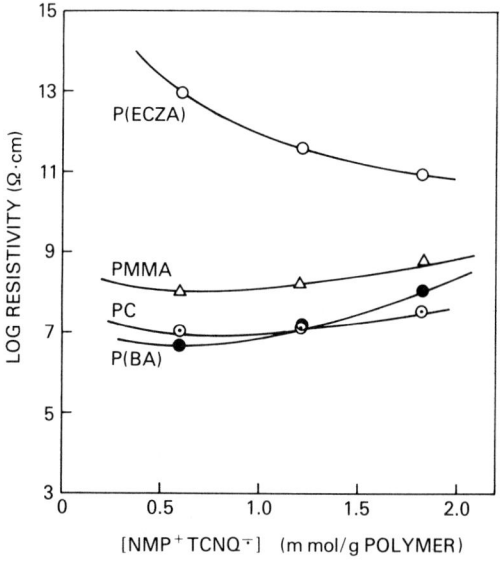

Figure 2. Electric resistivity of composite films as a function of $NMP^+TCNQ^{\bar{\cdot}}$ added to DMF casting solutions of polymers.

With a monomeric analog of PD, $TCNQ^o$ does not form $TCNQ^{\bar{\cdot}}$ but the neutral CT complex is formed. Apparently, in the case of the polymer, a number of donor moieties in the chain mutually participate in the interaction with $TCNQ^o$. With the polymer donor, K_2 should be greater than K_1, while with the monomeric analog, the situation is reversed. However, the CT interactions between $TCNQ^o$ and polymer are more strongly influenced by the solvent. The solvent, DMF, used in the present system predominates over the polymer donors in the CT interaction with $TCNQ^o$. However, the resulting $TCNQ^{\bar{\cdot}}$ concentration in the solution was found to decrease with an increase of $TCNQ^o$ concentration, independent of the donor polymer. On the other hand, it is anticipated that CT interaction of the polymer donor with $TCNQ^o$ can be restored when films are cast from the solution by removing DMF. Through a series of experimental manipulation, we confirmed that the effect of $TCNQ^o$ concentration on the interaction with DMF is reproduced in the cast film.

The polymer donor participation in the CT interaction with $TCNQ^o$ in the composite films is evident from Fig. 3; $TCNQ^{\bar{\cdot}}$ concentrations produced in the cast films are all different among the polymers, and some of the typical donor polymers show a steady increase of $[TCNQ^{\bar{\cdot}}]$ with an increase of $TCNQ^o$ concentration, whereas the DMF contribution to the $[TCNQ^{\bar{\cdot}}]$ should show a reverse trend (8). If there is no or negligible CT interaction between polymers and $TCNQ^o$ in the cast films, $TCNQ^{\bar{\cdot}}$ produced in the films should be the sole result of DMF participation in the interaction and thus the total $TCNQ^{\bar{\cdot}}$ concentrations in the films should be all the same at a given concentration of $TCNQ^o$, since the volume of DMF and the amount of polymers are kept constant in the casting solution. Consequently, the finding in Fig. 3 strongly suggests that CT interaction takes place between $TCNQ^o$ and polymers in the solid dispersion to an extent which depends on the donor strength of the polymers: $P(ECZA) > P(NA) > P(BA) \geq PMMA$. Considering such a chemical change of $TCNQ^o$ in the solid dispersion, it is most conceivable that the stoichiometry ($TCNQ^o/TCNQ^{\bar{\cdot}}=1$) of the complex salt in the present conductive system must be changed to less than 1, resulting in a mixture of simple and complex salts.

It is, therefore, important to examine the effect of compositional variation of TCNQ species on the conductivity. It is seen in Fig. 4 that the addition of a small amount of $TCNQ^o$ to DMF casting solution of the complex salt dramatically reduced the resistivity of composites with relatively strong donor matrices P(ECZA), P(NA) and P(ANT). With the weak donor matrix, P(BA), this effect was minor, and with the essentially non-donor matrix, PMMA, the resistivity was, in fact, even increased slightly by the addition of $TCNQ^o$ to the complex salt system. However, the initially decreasing resistivity on the addition of a small amount of $TCNQ^o$ shows an upward tendency with a further addition of $TCNQ^o$ beyond an optimum point. It is of interest, therefore,

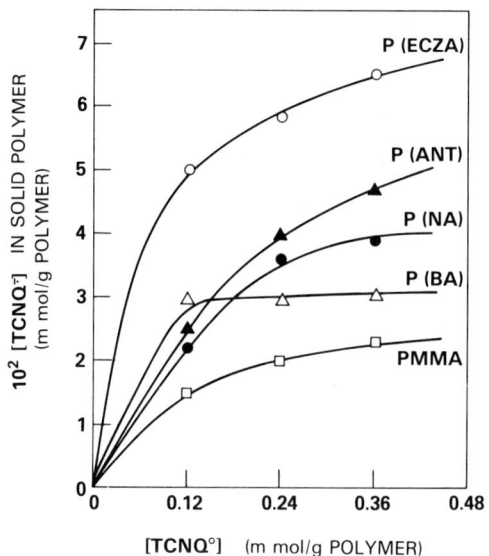

Figure 3. $TCNQ^-$ concentration in films cast from solutions of 23.5 mg polymer per mL DMF as a function of added TCNQ. (Reproduced with permission from Ref. 8. Copyright 1982, J. Polym. Sci. Polym. Chem. Ed.)

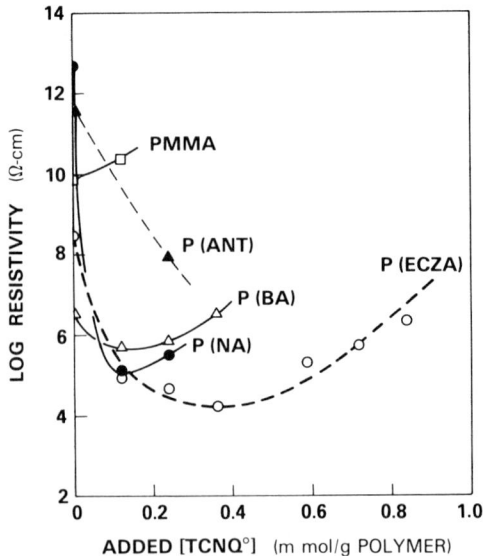

Figure 4. Electric resistivity of TCNQ -doped $Et_3NH^+(TCNQ)_2^-$/polymer composites as a function of TCNQ added to DMF casting solutions. Key (mmol $Et_3NH^+(TCNQ)_2^-$ per gram polymer): ———, 0.12; and ---, 0.24. (Reproduced with permission from Ref. 8. Copyright 1982, J. Polym. Sci. Polym. Chem. Ed.)

to identify the TCNQ salt stoichiometry of individual systems from a plot (8) of $[TCNQ^o]/[TCNQ^{\cdot -}]$ in the composites vs. $[TCNQ^o]$ added to the complex salt dissolved in DMF, and correlate the compositions to their resistivities. Without the addition of $TCNQ^o$, the actual change in the $[TCNQ^o]/TCNQ^{\cdot -}]$ of the complex salt (due to the CT interaction) in some composite was down to as low as 0.4. The resistivity was, however, decreased with increasing $TCNQ^o$ doping as long as doping brings the composition of the dispersed salt closer to $[TCNQ^o]/[TCNQ^{\cdot -}] = 1$. Further doping of $TCNQ^o$ changes the stoichiometry to > 1 where the resistivity turns up again, probably because of the lowered carrier mobility.

In a solid dispersion of polymer donor and TCNQ complex salt, one can expect polymers to form a CT complex with the complex salt by analogy with the preceding equilibria:

$$PD + R_3NH^+(TCNQ)_2^{\cdot -} \rightleftharpoons [PD^+ \ldots TCNQ^{\cdot -}] + R_3NH^+ TCNQ^{\cdot -}$$

$$[PD \ldots TCNQ^{\cdot -}] + n\ R_3NH^+(TCNQ)_2^{\cdot -} \rightleftharpoons (m+1)\ PD^+ TCNQ^{\cdot -}$$
$$+ m\ R_3NH^+ TCNQ^{\cdot -} + (n-m)\ R_3NH^+(TCNQ)_2^{\cdot -}$$

Thus, the CT interaction is the driving force for binding of TCNQ species to the polymer chain. The binding of $TCNQ^{\cdot -}$ to a polymer donor site is similar to that in polycation-$TCNQ^{\cdot -}$ system. A consequence is that in a strongly interacting polymer donor matrix, more $TCNQ^o$ (of the complex salt) is solubilized by binding to the polymer up to saturation, and further addition of the complex salt beyond the saturation level alters the equilibria such that aggregation and microcrystallization of the salt to occur, particularly when the $TCNQ^o$ is externally added to the system. In a polymer matrix of a low donor strength, microcrystallite formation occurs at a relatively low salt concentration because so little of the salt becomes associated with the polymer. Thus, it is anticipated that the lowest resistivities of the complex salt composites appear at relatively higher salt concentrations in such strong donor matrices as (P(ECZA), P(ANT) and P(NA), where SEM micrographs showed a densely packed filamentary dispersion of the salt microcrystallites.

The mode of binding of $NMP^+ TCNQ^{\cdot -}$ to the polymer should be different from that of $Et_3NH^+(TCNQ)_2^{\cdot -}$, because of the absence of $TCNQ^o$, however the solubility of $NMP^+ TCNQ^{\cdot -}$ in matrix polymers is much higher than that of $Et_3NH^+(TCNQ)_2^{\cdot -}$, probably due to an effect of a strong association of the cation with polymers. If the association is an intensive one, the donor exchange can take place between monomeric donor (D) and polymeric donor (PD):

$$D^+TCNQ^- + PD \rightleftharpoons PD^+ TCNQ^- + D$$

Under the condition of donor exchange reaction, it is unlikely to form an orderly packed homogeneous stack of TCNQ salt crystals, as in the case of $TCNQ^o$ binding to the polymers through CT interaction, though the molecules can form a larger aggregate. In order for the composites to become conductive, the TCNQ salt, initially dissolved in DMF solution before casting, should subsequently recrystallize as pure microcrystals. To realize this condition with NMP^+TCNQ^-, two approaches are possible; one is to reduce the association through choosing weak donor polymers as matrices, and another is to dope the system with $TCNQ^o$ so that TCNQ salt association with the polymer should be reduced through a competitive reaction with $TCNQ^o$, thereby facilitating salt crystallization.

The effect of $TCNQ^o$ doping of the NMP^+TCNQ^-/polymer system (Fig. 2) was somewhat dramatic; the DMF casting solution became very stable in the presence of $TCNQ^o$, microcrystallization of the salt was noticeable even with 30 mole % $TCNQ^o$ doping with respect to $TCNQ^-$ concentration, and the resistivity drop under this doping condition was four orders of magnitude lower than the undoped one. The resistivity decreased further with an additional increase of $TCNQ^o$ concentration, reaching a minimum where $[TCNQ^o][TCNQ^-] \simeq 1$, and then tended to gradually increase with further addition of $TCNQ^o$. As expected, the matrix effect was greater with low donor strength (contrasting to the case of $Et_3NH^+(TCNQ)_2^-$); weak donor polymers such as P(BA), PMMA, and PC are all far more effective than P(NA) and P(ECZA). With the former group, as shown in Fig. 5, the resistivity (12) of the composites attained was ca. 25 (ohm-cm) at less than 30 wt % total TCNQ salt concentration in the composites, and ca. 10 (ohm-cm) at about 45 wt % total TCNQ salt; with the latter group the resistivity was ca. 10^4 (ohm-cm) and ca. 10^3 (ohm-cm) under the respective TCNQ salt concentrations.

Controlling the rate of solvent evaporation in the film casting had no significant effect on the film morphology of NMP^+TCNQ^-/polymer composites, while it had a tremendous impact on the morphology of $Et_3NH^+(TCNQ)_2^-$/polymer composites; a slow evaporation of the solvent, most favorably, at 12 mm Hg, for example, allowed conductivity increases of 5 orders of magnitude in some cases. The reason for the different mode of microcrystallization is not clear but may be related to the rates of molecular association and subsequent crystallization of the colloidal dispersion. Therefore, if the system has a slow equilibrium, then the aging might become an important process for crystallization. The film resistivity of $Et_3NH^+(TCNQ)_2^-$/polymer composites is closely related to the degree of microcrystallite packing in the films. We can better illustrate this relation with the film morphology of NMP^+TCNQ^-/polymer composites. One interesting feature of these composites revealed by SEM pictures

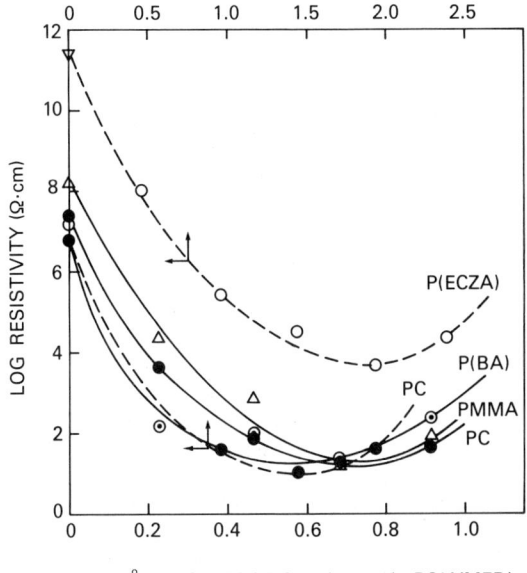

Figure 5. Electric resistivity of TCNQ-doped NMP^+TCNQ^-/polymer composites as a function of TCNQ added to DMF casting solutions. Key (mmol NMP^+TCNQ^- per gram polymer): ——, 0.61; and ---, 1.22.

(Fig. 6) is a unique film morphology showing that in the weak donor matrix polymers, P(BA) and PC (c and d), the $TCNQ^0$-doped NMP^+TCNQ^- salt was recrystallized as a fully interconnected fine filamentary structure. The surface is fully covered along the axis by the insulating matrix polymer in a flexible manner, as if the filamentary network of the conductor were protected from the environmental change. In the strong matrix polymer, P(ECZA) (a and b), $TCNQ^0$-doping effect on the microcrystallization of NMP^+TCNQ^- is extremely slight (b), and furthermore no microcrystallization occured without $TCNQ^0$ doping (a). With the matrices of P(BA) and PC, the optimum $TCNQ^0$-doped resistivities were ca. 25 and 10 (ohm-cm), respectively, while with the matrix P(ECZA), the resistivities were ca. 5×10^4 and 1×10^{12} (ohm-cm) with and without $TCNQ^0$ doping, respectively. Throughout the optimization effort, the present composite system produced excellent conducting materials characterized by flexible film formation and high stability; for over one year, in air at room temperature, the conductivity change was negligible.

The association behavior of NMP^+TCNQ^- was explored by visible spectra to look into the microcrystallization of the salt under the influence of matrix polymers. As in the solutions, for dimers and higher molecular aggregates to form in polymers, the salt concentrations must exceed a certain level at which a constant amount of free volume within the polymer matrix is fully occupied. In this case, matrix polymers act as solvents whose dielectric constant and complexing power (donor strength) determine the ion-pairing situation. Visible spectra of NMP^+TCNQ^- in DMF showed a noticeable difference from what was observed in acetonitrile whose polarity is about the same as to DMF; the absorbances at 385 nm (due to $TCNQ^-$) is much lower in DMF than in acetonitrile, while the absorbances at 845 nm (due to $TCNQ^-$) are nearly the same. It is inferred that NMP^+TCNQ^- salt ions are highly separated in DMF, while the ions form a contact pair in acetonitrile, and that the absorption at 385 nm is particularly sensitive to the changes in the ion-pairing situation of the TCNQ salt. This information in solution was applied to the composite films of NMP^+TCNQ^-/polymer system to evaluate the aggregation tendency of the salt as a function of salt concentration, where the absorbance ratio at 390 nm and 850 nm of the composite films, A_{390}/A_{850}, was used as a measure of the tight ion-pairing which can proceed to aggregation with a further increase of the salt. As shown in Fig. 7, up to a certain NMP^+TCNQ^- concentration (0.076 m mol per g polymer) in a fixed amount of polymers, the absorbance ratios stay nearly the same in the same matrix polymer, but the values are higher in parallel with the weakness of donor strength of polymers; P(BA)≲PMMA<PC<P(NA)<P(ECZA). The absorbance ratios increase gradually with a further increase of NMP^+TCNQ^- concentration up to 0.15 m mol per g polymer. Beyond that concentration, the increment of the ratios becomes smaller, then sharply increasing in P(ECZA), P(NA) and PC, whereas it

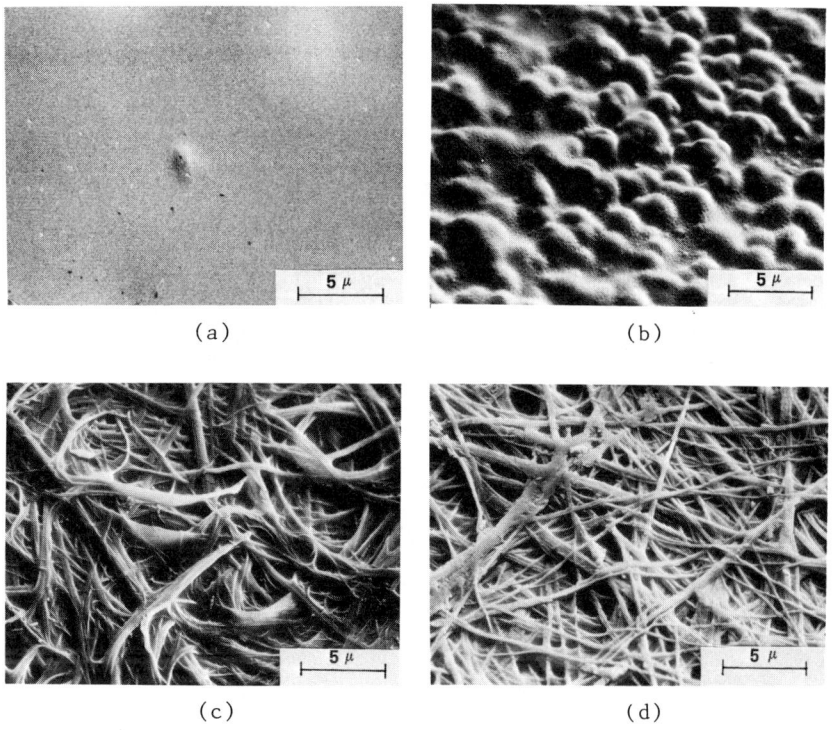

Figure 6. Scanning electron micrographs of conductive films with the following compositions: (mmol of NMP^+TCNQ^- and TCNQ per gram polymer) In P(ECZA); a, 0.61 and null; b, 0.61 and 0.69. In P(BA); c, 0.61 and 0.69. In PC; d, 1.22 and 1.38.

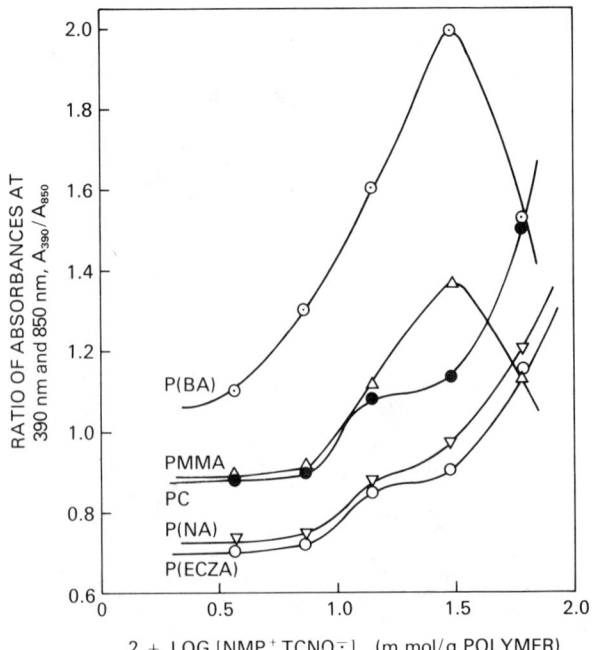

Figure 7. Dependence of the absorbance ratios at 390 nm and 850 nm (A_{390}/A_{850}) of NMP^+TCNQ^-/polymer composite films on NMP^+TCNQ^- concentrations in the casting solution.

becomes sharply decreasing after attaining a maximum (around 0.3 m mol per g polymer) in PMMA and P(BA). At such high NMP^+TCNQ^- concentrations, the absorption bands around 390 nm became broadened accompanied a new band around 370 nm which is probably due to NMP^+TCNQ^- aggregation. This behavior is typified with the case of P(BA) as shown in Fig. 8. A similar behavior was observed with dye molecules in polymer matrices (14). The implication of this result is that the salt molecule can distribute itself to follow a thermodynamic equilibrium at low concentrations, and once the volume is occupied, a further addition of molecules will force the formation of dimer or larger aggregates. From the observation of the absorbance ratio vs. NMP^+TCNQ^- concentration plot (Fig. 7), it appears that there exist three stages of equilibria in the salt aggregation in the polymer matrix; a loose ion-pairing, a tight ion-pairing, and an aggregation:

$$PD + NMP^+TCNQ^- \rightleftharpoons PD\ldots NMP^+ \ldots TCNQ^-$$
$$PD\ldots NMP^+ \ldots TCNQ^- + m\, NMP^+TCNQ^- \rightleftharpoons PD\ldots (NMP^+TCNQ^-)_{m+1}$$
$$PD\ldots (NMP^+TCNQ^-)_{m+1} + n\, NMP^+TCNQ^- \rightleftharpoons (NMP^+TCNQ^-)_{m+n+1} + PD$$

Such a matrix effect on the salt aggregation is reflected on the film morphology (Fig. 6) of the $NMP^+TCNQ^-/TCNQ^o$/polymer system; in P(BA), PMMA and PC, microcrystallization of the salt is well developed, while in P(ECZA) and P(NA), the salt crystallization is imperfect, resulting in poor conductivity.

Conclusions

The film conductivity and the stability of TCNQ salt/polymer composites are strongly dependent on the matrix polymers, more specifically on the electron-donor strength of the polymers which controls CT interaction with TCNQ salts. With $Et_3NH^+(TCNQ)_2^-$, $TCNQ^o$ is the major interacting species with polymers, whereas with NMP^+TCNQ^-, the cation seems to be responsible for the high solubilization in matrix polymers. Consequently, the donor strength of matrix polymers control the dispersion state of TCNQ salt in polymer matrix; high donor-strength polymers solubilize (or often destabilize) more TCNQ salts so that the microcrystallite network (conducting path) formation is suppressed. For an effective microcrystallization of the dissolved TCNQ salt, the polymer donor strength must be balanced with respect to the salt solubility or adjusted by the doping of $TCNQ^o$; with less soluble $Et_3NH^+(TCNQ)_2^-$, rather strong donor matrices are more effective, while with more soluble NMP^+TCNQ^-, the use of weak donor matrices is important. For a particular case of NMP^+TCNQ^-/polymer system, $TCNQ^o$ doping is indispensable for the stability and for the enhancement of conductivity. The resistivity of $TCNQ^o$-doped NMP^+TCNQ^-/polymer composites attained a minimum at $[TCNQ^o]/[TCNQ^-] \simeq 1$ in the composite film. The

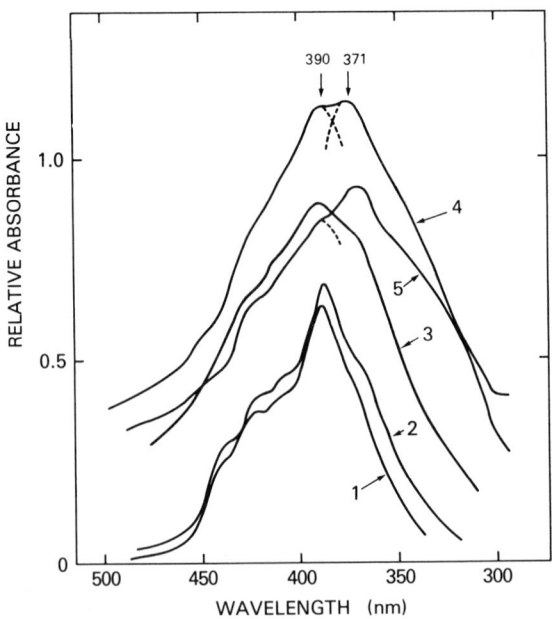

Figure 8. Spectral changes in the 390 nm region of NMP^+TCNQ^-/P(BA) composite films at varied NMP^+TCNQ^- concentrations: 1; 3.8×10^{-2}; 2, 7.6×10^{-2}; 3, 1.5×10^{-2}; 4, 3.0×10^{-2}; and 5, 6.1×10^{-2} mmol per gram P(BA).

electron micrographs of NMP^+TCNQ^-/$TCNQ^0$/polymer composite films clearly show the morphology dependence of the composite conductivity.

Literature Cited

1. J. H. Lupinski, K. D. Kopple and L. J. Hertz, J. Polym. Sci., Part C, 1967, 16, 1561.
2. A. Rembaum, A. M. Hermann, F. E. Stewart and F. Gutmann, J. Phys. Chem., 1969, 73, 513.
3. K. Mizoguchi, T. Suzuki, E. Tsuchida and I. Shinohara, Nippon Kagaku Kaishi, 1973, 1760.
4. J. M. Bruce and J. R. Herson, Polymer, 1967, 8, 619.
5. K. Nakatani, T. Sakata and H. Tsubomura, Bull. Chem. Soc. Japan, 1975, 48, 2205.
6. K. Mizoguchi, T. Kamiya, E. Tsuchida and I. Shinohara, J. Polym. Sci., Polymer Chem. Ed., 1979, 17, 649 and references therein.
7. O.-K. Kim and R. B. Fox, NRL Memorandum Report 1980, 4335, p. 103.
8. O.-K. Kim and R. B. Fox, J. Polym. Sci., Polym. Chem. Ed., 1982, 20, 2765.
9. L. R. Melby, R. J. Harder, W. R. Hertler, W. Mahler, R. E. Benson and W. E. Mochel, J. Am. Chem. Soc., 1982, 84, 3374.
10. L. R. Melby, Can. J. Chem., 1965, 43 1448.
11. R. C. Schulz and U. Geissler, Makromol. Chem., 1978, 179, 1955.
12. The conductivities (of crystalline compaction) of NMP^+TCNQ^- and $NMP^+(TCNQ)_2^-$ are known to be 2 and 0.7 ohm^{-1} cm^{-1}, respectively (Ref. 10).
13. K. W. Law and R. O. Loutfy, Macromolecules, 1981, 14, 587.

RECEIVED September 2, 1983

Synthesis and Properties of Conducting Films by Plasma Polymerization of Tetramethyltin

R. K. SADHIR

Research and Development Center, Westinghouse Electric Corporation, Pittsburgh, PA 15235

W. J. JAMES

Department of Chemistry and Graduate Center for Materials Research, University of Missouri—Rolla, Rolla, MO 65401

> Plasma polymerization of tetramethyltin has been investigated by using an inductively coupled tubular reactor. Plasma polymerized thin films of tetramethyltin, produced in the presence of oxygen, were reflective, having a conductivity in the range 10^2-10^4 $(\Omega cm)^{-1}$. A C/Sn ratio of <2 on the surface was necessary to achieve a conducting film. The transparent films consisted of an irregular network of C and Sn, but there was a threshold value of the C/Sn ratio below which the structure changed to β-tin with possibly some carbon in the interstices. At sufficiently low C/Sn ratios, the film exhibited metallic behavior which was evidenced from the decrease of conductivity with increasing temperature. The electron diffraction spectra, X-ray diffraction spectra, differential thermograms and SEM of the reflective films obtained in this manner confirmed the closely packed β-tin sturcutre of these films.

The preparation and properties of glow-discharge polymerized organotin films have been studied in the past [1] and such films have been used as insulating layers on microelectronic devices, as protective coatings, and as intermediate adhesive layers. [2,3] An up-to-date literature review on the production and properties of the organometallic films prepared by plasma polymerization has been reported by the authors in another paper. [4]

Kny et al. [5] have reported the preparation and properties of semiconducting organotin films having conductivites between 2 X 10^{-1} and 1 X 10^{-2} (ohm cm)$^{-1}$. Thermal treatment of such films increased the conductivity to a maximum value of 1 X 10^2 (ohm cm).$^{-1}$ Recently, Hyneck [6] prepared a transparent conductive film of tin(II) oxide on a substrate by glow discharge polymerization of a mixture of tetramethyltin and carbon dioxide. This paper describes the preparation of thin reflective organotin films having conductivities as high as 10^4 (ohm cm)$^{-1}$ by making use of a mixture of tetramethyltin and a non-polymerizing reactive gas, oxygen. The films produced by this method are found to have the sturcture of β-tin. This paper gives in detail the optimization of the conditions of film formation, the rate of film deposition, and structure elucidation using scanning electron microscopy (SEM), transmission electron microscopy (TEM), electron spectroscopy for chemical analysis (ESCA) and X-ray diffraction. The characterization of the film properties using differential thermal analysis (DTA), contact angle measurements, pull tests, water and air permeability measurements and conductivity measurement versus temperature have also been included.

Experimental

An inductively coupled glow discharge reactor was used for the preparation of the tetramethyltin (TMT) films. The details of the reactor are given in another paper.(4) The conditions used for the production of TMT films are given in Table I. The substrates were glass slides, polished brass, aluminum, stainless steel and polypropylene. The substrates were positioned along the length of the reactor. Prior to film deposition, each substrate was exposed to an argon plasma for thirty minutes. The vapors of TMT and a non-polymerizing reactive gas, oxygen, were introduced through different leak valves at a desired flow rate.

Film thicknesses were determined by the Nomarsky interference method(7) using a "Reichert Wien" polarization interferometer. The topological study of the films was effected by scanning electron microscopy (SEM). To study the competitive ablating effect of the surface of the films by oxygen, SEM micrographs were taken of the samples prepared at varying flow rates of oxygen. Transmission electron micrographs (TEM) and electron diffraction spectra were taken of thin films (less than 200 Å) deposited on KBr pellets by exposing them to the reactive glow discharge for 1300 secs. The KBr pellet was then dissolved in water and the ultrathin film floated off and picked up on a copper grid for TEM examination.

The PHI ESCA/Auger spectrometer 549 was used for determining the composition of these films. This spectrometer was operated at pressures below 5×10^{-6} μm Hg. X-ray photoelectron spectra (XPS) were recorded using a pass energy of 50 or 100 eV and a Mg K_α X-ray source. XPS were calibrated with solidified TMT vapors and SnO_2 as described by Kny et al.(8) The Sn 3 $d_{5/2}$, C 1s and O1s signals were used for determining the C/Sn and O/Sn ratios. For depth profiling, the sputtering was done with argon ions at a pressure of 5×10^{-5} torr with a voltage of 2 kV and a current of 30 mA. The sheet resistivity of the films was measured with a four point probe and was calculated using the following expression:

Table I. Conditions For Glow Discharge Film Formation

	Experiment 1	Experiment 2
Pressure before experiment	5 μm Hg	5 μm Hg
Ar ion cleaning	30 min; 40 μm Hg r.f. power 35 W	30 min; 40 μm Hg r.f. power 35 W
Starting monomer	TMT	TMT
Flow rate of TMT (STP)	2.8×10^{-3} cm^3 s^{-1}	2.66×10^{-3} cm^3 s^{-1}
Flow rate of oxygen (STP)	1.5×10^{-4} cm^3 s^{-1}	6.8×10^{-4} cm^3 s^{-1}
System Pressure before r.f. is applied	50–55 μm Hg	52–57 μm Hg
System pressure after r.f. is on	36–38 μm Hg	37–40 μm Hg
r.f. condition	31 W, 3.9 MHz	31 W, 3.9 MHz
Duration	15005 s	11800 s
Maximum deposition rate	0.30 Å s^{-1}	0.436 Å s^{-1}

$$\rho = \frac{V}{I} \cdot W \; \frac{\pi}{\ln 2} \; F\left(\frac{W}{S}\right)$$

where ρ = sheet resistivity (ohm cm)

V = voltage (volts)

I = current (amperes)

W = thickness of the film (cm)

$F\left(\frac{W}{S}\right)$ = correction factor (approaches unity as W approaches zero)

The conductivities were also measured at higher temperatures to establish the conducting or semiconducting behavior of the films.

Pull tests were carried out using an Instron machine. The procedure used for determining the pull strength is given in a subsequent paper.(4) Pull tests were also performed in aqueous environments. After the curing step, the rods were immersed in boiling water for 2 h. The lower jaw of the Instron tensile tester was modified to perform pull tests in an aqueous medium. A stainless steel container was fixed which moved with the crosshead of the Instron unit.

X-ray diffraction of the films were performed on G.E. XRD-5 diffractometer using Cu K_α radiation. A thick film (∼2500 Å) of tetramethyltin was deposited on glass for the X-ray diffraction study.

Water vapor permeabilities for plasma formed films on polypropylene were measured using a similar method reported by Yasuda et al.(9) The water vapor permeability was calculated using the following relationship.(10)

$$P = \frac{dp}{dt} \; \frac{V_s}{A} \; \frac{273}{273 + T} \cdot \frac{1}{P_v(76)} \; X \; L$$

P = permeability of water or air through the polymer film $\left(\frac{cm^3 (STP) \; cm}{cm^2 (sec) \; cm \; Hg}\right)$

$\frac{dp}{dt}$ = pressure change of water vapor (or air) permeates with time (cm Hg/sec)

V_s = volume of the system (cm^3)

A = area of the polymer film (cm^2)

T = temperature of the system (°C)

P_v = vapor pressure of liquid water (or air) at the experiment temperature (cm Hg)

L = thickness of the film (cm)

The contact angle of water on tetramethyltin films deposited on polypropylene films and glass substrates was determined by using a goniometer. For the TGA and DTA studies, thick deposits of the films were prepared on the reactor glass sleeves and then removed and used as powder samples. Alumina was used as the reference mateiral. The temperature was raised from room temperature to 750°C at the rate of 10°C per minute.

Results and Discussion

Highly conducting, reflective films were formed at a distance of 8-26 cm from the monomer inlet port, just preceeding to the position of the r.f. coil (Position of r.f. coil from the monomer inlet port 25 cm-40 cm). The rate of deposition was low in this metallic region, as shown in Figures 1 and 2. The SEM micrographs of the metallic films formed at 24 cm from the monomer inlet port under plasma conditions described in Exp. 1, Table I are shown at two magnifications in Figures 3(a) and 3(b). The surface was found to consist of closely packed small spheres. The sphere size ranges from 0.03 microns to 0.3 microns with an average diameter of 0.15 microns. Some of the small spheres has a diameter of 250-350Å. Havens et al.(11) from their low angle scattering study of glow-discharge-polymerized ethylene, calculated a D_{min} of 330 Å. Figures 3(c) and 3(d) show the scanning electron micrographs of a plasma polymerized tetramethyltin film, deposited at a distance of 20 cm from the monomer inlet port, prepared with a higher oxygen flow rate (Exp. 2, Table I). These electron micrographs do not show the agglomeration of the particles at a magnification of X6600. Even at this magnification it is observed that in some regions the film is partially removed and in other regions craters are formed. This can be attributed to the ablative effect of the oxygen. As suggested by Yasuda,(12) glow discharge polymerization can be represented by the competetive ablation and polymerization mechanism. Non-polymer-forming gas products are produced when reactive intermediates are formed as precursors to the polymeric thin films. Because polymer-forming species leave the gas (plasma) phase as the polymers are deposited, the major portion of the remaining gas consists of the product gas when a high conversion ratio of a starting material to a polymer is obtained. The characteristics of the product gas plasma are the predominant factors in determining the extent of the ablation process. During polymerization with hydrocarbons, hydrogen is produced as the product

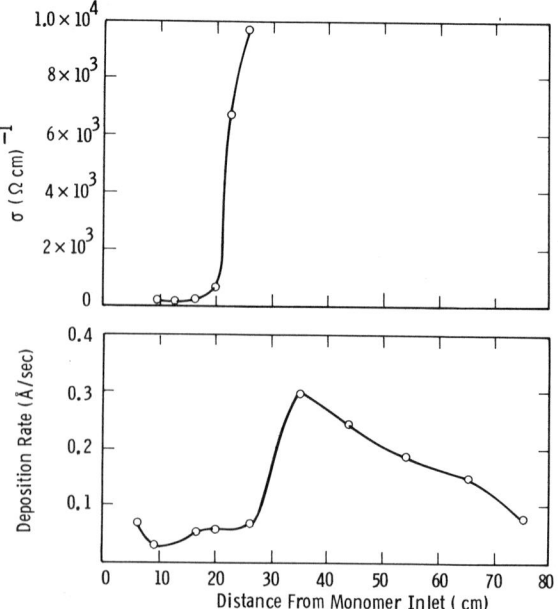

Figure 1. The rate of deposition and conductivity vs distance in centimeters from the TMT inlet port in the reactor. (Ex. 1, Table 1).

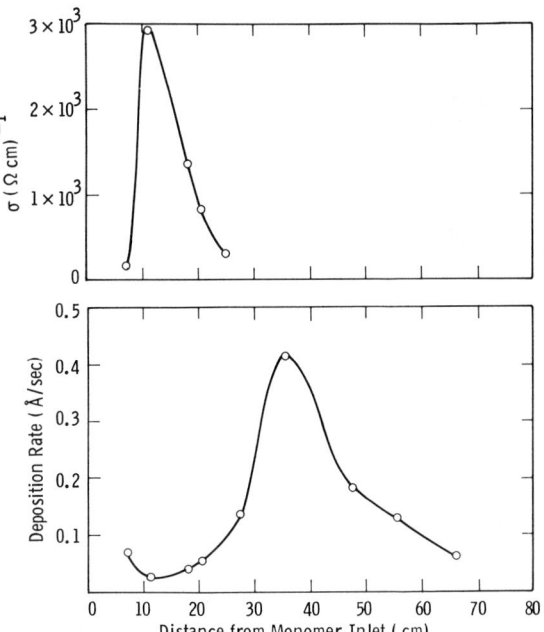

Figure 2. The rate of deposition and conductivity vs distance in centimeters from the TMT inlet port in the reactor (Ex. 2, Table 1).

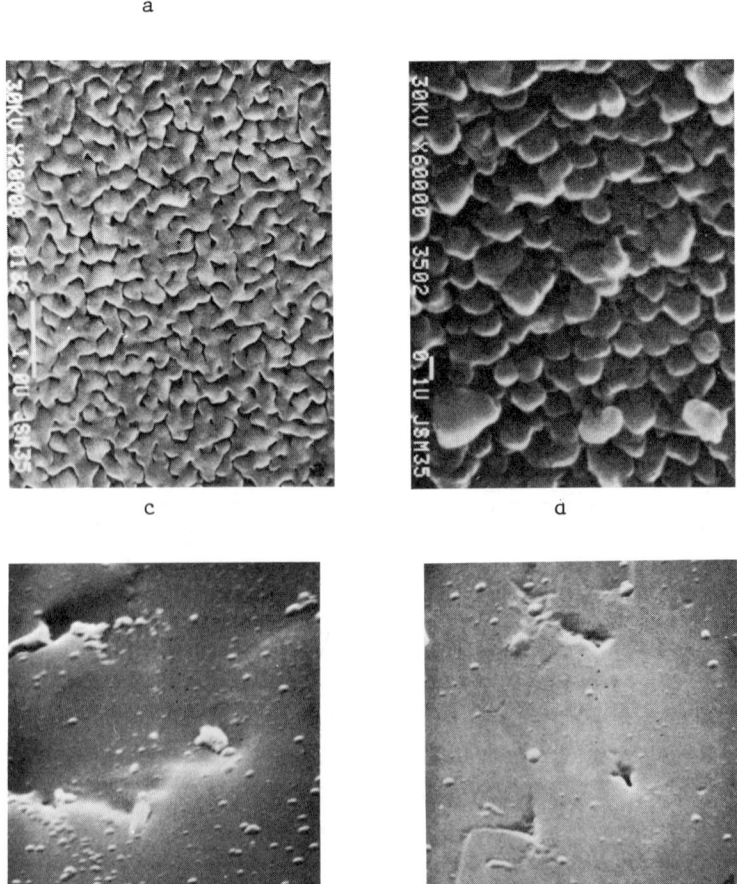

Figure 3. SEM micrographs of TMT films: (a) TMT metallic film on glass, (b) TMT metallic film on glass at a higher magnification (flow rate of O_2 1.5×10^{-4} cm^3 s^{-1}(STP)), (c) TMT metallic film on glass deposited at a higher flow rate of O_2 (6.8×10^{-4} cm^3 s^{-1} (STP)), (d) TMT metallic film on glass deposited at a higher flow rate of O_2 (6.8×10^{-4} cm^3 s^{-1} (STP)). (Magnifications: (a) 13,200X; (b) 40,000X; (c), (d) 6600X.) (Reproduced with permission from Ref. 20. Copyright 1981, Elsevier Sequoia .)

gas and does not show significant ablation effect. Pronounced ablative effects can be seen during the glow discharge polymerization of fluorine- or oxygen containing compounds.(13) In our experiments we introduced oxygen so as to extract the carbon by producing CO and CO_2, thus increasing the conductivity of the films. A low flow rate of oxygen was maintained to avoid having an appreciable reaction of oxygen with tin to form SnO_2. Excessive use of the oxygen led to an attack of the surface as evidenced by SEM micrographs in Figures 3(c) and 3(d). The tin mirror was obtained using TMT as the monomer and an oxygen flow rate of $6.8 \times 10^{-4} cm^3 s^{-1}$ at standard temperature and pressure (STP). The remaining parameters are given in Table I.

The sheet conductivity of samples placed at different positions within the reactor under two different conditions are shown in Figures 1 and 2. The conductivities on the tin mirrors prepared in Exp. 1 were in the range $(2.1 \times 10^2) - (9.7 \times 10^3)$ $\Omega^{-1} cm^{-1}$. The maximum conductivity obtained in yet another experiment under similar conditions was 3.75×10^4 $\Omega^{-1} cm^{-1}$ at a sample position 24 cm from the monomer inlet in the reactor. The samples obtained between 22 and 26 cm from the inlet exhibited metallic behavior, having conductivities of about 10^4 $\Omega^{-1} cm^{-1}$. Between 8 and 22 cm the conductivities ranged between 1.7×10^2 and 6.9×10^2 Ω^{-1} and showed semiconducting behavior. At all other positions insulating films were obtained. A carbon-to-tin ratio of 2.0 seemed to be a necessary, but not the only condition for conducting properties since insulating films with a metalloid appearance also exhibited carbon-to-tin ratios lower than 2.0. The conductivity of the films prepared at higher flow rates of oxygen (Exp. 2) are shown in Figure 2. The conductivities range between 3×10^2 $\Omega^{-1} cm^{-1}$ and 3×10^3 $\Omega^{-1} cm^{-1}$. Under these conditions higher conductivity was obtained near the monomer inlet port (about 10 cm from the monomer inlet). Thin films of tin oxide which formed on the surfaces of the samples do not explain the observed conductivity. The conductivity of SnO_2 films at room temperature has been reported to be in the range of $10^{-2} - 10^2$ $\Omega^{-1} cm^{-1}$. The high conductivities obtained in our experiments suggest the films to be mainly β-tin. Thick deposits of the films were prepared on glass sleeves using plasma conditions given in Exp. 1, Table I for DTA. The powder was scrapped from the films deposited between 8 cm to 26 cm from the monomer inlet port in the reactor. From the DTA curve of this powder an endothermic peak at 232°C due to melting of β-Sn was obtained which also confirmed the presence of β-Sn in the plasma polymerized TMT film.

The X-ray diffraction patterns and the TEM micrographs show a crystalline structure for these films. Major peaks of β-Sn corresponding to the (203) and (101) indices were obtained at angles 2θ of 30.7° and 31.2° (Figure 4). Figures 5(a) and 5(b) are the transmission electron micrographs of the metallic film, showing the presence of tin crystals. Figure 5(c) is a TEM micrograph of a carbon-enriched film. Figure 5(d) shows the

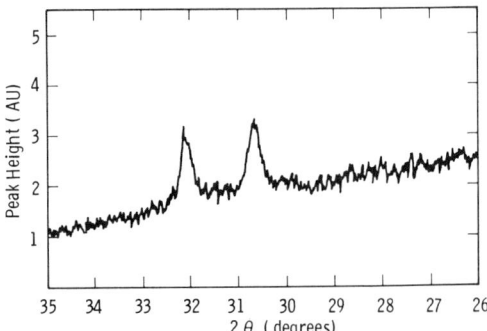

Figure 4. X-ray diffractogram of metallic TMT film.

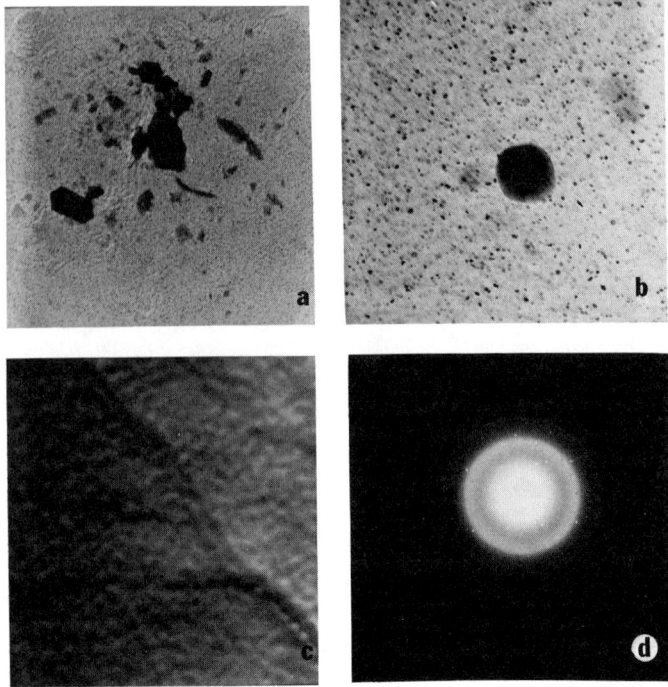

Figure 5. TEM micrographs and an electron diffractogram of TMT film prepared with an oxygen flow rate of 1.5×10^{-4} cm^3 s^{-1} (STP) (Position of the deposited films from the monomer inlet port in the reactor for a, b and d = 24 cms and c = 65 cms). (a) TMT metallic film, (b) TMT metallic film at a higher magnification; (c) TMT insulating type film; (d) electron diffraction pattern of a metallic TMT film (Magnification: (a) 11000X; (b), (c) 51060X). (Reproduced with permission from Ref. 20. Copyright 1981, Elsevier Sequoia .)

electron diffraction pattern of the metallic film. Electron diffraction of the metallic TMT film shows sharp rings suggesting a crystalline structure. A regular pattern of tiny dots all over the diffraction pattern are observed. The scattering angles and orientation of the single crystal reflections are found to correspond to β-Sn. Kny et al.(14) produced semiconducting organotin films which exhibited no crystalline structure when examined by TEM, SEM and X-ray diffraction. Electron and X-ray diffraction of their samples produced diffused rings similar to those for amorphous materials. They obtained the β-Sn structure after exposing the films to an electron beam for several minutes. The films prepared in our study were crystalline and exhibited the β-Sn structure.

The films prepared were very stable in air for several months. The conductivity showed no change during that time. Heat treatment of the film at 63°C did not show appreciable change in the room temperature conductivity (Figure 6). With an increase in temperature, the conductivity decreased which confirmed the metallic nature of the films. Prolonged heating at higher temperature increased the conductivity; heating at 91°C for 300s increased the conductivity from 8.59×10^3 to 1.29×10^4 $\Omega^{-1}cm^{-1}$ and 125°C for 1500s increased the conductivity from 1.30×10^4 to 2.40×10^4 $\Omega^{-1}cm^{-1}$ (Figure 7). After the heating was stopped, the conductivity values decreased to some extent but did not reach the value from which the heating was started. Kny et al.(14) have reported that the rate of increase in conductivity at 160°C is more than the rate of increase in conductivity at 128°C. They showed that by heating of the film at 154 and 200°C, the conductivity was decreased appreciably from the original condition. In our studies, we did not observe this behavior. This may come about because of the difference in the film composition.

The film composition was determined by ESCA. During ESCA analysis, the absolute value of the signal is influenced by instrumental parameters and the surface roughness of the samples. The elemental ratio of carbon to tin and of oxygen to tin were influenced less by these parameters. It has been reported in an earlier paper(8) that oxygen to tin and carbon to tin ratios on a series of samples gave reproducible results (variation of ±8%). Hence it was decided in this study to use the elemental ratios of carbon to tin and oxygen to tin for characterization of film. Using peak heights in XPS instead of peak areas speeded up the analysis and evaluation of data. A comparison of peak height ratios with peak area ratios on a selected series of samples showed that only a very small error was introduced by using peak heights. The XPS ratio of carbon to tin was calibrated with TMT vapors solidified at the sample stage at -112°C. A conversion factor $f_{C/Sn}$ of 19.9 was determined. The XPS ratio of oxygen to tin was determined with finely powdered SnO_2 spread onto indium foil. A conversion factor of $f_{O/Sn}$ of 6.0 resulted from this procedure. All the values given for the ratios of carbon to tin and of oxygen to tin in the following discussion refer to atomic ratios calibrated with the above procedure.

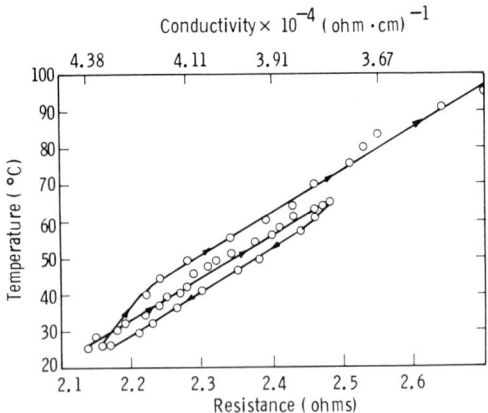

Figure 6. Change in sheet conductivity with increase in temperature.

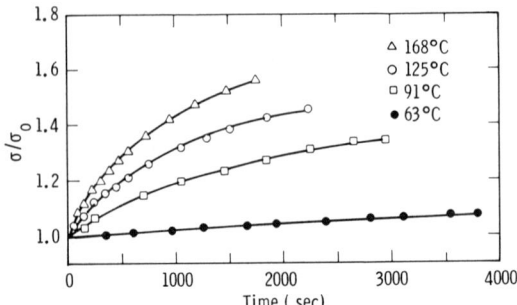

Figure 7. Increase in sheet conductivity with time at higher temperature (σ is the conductivity measured at higher temperature, after heating the film for certain time at that temperature; σ_0 is the conductivity of the film measured at the same temperature at zero time). (Reproduced with permission from Ref. 20. Copyright 1981, Elsevier Sequoia.)

The carbon-to-tin and oxygen-to-tin atomic ratios of various samples along the length of the reactor are shown in Figure 8. The original composition of the monomer was not retained in the plasma polymerized TMT films. Tin enriched films were obtained near the monomer inlet port and carbon enriched films were obtained towards the end of the reactor. There appeared to be a correlation between the carbon-to-tin atomic ratio and the conductivity of the films. Lower carbon-to-tin ratios were obtained from the samples exhibiting higher conductivities. The composition of the film on the surface is different from the composition of the bulk (Figure 9). After the deposition of the film, when the reaction was stopped, there remained some free radical traps and these active sites were capable of reaction with the environment. A thin layer of SnO_2 might be formed(5) and CO_2 could be absorbed on the surface. The actual composition of the film was obtained by first removing 20-25 Å of film by sputtering with argon ions. Preferential removal of fluorine relative to carbon during argon sputtering and taking ESCA spectrum has been reported in plasma polymerized fluorinated polymers.(15,16) Reduction of metal oxide surfaces has also been reported after argon sputtering.(17) Though in our experiments, we have used mild conditions of sputtering and low pass energy for ESCA to minimize the preferential sputtering effects, the change in composition of the films after depth profiling can also be attributed to this fact.

The carbon-to-tin ratio in the bulk of the films obtained between 8 and 20 cm from the monomer inlet of the reactor varied between 0.2 and 0.5. Samples at positions between 20 and 26 cm gave no carbon ESCA signals. Figure 10 is the depth profie of such a metallic film, 1208 Å thick. A small amount of carbon was found on the surface. After 2 minutes of sputtering with argon ion at 2 kV and 30 mA emission current, the carbon signal totally disappeared and the tin signal increased by approximately 3 times its peak height on the surface.

The pull test results for glow-discharge-polymerized TMT films on aluminum 304, stainless steel and brass substrates are summarized in Table II. The films deposited onto the stainless steel substrate showed the maximum pull strength. In most instances, the failure of films on stainless steel occurred at the interface of the Chemlok adhesive and the film or on the back side (between the rod and the Chemlok adhesive) suggesting thereby that the actual adhesive and cohesive strengths of the film exceeded the values reported. For aluminum and brass, the failure appeared to be adhesive when observed visually or under a light microscope. However, when the pull region was examined by ESCA, a thin film residual containing tin (or tin and carbon) was present on the substrate surface indicating the failure mode to be cohesive. Pull tests of films deposited onto brass samples were also carried out after soaking of the samples in boiling

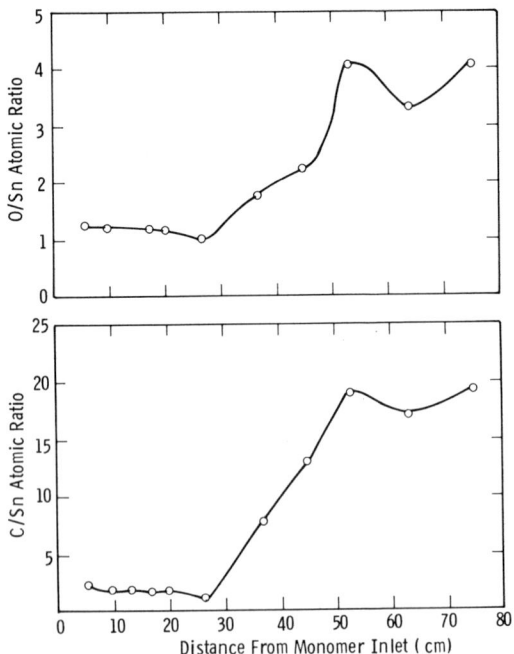

Figure 8. Carbon-to-tin and oxygen-to-tin atomic ratios vs distance in centimeters from the TMT inlet port in the reactor.

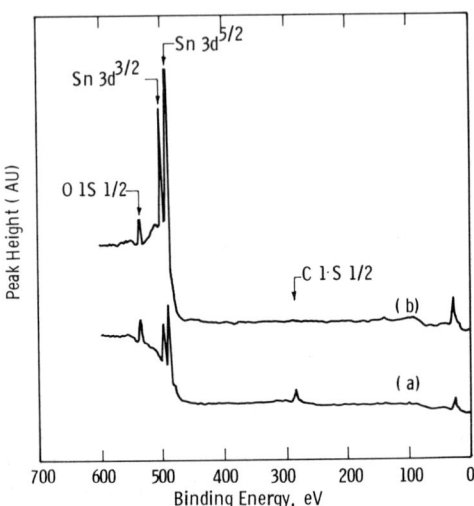

Figure 9. ESCA profile of metallic TMT film, (a) on the surface, (b) after 2 min. sputtering with Ar-ion.

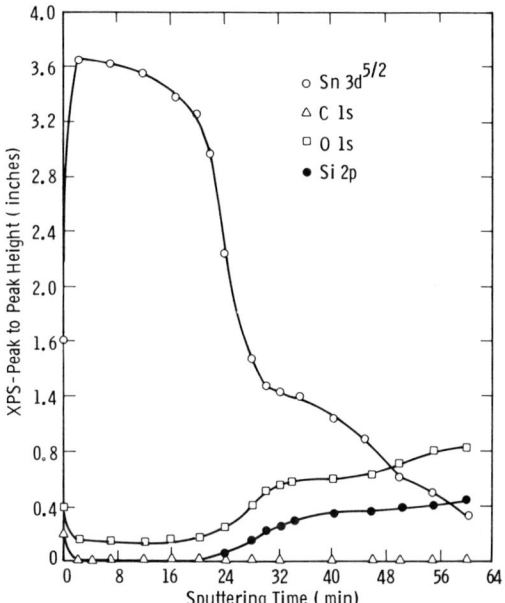

Figure 10. Effect of Ar-ion sputtering on Sn, C and O signals in ESCA. (Reproduced with permission from Ref. 20. Copyright 1981, Elsevier Sequoia.)

Table II. Pull Tests on Glow Discharge Polymerized Al (3000 series) Brass

S. No	Sample Position from the inlet of the monomer (cm)	Al	
		Failure Mode light microscope inspection	Pull Strength psi
1	9.0	Film-Al interface	2860 ± 400
2	17.0	"	2992 ± 510
3	20.5	Chemlok adhesive-rod interface	>3344 ± 325
4	26.5	Film-Al interface	3740 ± 225
5	33.0	Chemlok adhesive-film interface	>2552 ± 300
6	45.0	Film-Al interface	2728 ± 480
7	53.0	"	2090 ± 650
8	64.0	"	1760 ± 320

Tetramethyl Tin Films Deposited on and Stainless Steel

Brass		Stainless Steel	
Failure Mode light microscope inspection	Pull Strength psi	Failure Mode light microscope inspection	Pull Strength psi
Film-brass interface	2904 ± 235	Chemlok adhesive-film interface	>4004 ± 340
"	3168 ± 360	Chemlok adhesive-rod interface	>2640 ± 235
"	3476 ± 480	Chemlok adhesive-film interface	>3388 ± 405
Chemlok adhesive-rod interface	>2420 ± 125	"	>4840 ± 465
Film-brass interface	1320 ± 390	"	>3388 ± 130
"	968 ± 455	Chemlok adheisve-rod interface	>3586 ± 340
–	–	–	–
–	–	–	–

water for 2 h and were performed in hot water. The pull strength was reduced to about one-third of the value in air. Kny et al.(18) have reported pull strengths of glow-discharge-polymerized monobutyltrivinyltin films on aluminum and stainless steel. However, they did not introduce O_2 at a low flow rate simultaneously with the monomer and as a consequence did not experience in their systems the plasma etching introduced by an ablating gas such as O_2. Accordingly, the extent of atomic interfacial mixing was small and resulted in diminished adhesion.

It is noteworthy that the pull strengths are higher for the metallic tin-enriched coatings than for the coatings containing higher carbon-to-tin atomic ratios. Unfortunately, we were unable to extract from the ESCA studies irrefutable evidence of primary bonds at the film-substrate interface. On the basis of the higher pull strengths of films with low carbon-to-tin atomic ratios and the lower pull strengths of the films with a high carbon-to-tin ratio, it is evident that tin somehow plays an important role in the interfacial bonding mechanism. Indeed, the ESCA studies of Kny et al. did establish the presence of the bonding of tin to the surface of aluminum via an oxygen bridge bond.

A thin metallic film (1000 Å, carbon-to-tin ratio less than 0.3) was deposited onto polypropylene and the permeability determined. A water vapor permeability of 1.76×10^{-14} cm^3 (STP) cm cm^{-2} $s^{-1} cm^{-1} Hg^{-1}$ was obtained, which is of the order of values usually found for metal coatings such as copper. (Table III). For reference, the water vapor permeability of the polypropylene film is $(3.8 - 4.0) \times 10^{-9}$ cm^3 (STP)$cm^{-1} cm^{-2}$ $s^{-1} cm^{-1} Hg^{-1}$ (19)

The contact angle for water on the films deposited in the metallic region are given in Table IV. The contact angle of water on polypropylene is 106° which was reduced to 45° after the glow-discharge-polymerized TMT film was deposited. Similar values were obtained on glass. TMT films deposited by glow discharge in the absence of oxygen exhibited a contact angle of approximately 78°. These results suggest that the TMT film deposited in the presence of oxygen has the highest surface energy, which is in agreement with the increased metallic behavior of such films, i.e. high conductivities, luster etc.

Summary and Conclusions

Thin shiny metallic tin films can be produced by plasma polymerization of TMT in the presence of oxygen. Such films have a conductivity in the range of $10^2 - 10^4$ $\Omega^{-1} cm^{-1}$. A carbon-to-tin ratio of less than 2 on the surface was essential to achieve a conducting film. The transparent films were insulating, and might consist of an irregular network of carbon and tin, but there was a threshold value of the carbon-to-tin ratio below which the structure changed to β-tin with possibly some carbon in the interstices.

Table III. Water Vapor Permeability of Barriers

		$P \times 10^{10} \dfrac{cm^3(STP\ cm}{cm^2(sec)\ cm\ Hg}$
1	Polypropylene	40
2	Polypropylene + TMT film (without oxygen)	.806
3	TMT film (without oxygen)	.00165
4	Polypropylene + TMT film (in presence of oxygen)	.161
5	TMT film (in presence of oxygen)	.000176

Table IV. Contact Angle of the Metallic Films Deposited On Glass and Polypropylene Substrate

1	TMT deposited on glass (in absence of ablating gas)	78°
2	TMT deposited on glass (in presence of ablating gas)	48°
3	Polypropylene	106°
4	TMT deposited on polypropylene (in presnece of ablating gas)	46°

The flow rate of oxygen played an important role in rate of deposition of the film and their conductivities. Too high a flow rate appreciably ablated the surface of the film producing pinholes and defects. At sufficiently low carbon-tin ratios, the films exhibited metallic behavior which was evidenced from the decrease in conductivity with increasing temperature. The electron diffraction lines were sharp and corresponded to β-Sn. X-ray diffraction spectra of thicker metallic films showed major peaks at 30.7° and 31.2° corresponding to the (203) and (101) reflections of β-Sn. Differential thermograms of powders scraped from the walls of the glass sleeves inserted in the reactor showed an endothermic peak at 232°C attributed to the melting of β-Sn. All this evidence coupled with the SEM micrographs clearly proves that the reflective films obtained in the manner described possessed the closely packed β-Sn structure. Pull test results for all three substrates showed that the films enriched with tin have higher adhesive and cohesive strengths than those of higher carbon content. The metallic films acted as good water vapor permeability barriers in accord with what is expected for metal coatings.

Literature Cited

1. Bradley, A.; Hammes, J. P., J. Electrochem. Soc., 1963, 110, 15.
2. Tkachuk, B. V.; Kobtsev, Yu. D.; Laurs, E. P.; Mikhalchenko, V. I.; Marusshi, N. Ya., Iv. Vyssh. Ucheben. Zaved Fiz., 1972, 15, 117.
3. Tkachuk, B. V.; Marusshi, N. Ya; Laurs, E. P. Vysokomol. Soedin, Ser. A, 1973, 15, 2046.
4. Sadhir, R. K.; Saunders, H. E. and James, W. J., submitted for Publication in the same ACS Symposium Series.
5. Kny, E.; Levenson, L. L.; James, W. J.; and Auerbach, R. A.; J. Phys. Chem., 1980, 84, 1635.
6. Hynecek, J., U.S. Patent 4,140,814, February 20, 1979.
7. Nomarski, G.; Weill, R. A., Rev. Metall.(Paris), 1955, 55, 121.
8. Kny, E.; Levenson, L. L.; James, W. J.; Auerbach, R. A., Thin Solid Films, 1979, 64, 395.
9. Yasuda, H. K.; Stannett, V., J. Macromol. Sci.-Phys., 1969, B3(4), 589.
10. Yasuda, H. K., J. Appl. Polym. Sci., 1975, 19, 2529.
11. Havens, M. R.; Mayhan, K. G.; James, W. J., J. Appl. Polym. Sci., 1978, 22, 2793.
12. Yasuda, H. K., ACS Symp. Ser., 1978, 108, 37.
13. Yasuda, H. K., Contemporary Topics in Polymer Science, Ed. Shen, M. Plenum Publishing Corp., 1979, Vol. 3, 103.
14. Kny, E.; James, W. J.; Levenson, L. L.; Auerbach, R. A., Thin Solid Films, 1981, 85, 23.
15. Rice, D. W.; O'Kane, D. F., J. Electrochem. Soc., 1976, 123, 1308.

16. Yasuda, H. K.; Marsh, H. C.; Brandt, E. S.; Reilley, C. N., J. Polym. Sci., Polym. Chem. Ed., 1977, 15, 991.
17. Kim, G. S.; Baitinger, W. E.; Amy, J. W.; Winogard, N.; J. Electron. Spectr. Related Phen. 1974, 5, 351.
18. Kny, E.; Levenson, L. L.; James, W. J.; and Auerbach, R. A., J. Vac. Sci. Technol., 1979, 16(2), 359.
19. Lucari, J. J. and Brans, E. R. Mach. Des., 1967, 39(12), 192.
20. Sadhir, R. K., James, W. J.; and Auerbach, R. A., Thin Solid Films, 1982, 97, 17.

RECEIVED September 2, 1983

Plasma Polymerized Organometallic Thin Films: Preparation and Properties

R. K. SADHIR and H. E. SAUNDERS

Research and Development Center, Westinghouse Electric Corporation, Pittsburgh, PA 15235

W. J. JAMES

Department of Chemistry and Graduate Center for Materials Research, University of Missouri—Rolla, Rolla, MO 65401

>We review the recent work on plasma polymerization of organometallic monomers, and discuss the preparation and properties of organotin and organogermanium monomers. Various saturated and unsaturated monomers (such as tetraethyltin, tetravinyltin, hexamethylditin and tetramethylgermanium) have been polymerized in a glow discharge, and the mechanism of polymerization has been elucidated. Films were also prepared by glow discharge polymerization from the above mentioned monomers in the presence of oxygen. Films of varying conductivity were prepared by these two methods, the presence of oxygen giving rise to semiconducting films. The C/Sn and C/Ge ratios, as determined by ESCA, vary with the reactor operating parameters and with the location of the substrate within the reactor. Thin films prepared from tetramethylgermanium also show semiconducting and insulating behavior, depending upon their location within the plasma reactor.

Various methods have been used for dispersing metals in conventional polymeric systems, such as coevaporative techniques, (1) solution growth techniques (2) and ion implantation. (3) However, plasma polymerization has received relatively little attention in this field. The synthesis of organometallic films produced by plasma polymerization technqiues is an attractive prospect, since it can be envisaged that careful choice of an organometallic monomer and close control of the overall composition of the product by the control of the processing parameters would greatly extend the scope of these films in electrical, magnetic and optical applications.

In this paper we review the recent work on plasma polymerization of organometallic monomers and discuss the room temperature preparation, properties, and composition of glow discharge-polymerized tetraethyltin, hexamethylditin, tetravinyltin and tetramethylgermanium. A possible mechanism by which glow discharge polymerization of these monomers occurs is discussed.

Literature Review

There has been a considerable amount of work done on thin film deposition of pure organic vapors by glow discharge polymerization in the past.(4) Hollahan and Rosler(5) have discussed the preparation of inorganic thin films by the glow discharge process, and their uses in the electronics industry. In most cases, the substrate is heated to 200°C or above to form these inorganic films.

The initial work on plasma polymerization of organosilicon compounds was done by Tkachuk et al.(6-8) who prepared siloxane and organosiloxane films under different processing conditions, and studied rates of deposition. The films were characterized by IR and ESR. Some of the more recent publications show that the plasma polymerization of organosilicon monomers yields films exhibiting excellent properties, such as high thermal stability(9) and high dielectric constant.(10,11) Such films have found use as dielectrics for microelectronics, and in optical(12, 13) and biomedical applications.(14)

The technology of plasma formation of metal-containing polymers in the form of thin films dates from 1963, when Bradley and Hammes(15) prepared specimens from some forty different materials, and studied their electrical conductivities. Included in the study were organic compounds of iron, tin, titanium, mercury, selenium, and arsenic. The presence of a metal or transition element in the polymer did not lead to special electrical properties compared to the purely organic polymers studied.

Tkachuk and coworkers(16,17) have reported that organotin polymer films prepared in a glow discharge can be used as insulating layers on microelectronic devices, as protective coatings, and as intermediate adhesive layers. Subsequent thermal treatment of such coatings can produce thin films of different properties. For example, pyrolysis of organotin polymer films can produce tin oxide coatings. Kny et al.(18) have reported the production and properties of semiconducting films from tetramethyltin having conductivities between 2×10^{-1} and $10^{-2} \Omega^{-1} cm^{-1}$. Thermal treatment of such films increased the conductivity to a maximum value of $10^2 \Omega^{-1} cm^{-1}$. These workers have also determined the composition of such films by x-ray photoelectron spectroscopy (XPS) and Auger electron spectroscopy (AES).(19) Kny et al.(20) have also studied the polymer/substrate interface composition and adhesion of glow discharge-formed organotin polymers. The deposition of a transparent conductive layer of tin(II) oxide on a substrate by glow discharge polymerization of a mixture of tetramethyltin and carbon dioxide has been the subject of a U.S. patent.(21) Morosoff and Patel(22) investigated the preparation of plasma polymers containing the transition metals, iron and cobalt, and elucidated their chemical properties. The transition metals were introduced into the plasma as the volatile organometallic compounds, pentacarbonyliron and cyclopentadienyl-dicarbonylcobalt, and the resulting films were characterized by FTIR, ATR, ESCA and cyclic voltammetry. Anderson and Spear(23) prepared amorphous germanium carbide by decomposition of a mixture of germania and ethylene in a r.f. glow discharge at elevated temperatures.

Recently, plasma polymerized vinylferrocene films have been deposited on graphite and platinum electrodes,(24,25) and the characteristics of these electrodes have been studied. There have been a few publications(26-29) describing a method involving simultaneous plasma etching and polymerization in the same system, resulting in the synthesis of metal-containing fluoropolymers. In this method, the metal to be incorporated in the polymer film is used as a cathode in a capacitively coupled reactor, and a fluorinated hydrocarbon is used as a polymerizing/etchant gas. By using this technique, various polymeric films containing germanium, molybdenum, copper, tin and chromium have been obtained. XPS and MS were employed to determine the composition and structure of the films. Shuttleworth(30) has used two distinct routes to obtain a metal in the gas phase and subsequently incorporate it into the growing plasma polymer, namely, sputtering from a tin surface by use of an inert gas plasma and by the use of a volatile metal compound ($Cr(CO)_6$). The analysis of the polymer films was carried out by using ESCA and it was shown that these techniques produce a metal compound dispersed in the plasma polymer with a wide range of metal-polymer compositions. Liepins et al.(31) have reported the use of glow discharge on organometallic monomers such as tantalum (V) ethoxide, dimethyl mercury, diethyl mercury, tetramethyl lead, tetraethyl lead and trimetyl bismuth to deposit thin, metal-containing coatings on microspheres used as targets for fusion reactors.

Experimental

Glow discharge organometallic films were prepared in an inductively coupled plasma reactor(20) (Figure 1). The reactor consisted of a cylindrical pyrex glass tube (100 cm X 7 cm). The whole system was evacuated to the millitorr range by means of a rotary pump. The monomer gases or vapors were introduced by a leak valve at one end of the reactor at the desired flow rate, and the pressure was recorded by means of a thermocouple gauge. The generator for activating the plasma was operated at 3.9 MHz frequency.

Precleaned glass slides (2.5 X 2.5 X 0.045 cm) and aluminum plates (1 X 1 X 0.0625 cm) were used as substrates for film

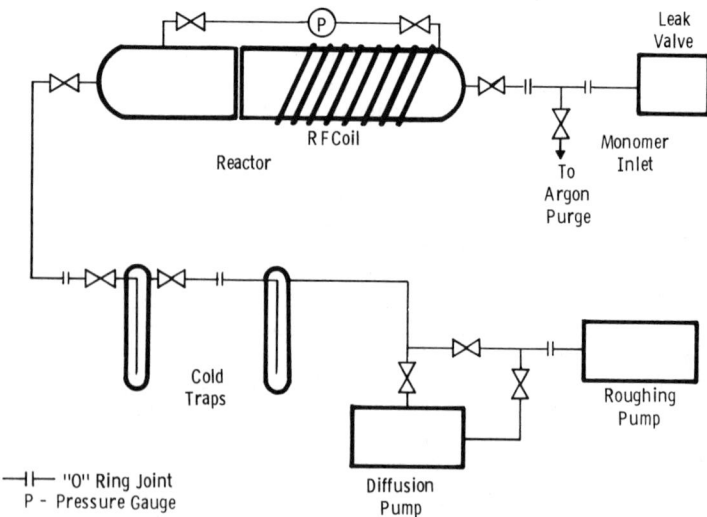

Figure 1. Tubular reactor for plasma polymerization.

deposition. These substrates were cleaned by exposure to a plasma of a nonreactive gas, argon, at a pressure of 50 millitorr prior to the deposition of the film. The starting materials for film formation were tetraethyltin (TET), hexamethylditin (HMDT), tetramethylgermanium (TMG) (Ventron Alfa), and tetravinyltin (TVT) (Pfaltz and Bauer). The gaseous monomer was introduced into the reactor through the leak valve. TET, HMDT, and TVT were heated to 130°C, 83°C, and 80°C, respectively at reduced pressures to transform the liquid monomer to the gaseous state. The feedline and the valve were wrapped with heating tape, and the temperature was adjusted by means of a Variac to prevent condensation of the monomer in the feedline and valve. Films were also prepared from these monomers in the presence of oxygen which was used to increase the conductivity of the resulting films. The parameters used for film deposition are given in Table I.

Film thicknesses were determined by the Nomarsky interference method(32) using a "Reichert Wien" polarization interferometer. The films were deposited by masking a portion of the glass slide so as to form a sharp step. This step was viewed under a polarization interferometer which produced a regular interference pattern with a displacement due to the difference in thickness. The distance between each interference line and the displacement was measured. The thickness was calculated by multiplying the ratio of these values by the wavelength. Conductivities of the films were determined by the four probe method at both ambient and elevated temperatures. The pull strength of the films on glass slides and aluminum substrates were measured. The coated and uncoated sides of the substrates were attached to stainless steel rods (1 cm diameter) with a thin layer of Chemlok-304 epoxy adhesive (Hughson Chemical Co.). After applying the epoxy, the rods were placed in a tray to maintain proper alignment while the epoxy was cured at 50°C for 24 hours. Pull testing was then performed at a strain rate of 0.125 cm/minute using an Instron Tensile Tester Model 02-002. Failure modes were determined by light microscopy and ESCA. The topological study of the films were effected by scanning electron microscopy (SEM). A Physical Electronics Industries ESCA-AES spectrometer model 549 was used to determine the composition of the films. The x-ray photoelectron spectroscopy (XPS) spectra were recorded using a pass energy of 50 eV with MgK$_\alpha$ x-ray source. Sputtering experiments were carried out using argon ions at 2 kV and 30 mA current.

Results and Discussion

The results of the rate of deposition and the composition of three organotin polymers are shown in Figure 2. The maximum rate of deposition was obtained with tetravinyltin (TVT). Figure 3 shows the rate of deposition of the organotin polymers deposited in the presence of oxygen as a function of the position in the reactor for conditions given in experiments II, IV and VI (Table I). A

Table I. Parameters for

Experiment	I	II	III
Starting Compound	TET	TET	TVT
Flow rate of monomer, (STP) cm³/sec	3.1 X 10^{-3}	8.4 X 10^{-3}	6.1 X 10^{-3}
Flow rate of oxygen gas, (STP) cm³/sec.	---	3.2 X 10^{-4}	---
r.f. power	32W 3.9 MHz	33W 3.9 MHz	33W 3.9 MHz
System pressure, mtorr	32-35	50-53	23-23
Duration, s	9800	7000	11400
Max. dep. rate, Å/s	0.21	0.31	0.95
Film appearance	transparent	reflective, shining, semi-transparent	transparent

Glow Discharge Polymerization

IV	V	VI	VII	VIII
TVT	HMDT	HMDT	TMG	TMG
6.3×10^{-3}	7.4×10^{-4}	2.5×10^{-3}	9.6×10^{-3}	1.2×10
3.5×10^{-4}	---	3.6×10^{-4}	---	3.6×10
33W	32W	32W	4W	4.5W
3.9 MHz	3.9 MHz	3.9 MHz	3.9 MHz	3.9 MHz
22-24	28-34	35-38	55-57	55-57
9050	6120	3800	2950	7600
1.38	0.34	0.75	1.42	0.78
reflective, seim-transparent	brown, reflective	brown, reflective	golden, semitransparent	reflecti metal-li

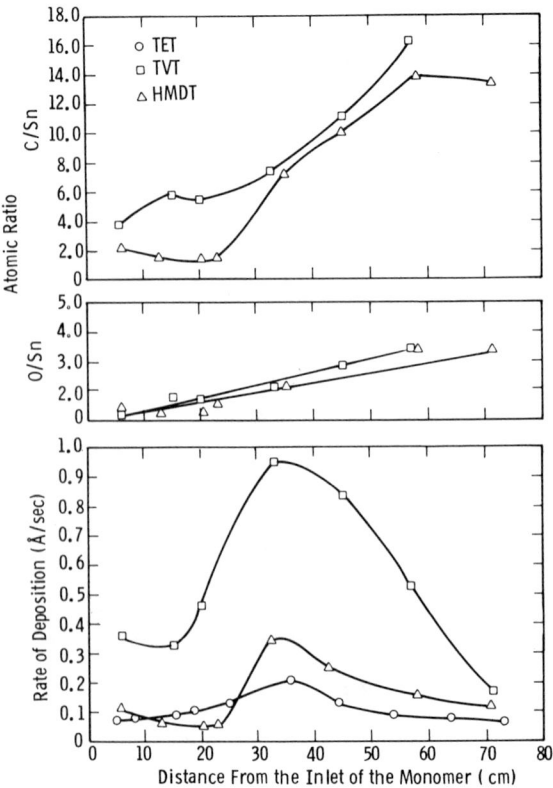

Figure 2. The rate of deposition, C/Sn and O/Sn atomic ratios vs distance in cm. from the monomer inlet in the reactor.

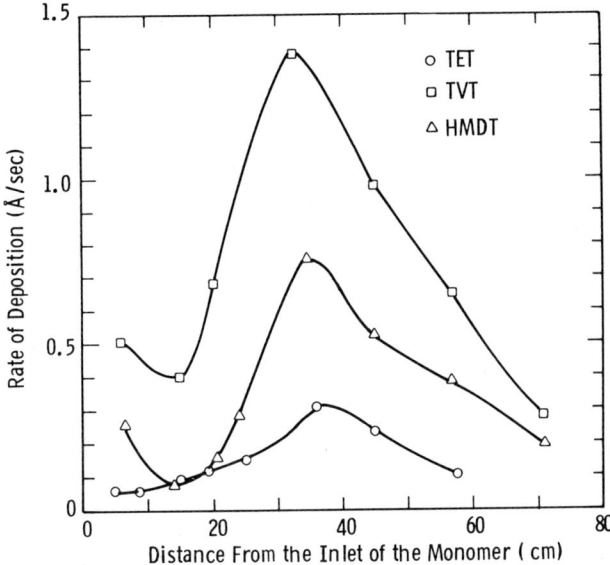

Figure 3. The rate of deposition (in presence of oxygen) vs distance in cm. from the monomer inlet in the reactor.

similar trend in the rate of deposition was observed, i.e., the maximum deposition rate was obtained with TVT. Another interesting feature observed during the glow discharge polymerization of TVT was the appreciable reduction in the pressure (65 millitorr to 23 millitorr) immediately following the application of the r.f. field. This occurred both in the presence and absence of oxygen. In the cases of TET and HMDT, only marginal changes in pressures were observed after applying the r.f. field. Yasuda(33) differentiates between plasma-induced polymerization and plasma state polymerization. Plasma-induced polymerization is essentially conventional polymerization which does not produce a gas-phase byproduct. From our findings it appears that under given conditions, the glow discharge polymerization of TVT proceeds mainly by plasma-induced polymerization as would be expected. Figure 4 shows the rate of deposition and the composition of the organogermanium films prepared both in the presence and absence of oxygen. In all these cases (see Figures 2 through 4) the maximum rate of deposition was observed at about 25 to 35 cm from the monomer inlet, i.e. where the r.f. coil was located.

The ESCA examination of the films revealed that the stoichiometry with regard to the starting monomer was not retained. The C/Sn and C/Ge atomic ratios were determined by considering the relative heights of C1s peak, Sn $3d^5$ peak and Ge Auger 422 peak. The correction factors for varying atomic sensitivity of the various peaks of C, Sn, Ge and O were determined by taking the XPS of the solidified TMT, SnO_2 and TMG. The procedure is described elsewhere in detail.(19) In our study, most of the significant peaks generated were sharp. Hence, the calculation of atomic ratios did not differ appreciably when either peak heights or peak areas were used. For convenience, atomic ratios were calculated using peak heights. As expected, the C/Sn and C/Ge ratios varied along the length of the reactor (Figures 2 and 3). Films of higher tin and germanium contents were deposited from 6 to 24 cm from the monomer inlet of the reactor, just preceding the r.f. coil, and carbon enriched films were deposited at the portion of the reactor most distant from the monomer inlet. Although various C/Sn ratios were obtained along the length of the reactor for plasma polymerized TET, TVT and HMDT (varying from 1.5 to 16.4), the general trend of the C/Sn ratios of the monomers (C/Sn ratio for TET and TVT = 8/1 and for HMDT = 3/1) was retained in the polymers i.e. the C/Sn ratios in the plasma polymerized TET and TVT were higher than those of HMDT. Figures 5 through 7 show ESCA depth profiles of the films. The sputtering was carried out by argon ions at 2 kV and 30 mA currents. The compositions of the films were determined after every 2-3 minutes of sputtering until reaching substrates. The peak heights of each component were plotted against the sputtering time for all the organotin films and are shown in Figures 5-7. The Sn content on the surface is less than in the bulk whereas oxygen is found to be less in the bulk of the films. The

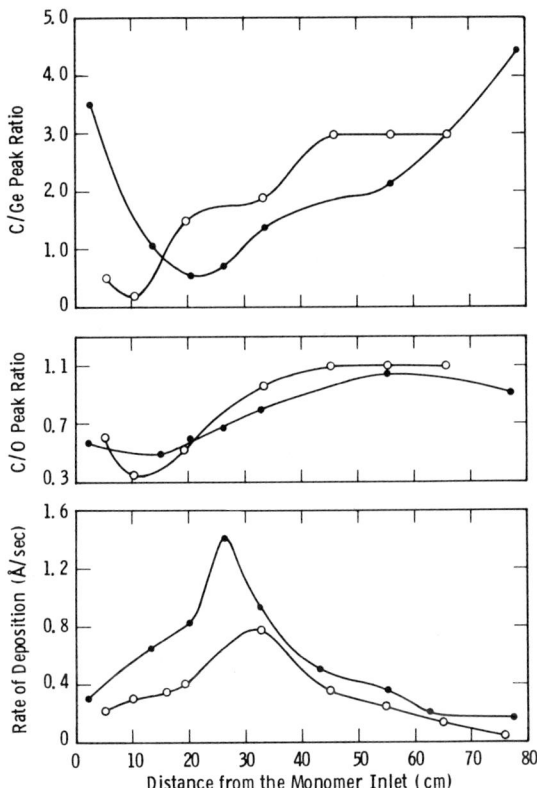

Figure 4. The rate of deposition, C/Ge and C/O atomic ratios vs distance in cm. from the momomer inlet in the reactor.

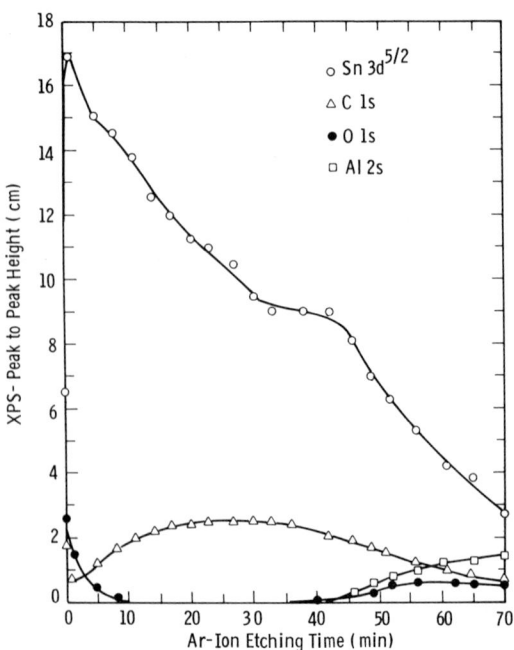

Figure 5. Effect of Ar-ion sputtering on TET film.

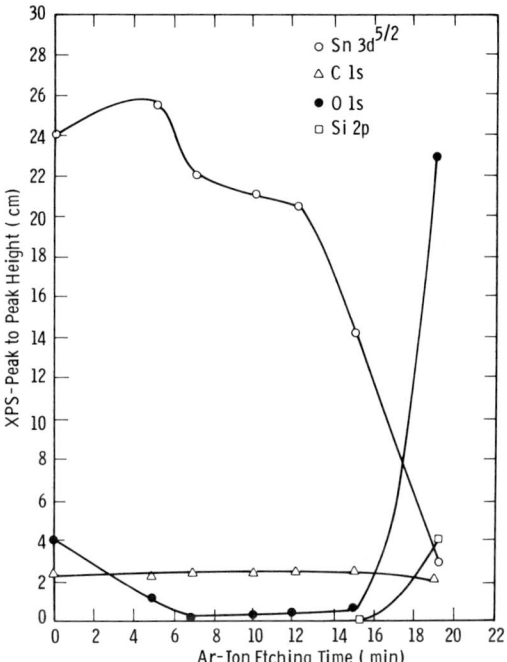

Figure 6. Effect of Ar-ion sputtering on HMDT film.

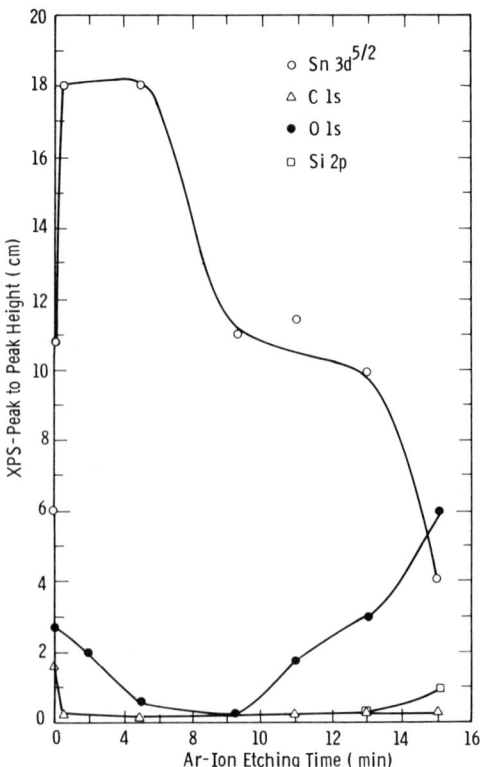

Figure 7. Effect of Ar-ion sputtering on TVT film.

difference in composition of the films on the surface and bulk can be attributed to the adsorbed carbon dioxide and the formation of tin oxides on the surface after exposure of the freshly prepared films to the air. It has been cited in the literature that an initially fluorine-rich plasma polymerized tetrafluoroethylene surface becomes fluorine poor on argon ion sputtering and this has been attributed to preferential sputtering of fluorine.(34) It has been found that the very act of taking ESCA spectra of fluorinated polymers results in loss of fluorine relative to carbon.(35) It has also been shown that argon sputtering leads to the reduction of metal oxide surfaces.(36) Though in our experiments, we have used mild conditions of sputtering and low pass energy for ESCA to minimize the preferential sputtering effects, the change in composition of the films after depth profiling can also be attributed to this fact. Because of their variable C/Sn and C/Ge ratios, these films sputtered at different rates.

The films of TET, TVT and TMG prepared in experiments I, III, and VII behave as insulators. Glow discharge HMDT film prepared in experiment V exhibited a conductivity of $\sim 10^{-4}$ (ohm cm)$^{-1}$. Films prepared from all these monomers in the presence of oxygen gas (experiments II, IV, VI and VIII) showed enhanced conductivity, e.g., the conductivity for polymerized TET ranges from 0.12 to 7.9 X 10^1 $(\Omega cm)^{-1}$, and for plasma polymerized HMDT from 0.8 to 2.2 X 10^1 $(\Omega cm)^{-1}$. This enhanced conductivity can be attributed to the formation of CO or CO_2, thus decreasing the carbon content of the polymer. A similar effect was observed by Coburn and Kay(37) in enhancing the etching capability of CF_4 by adding oxygen. They have described oxygen as a carbon scavenger which forms stable molecules such as CO and CO_2. These molecules do not participate in polymerization or etching. The glow discharge polymerized TVT (experiment IV) showed a conductivity of 1.6 X 10^{-3} $(\Omega cm)^{-1}$. The C/Sn ratio of the latter is 1.7 to 2.0. After 2 minutes of sputtering the carbon signal disappears and the remaining film is essentially β-tin. Similar behavior was noted for glow discharge polymerized tetramethyltin films prepared by the authors,(38) where the C/Sn on the surface of the film was 1.2. However, the sheet conductivity of these films was very high ($\sim 10^4$ $(\Omega cm)^{-1}$), approaching that of bulk β-tin. The lower conductivity of the HMDT glow discharge polymerized films is likely due to the higher carbon content on the surface.

Figure 8 shows the SEM micrographs of glow discharge-polymerized TVT and HMDT. The surfaces of the films exhibiting semiconducting behavior are smooth and featureless, while the transparent insulating films are revealed to have spherical globules on their surface.

The films were stable for several months at ambient conditions with no change in conductivity. Heat treatment at 55°C of the films of TET and TVT prepared in experiments II and IV did not change the conductivity appreciably. In the case of

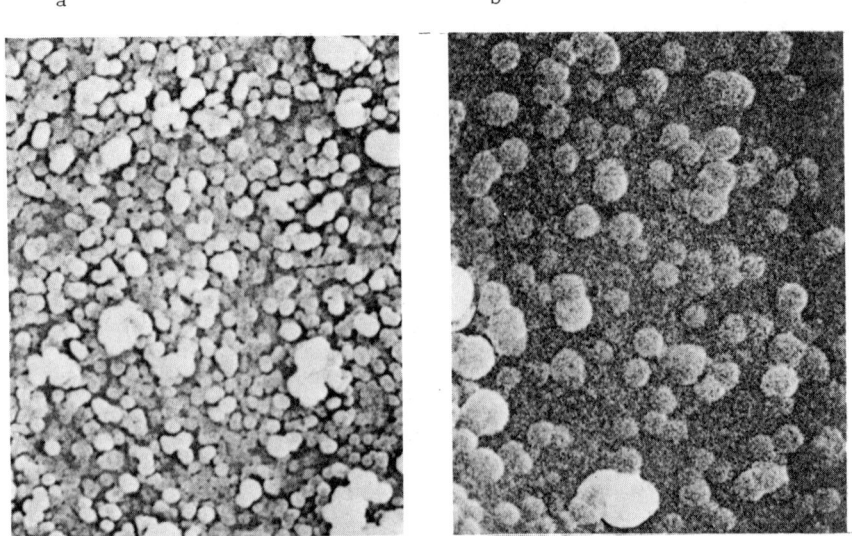

Figure 8. SEM micrographs: (a) HMDT film on glass (X 10,000); (b) TVT film on glass (X 10,000).

polymerized HMDT, prolonged heating at 76°C and above caused the conductivity to change irreversibly. The room temperature conductivity of HMDT films increased from 2.2×10^1 $(\Omega cm)^{-1}$ to 4.5×10^1 $(\Omega cm)^{-1}$ (i.e. $\sigma/\sigma_0 \cong 2.0$) after heating for 2600 secs at 90°C (Figure 9). The TET and HMDT glow discharge-polymerized films showed semiconducting behavior, i.e., as the temperature increased the conductivity increased. Figure 10 shows the relationship between conductivity and temperature for HMDT films prepared in Experiment VI. The conductivity of polymerized HMDT increased marginally with a rise in temperature upto 53°C and returned to the room temperature conductivity after cooling. However, after heating at 76°C, the conductivity increased from 1.4×10^1 to 2.22×10^1 $(\Omega cm)^{-1}$ and did not return to its original value after cooling. Similar behavior was observed for polymerized TET.

The adhesion of these films to glass and aluminum substrates was determined by a pull test technique. The conducting glow discharge TET films deposited on glass gave an average pull strength of 278 kg/cm^2. These films adhered better to glass than to metal, average pull strengths of 130 kg/cm^2 and 150 kg/cm^2 being obtained for films of TVT and HMDT, respectively on aluminum substrates. The films of higher tin content showed better adhesion than the carbon enriched films. A visual examination of the pulled samples suggested adhesive failure but ESCA examination of the failure area showed the presence of a thin organotin film, suggesting cohesive failure. The higher pull strengths observed in this study also suggest a degree of chemical bonding at the film-substrate interface. Pull tests and other factors suggest that plasma etching during glow discharge polymerization plays a role in the enhanced adhesion of the film to the substrate, as compared to deposition by conventional polymerization. The collective mechanism which involves absorption-desorption and bond reformation and interdiffusion of the polymer and metal phase are referred to as AIM (Atomic Interfacial Mixing).(39)

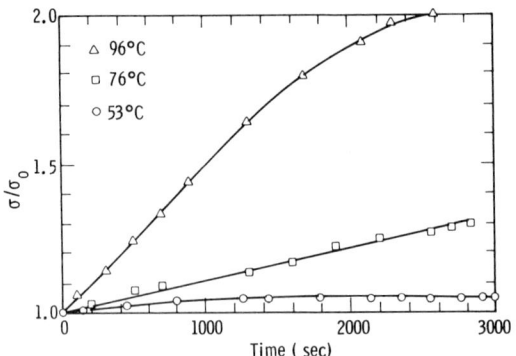

Figure 9. Increase in room temperature conductivity of HMDT film with time at higher temperature.

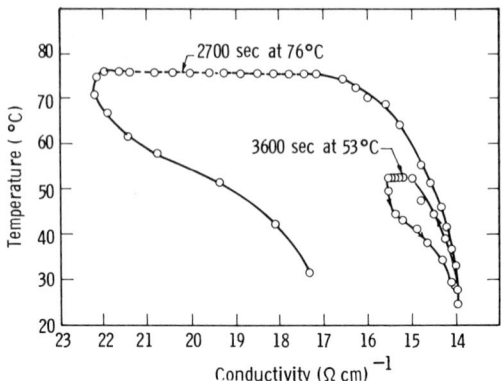

Figure 10. Change in sheet conductivity of HMDT film with an increase in temperature.

Literature Cited

1. Boonthanom, N.; White, M., Thin Solid Films, 1974, 24, 295.
2. Rao, T. V.; Chopra, K. L., J. Chem. Phys., 1978, 68, 1484.
3. Osaki, S.; Ishida, T., J. Polym. Sci., Polym. Phys. Ed., 1973, 11, 801.
4. Yasuda, H., in "Thin Film Processes"; Vossen, J.; Kern, W., Ed.; Academic Press, New York, 1979.
5. Hollahan, J.; Rosler, R. S., in "Thin Film Processes"; Vossen, J.; Kern, W. Ed; Academic Press, New York, 1979.
6. Tkachuk, B. V.; Kolotyrkin, V. M. Vysokomol. Soedin, 1968, Ser. B 10, 24.
7. Tkachuk, B. V.; Bushin, V. V., Ukr. Khim. Zh., 1967, 33, 224.
8. Tkachuk, B. V.; Takamura, M.; Kojima, K., Mem. Inst. Sci. Ind. (Osaka Univ.), 1967, 24, 65.
9. Fink, W., J. Paint Technol. 1970, 42, 220.
10. Maisonneuve, M.; Sequi, Y.; Bui, A., Thin Solid Films, 1976, 33, 35.
11. Wrobel, A. M., Conf. Org. Solid State, Gustrov DGR, 1973, pp. 59.
12. Tien, P. K.; Riva-Senseverian, S.; Martin, R. J.; Smolinsky, G., Appl. Phys. Let., 1974, 24, 547.
13. Varshney, S. K.; Beatty, C. L., ACS Organic Coatings and Applied Polymer Sci. Proceedings, 1982, 46, 127.
14. Nichols Hahn, A. W.; Easley, J. R.; Mahyan, K. G.; J. Bio. Mat. Res., 1979, 13, 299.
15. Bradley, A.; Hammes, J. P.; J. Electrochem. Soc., 1963, 110, 15.
16. Tkachuk, B. V.; Kobtsev, Yu. D.; Laurs, G. P.; Miklalchenko, V. I.; Marussi, N. Ya., Izv. Vyssh. Uchebn. Zaved., Fiz, 1972, 15, 117.
17. Tkachuk, B. V.; Marussi, N. Ya.; Laurs, E. P., Vysokomol. Soedin., 1973, Ser. A, 15, 2046.
18. Kny, E.; James, W. J.; Levenson, L. L.; Auerbach, R. A., Thin Solid Films, 1981, 85, 23.
19. Kny, E.; Levenson, L. L.; James, W. J.; Auerbach, R. A., Thin Solid Films, 1979, 64, 395.
20. Kny, E.; Levenson, L. L.; James, W. J.; Auerbach, R. A., J. Vac. Sci. Technol., 1979, 16(2), 359.
21. Hynecek, J., U.S. Patent 4,140,814, February 20, 1979.
22. Morosoff, N.; Patel, D. L.; Crubliss, A. L.; Lugg, P. S.; ACS Organic Coatings and Applied Polymer Science Proceedings, 1982, 47, 318.
23. Anderson, D. A.; Spear, W. E., Philos. Mag. 1976, 35, 1A.
24. Nowak, A. J., Schultz, F. A.; Umana, M.; Lam, R.; Murray, R. W., Anal. Chem., 1982, 52, 315.
25. Dautartos, M. F.; Evans, J. W., J. Electroanal. Chem., 1980, 109, 301.
26. Kay, E.; Dilks, A. in "Plasma Polymerization", ACS Symposium Series, 1979, 108.
27. Kay, E.; Dilks, A.; Seybold, D., J. Appl. Phys. 1980, 51(11), 5678.

28. Kay, E.; Dilks, A., J. Vac. Sci. Technol., 1981, 18(1), 1.
29. Kay, E.; Dilks, A., Thin Solid Films, 1981, 78, 309.
30. Shuttleworth, D., J. Phys. Chem., 1980, 84, 1629.
31. Liepins, R.; Campbell, M.; Clements, J. S.; Hammond, J.; Fries, R. J., J. Vac. Sci. Technol., 1981, 18(3), 1218.
32. Nomarski, G.; Weiff, A. R., Rev. Met., Paris, 1955, 55, 121.
33. Yasuda, H., in "Thin Film Processes" Academic Press, 1978, 361.
34. Rice, D. W.; O'Kane, D. F., J. Electrochem. Soc., 1976, 123, 1308.
35. Yasuda, H.; Marsh, H. C.; Brandt, E. S.; Reilly, C. N., J. Polym. Sci., Polym. Chem. Ed., 1977, 15, 991.
36. Kim, G. S.; Baitinger, W. E.; Amy, J. W., Winogard, N.; J. Electron. Spectr. Related Phen., 1974, 5, 351.
37. Coburn, J. W.; Kay, E.; IBM J. Res. Develop., 1979, 23, 33.
38. Sadhir, R. K.; James, W. J., Auerbach, R. A., Thin Solid Films, 1982, 97, 17.
39. Yasuda, H. K.; Hale, E. B.; James. W. J., J. Adhesion, in press.

RECEIVED September 14, 1983

44

Polyacetylene, $(CH)_x$: An Electrode-Active Material in Aqueous and Nonaqueous Electrolytes

R. B. KANER, A. G. MACDIARMID, and R. J. MAMMONE

Department of Chemistry, University of Pennsylvania, Philadelphia, PA 19104

A potentially large, new, unexpected field of conducting polymers is opened up by the discovery that $(CH)_x$ can be doped electrochemically, in <u>aqueous</u> solution to the metallic regime without the inclusion of oxygen. For example, use of $(CH)_x$ film as the anode in an electrochemical cell employing $NaAsF_6$ in 52% aqueous HF as an electrolyte gives flexible, golden films (σ =10-100 $ohm^{-1}cm^{-1}$). The energy density, power density and other characteristic properties of four different types of rechargeable batteries using $(CH)_x$ electrodes are described. The batteries are illustrated below by equations representing their discharge reactions: <u>Type I</u>: $(CH)_x$ cathode/Li anode: (i) $[CH^{+y}(ClO_4)_y^-]_x$ + xyLi → $(CH)_x$ + $xyLiClO_4$, and (ii) $(CH)_x$ + xyLi → $[Li_y^+CH^{-y}]_x$; <u>Type II</u>: $(CH)_x$ cathode/$(CH)_x$ anode: (i) $(CH)_x$ + $[Li_y^+CH^{-y}]_x$ → $2[Li_{y/2}^+CH^{-y/2}]_x$, and (ii) $[CH^{+y}(ClO_4)_y^-]_x$ + $[Li_y^+CH^{-y}]_x$ → $2(CH)_x$ + $xyLiClO_4$.

<u>Aqueous Electrochemistry of $(CH)_x$</u>

When <u>trans</u>-$(CH)_x$ film is partially oxidized (i.e. p-doped) by bromine, iodine, arsenic pentafluoride, etc. its conductivity increases by ~ eight orders of magnitude and it is converted to an "organic metal" having all the electronic properties of a conventional metal (<u>1-3</u>). Until recently it had been believed that all p-doped material was very unstable in the presence of water. When the present study was undertaken there were only two apparent exceptions to this water instability.
    The first exception involved the electrochemical oxidation of a piece of $(CH)_x$ film when it was placed in an aqueous 0.5 M

solution of KI and was attached to the positive terminal of a
9V dry cell, the other terminal being attached to a platinum
counter electrode immersed in the solution. Doping took place
in a few minutes to give $(CHI_{0.07})_x$ having a conductivity in the
metallic regime (4). The sum of the elemental analyses for C, H
and I was 99.8%. This showed that no reaction with water, to
incorporate oxygen into the $(CH)_x$, had taken place, at least
during the time needed for doping. As time proceeded, it was
believed that the analysis must have been in error and that
oxygen surely must have been incorporated during the doping process since it has been shown that $(CHI_{3y})_x$ is a polycarbonium
ion (1-3), $[CH^{+y}(I_3^-)_y]_x$ and carbonium ions are known to react
readily with water especially in solutions whose pH is close to
neutral. However, Guiseppi-Elie and Wnek (5) have recently demonstrated a second exception by showing that $(CHI_{0.21})_x$ exhibits
unexpected stability when immersed in solutions of varying pH
values and varying chloride ion concentrations.

In order to determine whether this unexpected water stability was restricted only to the case where the dopant anion was
I_3^-, further electrochemical doping experiments in aqueous solutions were performed. A piece of $(CH)_x$ film (2cm x 2cm) and a
piece of platinum foil were placed in a saturated solution ~ 0.5M
$NaAsF_6$ in 52% aqueous HF. The $(CH)_x$ was attached to the positive electrode and the platinum to the negative electrode, respectively, of a d.c. power supply. A constant potential of 1.0V
was applied between the electrodes for ~ 30 minutes and the
film was then washed in 52% HF and pumped in the vacuum system
for 18 hours. In several different experiments, carried out under slightly different conditions, flexible, golden films having
good metallic conductivity (σ = 10 to 100 $ohm^{-1}cm^{-1}$) were obtained. Elemental analysis showed that the films contained no oxygen. The fluorine content varied from one preparation to another,
e.g. $[CH^{+0.026}(AsF_{5.1})_{0.026}^-]_x$, (C+H+As+F = 100.2%) and
$[CH^{+0.029}(AsF_{4.7})_{0.029}^-]_x$, (C+H+As+F = 100.3%). The nature of
the dopant species and the cause of the variable fluorine content is currently being investigated. It is believed that the
dopant probably consists of a mixture of the $(AsF_6)^-$ and $(AsF_4)^-$
ions. The fact that $(CH)_x$ can be doped to the metallic regime
either with iodine or with arsenic-fluorine species without the
inclusion of oxygen suggests the possibility of an extensive
aqueous chemistry not only for $(CH)_x$ but also possibly for other
conducting polymers.

The low reactivity of the $(CH^{+y})_x$ ion in these cases may be
related to the fact that the positive charge is believed to be
delocalized over a positive soliton, consisting of approximately
15 CH units (6-7). Thus the carbon atoms in a CH unit would be
less susceptible to nucleophilic attack by OH^- or H_2O than if
the charge were localized on only one carbon atom. It seems not
unlikely that the size of a positive soliton may vary with the
size and polarizability of the counter anion. Moreover, the

ease of hydrolysis of the $(CH^{+y})_x$ may, therefore, also be affected by the nature of the counter anion.

Rechargeable Batteries Using $(CH)_x$ Film as Electrodes

We have shown previously that $(CH)_x$ film can be oxidized and reduced in reversible electrochemical reactions (8-9). This makes it an interesting material to study as a potentially useful electroactive electrode material. Four different kinds of batteries employing $(CH)_x$ electrodes have been studied. The dopant concentrations given for the four types of sealed battery cells discussed below are based on the coulombs passed during the charging and/or discharging processes and on the weight of the $(CH)_x$ film employed.

Type I: Polyacetylene Cathode in Conjunction with a Lithium Anode: (i) p-doped $(CH)_x$ cathode and Li anode. Cells were constructed from cis-polyacetylene film (thickness ~ 0.1mm; ~ 3.5mg/cm^2) and lithium metal immersed in an electrolyte of 1.0M LiClO$_4$ in propylene carbonate (P.C.). The overall charging process in which the $(CH)_x$ is oxidized and the Li$^+$ is reduced is represented by the equation :

$$(CH)_x + xyLi^+(ClO_4)^- \rightarrow [CH^{+y}(ClO_4)^-_y]_x + xyLi \quad (1)$$

where y ≤ 0.07. Reversal of the above reaction may be obtained on discharging at a fixed applied potential of 2.5V, the potential (vs. Li) of parent, unoxidized $(CH)_x$ in this electrolyte(10).

Significant oxidation of the $(CH)_x$ occurs only at an applied potential greater than 3.1V (10). After the onset of oxidation, the open circuit voltage, V_{oc}, rises rapidly with increasing oxidation up to oxidation levels of ~ 1% and then increases more slowly (11). The relationship between cell potential and degree of oxidation under various conditions has been studied at oxidation levels up to 7%. Diffusion of $(ClO_4)^-$ ions from the exterior to the interior of a 200Å $(CH)_x$ fibril following a charge cycle causes the V_{oc} to fall on standing. Conversely, after a partial discharge cycle, the V_{oc} rises as $(ClO_4)^-$ ions diffuse from the interior to the exterior of a fibril (11).

Coulombic efficiencies, $(Q_{discharge}/Q_{charge})100$ where Q_{charge} refers to the total coulombs involved in a given charge process and $Q_{discharge}$ refers to the total coulombs involved in a discharge process to 2.5V, have been determined for several different levels of oxidation (11). Corresponding energy efficiencies have also been measured. They are (oxidation, coulombic efficiency, energy efficiency): 1.54%, 100.0%, 80.8%; 2.01%, 99.2%, 79.7%; 2.17%, 100.1%, 81.5%; 2.51%, 95.8%, 78.2%; 4.0%, 89.5%, 72.8%; 6.0%, 85.7%, 68.2%.

Studies have been made of the change in voltage during constant current discharges of a 7% oxidized film at 0.1mA

(19.5A/kg), 0.55mA (107A/kg) and at 1.0mA (195A/kg). Values in parentheses refer to the discharge current normalized per kg of $[CH(ClO_4)_{0.07}]_x$ employed. The corresponding energy density values upon discharge to 2.5V and 3.0V, between which potentials the voltage begins to drop rapidly are 255Whr/kg and 217Whr/kg, respectively. The energy density values are based only on the mass of the electroactive material involved and are calculated using the weight of the $[CH(ClO_4)_{0.07}]_x$ employed and the weight of Li consumed in the discharge reaction (the reverse reaction to that given by the above equation for the charging reaction). The average power densities for the 0.1mA, 0.55mA and the 1.0mA discharges are 70W/kg, 354W/kg and 591W/kg, respectively (11).

Maximum power densities, $P_{max.}$, were obtained at the beginning of a discharge cycle using an external load whose resistance was matched to the internal resistance of the cell. Values were obtained by measuring V_m and/or I_m (V_m and I_m = voltage and current respectively at the very beginning of a discharge cycle) and R_ℓ (the external load resistance) by use of the relationships $P_{max.} = V_m I_m$, $P_{max} = V_m^2/R_\ell$ and/or $P_{max} = I_m^2 R_\ell$. The values obtained, $\sim$ 30,000W/kg, were relatively independent of either the extent of oxidation of the $[CH(ClO_4)_y]_x$ or the absolute weight of the $[CH(ClO_4)_y]_x$ employed (11).

All of the above electrochemical characteristics of the cell are extremely sensitive to the method of cell construction, presence of impurities (especially oxygen), relative ratio of electrolyte to $(CH)_x$, method of charging, etc.

(ii) neutral $(CH)_x$ cathode and Li anode. Selected electrochemical characteristics of a cell constructed from cis-polyacetylene film (thickness $\sim$ 0.1mm; $\sim$ 3.5mg/cm^2) and lithium metal immersed in an electrolyte of 1.0M LiClO$_4$ in tetrahydrofuran (THF) have been studied. Cells were constructed in a similar manner to those discussed in the preceeding section (11). A spontaneous electrochemical reaction occurs when the $(CH)_x$ and Li electrodes are connected by a wire external to the cell. The Li is oxidized and the $(CH)_x$ is reduced during the process according to the reactions given below:

Anode Reaction $\quad xyLi \rightarrow xyLi^+ + xye^-$ (2)

Cathode Reaction $\quad (CH)_x + xye^- \rightarrow (CH^{-y})_x$ (3)

giving the overall net reaction:

$$xyLi + (CH)_x \rightarrow [Li_y^+(CH^{-y})]_x \quad (4)$$

where y ≤ 0.1. It should be noted that the reaction given by Equation 4 is the discharge reaction of a voltaic cell and that the cell, in its completely charged state consists of parent, neutral $(CH)_x$ which appears to be stable indefinitely in the electrolyte. Reversal of the reaction given by Equation 4 may be obtained on charging at a fixed applied potential of 2.5V, the potential (vs. Li) of the parent, neutral $(CH)_x$ in this electrolyte.

Although reduction occurs spontaneously, most studies were carried out at constant applied potentials at various selected values in order to study the system in a controllable manner. Significant reduction of the $(CH)_x$ occurs only at an applied potential less than 1.7V (10). After the onset of reduction, the open circuit voltage, V_{oc}, falls rapidly with increasing reduction up to a reduction level of ~ 1% and then decreases more slowly.

The relationship between cell potential and degree of reduction has been studied up to 10% reduction levels. The reduction process was stopped at intervals and the V_{oc} value (vs. Li) was measured immediately. The cell was then allowed to stand for a period of 24 hours in order to permit partial equilibration of the Li^+ ions within the ~ 200Å $(CH)_x$ fibrils. An increase in potential was observed on standing. This is caused by a decrease in the degree of reduction on the outside of the $(CH^{-y})_x$ fibrils as the counter Li^+ ions diffuse towards the center of the fibrils together with their attendant negative charge on the polyacetylene. Exactly the opposite effect is observed after a partial electrochemical oxidation (charge reaction) of $(CH^{-y})_x$ to a less reduced state. In this case, the V_{oc} falls on standing.

Coulombic efficiencies, $(Q_{discharge}/Q_{charge})100$ where $Q_{discharge}$ refers to the total coulombs involved in a given discharge (reduction) process and Q_{charge} refers to the total coulombs involved in a charge (oxidation) process to 2.5V have been determined for several different levels of reduction. Values of ~ 100% were found up to 6% reduction levels. Somewhat smaller values were found at higher levels of reduction. These high coulombic efficiencies are undoubtedly related, at least in part, to the excellent chemical stability of the partly reduced polyacetylene in the electrolyte. For example, studies show that the V_{oc} of a 7% reduced polyacetylene electrode remains remarkably constant at ~ 1.05V for 40 days as shown in Figure 1 (lower curve). When reoxidized back to neutral $(CH)_x$ the cell displayed a stable V_{oc} of ~ 2.04V for at least 40 days as shown in Figure 1 (upper curve).

Studies have been made of the change in voltage during constant current discharges of 0.1mA (19.8A/kg), 0.5mA (98.8A/kg) and 1.0mA (197.6A/kg) to 6% reduction of the $(CH)_x$, i.e., to a composition of $[Li^+_{0.06}(CH)^{-0.06}]_x$ in each case. The results are shown in Figure 2. The weight of the $(CH)_x$ employed (4.9mg, ~ 1.5cm^2) and the weight of the Li consumed in the discharge reaction were used to calculate the normalized discharge rates in Amps/kg given above. Even though each constant current discharge involved the same number of coulombs and hence resulted in the same average percent reduction, the final discharge vol-

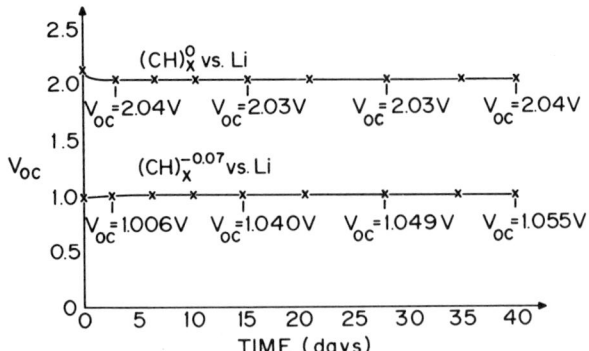

Figure 1. Open circuit voltage, V_{oc}, vs. time characteristics of a Li/1M LiClO$_4$, THF/[Li$^+_{0.07}$(CH$^{-0.07}$)]$_x$ cell (lower curve), and of a Li/1M LiClO$_4$, THF/(CH)$_x$ cell (upper curve).

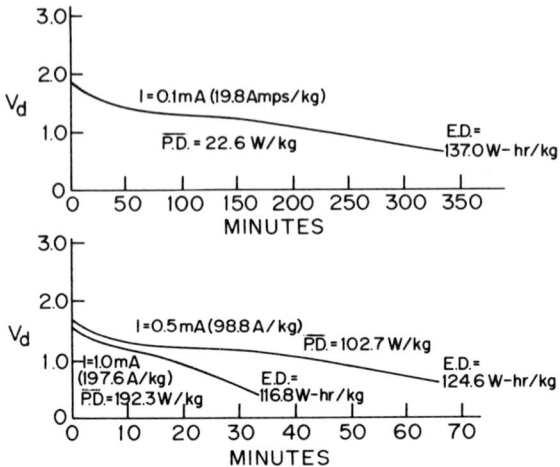

Figure 2. Constant current discharge characteristics of a (CH)$_x$/1M LiClO$_4$, THF/Li cell. 4.9 mg of (CH)$_x$ was employed as the cathode active material.

tages, V_d, decreased as the discharge currents increased, e.g., V_d=0.62V at 0.1mA, V_d=0.52V at 0.5mA and V_d=0.35V at 1.0mA. This is due to the fact that the diffusion equilibrium involving migration of the Li^+ ions from the exterior to the interior of the $(CH)_x$ fibrils becomes less complete the more rapidly the electrochemical reduction process is carried out.

The energy density (E.D.) and <u>average</u> power density (P.D.) values obtained in each of the above discharges are, respectively, 0.1mA, 137.0 Whr/kg and 22.6 W/kg; 0.5mA, 124.6 Whr/kg and 102.7 W/kg; 1.0mA, 116.8 Whr/kg and 192.3 W/kg.

<u>Maximum</u> power density of a $Li/(CH)_x$ cell using 4.5 cm^2 (~ 15mg) of $(CH)_x$ film was obtained by discharging it through an external resistor having the same resistance as the internal resistance of the cell (~ 15 ohms). The current vs. time discharge characteristics are shown in Figure 3 (curve 1). The cell was then recharged in a series of constant potential charging steps to a final voltage of 2.5V. The coulombic efficiency was ~ 100%. Measurement of the <u>initial</u> current 54mA, Figure 3 (curve 2) gave a <u>maximum</u> power density of ~ 2,900W/kg using the relationship $P_{max} = I_m^2 R$ where I_m = current at the beginning of the discharge. The circuit diagram is also shown in Figure 3. Power densities, calculated using the same relationship, after ~ 30 seconds and ~ 100 seconds of discharge were 2,500W/kg and 1,200W/kg respectively. After 1 minute the current was 45mA, after 2.5 minutes it was 24mA and after 5 minutes it had fallen to 14mA.

The present studies indicate that cells involving neutral and/or partly reduced polyacetylene have excellent stability and exhibit interestingly large energy and power densities even at relatively small levels of reduction of the polyacetylene.

<u>Type II: Polyacetylene Cathode in Conjunction with a Polyacetylene Anode</u>: (i) neutral $(CH)_x$ cathode + n-doped $(CH)_x$ anode. Since both neutral and reduced $(CH)_x$ have good stability in an electrolyte of 1M $LiClO_4$ in tetrahydrofuran a voltaic cell can be constructed using $(CH)_x$ as the cathode and $(CH^{-y})_x$ as the anode. During discharge the $(CH^{-y})_x$ gives up an electron to the $(CH)_x$ producing the net overall reaction:

$$[Li_y^+(CH^{-y})]_x + (CH)_x \rightarrow 2[Li_{y/2}^+(CH^{-y/2})]_x \qquad (5)$$

where Li^+ acts as the counter cation to stabilize the polycarbanion species. A cell of this type using 7% electrochemically reduced $(CH)_x$ for the anode and neutral $(CH)_x$ for the cathode has an open circuit voltage, V_{oc}, of ~ 1.0V and a short circuit current, I_{sc}, of ~ 3mA/cm^2 of $(CH)_x$. The cell has excellent stability; during a five month period its V_{oc} remained constant at 0.99V as shown in Figure 4. It is fully rechargeable with coulombic efficiencies of ~ 100%. It is the first stable, rechargeable battery developed in which both the cathode and anode active materials are organic polymers.

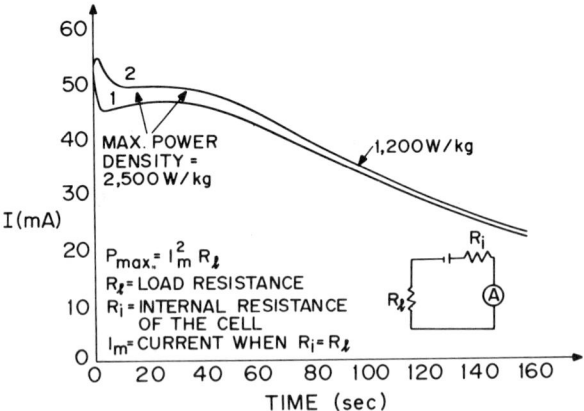

Figure 3. Current vs. time discharge through a matched load resistor for a Li/1M LiClO$_4$, THF/(CH)$_x$ cell employing 15 mg of (CH)$_x$. The initial current (54 mA) during the second discharge gave a maximum power density of 2900 W/Kg.

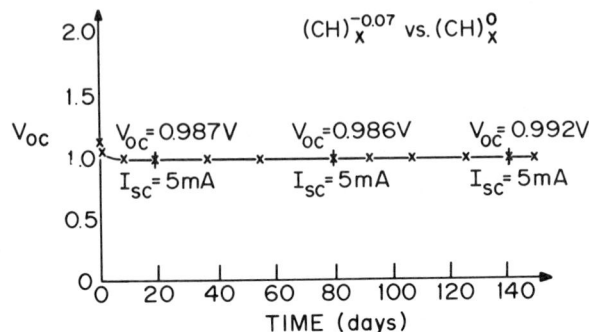

Figure 4. Open circuit voltage, V_{oc}, and short circuit current, I_{sc}, characteristics of a [Li$^+_{0.07}$(CH$^{-0.07}$)]$_x$/ 1M LiClO$_4$, THF/(CH)$_x$ cell.

(ii) p-doped $(CH)_x$ cathode + n-doped $(CH)_x$ anode. A number of cells of this type have been constructed by placing two pieces of $(CH)_x$ film (attached mechanically to Pt or Ni wire) in an electrolyte consisting of ~ 1M $LiClO_4$ or $(Bu_4N)^+(PF_6)^-$ dissolved in propylene carbonate or sulfolane. The electrodes are then attached to the positive and negative terminals respectively of a ~ 2.5V d.c. power source. The $(CH)_x$ attached to the positive terminal is oxidized and the $(CH)_x$ attached to the negative terminal is reduced to the same extent. The overall charging reaction, exemplified with $(Bu_4N)^+(PF_6)^-$ for 6% charging is:

$$2(CH)_x + 0.06x(Bu_4N)^+(PF_6)^- \rightarrow$$
$$[CH^{+0.06}(PF_6)^-_{0.06}]_x + [(Bu_4N)^+_{0.06}CH^{-0.06}]_x \qquad (6)$$

On connecting the two terminals, the cell discharges with a reversal of the above reaction. The basic electrochemical reaction during discharge is:

$$(CH^{+0.06})_x + (CH^{-0.06})_x \rightarrow 2(CH)_x \qquad (7)$$

Cells of this type have an open circuit voltage of ~ 2.4V for 6% doping and have a maximum short circuit discharge current of ~ 100mA/cm^2 of $(CH)_x$. They represent the ideal type of all polymer cells because of their relatively large voltage, but their shelf life is not as good as the $(CH^{-y})_x/(CH)_x$ cells described before.

The present results suggest that electrochemical studies not only of $(CH)_x$ but also of other conducting polymers represent an extensive area for further research not only of fundamental scientific interest but also of possible potential technological value.

Acknowledgments

The work involving aqueous electrochemistry, (R.J.M.), was supported by the Office of Naval Research and the Defense Advanced Research Projects Agency (through a grant monitored by the Office of Naval Research). Studies involving nonaqueous rechargeable batteries, (R.B.K.), were supported by the Department of Energy, Contract No. DE-AC02-81-ER10832. The authors wish to thank Dr. Mahmoud Aldissi and Dr. MacRae Maxfield for many helpful discussions.

Literature Cited

1. MacDiarmid, A. G.; Heeger, A. J. Synth. Met. 1979/80, 1, 101-118.

2. Heeger, A. J.; MacDiarmid, A.G. in "The Physics and Chemistry of Low Dimensional Solids"; Alcacer, L., Ed.; D. Reidel Publishing Co.: Dordrecht, Holland, 1979; pp. 353-391.
3. Etemad, S.; Heeger, A. J.; MacDiarmid, A. G. in "Annual Review of Physical Chemistry"; Rabinovitch, B. S., Ed.; Annual Reviews, Inc.: Palo Alto, CA, 1982; Vol. 33, pp. 443-469.
4. Nigrey, P. J.; MacDiarmid, A.G.; Heeger, A. J. J. Chem. Soc., Chem. Commun. 1979, 594-5.
5. Guiseppi-Elie, A.; Wnek, G. E. J.Chem. Soc., Chem. Comm. 1983, 63-65.
6. Su, W. P.; Schrieffer, J. R.; Heeger, A. J. Phys. Rev. Lett. 1979, 42, 1698-1701.
7. Su, W. P.; Schreiffer, J. R., Heeger, A. J.: Phys. Rev. B. 1980, 22, 2099-2111.
8. Nigrey, P. J.; MacInnes, D., Jr.; Nairns, D. P.; MacDiarmid, A. G.; Heeger, A. J. in "Conducting Polymers"; Seymour, R. B., Ed.; Plenum: New York, 1981; pp. 227-233.
9. Nigrey, P. J.; MacInnes, D., Jr.; Nairns, D. P.; MacDiarmid, A. G.; Heeger, A. J. J. Electrochem. Soc. 1981, 128, 1651-1654.
10. Kaufman, J. H.; Kaufer, J. W.; Heeger, A. J.; Kaner, R. B.; MacDiarmid, A. G. Phys. Rev. B. 1982, 26, 2327-2330.
11. Kaneto, K.; Maxfield, M.; Nairns, D. P.; MacDiarmid, A. G.; Heeger, A. J. J. Chem. Soc., Faraday Trans. I., 1982, 78, 3417-3429.

RECEIVED September 2, 1983

INDEXES

Author Index

Acampora, L. A., 461
Ainger, F. W., 313
Allara, D. L., 135
Anderson, C. C., 119
Babie, W. T., 41
Bednar, B., 129,201
Boudreaux, D. S., 433
Bowden, M. J., 135,153
Bowmer, T. N., 153
Bredas, J. L., 433
Brettle, J., 313
Bush, R. W., 373
Chance, R. R., 433,447
Chow, M-F., 41
Clecak, N., 25
Cotts, P. Metzger, 227
Crivello, J. V., 3
Dandreaux, G. F., 507
Davis, Gary C., 259
Day, David R., 423
Devaty, J., 129,201
Dix, I., 313
Downing, J., 25
Druy, M. A., 473
Dugger, D. L., 461
Duran, J., 239
Elsenbaumer, R. L., 447
Emma, T., 461
Enns, J. B., 345
Falcone, D. R., 135
Fox, Neil S., 367
Frackoviak, J., 135
Frommer, J. E., 447
Fujiwara, Shozo, 177
Furuta, Akihiro, 191
Galvin, M. E., 507
Goldstein, Lesley J., 409
Goosey, M. T., 313
Helbert, J. N., 65,91
Hiraoka, H., 55
Imamura, Saburo, 103
Isobe, Asao, 167
Ito, Hiroshi, 11
James, W. J., 533,555
Jopson, Harriet, 497
Kaner, R. B., 575
Kelley, L. C., 135
Kim, Oh-Kil, 515
Kralicek, J., 129,201
Krasicky, P. D., 119
Kyle, D. R., 373
Lai, Juey H., 213

Lee, Kang I., 497
MacDiarmid, A. G., 575
Malkiewicz, C., 91
Mammone, R. J., 575
Marchetti, J. R., 379
McKean, D. R., 25
Michl, J., 25
Mih, Winston C., 273
Miller, R. D., 25
Mohammed, J., 461
Moreau, W. M., 41
Morgan, C. R., 373
Murai, Fumio, 167
Namaster, Y., 119
Nonogaki, Saburo, 167
Obendorf, S. K., 119
Ohnishi, Yoshitake, 177,191
Ors, J. A., 325,345
Ouano, A. C., 79
Pantelis, P., 399
Pittman, C. U., Jr., 91
Raghava, R. S., 379
Rodriguez, F., 119
Royce, B. S. H., 325
Rubner, M. F., 461,487
Sadhir, R. K., 533,555
Samuelson, L., 461
Sandman, D. J., 461
Sanjana, Z. N., 379
Saunders, H. E., 555
Schmidt, M. A., 91
Seymour, R. J., 473
Shiraishi, Hiroshi, 167
Silbey, R., 433
Small, R. D., 325
Sugawara, Shungo, 103
Tamamura, Toshiaki, 103
Tanigaki, Katsumi, 177,191
Tompkins, T. L., 25
Tripathy, S. K., 461,473,487
Viswanathan, N. S., 239
Volksen, W., 227
Vroom, W. I., 135
Walker, C. C., 65
Wallace, E., Jr., 91
Welsh, L. W., Jr., 55
Wilson, C. Grant, 11,25
Wischmann, K. B., 305
Wnek, G. E., 507
Wong, Ching-Ping, 285
Zachoval, J., 129

Subject Index

A

Ablation process, TMT plasma polymerization, 537
Absorbance rate, solvent, cure in acrylated epoxy systems, 331
Absorbance ratio, TCNQ salt-polymer composites, 526,528f
Absorbance shift, conducting polymer solution, 457
Absorbance spectra
  diazoketone derivatives, 33f
  diazoketone sulfonate derivative, 37f
  DQ and PDQ, 45f
  mid-UV sensitizers, 25-39
  novolac and PMPS film, 147f,148f
  PBOCST and PHOST, 17
  poly(2,2'-bithiophene), 478f
  poly(olefin sulfones), 58f,59f
Absorption, optical, AsF_5-doped PPS, 458f
Abstraction reaction, cross-linking negative resists, 181
Acetylacetonate complexes, magnetic recording media, 415
Acid-catalyzed reactions, various, 6,8
Acoustic impedance, piezoelectric PVDF, 405
Acrylate systems, modified, curing behavior, 325-43
Acryloyl chloride, photosensitive polyimide siloxane, 262
Actinic sensitivity, diazoquinone-novolac resists, 41-53
Adhesion
  electron resists, 104t
  organometallic plasma polymerization, 571
  ultrapure polyimides, 252
Adhesive, Chemlok, TMT plasma polymerization, 545
Adsorption inhibitors, encapsulants, 315
Agglomeration, TMT plasma polymerization, 537
Aggregation, TCNQ salt-polymer composites, 529
Alkanolamine, epoxy molding compounds, 279
Alkoxide ion, epoxy molding compounds, 276
Alkylsulfonate derivatives, diazoketone, mid-UV sensitizers, 38
Aluminum
  corrosion, inhibited encapsulants, 314-16
  TMT plasma polymerization, 546t
Ammonium salt, catalyst, epoxy molding compounds, 279
Ammonium tungstate, inhibited encapsulants, 317,321t
Anhydride-cured epoxy reaction mechanism, catalysts, 275-78
Anode reaction, PA-Li battery, 578
Applied field effects, ultrapure polyimides, 250f
Applied voltage changes, poly(2,2'-bithiophene), 479
Aqueous electrochemistry, PA, 575
Argon ion sputtering
  effect on ESCA, 547f
  organometallic plasma polymerization, 564,566f,567f,568f
  TMT plasma polymerization, 545
Arsenic
  organometallic plasma polymerization, 556
  in PPS film, 452f
Arsenic-fluorine, PA in electrolytes, 576
Arsenic pentafluoride
  doping of PA, 575
  PPSe and PPTe, 469
Arsenic pentafluoride-arsenic trifluoride synergism, gas phase, 450
Arsenic trifluoride, conducting polymer solution, 447-58
Arsenic trifluoride-arsenic pentafluoride synergism, gas phase, 450
Association behavior, TCNQ salt-polymer composites, 526
Atomic interfacial mixing, organometallic plasma polymerization, 571
Aza substitution, mid-UV sensitizers, 31

B

Band gap
  polyenes and diphenylpolyenes, 439f
  redox properties of conjugated polymers, 438t
Battery electrodes
  conjugated polymers, 433
  PA film, 577-83
Benzene derivatives, substituted, 36t

INDEX

Benzophenone moiety, mid-UV sensitizers, 33
Benzopinacol, UV-thermal cure encapsulants, 374
Binding, TCNQ salt-polymer composites, 523
Blend composition, effect, PA-PB blends, 491t
Block polymer, conductive PA hybrids, 509,512
Brass, TMT plasma polymerization, 546t
Bromine, doping of PA, 575
Bubble memory fabrication, ultrapure polyimides, 239
Buffer inhibitors, encapsulants, 315
Bulk polymerization rates, acrylated epoxy systems, 332f

C

Calcium citrate, inhibited encapsulants, 317,321t
Carbon, quaternary, resist dry-etch susceptibility, 98
Carbon disulfide-tricyclohexylphosphine adduct, catalyst, 281t
Carbon-germanium ratio, organometallic plasma polymerization, 564
Carbon-oxygen ratio, organometallic plasma polymerization, 564
Carbon scavenger, organometallic plasma polymerization, 569
Carbon-tin ratio
 vs. distance, TMT plasma polymerization, 546f
 organometallic plasma polymerization, 564
 TMT plasma polymerization, 543
Carbonyl carbon, formation, deep-UV photolithography, 60,61f
Carboxyl group
 epoxy molding compounds, 275
 magnetic recording media, 410,415
Carboxylic acid, photosensitive polyimide siloxane, 262
Carboxylic acid-isocyanatoethyl methacrylate ratio, photosensitive polyimide siloxane, 264t
Catalyst
 epoxy molding compounds, 273-82
 magnetic recording media, 411t,412
 photosensitive polyimide siloxane, 262
 Shirakawa
  conductive PA hybrids, 503
  PA-PB blends, 489
Catalytic cure rate, ligands, 413
Catalytic reactions, photoinitiators and resist design, 14-19
Cathode reaction, PA-Li battery, 578

Cationic reactions
 depolymerization, 7
 depropagation, 162
 dimerization, 144
 photoinitiated, 3-9
 radiation-induced degradation of PMPS, 144,162
Cell construction, PA-Li battery, 578
Cell potential, PA-Li battery, 577,579
Chain dimensions, polyimide materials, 233
Chain extension, removable polyurethane encapsulants, 308
Chain reaction
 cross-linking negative resists, 198
 path, 179
Chain scission, negative e-beam resist, 204
Chain-scission yield, methacrylate copolymers with e-beam radiation, 119-26
Charge carrier reactivities, doped PA, 512
Charge-transfer (CT)
 PPSe and PPTe, 469
 radiation-induced degradation of PMPS, 162
 TCNQ salt-polymer composites, 515-30
Charging process, PA-Li battery, 577
Charlesby's gel formation theory, cross-linking negative resists, 178,198
Chemical amplification, photoinitiators and resist design, 11-22
Chemlok adhesive, TMT plasma polymerization, 545
α-Chloro group, resist dry-etch susceptibility, 96
Chromophores, lithographically important, 35
Circuit pattern, UV-thermal cure encapsulants, 376f
Cleavage reactions, photoinitiators and resist design, 12
CNDO model, mid-UV sensitizers, 27
Coating, UV-thermal cure encapsulants, 375
Coaxial tube transducer forms, PVDF, 406f
Cobalt-modified powders, magnetic recording media, 409
Cobalt naphthanate, catalyst, magnetic recording media, 412
Cobalt octoate, catalyst, magnetic recording media, 412
Cobilt evaluations, positive photoresist lithographic performance, 67-71
Colorants, PCB insulators, 369

Component ratio, negative
  resists, 197f
Component vs. sensitivity ratio,
  cross-linking negative
  resists, 182-87
Composite resist
  conductive PA hybrids, 509-511
  UV spectra, 58f,59f
Condensation polymerization, 6,9
Conducting polymer
  design, 447
  poly(2,2'-bithiophene), 473
  PPSe and PPTe, 461-71
  solutions, 447-58
  TMT, 533-52
Conductive polyimide (CPI) gate
  transistors, 425
Conductivity
  AsF_5-doped PPS in AsF_3, 454
  conducting polymer solution, 450
  conductive PA hybrids, 510,511f
  CPI-gate transistors, 428
  vs. distance, TMT plasma
    polymerization, 533f,539f
  doped polymers, 449t
  EPDM-PA blends, 501,502f
  iodine doping, 505t
  Kraton-PA blends, 505f
  organometallic plasma
    polymerization, 559,571,572f
  piezoelectric PVDF, 402
  poly(2,2'-bithiophene), 475
  stretch-enhanced, PA-PB blends, 490
  TCNQ salt-polymer composites, 516
  vs. temperature, TMT plasma
    polymerization, 544f
  vs. time, AsF_5 doping of a PPS
    film, 451f
  vs. time, TMT plasma
    polymerization, 544f
  TMT plasma polymerization, 541
Conductor, organic, novel
  phase, 447-58
Conjugated bridge, conducting polymer
  solution, 454
Conjugated polymer, electrochemical
  properties, 433-44
Contact angle, TMT plasma
  polymerization, 537,551t
Contamination, electron resists, 104t
Conventional polysilicon gate
  process, 426
Copolymer
  cross-linking negative
    resist, 192-98
  negative e-beam resist, 207f
  sensitivity and component
    ratio, 177-88,192-98
Corrosion of aluminum,
  inhibited, 314-16

Corrosion inhibition,
  encapsulants, 313
Coulombic efficiency, PA-Li
  battery, 577,579
Covalent bond formation, conducting
  polymer solution, 453
Creep, casting resin, 308
Creep compliance vs. time, linear
  polyurethane encapsulants, 310f
p-Cresol, inhibited
  encapsulants, 317
Critical dimension
  Nanoline, 66-73
  positive photoresist lithographic
    performance, 66,68f,74f,75f
Cross-linked epoxy resin matrix, 315
Cross-linking
  acrylated epoxy systems, 331
  conducting polymer solution, 453,457
  e-beam lithography, 109
  EPDM-PA blends, 501
  magnetic recording media, 410
  methacrylate copolymers, 123
  negative electron
    resists, 177-88,191-200
  PFCPM positive electron resist, 130
  photoinitiators, 12
  photosensitive polyimide
    siloxane, 266
  polyimide materials, 233
  removable polyurethane
    encapsulants, 308,311
  resist dry-etch susceptibility, 99
  UV-thermal cure encapsulants, 373
Crystal form content vs. machine draw
  ratio, piezoelectric PVDF, 403t
Crystallinity
  EPDM-PA blends, 501
  Kraton-PA blends, 503
  PA-PB blends, 494
  poly(2,2'-bithiophene), 484
  TMT plasma polymerization, 541
Cured film, ultrapure
  polyimide, 248-52
Cured molecular parameters, polyimide
  materials, 231,234t
Curing
  epoxy molding compounds, 274
  modified acrylate systems, 335-43
  rubber-modified epoxy
    photopolymers, 348-52
  ultrapure polyimides, 252,254f,256
  UV solder mask as PCB insulator, 371
  UV-thermal cure encapsulants, 375
Curing rate
  acrylated epoxy systems, 330
  photosensitive polyimide
    siloxane, 266
Current density, poly(2,2'-
  bithiophene) film growth, 479-84

INDEX

α-Cyano group, resist dry-etch
  susceptibility, 96

D

Debye ring patterns, PA-PB blends, 494
Deep-UV photolithography, negative
  tone e-beam resists, 57,60,62f
Degradation
  negative e-beam resist, 204
  photosensitized, poly(olefin
    sulfones), 56
  PPSe and PPTe, 467
  radiation-induced
    methacrylate copolymers, 126f
  PMPS, 135-50,153-65
Density, piezoelectric PVDF, 403
Depolymerization
  catalytic, photoinitiators, 19
  cationic, photoinitiators, 7
  radiation-induced degradation of
    PMPS, 144
Deposition rate
  organometallic plasma
    polymerization, 559,562f,563f,565f
  TMT plasma
    polymerization, 533f,537,539f
Depropagation, radiation-induced
  degradation of PMPS, 158,161f
Developer, effect, photoinitiators and
  resist design, 13,17
Development, positive photoresist
  lithographic
  performance, 67,69t,70t,72f
Device reliability, inhibited
  encapsulants, 319f
Dialkylphenacylsulfonium salts (III),
  photolysis, 4
Diaryliodonium salts
  photolysis, 3
  photosensitizers, 5
Diazoketone derivatives
  absorption spectra, 33f,37f
  experimental and calculated
    spectra, 25f
  mid-UV sensitizers, 25-39
Diazonaphthoquinone (DQ)
  conversions vs. pyrene, 46t
  mid-UV phototosensitization, 41-53
  UV spectra, 44
Diazoquinone-novolac resists, 41-53
Diazoquinone positive
  photoresists, 41-53
Dielectric photopolymer, acrylated
  epoxy systems, 325-43
Dielectric passivation layers,
  ultrapure polyimides, 240
Differential scanning calorimetry
  (DSC)
  PPSe and PPTe, 465-69

Differential scanning calorimetry
  (DSC)--Continued
  rubber-modified epoxy
    photopolymers, 353,354f
Differential thermal analysis (DTA),
  TMT plasma polymerization, 541
Diffractometer tracing, PPSe and
  PPTe, 466f
Diffusing moisture, encapsulants, 315
Dilatometer, thermal expansion of
  PWB, 382
α,α-Dimethoxy-α-phenacetophenone,
  UV-thermal cure
  encapsulants, 374
Diphenyl acetylene (DPAE), plasma
  developable e-beam resists, 214
Diphenyl acetylene-poly(trichloroethyl
  methacrylate) (DPAE-PTCEM)
  resist, 218,221,222f
Diphenylpolyenes (DPP)
  band gap, 439f
  VEH computed IP, 441f
Dipole-dipole interaction, diazo-
  quinone resists, 44
Discharge rates, PA-Li battery, 579
Diselenide links, PPSe and PPTe, 464
Dissolution kinetics, effects of
  molecular dynamics, 79-90
Dissolution rate
  mid-UV photosensitization, 45f
  novolac-based positive e-beam
    resist, 169,170,171f
Dopants, exposure, PPSe and PPTe, 469
Doped polyacetylene
  anode with neutral PA cathode, 581
  cathode and anode, 583
  cathode with Li anode, 577
Doping
  AsF_5, PPS, 453
  iodine, EPDM-PA blends, 501
  Kraton-PA blends, 503
  PA hybrids, 510,511f
  PA-PB blends, 490
  PA, 575
  TCNQ, 523
Doping experiments, PA, 576
Dose dependence, radiation-induced
  degradation of PMPS, 142-45
Drain current, CPI-gate
  transistors, 430f
Dry-etch durability, electron
  resists, 104t
Dry-etch resistance, cross-linking
  negative resists, 192,194f
Dry etching
  durable resists for e-beam
    lithography, 107-113
  vinyl polymer resist, 91-100
Dynamic mechanical properties, linear
  polyurethane encapsulants, 310f

E

Efficiency, acrylate-based mixtures, 330t
Elasticity
 EPDM-PA blends, 499
 Kraton-PA blends, 503
Elastomer
 removable polyurethane encapsulants, 309t
 silicone, TGA, 285-303
Electrical conductivity, enhancement in PA-PB blends, 487-96
Electrical properties
 CPI-gate transistors, 428
 PPSe and PPTe, 469
Electrochemical doping, 433-44
 PA in electrolytes, 576
Electrochemical reaction, spontaneous, PA-Li battery, 578
Electrochemical synthesis, poly(2,2'-bithiophene), 473
Electrochemistry, aqueous, PA, 575
Electrochromic display, poly(2,2'-bithiophene), 479
Electrode, battery, conjugated polymers, 433
Electromigration, inhibited encapsulants, 317
Electron affinity, conjugated polymers, 442f,443
Electron-beam delineation, novolac based positive e-beam resist, 170
Electron-beam dosage, resist dissolution kinetics, 85f,89f
Electron-beam lithography, resists, 103-16
Electron-beam radiation
 chain-scission of methacrylate copolymers, 119-26
 PFCPM positive e-beam resist, 132
 photoinitiators, 17,19
Electron-beam resist
 dry-etch susceptibility, 99
 negative
  e-beam lithography, 109-16
  poly(butadiene-co-glycidyl methacrylate), 201-210
  poly(chloromethylstyrene-co-2-vinylnaphthalene), 191-200
 plasma developable, 213-22
 positive
  deep-UV photolithography, 57,60,62f
  e-beam lithography, 105-107
  novolac-based, 167-75
  PFCPM, 129-33
  requirements, 104t
Electron diffraction
 poly(2,2'-bithiophene), 484

Electron diffraction—Continued
 PPSe and PPTe, 464
 TMT plasma polymerization, 534,542f,543
Electron paramagnetic resonance (EPR)
 AsF_3 solution, 456f
 conducting polymer solution, 454,455t
Electron spectroscopy
 calculations, 25-39
 conducting polymer solution, 457
 PPSe and PPTe, 464
Electron spectroscopy for chemical analysis (ESCA)
 deep-UV photolithography with poly(olefin sulfones), 57,61f
 depth profile, organometallic plasma polymerization, 564
 profile, TMT plasma polymerization, 546f
 TMT plasma polymerization, 543
Elongation, removable polyurethane encapsulants, 309t
Encapsulation
 epoxy molding compounds, 273-82
 IC device, 285-303
 inhibited, 313-21
 removable polyurethane, 305-12
 solventless, UV-thermal cure, 373
Energy density, PA-Li battery, 581
Energy transfer mechanism, mid-UV photosensitization, 43
Epoxide, catalyst, molding compounds, 275
Epoxy
 photoinitiators and resist design, 12
 TCE vs. temperature, 382t
 thermal expansion of PWB, 380
Epoxy curing mechanism, epoxy molding compounds, 274
Epoxy group, e-beam lithography, 109
Epoxy laminates, various, TCE as function of temperature, 384-95
Epoxy photopolymers, rubber-modified, 345-64
Epoxy reaction mechanism
 anhydride cured, catalysts, 275-78
 phenolic cured, catalysts, 278-81
Epoxy-rubber mixtures, modified epoxy photopolymers, 347t
Equilibria, TCNQ salt-polymer composites, 529
Equilibrium molecular parameters, polyimide materials, 230,232f
Etching
 cross-linking negative resists, 194f
 plasma, PFCPM positive e-beam resist, 132
 reactive ion with oxygen, 112

INDEX

Etching--Continued
  resist dry-etch susceptibility, 94f
Ethylene oxide, and PA, 510
Ethylene-propylene-diene terpolymer-polyacetylene (EPDM-PA), 497-505
Exchange interaction, diazoquinone, 44
Expansion, thermal coefficient (TCE), PWB, 380
Expansion coefficient, removable polyurethane encapsulants, 309t
Exposure
  vs. critical dimension, 68f
  double, 173,174f
  novolac-based positive e-beam resist, 170,172f
  positive photoresist lithographic performance, 67,68f

F

Ferric acetylacetonate, catalyst, magnetic recording media, 412
Fiberglass, thermal expansion of PWB, 390
Field-effect transistor (FET)
  conductive polyimide gates, 425-31
  CPI-gate transistors, 427
Filler
  interactions, silicone elastomers, 293,294
  TGA, discussion, 291-98
  UV solder masks, 369
Film
  cured, ultrapure polyimides, 248-52
  glow discharge, 535t
  growth
    photosensitive polyimide siloxane, 261
    poly(2,2'-bithiophene), 479
  morphology, TCNQ salt-polymer composites, 517,526
  plasma polymerized, 555-70
  thin organometallic, 555-70
  tin-enriched, 545
Film thickness
  vs. IR absorbance, 142,143f
  mid-UV photosensitization of diazo- quinone resists, 51t
  organometallic plasma polymerization, 559
  photoinitiators and resist design, 13
  photosensitive polyimide siloxane, 266
  plasma developable e-beam resists, 222f
  poly(olefin sulfones), 60,63f
  TMT plasma polymerization, 534
  vs. viscosity, ultrapure polyimides, 247f

First derivative thermogravimetry spectra, silicone, 298f
Fischer mechanism, epoxy molding compounds, 276
Flame-retardant epoxy resin system, 380
Flexibility, UV-thermal cure encapsulants, 375
Fluorine
  e-beam lithography, 105
  organometallic plasma polymerization, 569
  PA-Li battery, 576
Fluorine-arsenic, PA in electrolytes, 576
Fock operator, redox properties of conjugated polymers, 435
Fourier transform IR-photoacoustic spectroscopy (FTIR-PAS)
  acrylated epoxy systems, 326,328,329f
  rubber-modified epoxy photopolymers, 351f
  silicone extractables, 300f
  TGA of silicone elastomers, 296
Free-radical polymerization, UV-thermal cure encapsulants, 373

G

Gas chromatography-mass spectrometry (GC-MS)
  PPSe and PPTe, 465-69
  silicone elastomers, 296
  silicone extractables, 301f
Gas phase AsF_3-AsF_5 synergism, 450
Gas phase ionization potentials (IP), conjugated polymers, 438t
Gel dose, copolymer, cross-linking negative resists, 181
Gel formation, conducting polymer solution, 454
Gel formation theory, Charlesby's, cross-linking negative resists, 178,198,264
Gel permeation chromatography
  conductive PA hybrids, 513
  silicone fluids, 295f
Gelation, rubber-modified epoxy photopolymers, 358
Generic RTV silicones, TGA, 291-98
Geometry, mask, mid-UV photosensitization, 51t
Germanium-carbon ratio, organometallic plasma polymerization, 564
Glass adhesion, organometallic plasma polymerization, 571
Glass-epoxy-Kevlar laminate, TCE as a function of temperature, 393f

Glass transition temperature
  cross-linking negative
    resists, 187,198
  e-beam lithography, 105,113
  plasma developable e-beam
    resists, 218
  resist dry-etch
    susceptibility, 97t,99t
  rubber-modified epoxy
    photopolymers, 353-358
  thermal expansion of PWB, 390
Glow discharge
  inductively coupled, TMT plasma
    polymerization, 534
  organometallic
    polymerization, 557,560t
  TMT film formation, 535t
Glycidyl methacrylate, photosensitive
  polyimide siloxane, 262
Goniometer, TMT plasma
  polymerization, 537
Graft polymer, conductive PA
  hybrids, 509,510
Graphite particles, conductive, CPI
  gate transistors, 431

H

Halogens, e-beam
  lithography, 107,109,112
Hamiltonians, conjugated polymers, 434
Header test vehicle, inhibited
  encapsulants, 318f
Hermetic chip carrier (HCC), 380
Heterocyclics, epoxy molding
  compounds, 279
Hexamethylditin (HMDT), 559,564
  SEM, 569
High sensitivity resists, e-beam
  lithography, 105-107
Humidity, effect, PCB insulators, 371
Hydrocarbons, radiation-induced
  degradation of PMPS, 155-58
Hydrogen, radiation-induced degrada-
  tion of PMPS, 155-58
Hydrogen abstraction, 156
  methacrylate copolymers, 125
Hydrogen bonding, epoxy molding
  compounds, 275
α-Hydrogen subtraction, resists for
  e-beam lithography, 113
Hydrolysis, acid-catalyzed,
  polymerization photoinitiators, 8
Hydrolytic stability, UV-thermal cure
  encapsulants, 376
Hydrostatic coefficient, piezoelectric
  PVDF, 402
Hydroxyl-alkyl products, epoxy
  molding compounds, 278

I

Image bias vs. prebake temperatures,
  ultrapure polyimides, 255f
Imidazole, epoxy molding
  compound, 281t
Imidization
  vs. temperature, ultrapure
    polyimides, 254f
  ultrapure polyimides, 249f
Impurities, PA-Li battery, 578
In-situ polymerization, conductive PA
  hybrids, 509
INDO/S model, mid-UV sensitizers, 27
Inductive effect, mid-UV
  sensitizers, 29-39
Inductively coupled glow discharge
  reactor, TMT, 534
Inductively coupled plasma reactor,
  organometallic plasma
  polymerization, 557
Ingot reactor, solid, ultrapure
  polyimides, 247f
Inorganic bases, catalyst, epoxy
  molding compounds, 279
Inorganic crystallites, TEM,
  poly(2,2'-bithiophene), 479
Insulator
  IC devices, 240
  organometallic plasma
    polymerization, 569
  PPSe and PPTe, 469
  UV solder masks, 367-71
Integrated circuit (IC) device
  encapsulants, silicone
  elastomers, 285-303
Interstitial distances, resist dissolution kinetics, 87
Intramolecular bridging, conducting
  polymer solution, 457
Intrinsic viscosity, polyimide
  materials, 236
Iodine
  doping
    conductive PA hybrids, 510,511f
    EPDM-PA blends, 501
    Kraton-PA blends, 503
    PA, 575
    PA-PB blends, 490,494
    PA in electrolytes, 576
    PPSe and PPTe, 469
  stain, TEM of PA-PB blends, 494
Ion implantation, CPI-gate
  transistors, 427
Ion-pairing, TCNQ salt-polymer
  composites, 526
Ionization potential (IP)
  conjugated polymers, 442f
  gas phase, conjugated polymers, 438t

INDEX

Ionization potential (IP)--Continued
  VEH, conjugated polymers, 437t
Ionizing radiation sensitivity, negative e-beam resist, 204,206f,207f
IR absorbance
  and film thickness, 142,143f
  and PMPS concentration, 145,147f
IR spectra
  conducting polymer solution, 455
  conductive PA hybrids, 512
  EPDM-PA blends, 499
  Kraton-PA blends, 503
  PBOCST film, 14
  photosensitive polyimide siloxane, 267f
  piezoelectric PVDF, 402
  PMPS, 139,140f
  poly(2,2'-bithiophene), 478f
  PPS, 456f
Iron, organometallic plasma polymerization, 556
Irradiation
  acrylated epoxy systems, 3
  rubber-modified epoxy photopolymers, 358-62
Irradiation dose
  negative e-beam resist, 204
  degradation of PMPS, 162-64
Irradiation time vs. temperature, degradation of PMPS, 141f
Isocyanate-carboxyl group, magnetic recording media, 415
Isocyanate-glycol ether reaction, magnetic recording media, 410
Isocyanatoethyl methacrylate (IEM)
  modified polyimide siloxane, 261
  photosensitive polyimide siloxane, 262
  ratio, carboxylic acid, 264t
Isomerization, radiation-induced degradation of PMPS, 158,159f,162,164f
Itaconate copolymers, chain-scission of methacrylate copolymers, 123,126f

K

Kevlar
  laminates, various
    TCE vs. temperature, 384-93
    Young's modulus vs. temperature, 383f
  TCE vs. temperature, 382t
  thermal expansion of PWB, 380,390
Kinetics
  radiation-induced degradation of PMPS, 135-50
  TGA of silicone elastomers, 302
Kraton-polyacetylene, 497-505

L

Laminates, various, TCE vs. temperature, 382-94
Leadless chip carrier (LCC), 380
Leakage current, ultrapure polyimides, 249f
Lewis acids and bases, catalyst, epoxy molding compounds, 275,276,279,281t
Ligand, effect, magnetic recording media, 413
Light scattering measurement, polyimide materials, 234f
Linear thermal coefficient of expansion (TCE), PWB, 380
Lithium anode
  with neutral PA cathode, 578-81
  with PA cathode, 577-81
Lithography
  diazoquinone resists, 41-53
  e-beam resists, 103-16
  mid-UV
    photosensitization, 43,47,48f,51
  negative e-beam resist, 204,208-10
  positive photoresist, 65-76
  semiempirical quantum mechanics, 35
Low-density polyethylene-polyacetylene (LDPE-PA) composite, 509

M

Machine draw ratio vs. crystal form content, piezoelectric PVDF, 403
Magnetic bubble devices, ultrapure polyimides, 239-57
Magnetic circular dichroism (MCD), mid-UV sensitizers, 34
Magnetic field, magnetic recording media, 410
Magnetic recording media, polymeric reactions, 409-19
Mask geometry, diazoquinone resist, 51t
Mass spectrometry (MS)
  PPSe and PPTe, 465-69
  silicone elastomers, 296,302t
  silicone extractables, 301f
Matrix effect, TCNQ salt-polymer composites, 529
Mechanical properties
  linear polyurethane encapsulants, 310f
  UV-thermal cure encapsulants, 375
Mechanical treatment, effect, PA-PB blends, 491t
Membrane osmometry, polyimide materials, 230
Mercury, organometallic plasma polymerization, 556

Metal adhesion, organometallic plasma polymerization, 571
Metal complexes, order of reactivity, 419
Metal-containing polymers, plasma polymerization, 556
Metal-gate field-effect transistor, CPI-gate transistors, 427
Metal oxide semiconductor (MOS)
  conventional, 426
  parasitic, 317
Metal powders, magnetic recording media, 409
Metal salt complexes, catalyst, epoxy molding compounds, 276
Metallic behavior, TMT plasma polymerization, 541
Metallic conductivity, PA-Li battery, 576
Methacrylate copolymers, chain-scission, 119-26
Methacrylic polymers, e-beam lithography, 105
α-Methyl group
  chain-scission of methacrylate copolymers, 125
  resist dry-etch susceptibility, 98
Methylene chloride
  conductive PA hybrids, 512
  PPSe and PPTe, 469
Michler's ketone, photosensitive polyimide siloxane, 264
Microcircuits, inhibited encapsulants, 313-21
Microcrystallite, TCNQ salt-polymer composites, 523,524
Microelectronic encapsulation, epoxy molding compounds, 273-82
Microprobe analyses, PPS films doped with AsF_5, 450
Microstructure, PA-PB blends, 487-96
Mid-UV photosensitization, diazoquinone positive photoresists, 41-53
Mid-UV sensitizers, semiempirical quantum mechanics, 25-39
MNDO (modified neglect of differential overlap) semiempirical procedure, 436
Model, thermal expansion of PWB, 381
Moisture
  diffusing, inhibited encapsulants, 315
  resistance, UV solder masks as PCB insulators, 368
Molding resin, inhibited encapsulant, 320
Molecular design, cross-linking negative electron resists, 191-200
Molecular dynamics, resist dissolution kinetics, 79-90
Molecular orbital theory, mid-UV sensitizers, 29-39
Molecular parameters, polyimide materials, 230,232f
Molecular weight
  chain-scission yields of methacrylate copolymers, 121t,124f
  cross-linking negative resists, 195,197f
  e-beam lithography, 109,114f
  PFCPM as positive e-beam resist, 130
  polyimide materials, 230,236
  resist dissolution kinetics, 82
Monomer
  magnetic recording media, 419
  plasma developable e-beam resists, 215,216f
  UV solder masks as PCB insulators, 368
Monomer-polymer compatibility, plasma developable e-beam resists, 215
Morphology
  poly(2,2'-bithiophene) films, 479
  rubber-modified epoxy photopolymers, 358-62
Multilevel photoresists, polymerization photoinitiators, 5

N

Nanoline critical dimension, 66-73
Naphthalene ring, e-beam lithography, 112
Negative electron beam resist
  copolymer, sensitivity and component ratio, 177-88
  deep-UV photolithography, 57,60,62f
  e-beam lithography, 109-16
  onium salt photoinitiators, 7
  poly(butadiene-co-glycidyl methacrylate), 201-210
  poly(chloromethylstyrene-co-2-vinylnaphthalene), 191-200
  polymerization photoinitiators, 6
  SEM, 13
Negative mode, photoinitiators, 14-19
Negative patterning, photoinitiators and resist design, 20
Nitrogen atmosphere, TGA of PPSe and PPTe, 467
Nitrogen bases, catalyst, epoxy molding compounds, 279
Nonchain reaction
  cross-linking negative resists, 198
  paths, 179,180f
Nonpolar developer, photoinitiators, 17

INDEX 597

Novel vinyl polymers, resist dry-etch susceptibility, 93,96-99
Novolac-based positive e-beam resist, 167-75
Novolac-diazoquinone resists, 41-53
Novolac film, absorbance spectra, 147f,148f
Novolac resin
  composite photoresists, 57-64
  polyfunctional epoxy, 12
  positive e-beam resist, 169
  radiation-induced degradation of PMPS, 145-50

O

Octadecyl methacrylate (ODMA), plasma developable e-beam resists, 214
Octadecyl methacrylate-poly(trichloroethyl methacrylate) (ODMA-PTCEM) resist, plasma developable e-beam resists, 218
OH-terminated silicone fluids, TGA, 291-98
Olefin, radiation-induced degradation of PMPS, 155-58
Oligomer, UV solder masks as PCB insulators, 368
Oligomerization, radiation-induced degradation of PMPS, 158,161f
Onium salt, 4
  photoimaging processes, 5
  photoinitiators, 11-22
  rubber-modified epoxy photopolymers, 345-364
Optical absorption, AsF_5-doped PPS, 458f
Optical micrograph, CPI gate transistors, 430f
Optical response, poly(2,2'-bithiophene), 479
Optical transition energies, DPP, 440
Organic conductors, novel phase, 447-58
Organic phosphine, catalyst, epoxy molding compounds, 281t
Organic transition metal oxides, magnetic recording media, 409
Organometallic catalyst, epoxy molding compounds, 281t
Organometallic plasma-polymerized thin films, 555-70
Organosilicon, plasma polymerization, 556
Osmium tetroxide, TEM, rubber-modified epoxy photopolymers, 358
Oxidation
  inhibitor, encapsulants, 314
  PA-Li battery, 577

Oxidation--Continued
  potentials, VEH theory vs. experimental, 443t
Oxygen
  organometallic plasma polymerization, 559,569
  PA in electrolytes, 576
  PA-Li battery, 578
  reactive ion etching, 112
  TMT plasma polymerization, 537
Oxygen-carbon ratio, organometallic plasma polymerization, 564
Oxygen-tin ratio
  vs. distance, TMT plasma polymerization, 546f
  organometallic plasma polymerization, 564
  TMT plasma polymerization, 543

P

Parasitic metal oxide semiconductor, inhibited encapsulants, 317
Pariser-Parr-Pople model, mid-UV sensitizers, 27
Particle size distribution, rubber-modified epoxy photopolymers, 361f,363f
Passivation
  conventional MOS transistors, 426
  ultrapure polyimides, 239-57
Patterning
  conventional and photosensitive polyimides, 260s
  CPI-gate transistors, 428
  e-beam lithography, 107,108f,110f,115f
  novolac-based positive e-beam resist, 169
  poly(butadiene-co-glycidyl methacrylate), 209
Percolation threshold, conductive PA hybrids, 510
Permeability, water vapor
  TMT plasma polymerization, 536,551t
  UV-thermal cure encapsulants, 375
Perturbation substituents, mid-UV sensitizers, 30f
Phenolic cured epoxy reaction mechanism, catalysts, 278-81
Phenyl end groups, effect, redox properties of conjugated polymers, 440
Phenyl-sulfur stretch, conducting polymer solution, 455
N-Phenylmaleimide, photosensitive polyimide siloxane, 266
Phosphine, catalyst, epoxy molding compounds, 279,281t

Phosphoric-acetic-nitric solution (PAN etch), CPI gate transistors, 427
Photochemistry, diazoquinone resists, 43-46
Photocross-linking
  deep-UV photolithography, 56
  photosensitive polyimide siloxane, 264
Photoenergy transfer, deep-UV photolithography, 57,59f
Photogenerated acids, photoinitiators and resist design, 12
Photoimage
  deep-UV photolithography, 60,62f
  process based on onium salts, 5
Photoinitiated cationic polymerization, application, 3-9
Photoinitiation system, UV solder masks as PCB insulators, 368
Photoinitiator (PI), 326
  acrylated epoxy systems, 331-40
  cationic polymerization, 4-5
  resist design, 17
  semiconductor manufacturing, 11-22
  UV-thermal cure encapsulants, 374
Photolabile group, photosensitive polyimide siloxane, 262
Photolithography, deep-UV with poly(olefin sulfones), 55-64
Photolysis, photoinitiators, 3-4
Photomask fabrication, 104t
Photopolymer dielectrics, acrylated epoxy systems, 325-43
Photopolymer shelf stability, photosensitive polyimide siloxane, 264
Photoresist
  deep-UV photolithography, 55-64
  diazoquinone positive, 41-53
  dissolution kinetics and molecular dynamics, 79-90
  dry-etch susceptibility, 91-99
  lithographic performance, 65-76
  mid-UV sensitizers, 25-39
  novel vinyl polymers, 91-99
  photoinitiators
    cationic, 3-9
    semiconductor manufacturing, 11-23
  poly(olefin sulfone), 55-64
Photosensitive polyimide, CPI gate transistors, 426
Photosensitive polyimide siloxane, electronic applications, 259-69
Photosensitization
  mid-UV, diazoquinone positive photoresists, 41-53
  polymerization photoinitiators, 4-5
Photosensitized degradation, poly(olefin sulfones), 56

Picrylsulfonic acid, PPSe and PPTe, 470
Piezoelectric coefficients, PVDF, 402
Piezoelectric poly(vinylidene fluoride) (PVDF), 399-406
Plasma developable electron resists, 213-22
Plasma etching
  negative e-beam resist, 209
  oxygen
    CPI-gate transistors, 427
    e-beam lithography, 112
  PFCPM as positive e-beam resist, 132
  plasma developable e-beam resist, 218
  vinyl resist polymer composition, 91-100
Plasma polymerization
  organometallic thin films, 555-70
  TMT, 533-52
Plastic encapsulated device (PED), 313
Polar solvent, removable polyurethane encapsulants, 311
Polarity
  photoinitiators, 17
  resists for e-beam lithography, 116
Polarization interferometer
  organometallic plasma polymerization, 559
  TMT plasma polymerization, 534
Poly(acetylene) (PA)
  cathode
    Li anode, 577-81
    PA anode, 581-83
  conductive hybrids, 507-13
  conjugated polymers, 443t
  electrode-active material, 575-83
  EPDM and Kraton blends, 497-505
  redox properties, 442f
  VEH computed IP, 438t
Polyacetylene-polybutadiene (PA-PB) blends, effect of microstructure, 487-96
Polyaldehyde, photoinitiators and resist design, 19
Polyamic acid
  control of viscosity, 246
  solution characterization, 227-36
  ultrapure polyimides, 244-46
Poly(2,2'-bithiophene), electrochemical synthesis, 473-84
Poly(p-t-butoxycarbonyloxystyrene) (PBOCST), photoinitiators and resist design, 14,20
Poly(chloromethylstyrene-co-2-vinylnaphthalene), cross-linking negative electron resists, 191-200
Polydimethylsiloxane, 112,285-303
Polyelectrolytes, polyimide materials, 230,236

INDEX

Polyene
  band gap, 439f
  VEH computed IP, 441f
Poly(ethylene oxide)-polyacetylene
  (PEO-PA) grafts, 509
Poly(p-hydroxystyrene) (PHOST),
  photoinitiators and resist
  design, 14
Polyimide
  gate transistors, 425-31
  photosensitive siloxane, 259-272
  solution characterization, 227-237
  ultrapure, 239-258
Polymer
  block, conductive PA
    hybrids, 509,512
  conducting
    design, 447
    PA-PB blends, 487-96
    poly(2,2'-bithiophene), 473
    PPSe and PPTe, 461-71
    TCNQ salt dispersion, 515-30
  conjugated, electrochemical
    properties, 433-44
  cross-linking negative resists, 193t
  encapsulation, 273
  magnetic recording media, 419
  metal-containing, plasma
    polymerization, 556
  negative e-beam resists, 106,111t
  PA copolymers and blends, 507-13
  preparation
    conductive PA hybrids, 509
    EPDM and Kraton PA blends, 498
    methacrylate copolymers, 120-22
    novolac-based positive e-beam
      resist, 168
    PA-PB blends, 488
    PFCPM as positive e-beam resist, 1
    photosensitive polyimide
      siloxane, 259
    plasma developable e-beam
      resists, 214
    poly(butadiene-co-glycidyl
      methacrylate), 202,203t
    polyimide materials, 229,232t
    PPSe and PPTe, 462-64,470
    radiation-induced degradation of
      PMPS, 138,154
    TCNQ salt-polymer composites, 516
Polymer-filler interaction, silicone
  elastomers, 294
Polymer patterns, deep-UV
  photolithography, 60,62f
Polymer-transition metal bonds, conductive PA hybrids, 503
Polymeric reactions, magnetic recording media, 409-19
Polymerization
  acid-catalyzed condensation, 6

Polymerization--Continued
  acrylated epoxy systems, 331
  PFCPM as positive e-beam resist, 130
  photoinitiated cationic, 3-9
  plasma, TMT, 533-52
  polyimide materials, 230
Polymethacrylonitrile (PMCN), resist
  dry-etch susceptibility, 93
Poly(2-methyl-1-pentene sulfone)
  (PMPS)
  absorbance spectra, 147f,148f
  radiation-induced
    degradation, 135-50
Polynuclear aromatics, photoinitiators
  and resist design, 17
Poly(olefin sulfones), deep-UV
  photolithography, 55-64
Poly(p-phenylene) (PPP)
  redox properties, 443t
  VEH computed IP, 438t
Poly-p-phenylene selenide
  (PPSe), 461-71
Poly(p-phenylene sulfide) (PPS)
  conducting polymer solution, 453
  polymeric organic conductor, 448
Poly-p-phenylene telluride
  (PPTe), 461-71
Polyphthalaldehyde, polymerization
  photoinitiators, 8
Polypyrrole (PPY)
  redox properties, 437t,443t
  VEH computed IP, 438t
Polysilicon
  conventional MOS transistors, 426
  gate process, 426
Polystyrene (PS)
  cross-linking negative resists, 198
  e-beam lithography, 112
  living, conductive PA hybrids, 509
Polystyrene-polyacetylene (PS-PA)
  block polymer, 512
Poly(tetrafluoro-chloropropyl
  methacrylate) (PFCPM), 129-33
Polythiophene (PTP)
  redox properties, 443t
  VEH computed IP, 438t
Poly(trichloroethyl methacrylate)
  (PTCEM), 214,218
Polyurethane
  encapsulant
    dynamic mechanical
      properties, 310f
    removable, 305-12
  magnetic recording media, 410
Polyurethane-prepolymerized isocyanate
  reaction, 415
Poly(vinyl alcohol) (PVA), monomer
  sublimation barrier, 217
Poly(vinyl pyrroidone) (PVP), monomer
  sublimation barrier, 217

Poly(vinylidene fluoride) (PVDF)
  piezoelectric, 399-406
  tube applications, 405
Positive e-beam resist
  deep-UV photolithography, 57
  e-beam lithography, 105-109
  novolac-based, 167-75
  PFCPM, 129-33
  photoinitiators and resist
    design, 14
Positive mode, photoinitiators and
  resist design, 14-19
Positive photoresist
  diazoquinone, mid-UV
    photosensitization, 41-53
  dry-etch susceptibility, 91-99
  lithographic performance, 65-76
Postexposure baking
  novolac-based positive e-beam
    resist, 170
  polyimide materials, 233
  radiation-induced degradation of
    PMPS, 140f,144,146f
Postirradiation polymerization, crosslinking negative resists, 195
Power density, PA-Li battery, 578,581
Prebake temperature
  vs. image bias, ultrapure
    polyimides, 255f
  resist dissolution kinetics, 81f
Prebake time, resist dissolution
  kinetics, 83f
Prebaking, poly(butadiene-co-glycidyl
  methacrylate), 209
Preimidization, photosensitive
  polyimide siloxane, 262
Prepolymerized isocyanate reaction,
  polyurethane, magnetic recording
  media, 415
Printed circuit board (PCB), UV solder
  masks, 367-72
Printed wiring board (PWB),
  TCE, 379-95
Proton NMR spectra, radiation-induced
  degradation of PMPS, 156,157f
Puddle coating, positive photoresist
  lithographic performance, 71
Pull strength
  organometallic plasma
    polymerization, 559
  TMT plasma
    polymerization, 536,545,546t
Pyrene
  diazoquinone resists, 41-53
  lifetime, mid-UV
    photosensitization, 42
Pyrene-sensitized diazonaphthoquinone
  derivative (PDQ), UV spectra, 44
Pyridine N-oxide, deep-UV
  photolithography, 56

Pyroelectric coefficient, piezoelectric PVDF, 402
Pyrolysis, CPI-gate transistors, 428
Pyromellitic dianhydride (PMDA),
  condensation and
  cyclodehydration, 228f

Q

Quartz, TCE as a function of
  temperature, 382t
Quartz laminates, various
  TCE vs. temperature, 395f
  TCE of PWB, 394
Quaternary phosphonium salts, epoxy
  molding compounds, 281t

R

γ-Radiation, EPDM-PA
  blends, 501
Radiation chemical yield, chain-
  scission of methacrylate
  copolymers, 122,124f
Radiation dose
  cross-linking negative resists, 178
  degradation of PMPS, 155-58
  resist dissolution
    kinetics, 82-84,85f
Radiation-induced degradation
  chain-scission of methacrylate
    copolymers, 126f
  itaconate copolymers, 125
  PMPS, 135-50,153-65
Radiation product yield, radiation-
  induced degradation of
  PMPS, 158,160f
Radiation sensitivity,
  poly(butadiene-co-glycidyl
  methacrylate), 204,206f,207f
Radical cations, resists for e-beam
  lithography, 116
Radical scavenger, novolac-based
  positive e-beam resist, 173
Radiolysis products, radiation-
  induced degradation of
  PMPS, 153-65
Reactive ion etching with oxygen,
  e-beam lithography, 112
Reactivity
  order, magnetic recording media, 419
  PA in electrolytes, 576
Recording media, magnetic, 409-19
Recrystallization, ultrapure
  polyimides, 243
Reduction potential
  DPP, 443
  VEH theory vs. experimental, 443t
Reflective films, TMT plasma
  polymerization, 537

INDEX

Reinitiation, polymerization photoinitiators, 7
Relative dose vs. critical dimension, positive photoresist, 74f,75f
Relative permittivity, piezoelectric PVDF, 402
Relative process latitude, positive photoresist, 67,69t,70t
Reliability, device and inhibited encapsulants, 319f
Resin
  inhibited encapsulants, 320
  novolac, photoinitiators and resist design, 12
Resist
  design, photoinitiators, 11-22
  diazoquinone-novolac, mid-UV actinic sensitivity enhancement, 41-53
  dissolution kinetics and molecular dynamics, 79-90
  dry developing, photoinitiators and resist design, 20
  e-beam lithography, 103-16
  negative e-beam
    deep-UV photolithography, 60,62f
    polymerization photoinitiators, 6
  novolac-based positive e-beam, 169
  photoactive compound (PAC), 42
  plasma developable, 213-22
  positive e-beam
    novolac-based, 167-75
    PFCPM, 129-33
  positive tone, photoinitiators and resist design, 14
  semiempirical quantum mechanical calculations, 25-39
  theoretical predictions, 191-200
  two-layer diazoquinone, 49,50f,52f
  vinyl polymer, dry-etch susceptibility, 91-100
Resist pattern
  deep-UV photolithography, 60,62f
  e-beam lithography, 110f
Resist process optimization, ultrapure polyimides, 253
Resist profile
  cross-linking negative resists, 199f
  mid-UV sensitizers, 25f
Resistivity
  TCNQ salt-polymer composites, 518f,519f,522f,525f
  TMT plasma polymerization, 534
  UV-thermal cure encapsulants, 376
Resolution
  cross-linking negative resists, 198
  electron resists, 104t
  PFCPM positive e-beam resist, 132
Resonance effect, mid-UV sensitizers, 29-39
Rinse composition, photoinitiators and resist design, 13
RTV silicones, generic, TGA, 291-98
Rubber-modified epoxy photopolymers, 345-64
Rubber modifier, effect on cure degree, 348
Rubber particle distribution, modified epoxy photopolymers, 358

S

Salt concentration, TCNQ salt-polymer composites, 517
Scanning electron microscopy (SEM), 569
  As in PPS film, 452f
  diazoquinone resists, 47,51,52f
  mid-UV sensitizers, 39f
  novolac-based positive e-beam resist, 172f
  organometallic plasma polymerization, 559,570f
  PA-PB blends, 492-96
  photoinitiators and resist design, 13
  photosensitive polyimide siloxane, 268
  plasma developable e-beam resist, 219f,220f,222f
  poly(2,2'-bithiophene), 482f,483f
  positive photoresist, 66
  resist dissolution kinetics, 83f
  TCNQ salt-polymer composites, 527f
  TMT plasma polymerization, 534,537,540f
β-Scission, methacrylate copolymers, 125
Selenium, organometallic plasma polymerization, 556
Self-alignment gate process, CPI-gate transistors, 426
Semiconductor encapsulation, epoxy molding compounds, 274
Semiconductor manufacturing, photoinitiators, 11-22
Semiempirical quantum mechanical models, mid-UV sensitizers, 25-39
Sensitivity
  vs. component ratio, cross-linking negative resists, 182-87
  cross-linking negative resists, 192-98,199f
  electron resists, 104t
  ionizing radiation, negative e-beam resist, 204,206f,207f
  novolac-based positive e-beam resist, 169
Sensitizer
  mid-UV, 25-39

Sensitizer—Continued
  photosensitive polyimide
    siloxane, 264
Sheet resistivity, TMT plasma
  polymerization, 534
Shirakawa catalyst
  conductive PA hybrids, 503
  PA-PB blends, 489
Silicon chip test circuit, inhibited
  encapsulants, 316
Silicone elastomers, TGA, 285-303
Silicone resins, e-beam
  lithography, 112
Silicone—See also
  Polydimethylsiloxane
Sodium ion mobility, ultrapure
  polyimides, 250f
Sodium naphthalenide, PPSe and
  PPTe, 469
Softening, removable polyurethane
  encapsulants, 309t
Solder masks, UV, PCB
  insulator, 367-72
Soldering, effect, UV solder masks as
  PCB insulators, 371
Solid ingot synthesis, ultrapure
  polyimides, 247f
Solvation
  AsF_3 in conducting polymer
    solution, 457
  AsF_5-doped PPS, 453
Solvent absorption (SA), acrylated
  epoxy systems, 326,331,335t
Solvent evaporation, TCNQ salt-polymer
  composites, 524
Solvent extraction (SE), acrylated
  epoxy systems, 326,339
Solvent resistance, UV-thermal cure
  encapsulants, 376
Solvent size effect, resist dissolution kinetics, 87,88f
Solvolysis, epoxy molding
  compound, 276
Spectral shift, TCNQ salt-polymer
  composites, 530f
Spin coating, mid-UV photosensitization of diazoquinone resists, 52f
Spin-spray coating, positive
  photoresist lithographic
  performance, 71
Sputter etching, e-beam
  lithography, 107,108f
Sputtering, organometallic plasma
  polymerization, 559
Stability, RTV silicone material, 296
Stainless steel, TMT plasma
  polymerization, 546t
Stannous salts, catalyst, epoxy molding compounds, 281
Stereochemistry effect, resist dissolution kinetics, 84,86f,87

Stoichiometry, TCNQ salt, 523
Stress-strain behavior, Kraton-PA
  blends, 504f
Stretch elongation, effect, PA-PB
  blends, 490
Styrene-based resist, cross-linking
  negative resists, 198
Styrene-diene triblock copolymer-
  polyacetylene (Kraton-PA), 497-505
Sublimation
  barrier, plasma developable e-beam
    resists, 217
  high vacuum, ultrapure
    polyimides, 243
  rate, plasma developable e-beam
    resists, 215,216f
Substituent effect, mid-UV
  sensitizers, 29-39
Sulfonamide derivates, diazoketone,
  mid-UV sensitizers, 38
Sulfonate derivatives, diazoketone,
  absorbance spectra, 37f
Sulfonyl chloride diazonaphthoquinone,
  conversions, 46t
Sulfonyl group, mid-UV, 33,36t
Sulfur dioxide, radiation-
  induced degradation of
  PMPS, 144,155-58
Surface frying, resist dry-etch
  susceptibility, 93,95f
Surface polymerization rates, acrylated epoxy systems, 332f
Swelling
  cross-linking negative resists, 195
  negative e-beam resist, 209
  photoinitiators and resist
    design, 13,17
  photosensitive polyimide
    siloxane, 266
  poly(butadiene-co-glycidyl
    methacrylate), 209
  removable polyurethane
    encapsulants, 311

T

Temperature
  CPI-gate transistors, 428
  epoxy molding compounds, 278
  vs. imidization level, ultrapure
    polyimides, 254f
  vs. irradiation time, PMPS
    degradation, 141f
  magnetic recording media, 411t
  negative e-beam resist, 202-204
  organometallic plasma
    polymerization, 571
  photoinitiators and resist
    design, 19
  photosensitive polyimide
    siloxane, 266

INDEX

Temperature—Continued
  poly(butadiene-co-glycidyl methacrylate), 202-204
  positive photoresist lithographic performance, 71
  radiation-induced degradation of PMPS, 139,158,160f
  resist dissolution kinetics, 81f
  vs. sheet conductivity, TMT plasma polymerization, 544f
  vs. thermal expansion, various laminates, 382-94
Tensile property, UV-thermal cure encapsulants, 375
Tensile strength, removable polyurethane encapsulants, 309t
Tertiary amine, catalyst, epoxy molding compounds, 276
7,7,8,8-Tetracyanoquinodimethane (TCNQ) salt dispersion, conductive composites, 515-30
Tetraethyltin (TET), 559,564
Tetrahydrofuran (THF), 581
  PPSe and PPTe, 469
Tetramethylgermanium (TMG), 559
Tetramethyltin (TMT), plasma polymerization, 533-52
Tetrathiafulvalene (TTF), e-beam lithography, 113
Tetravinyltin (TVT)
  organometallic plasma polymerization, 559
  SEM, 569
Thermal analysis, poly(butadiene-co-glycidyl methacrylate), 202-204
Thermal coefficient of expansion (TCE), PWB, 379-95
Thermal degradation mechanism, silicone, 292f
Thermal diffusion, CPI-gate transistors, 427
Thermal expansion vs. temperature, various laminates, 382-94
Thermal initiator, UV-thermal cure encapsulants, 374
Thermal properties, PPSe and PPTe, 465-69
Thermal stress resistance, UV-thermal cure encapsulants, 376
Thermal treatment
  photosensitive polyimide siloxane, 266
  polyimide materials, 233
Thermogravimetric analysis (TGA)
  acrylated epoxy systems, 326,335
  description, 286-291
  first derivative, silicone, 298f
  PPSe and PPTe, 465-69
  resist dry-etch susceptibility, 97t,99t

Thermogravimetric analysis (TGA)—Continued
  rubber-modified epoxy photopolymers, 349f
  silicone elastomers, 285-303
  ultrapure polyimides, 251f
Thermomechanical spectra (TS), rubber-modified epoxy photopolymers, 355f,356f,357f,359f
Thermoplastics, removable encapsulants, 308,309t
Thickness
  photosensitive polyimide siloxane, 266
  plasma developable e-beam resists, 222f
  poly(2,2'-bithiophene), 476
  poly(butadiene-co-glycidyl methacrylate), 209
  ultrapure polyimides, 249f
  vs. viscosity, ultrapure polyimides, 247f
Thiophene ring, poly(2,2'-bithiophene), 475
Time
  vs. conductivity, AsF_5 coping of a PPS film, 451f
  poly(2,2'-bithiophene), 477f
  vs. creep compliance, linear polyurethane encapsulants, 310f
  prebake, resist dissolution kinetics, 83f
  vs. sheet conductivity, TMT plasma polymerization, 544f
Tin
  organometallic plasma polymerization, 556
  TMT plasma polymerization, 545,550
Tin-carbon ratio
  organometallic plasma polymerization, 564
  TMT plasma polymerization, 543
Tin-oxygen ratio
  vs. distance, TMT plasma polymerization, 546f
  organometallic plasma polymerization, 564
  TMT plasma polymerization, 543
Titanium, organometallic plasma polymerization, 556
Toluene extraction
  EPDM-PA blends, 499
  SEM PA-PB blends, 492-96
Topology, mid-UV photosensitization of diazoquinone resists, 51
Torsion pendulum (TP) data, rubber-modified epoxy photopolymers, 353-58
Transfer molding resins, inhibited encapsulants, 321

Transistors, polyimide gate,
  fabrication, 425-31
Transition, calculated, mid-UV
  sensitizers, 32f
Transition energy, optical, DPP, 440
Transition metal, organometallic
  plasma polymerization, 557
Transmission electron micrograph (TEM)
  EPDM-PA blends, 499,500f
  Kraton-PA blends, 504f
  PA-PB blends, 495f
  poly(2,2'-bithiophene), 479,481f
  rubber-modified epoxy
    photopolymers, 358,360f
  TMT plasma
    polymerization, 534,541,542f
Transmission spectra, rubber-modified
  epoxy photopolymers, 364f
Triarylsulfonium salts, photolysis, 3
Trichloroethyl ester group, resist
  dry-etch susceptibility, 96
Trichloroethyl methacrylate (TCEM),
  plasma-developable e-beam
  resists, 214
Tricyclohexylphosphine adduct,
  catalyst, epoxy molding
  compounds, 281t
Tube production, piezoelectric
  PVDF, 400
Tubular reactor, plasma
  polymerization, 558f
Two-layer resist, mid-UV
  photosensitization, 49,50f,52f
Two-step procedure, removable
  polyurethane encapsulants, 306

U

Ultrapure polyimides, 239-57
Urea derivatives, catalyst, epoxy
  molding compounds, 281t
Urethane system, magnetic record-
  ing media, 411t
UV absorption shift, mid-UV
  sensitizers, 34
UV-curable conformal coatings, 373-76
UV photolithography, deep
  negative tone e-beam
    resists, 57,60,62f
  with poly(olefin sulfones), 55-64
UV radiation
  photoinitiators and resist
    design, 19
  photosensitive polyimide
    siloxane, 266
UV solder masks, insulators for
  PCB, 367-72
UV spectra, diazoquinone resists, 44

V

Valence effective Hamiltonian (VEH)
  technique, 434
Vapor development
  PMPS, novolac-based positive e-beam
    resist, 170,171
  radiation-induced degradation of
    PMPS, 149f,150f
Varcum resin, deep-UV
  photolithography, 57
Variation rate, plasma developable
  e-beam resists, 225,216f
Vibrational spectroscopy, PPSe and
  PPTe, 464
$\underline{N}$-Vinyl carbazole (NVC), plasma-
  developable e-beam resists, 214
$\underline{N}$-Vinyl carbazole-poly(trichloroethyl
  methacrylate) (NVC-PTCEM)
  resist, 218,219f,220f
Vinyl group, e-beam lithography, 109
Vinyl polymer, resist dry-etch
  susceptibility, 91-100
Viscosity
  acrylate-based mixtures, 330t
  vs. film thickness, ultrapure
    polyimides, 247f
  magnetic recording media, 413
  polyimide materials, 231
  vs. weight percent, ultrapure
    polyimides, 247f
Visible spectra
  poly(2,2'-bithiophene), 476,478f
  TCNQ salt-polymer composites, 526
Volatile components, acrylated epoxy
  systems, 335
Volatile products, radiolysis of
  PMPS, 154
Voltammagram,
  poly(2,2'-bithiophene), 480f
Volume resistivity
  removable polyurethane
    encapsulants, 309t
  UV-thermal cure encapsulants, 376

W

Wafer geometry, measured, diazoquinone
  resists, 51t
Wafer process
  CPI-gate transistors, 427
  positive photoresist, 67
Wafer writing, electron resists, 104t
Water, contact angle, TMT plasma
  polymerization, 537
Water vapor permeability
  TMT plasma polymerization, 536,551t
  UV-thermal cure encapsulants, 375

Weight percent vs. viscosity,
  ultrapure polyimides, 247f
Wet etching,
  poly(butadiene-co-glycidyl
  methacrylate), 209

X

X-ray diffraction
  EPDM-PA blends, 501
  Kraton-PA blends, 503
  PA-PB blends, 490,494
  piezoelectric PVDF, 402
  PPSe and PPTe, 464
  TMT plasma polymerization, 541,542f

X-ray photoelectron spectroscopy (XPS)
  organometallic plasma
  polymerization, 559
  TMT plasma polymerization, 534,543
X-ray radiation, photoinitiators and
  resist design, 19
X-ray resist, dry-etch
  susceptibility, 99

Z

Zinc octoate, catalysts, magnetic
  recording media, 412
Zone refining, ultrapure
  polyimides, 243

Production and indexing by Susan Robinson
Jacket design by Anne G. Bigler
Elements typeset by Hot Type Ltd., Washington, D.C.
Printed and bound by Maple Press Co., York, Pa.

RECENT ACS BOOKS

"Radionuclide Generators: New Systems
for Nuclear Medicine Applications"
Edited by F. F. Knapp, Jr., and Thomas A. Butler
ACS SYMPOSIUM SERIES 241; 240 pp.; ISBN 0-8412-0822-0

"Polymer Adsorption and Dispersion Stability"
Edited by E. D. Goddard and B. Vincent
ACS SYMPOSIUM SERIES 240; 477 pp.; ISBN 0-8412-0820-4

"Assessment and Management of Chemical Risks"
Edited by Joseph V. Rodricks and Robert C. Tardiff
ACS SYMPOSIUM SERIES 239; 192 pp.; ISBN 0-8412-0821-2

"Chemical and Biological Controls in Forestry"
Edited by Willa Y. Garner and John Harvey, Jr.
ACS SYMPOSIUM SERIES 238; 406 pp.; ISBN 0-8412-0818-2

"Chemical and Catalytic Reactor Modeling"
Edited by Milorad P. Dudukovic and Patrick L. Mills
ACS SYMPOSIUM SERIES 237; 240 pp.; ISBN 0-8412-0815-8

"Multichannel Image Detectors Volume 2"
Edited by Yair Talmi
ACS SYMPOSIUM SERIES 236; 333 pp.; ISBN 0-8412-0814-X

"Efficiency and Costing: Second Law Analysis of Processes"
Edited by Richard Gaggioli
ACS SYMPOSIUM SERIES 235; 262 pp.; ISBN 0-8412-0811-5

"Xenobiotics in Foods and Feeds"
Edited by John W. Finley and Daniel E. Schwass
ACS SYMPOSIUM SERIES 234; 432 pp.; ISBN 0-8412-0809-3

"Nonlinear Optical Properties of Organic and Polymeric Materials"
Edited by David J. Williams
ACS SYMPOSIUM SERIES 233; 252 pp.; ISBN 0-8412-0802-6

"Rings, Clusters, and Polymers of the Main Group Elements"
Edited by Alan H. Cowley
ACS SYMPOSIUM SERIES 232; 182 pp.; ISBN 0-8412-0801-8

"Archaeological Chemistry--III"
Edited by Joseph B. Lambert
ADVANCES IN CHEMISTRY SERIES 205; 324 pp.; ISBN 0-8412-0767-4

"Molecular-Based Study of Fluids"
Edited by J. M. Haile and G. A. Mansoori
ADVANCES IN CHEMISTRY SERIES 204; 524 pp.; ISBN 0-8412-0720-8